Fabrication, Structure and Reactivity of Anchored Nanoparticles

Seminaris CampusHotel,
Berlin, Germany
10–12 April 2013

FARADAY DISCUSSIONS

Volume 162, 2013

RSC Publishing

The Faraday Division of the Royal Society of Chemistry, previously the Faraday Society, founded in 1903 to promote the study of sciences lying between Chemistry, Physics and Biology.

Faraday Discussions (Print ISSN 1359-6640, Electronic ISSN 1364-5498) is published 8 times a year by the Royal Society of Chemistry, Thomas Graham House, Science Park, Milton Road, Cambridge, UK CB4 0WF. Volume 162 ISBN-13: 978-1-84973-689-3

2013 annual subscription price: print+electronic £765, US $1428; electronic only £727, US $1356. Customers in Canada will be subject to a surcharge to cover GST. Customers in the EU subscribing to the electronic version only will be charged VAT. All orders, with cheques made payable to the Royal Society of Chemistry, should be sent to RSC Order Department, Royal Society of Chemistry, Thomas Graham House, Science Park, Milton Road, Cambridge, CB4 0WF, UK. Tel +44 (0) 1223 432398; E-mail orders@rsc.org

If you take an institutional subscription to any RSC journal you are entitled to free, site-wide web access to that journal. You can arrange access *via* Internet Protocol (IP) address at www.rsc.org/ip. Customers should make payments by cheque in sterling payable on a UK clearing bank or in US dollars payable on a US clearing bank.

PRINTED IN THE UK

MIX
Paper from responsible sources
FSC® C013604

Faraday Discussions documents a long-established series of *Faraday Discussion* meetings which provide a unique international forum for the exchange of views and newly acquired results in developing areas of physical chemistry, biophysical chemistry and chemical physics.

Fabrication, Structure and Reactivity of Anchored Nanoparticles

Faraday Discussions

www.rsc.org/faraday_d

A General Discussion on Fabrication, Structure and Reactivity of Anchored Nanoparticles was held in Berlin, Germany on 10th, 11th and 12th April 2013.

CONTENTS

ISSN 1359-6640; ISBN 978-1-84973-689-3

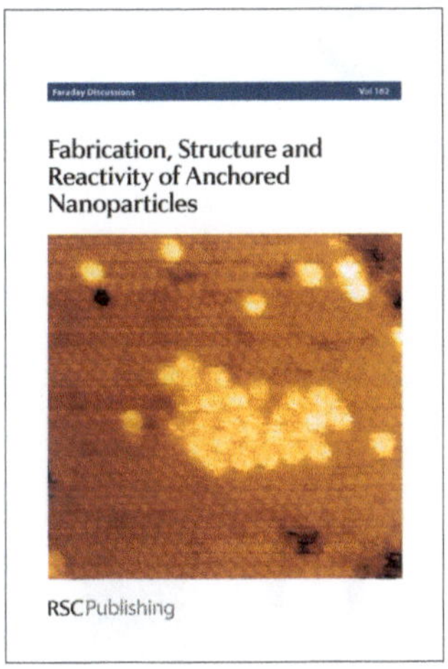

Cover
See Bowker, Davies *et al.*, *Faraday Discuss.*, 2013, **162**, 201–212.

STM image after exposure of a clean Fe_3O_4 (111) single crystal surface to molybdenum oxide followed by annealing in oxygen at 10^{-7} mbar. The Mo oxide layer is seen as the bright islands consisting of Mo_3O_9 units organised in a (2 × 2) structure with respect to the iron oxide lattice underneath. The image size is 17 × 17 nm.

Image reproduced by permission of Michael Bowker from *Faraday Discuss.*, 2013, **162**, 201.

INTRODUCTORY LECTURE

PAPERS AND DISCUSSIONS

Sponsors

FCI
FONDS DER
CHEMISCHEN
INDUSTRIE

BASF
The Chemical Company

Exhibitor

SPECS™

Co-sponsors

GECATS
GERMAN
CATALYSIS
SOCIETY

DBG

GDCh

CONCLUDING REMARKS

ADDITIONAL INFORMATION

Anchored metal nanoparticles: Effects of support and size on their energy, sintering resistance and reactivity

Charles T. Campbell* and Jason R. V. Sellers

Received 21st May 2013, Accepted 30th May 2013
DOI: 10.1039/c3fd00094j

Many catalysts consist of metal nanoparticles anchored to the surfaces of oxide supports. These are key elements in technologies for the clean production and use of fuels and chemicals. We show here that the chemical reactivity of the surface metal atoms on these nanoparticles is closely related to their chemical potential: the higher their chemical potential, the more strongly they bond to small adsorbates. Controlling their chemical potential by tuning the structural details of the material can thus be used to tune their reactivity. As their chemical potential increases, this also makes the metal surface less noble, effectively pushing its behavior upwards and to the left in the periodic table. Also, when the metal atoms are in a nanoparticle with higher chemical potential, they experience a larger thermodynamic driving force to sinter. Calorimetric measurements of metal vapor adsorption energies onto clean oxide surfaces in ultrahigh vacuum show that the chemical potential increases with decreasing particle size below 6 nm, and, for a given size, decreases with the adhesion energy between the metal and its support, E_{adh}. The structural factors that control the metal/oxide adhesion energy are thus also keys for tuning catalytic performance. For a given oxide, E_{adh} increases with $(\Delta H_{sub,M} - \Delta H_{f,MOx})/\Omega_M^{2/3}$ for the metal, where $\Delta H_{sub,M}$ is its heat of sublimation, $\Delta H_{f,MOx}$ is the standard heat of formation of that metal's most stable oxide (per mole of metal), and Ω_M is the atomic volume of the bulk solid metal. The value $\Delta H_{sub,M} - \Delta H_{f,MOx}$ equals the heat of formation of that metal's oxide from a gaseous metal atom plus $O_2(g)$, so it reflects the strength of the chemical bonds which that metal atom can make to oxygen, and $\Omega_M^{2/3}$ simply normalizes this energy to the area per metal atom, since E_{adh} is the adhesion energy per unit area. For a given metal, E_{adh} to different clean oxide surfaces increases as: MgO(100) $\approx$ TiO_2(110) $\leq$ α-Al_2O_3(0001) < CeO_{2-x}(111) $\leq$ Fe_3O_4(111). Oxygen vacancies also increase E_{adh}, but surface hydroxyl groups appear to decrease E_{adh}, even though they increase the initial heat of metal adsorption.

1. Introduction

Metal nanoparticles dispersed across the surfaces of oxide and carbon supports form the basis for many catalysts and electrocatalysts of importance to future energy technologies, pollution prevention and environmental protection, all necessary for any sustainable technological infrastructure that maintains a high quality of life. The catalytic activity per surface metal atom and selectivity can vary with particle size when the metal particles are below about 7 nm in diameter,[1–18] in some cases

Department of Chemistry, University of Washington, Seattle, WA 98195-1700, USA. E-mail: campbell@chem.washington.edu

dramatically.[7,8,19–28] Measurements have even shown that the activity depends on the nanoparticle shape for particles of the same size.[29] The activity can also depend strongly on the choice of support, even for particles of the same size,[1,2,7–9,11,13,25,28,30–42] the extent of reduction if an oxide support,[28,43–47] and oxidation if a carbon support.[39,40,48] The proceedings of this Faraday Discussion include many examples of the effects listed above.

In general, these effects are poorly understood, largely due to the difficulty in structurally characterizing high-area catalysts with sufficient detail, resulting in an inability to analyze thoroughly any structure–reactivity relationships that control the effects, although the proceedings of this Faraday Discussion contain many papers that demonstrate that this situation is improving rapidly. Furthermore, the metal nanoparticles often sinter rapidly under catalytic reaction conditions, resulting in particle growth and loss of metal area during activity measurements, and thus challenging the ability to even measure area-specific activity let alone correlate it to size and support. The relationships of sintering kinetics to catalyst structure are also poorly understood.

In order to better understand structure–reactivity relationships in such catalysts and their sintering kinetics, a number of researchers including our own group have adopted a model catalysts approach whereby structurally well-defined samples of metal nanoparticles on oxides are prepared by vapor deposition of the metal onto single-crystal oxide surfaces.[6,21–25,27,28,43,44,49–93] The deposited metal atoms generally nucleate small particles, whose size and number density can be controlled by the amount of deposited metal and the temperature. The aim of this approach is to correlate the metal's chemisorption, catalytic, sintering and/or electronic properties with the lateral diameter, thickness and composition of the nanoparticles and with the structure and extent of reduction of the oxide support upon which they sit.

Since we recently lost one of the most significant figures in this field, Professor D. Wayne Goodman, we wish to specifically recognize his contributions to this research area, especially his pioneering development of this model catalyst approach for studying structure–reactivity relationships in anchored nanoparticle catalysts. One can see the large impact of his contributions by looking at the very high frequency of citations to his papers in the paragraphs above. His creative approaches to establishing surface structure–function relationships in this area, and in the area of bimetallic catalysts, have paved the way for many others who later became active in this arena. His papers and talks in these areas eloquently outlined how one can approach industrially-important fundamental questions in catalysis using this model catalyst approach. Many researchers worldwide picked up on the infectious enthusiasm and clear-headed logic he projected, subsequently shifting their own research in directions he had set. Many of us who did that were also lucky enough to get to know him personally, and thus discovered his exceptionally enjoyable personality, intense competitiveness and unparalleled sense of humor.

In this paper, we review studies of well-defined model metal nanoparticle catalysts supported on clean, single-crystalline oxide surfaces, structurally characterized using a variety of ultrahigh vacuum surface science techniques. The main goal here is to relate the reactivity of late transition metal nanoparticles and the rates at which they sinter with the energies of metal atoms in these supported nanoparticles measured using unique adsorption calorimetry tools. To better understand how the choice of oxide support material affects the energies of metal atoms in nanoparticles, we also analyze temperature programmed desorption (TPD) measurements of metal desorption energies from metal nanoparticles anchored to clean oxide surfaces (from D. W. Goodman's group) as well as measurements of metal/oxide adhesion energies obtained from microscopic measurements of metal particle and molten metal droplet shapes.

We find that the concept of metal atom chemical potential is very powerful for predicting how chemisorption bond strengths and sintering rates depend on metal particle size for a given support, and on the support material for a given particle size of late transition metal. We further show how this chemical potential depends on the structural properties of the catalyst including the metal particle size and the composition and extent of reduction of the oxide support surface.

2. The chemical potential of metal atoms in anchored nanoparticles

We will show below that in order to understand how anchored metal nanoparticles function with respect to surface reactivity, catalytic activity and sinter resistance, we must first understand how the chemical potential of the metal atoms in a nanoparticle depends on the structural details of the material. These structural details include: (1) the nature of the metal, (2) the composition and structure of the support surface upon which they sit, (3) the size of the metal nanoparticle, and (4) the shape of the metal nanoparticle. A metal atom's chemical potential is so important because it directly reflects how aggressive that atom will be in its chemical bonding to adsorbates, but at the same time it is also a direct measure of how unstable it is with respect to sintering into very large particles. We will limit this discussion of structural effects on metal chemical potential (μ, or partial molar Gibb's free energy) to a description of only the internal energy differences that arise due to these structural differences, since entropic differences are small and nearly constant in these respects, and since the difference between enthalpy (H) and internal energy (U) is also nearly constant from system to system. We will often refer to these simply as "energy".

The internal energy of a free metal (M) particle in vacuum relative to the bulk solid metal is:

$$U_{\mathrm{M}} = \bar{U}_{\mathrm{M,bulk}} n_{\mathrm{M}} + \gamma_{\mathrm{M}} A_{\mathrm{M}} = \gamma_{\mathrm{M}} A_{\mathrm{M}}, \tag{1}$$

where $\bar{U}_{\mathrm{M,bulk}}$ is the molar internal energy of the pure, bulk metal (which we set to the zero reference energy), n_{M} is the number of moles of metal atoms in the particle, γ_{M} is the surface energy of the metal–vacuum interface (which we will show depends on particle diameter below 6 nm), and A_{M} is the particle's surface area. When this same particle adheres to a support without any change in A_{M}, the system's total energy decreases by the adhesion energy (E_{adh}) times the total area of metal/support contact (A):

$$\Delta U = -E_{\mathrm{adh}} A, \tag{2a}$$

so that the system's total energy is given by:

$$U = U_{\mathrm{M}} + U_{\mathrm{Sup}} - E_{\mathrm{adh}} A = \gamma_{\mathrm{M}} A_{\mathrm{M}} + U_{\mathrm{Sup}} - E_{\mathrm{adh}} A, \tag{2b}$$

where U_{Sup} is the internal energy of the particle-free support before adhesion. The metal's chemical potential (μ_{M}, relative to the bulk metal reference state) is the partial derivative of this with respect to n_{M} (neglecting entropy differences, which are relatively small). Both A_{M} and A depend on n_{M} (typically varying as $n_{\mathrm{M}}^{2/3}$), so μ_{M} will have a term that increases with γ_{M} and a second term that decreases with E_{adh}. As the adhesion energy increases, the metal's chemical potential decreases. So, to understand how the chemical potential of metal nanoparticles depends on a nanomaterial's structural properties, we must first understand how the adhesion energy depends on these properties. Note that both γ_{M} and E_{adh} can depend on particle size for particles below ~6 nm in effective diameter.[94,95] We address in the next section how E_{adh} depends on the nature of the metal and of the support surface in the large-particle limit.

3. How metal/oxide adhesion energies depend on the metal and the oxide

Metal/oxide adhesion energies have mainly been measured for molten metal drops on flat oxide surfaces, using sessile drop/contact angle techniques, often under controlled inert gas or H_2 atmospheres (although not under conditions of proven surface cleanliness). Two extensive compilations of such values have appeared,[51,96] with much of these data coming from Chatain *et al.*[97–100] There are similar measurements for facetted solid particles which instead of the contact angle use a ratio of distances that is similar to an aspect ratio (height/width) but which is specific to the Wulff construction of the particle shape (the ratio of the distance from the particle's center of symmetry or Wulff's point to the oxide surface divided by the distance from the Wulff's point to one exposed face of the particle).[101] Some older values for solid particles that were collected on single crystal oxide surfaces but not under conditions of proven surface cleanliness have been summarized by Li.[96] The E_{adh} reported from such studies of molten drops and solid particles is directly proportional to the value for the bulk surface energy of the pure metal used in the data analysis, and many of the surface energy values originally used are now outdated. These E_{adh} values have been recalculated using the latest surface energies of the metals, as reported in ref. 102 and are presented in Table 1 along with the new surface energies used. More recently, such measurements have been extended to the clean conditions of ultrahigh vacuum (UHV) on single crystal oxide surfaces that were proven to be clean using surface spectroscopies. Table 2 lists many of those results.

Our group added another approach for determining adhesion energies, whereby we measure the heat of adsorption of metal atoms as they are vapor deposited onto the clean and ordered surface of a single-crystal oxide under UHV using single crystal adsorption calorimetry (SCAC). We derived the relationship between metal atom adsorption enthalpies on flat substrates and E_{adh} using a thermodynamic cycle.[51] That cycle proved that the integral heat of adsorption per mole ($q_{ad,int}$) at some multilayer coverage where the differential heat of adsorption has just reached the bulk heat of sublimation of the metal per mole ($\Delta H_{sub,M}$) is given by:

$$q_{ad,int} = \Delta H_{sub,M} - [(1 + f_r)\gamma_M - E_{adh}]/(n_M/A), \tag{3}$$

where n_M/A is the coverage in moles of metal atoms per unit area of substrate covered by the metal and f_r is the roughness factor of the resulting multilayer metal film. We use here the convention that both $q_{ad,int}$ and $\Delta H_{sub,M}$ are positive, and neglected entropy. For metals which don't wet, $q_{ad,int} < \Delta H_{sub,M}$.[51] This approach has also been applied to cases where the metal film is not continuous but instead the deposited metal assumes the form of large 3D particles.[95,107,113] Table 2 lists all the values we have determined in this way.

Heats of adsorption of late transition metals on clean oxide single crystal surfaces have also been determined by temperature programmed desorption (TPD) in UHV, in a series of beautiful measurements made by the group of the late D. W. Goodman. We recently discovered that the leading-edge analysis used to analyze their TPD data, while usually quite valid, fails for these particular types of systems due to the simultaneous increase in particle size with time (due to both sintering and selective removal of the smaller particles by desorption).[106] We therefore reanalyzed much of their TPD data to get the heat of adsorption *versus* coverage for each system.[106] We then used those heats together with eqn (3) to extract adhesion energies (and equilibrium contact angles, assuming the solid particle shape can be approximated as a spherical cap to approximate the effect of surface roughness).[106] Those results are also listed in Table 2. However, we list these values here not as we originally reported them, but instead as upper limits. This is because we originally treated A in eqn (3) as if it were the full area of the oxide support surface. We recognize now that these multilayer metal films could have agglomerated into 3D particles that cover much less than 100% of the oxide during the heating of TPD prior to desorption. This is especially likely to happen on oxides with low adhesion energies. Decreasing A in eqn (3) leads to lower values of E_{adh}, thus the values we reported[106] should instead be considered as upper limits.

Table 1 Adhesion energies of metals on oxide surfaces measured by sessile drop/contact angle techniques for the liquid metal drops (top section) and by the shape of the particle for solid metals (bottom section). All data are from ref. 96, but the values marked with an asterisk have been recalculated based on newer values for the surface energies of the pure metals reported in ref. 102. Metals are ordered by their bulk heat of sublimation ($\Delta H_{sub,M}$) divided by the area per metal atom ($\Omega_M^{2/3}$). Also listed is the heat of formation of the bulk oxide support (per mole of oxygen atoms), $\Delta H_{f,OxSup}$

Metal	Oxide	Apparent $\theta_{contact}$	γ_M/(J m^{-2})	E_{adh}/(J m^{-2})	$E_{adh}/2\gamma_M$	$\Delta H_{sub,M}$/(kJ mol^{-1})	$\Omega_M^{2/3}$/(Å^2)	$-\Delta H_{f,OxSup}$/(kJ (mol O)$^{-1}$)	Atmosphere
Liquid metal drops									
Pb	α-Al_2O_3(0001)	125 ± 7°	0.44[103]	0.19 ± 0.05*	0.22	195.2	6.01	558.6	Ar + H_2
In	α-Al_2O_3(0001)	124°	0.700	0.309*	0.22	243	8.79	558.6	Vacuum
Sn	α-Al_2O_3(0001)	125°	0.709	0.302*	0.21	301.2	9.03	558.6	Vacuum
Ga	α-Al_2O_3(0001)	118°	0.881	0.467*	0.27	271.96	7.27	558.6	Vacuum
Ag	α-Al_2O_3(0001)	130°	1.22	0.436*	0.18	284.9	6.63	558.6	Ar + H_2
Ge	α-Al_2O_3(0001)	104°	0.555	0.421	0.38	372	8.01	558.6	Vacuum
Al	α-Al_2O_3(0001)	103 ± 8°	1.14	0.89 ± 0.16*	0.39	330.9	6.51	558.6	Vacuum
Mn	α-Al_2O_3(0001)	103°	1.54	1.19*	0.39	283.3	5.39	558.6	Ar
Au	α-Al_2O_3(0001)	120°	1.51	0.77*	0.25	368.2	6.60	558.6	Vacuum
Si	α-Al_2O_3(0001)	80°	0.745	0.875	0.59	450.0	7.37	558.6	He
Pd	α-Al_2O_3(0001)	120°	2.00	1.00*	0.25	376.6	6.01	558.6	He
		104°*		1.52*	0.38				Vacuum
Cu	α-Al_2O_3(0001)	128°	1.76	0.676*	0.19	337.4	5.18	558.6	Vacuum, Ar + H_2
Fe	CaO	132°	2.42	0.806*	0.17	415.5	5.18	634.9	Ar
	BeO	127°		0.970*	0.20			609.4	He
	MgO	123°		1.11*	0.23			601.6	Vacuum
	UO_2	123°		1.11*	0.23			542.5	Ar + H_2
	ZrO_2	115°		1.37*	0.28			550.3	Ar + H_2
	α-Al_2O_3(0001)	111°		1.55*	0.32			558.6	Ar + H_2
		109°		1.63*	0.34				Ar + H_2
	Cr_2O_3	88°		2.52*	0.52			379.9	Ar
Co	ZrO2	118°	2.35	1.26*	0.27	426.7	4.96	550.3	Ar + H_2
	α-Al_2O_3(0001)	115°		1.36*	0.30			558.6	Ar + H_2
		113°		1.53*	0.30				Ar + H_2

Table 1 (*Contd.*)

Liquid metal drops									
Metal	Oxide	Apparent $\theta_{contact}$	γ_M/(J m^{-2})	E_{adh}/(J m^{-2})	$E_{adh}/2\gamma_M$	$\Delta H_{sub,M}$/(kJ mol^{-1})	$\Omega_M^{2/3}$/(Å^2)	$-\Delta H_{f,OxSup}$/(kJ (mol O)$^{-1}$)	Atmosphere
	CoO	70°		3.18*	0.68			237.9	Ar
Ni	CaO	135°	2.38	0.697*	0.15	430.1	4.93	634.9	—
	MgO	132°		0.788*	0.17			601.6	Vacuum
	BeO	127°		0.965*	0.20			609.4	He
	SiO_2	125°		1.02*	0.21			455.4	Vacuum
	ZrO_2	121°		1.16*	0.24			550.3	Ar + H_2
	UO_2	112°		1.49*	0.31			542.5	Ar
	α-Al_2O_3(0001)	111°		1.52*	0.32			558.6	Ar + H_2
		109°		1.61*	0.34				Vacuum, Ar, He, H_2, Ar + H_2
	TiO_2	105°		1.76*	0.37			472.0	Vacuum
	NiO	60°		3.57*	0.75			244.3	—
	CoO	58°		3.64*	0.76			237.9	Ar
Pt	α-Al_2O_3(0001)	98°	1.70	2.16*	0.43	565.7	6.10	558.6	Vacuum

Solid metal particles									
Metal	Oxide	Apparent $\theta_{contact}$	γ_M/(J m^{-2})	E_{adh}/(J m^{-2})	$E_{adh}/2\gamma_M$	$\Delta H_{sub,M}$/(kJ mol^{-1})	$\Omega_M^{2/3}$/(Å^2)	$-\Delta H_{f,OxSup}$/(kJ (mol O)$^{-1}$)	Atmosphere
Ag	α-Al_2O_3(0001)	125°	1.22	0.51 ± 0.07	0.21	284.9	6.63	558.6	—
Au	α-Al_2O_3(0001)	125°	1.51	0.63 ± 0.09	0.21	368.2	6.60	558.6	—
Cu	α-Al_2O_3(0001)	131°	1.79	0.61 ± 0.11	0.17	337.4	5.18	558.6	—
Ni	α-Al_2O_3(0001)	130°	2.38	0.84 ± 0.09	0.18	430.1	4.93	558.6	—
Fe	α-Al_2O_3(0001)	124°	2.42	1.06 ± 0.17	0.22	415.5	5.18	558.6	—

Table 2 Adhesion energies of solid metals measured on clean oxide surfaces under ultrahigh vacuum conditions. Methods used are based either on the particle shape (HRTEM and GISAXS) or on the integral heat of adsorption of metal vapor at multilayer coverage (SCAC and TPD). All data are from the citations listed, but the values marked with an asterisk have been recalculated based on newer values for the surface energies of the pure metals reported in ref. 102. Also listed is the heat of formation of the most stable bulk oxide of that metal (per mole of metal atoms), $\Delta H_{f,MOx}$. Metals are ordered by their bulk heat of sublimation divided by the area per metal atom. (Table 1 lists these values.)

Metal	Oxide	Particle size/nm	Apparent $\theta_{contact}$	E_{adh}/(J m^{-2})	$E_{adh}/2\gamma_{v/m}$	$\Delta H_{f,MOx}$/(kJ (mol M)$^{-1}$)	Method	Citation
Pb	MgO(100)	10 + (4.75 ML)	72°	0.57 ± 0.20*	0.65	−219	SCAC	104
Ag	MgO(100)	6.6	150 ± 30°	0.30 ± 0.30	0.12	−15.1	SCAC	105
	α-Al_2O_3(0001) or γ-Al_2O_3(111)	—	>70.8°	<1.6	<0.66		TPD	106
	$CeO_{1.9}$(111)	3.6	90°	2.3 ± 0.3	0.94		SCAC	95
	$CeO_{1.8}$(111)	3.6	90°	2.5 ± 0.3	1.0		SCAC	95
	Fe_3O_4(111)	3.6	90°	2.5 ± 0.3	1.0		SCAC	107
Au	MgO(100)	1–5	124°	0.67*	0.22	9.65	HRTEM	108
	TiO_2(110)	2–8	130 ± 5°	0.54 ± 0.10*	0.18 ± 0.03		GISAXS	159
	TiO_2(110)	>4	122 ± 5°	0.71 ± 0.11*	0.24 ± 0.04		HRSEM	160
	TiO_2(001)	—	>79.1°	<1.8	<0.58		TPD	106
	Disordered SiO_2	—	>72.8°	<2.0	<0.65		TPD	106
Pd	MgO(100)	10–15	116°	1.11*	0.28	−85.4	HRTEM	109
		2–8	109°	1.34*	0.34		GISAXS	110
	Disordered SiO_2	—	>65.8°	<2.8	<0.70		TPD	106
Cu	MgO(100)	10 + (7 ML)	41°	1.92 ± 0.60	0.55	−168.2	SCAC	111
	Disordered SiO_2	—	>71.4°	<2.3	<0.65		TPD	106
	α-Al_2O_3(0001) or γ-Al_2O_3(111)	—	>62.4°	<2.6	<0.74		TPD	106
Pt	MgO(100)	1–5	113°	1.51*	0.30	−54.3	GISAXS	112

Combining the adhesion energy data in Tables 1 and 2 reveals several important trends. As shown in Fig. 1, the adhesion energies for different metals on a given oxide (in this case α-$Al_2O_3(0001)$) increase nearly linearly with the surface energy of the metal, and with its sublimation energy ($\Delta H_{sub,M}$) divided by the area per metal atom (approximated here as the atomic volume of the bulk metal, Ω_M, to the 2/3 power). The adhesion energies for molten metal drops are larger than for solid particles by 50 to 100%, but both follow the same trend. We can summarize the results for solids in Fig. 1b by the relationship:

$$E_{adh} = m[(\Delta H_{sub,M}/N_A)/\Omega_M^{2/3}] + b, \quad (4)$$

where N_A is Avogadro's number. The slope and y-intercept for α-$Al_2O_3(0001)$ are $m = 0.15$ and $b = -0.43$ J m^{-2}, respectively.

The trends in Fig. 1 can be easily explained as follows. Both the heat of sublimation and the surface energy of a metal reflect the strength of metal–metal bonding in that pure bulk metal. The same electronic properties that determine the strength with

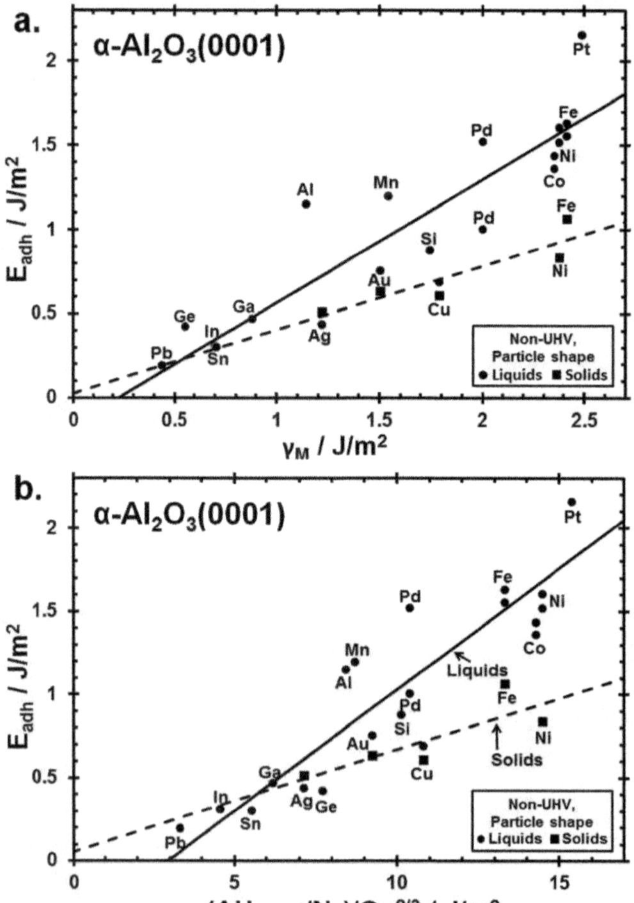

Fig. 1 Adhesion energies of different metals on α-$Al_2O_3(0001)$ plotted *versus* the surface energy of the metal (a) and *versus* the bulk metal's sublimation enthalpy per atom ($\Delta H_{sub,M}/N_A$) divided by the area per metal atom ($\Omega_M^{2/3}$) (b). Black points are based on measurements of particle shapes for molten metal drops (round) and solid particles (squares) without verification of surface cleanliness, from Table 1. Lines through the subsets of data are the best linear fits.

which a metal surface bonds to its own metal atoms also determine to a large extent the strength with which it bonds to other elements, and thus also to oxide surfaces. Similarly, noble gases bond very weakly to themselves, as reflected in very small sublimation energies and surface energies, and thus they also bond very weakly to all other elements and surfaces. Two important exceptions to such trends are the elements Au, which has relatively high sublimation energy but bonds very weakly to oxygen (as measured by the positive heat of formation of Au oxide[114]), and Pb, which has low sublimation energy but bonds relatively strongly to oxygen (as measured by the large negative heat of formation of Pb oxide[114]). Since the dominant bonding of the metal to the oxide is to the oxygen anions (rather than to the cations, see below), it is surprising how well Au and Pb fit the trends in Fig. 1.

Fig. 2a shows the same type of plot as Fig. 1, but now only using the E_{adh} values from Table 2 for different metals measured on clean MgO(100) under UHV conditions, where the oxide surface was proven to be clean. The fit for these clean-surface data is very poor (correlation coefficient = R^2 = 0.49). Note that the points for metals with high heat of oxide formation (Pb and Cu) fall left of this line, and those with low heats of oxide formation (Au and Ag) fall to its right.

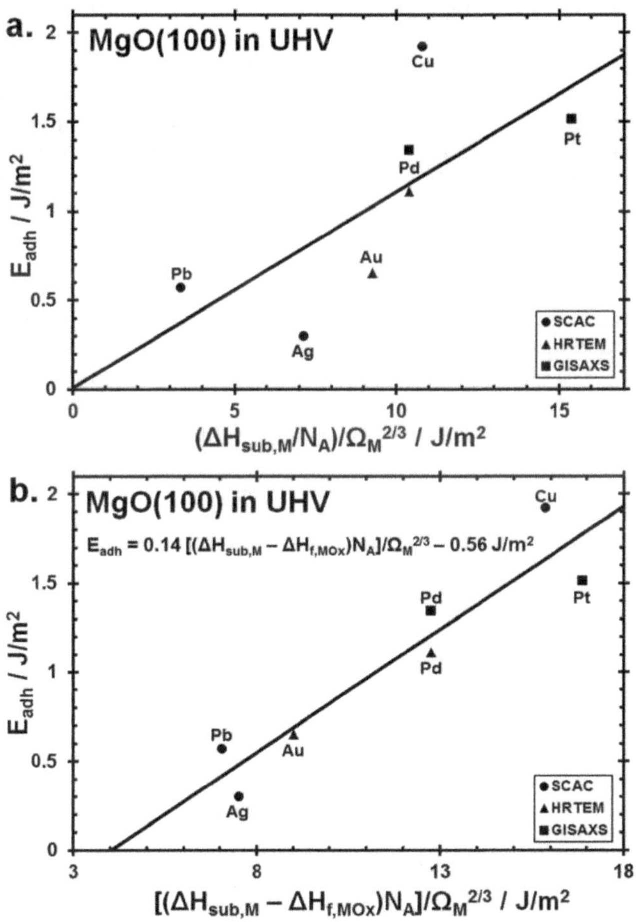

Fig. 2 Adhesion energies of different metals on clean MgO(100) measured in ultrahigh vacuum (from Table 2) plotted *versus* the bulk metal's value of $[\Delta H_{sub,M}/N_A]/\Omega_M^{2/3}$ (a) and *versus* $[(\Delta H_{sub,M} - \Delta H_{f,MOx})/N_A]/\Omega_M^{2/3}$, where $\Delta H_{f,MOx}$ is the standard heat of formation of the most stable bulk oxide of that metal (per mole of metal atoms) (b). The best-fit lines are also shown. Different symbols correspond to different measurement methods.

Several other models have been offered to explain trends in metal/oxide adhesion energies.[96,99,100,115–118] We suggested earlier that the adhesion energy on a given oxide correlates better with the sum of the magnitudes of the metal's heat of sublimation plus the standard heat of formation of the most stable bulk oxide of that metal (per mole of metal atoms), $\Delta H_{f,MOx}$,[115] which followed a similar suggestion by Chatain's group, where they used the enthalpy of formation of the alloy of the metal with the metal of the oxide support instead of the metal's heat of sublimation.[97–100] In Fig. 2b, we have reanalyzed the same data from Table 2 for E_{adh} of different metals measured on clean MgO(100) in UHV based on our earlier correlation with $\Delta H_{sub,M} - \Delta H_{f,MOx}$,[115] but now also dividing that by $\Omega^{2/3}$, as was done in Fig. 1 and 2a. As can be seen, the correlation for MgO(100) improves markedly if we add $-\Delta H_{f,MOx}$ per mol metal to $\Delta H_{sub,M}$ on the *x*-axis. These UHV data for MgO(100) are well fitted with the equation:

$$E_{adh} = m[(\Delta H_{sub,M} - \Delta H_{f,MOx})/N_A]/\Omega^{2/3} + b, \quad (5)$$

where $m = 0.14$ and $b = -0.56$ J m^{-2} (R^2 of 0.88). This quantity $\Delta H_{sub,M} - \Delta H_{f,MOx}$ equals the heat of formation of the metal's most stable oxide from a metal gas atom plus O_2(g), so it more directly reflects the strength of the chemical bonds which each metal can make to oxygen. Obviously, this is a better way to estimate how strongly different metal surfaces can bond to the oxygen anions on the surface of an oxide like MgO than $\Delta H_{sub,M}$ alone. Again, the denominator on this axis ($\Omega^{2/3}$) simply normalizes this energy to the number of metal atoms per unit area, as before.

Unfortunately, replotting the non-UHV data of Fig. 1 for α-Al_2O_3(0001) using this new *x*-axis of Fig. 2b results in a poorer fit ($R^2 = 0.24$ for the liquid points, *versus* 0.81 in Fig. 1b). These non-UHV points are less reliable since these surfaces could easily have surface impurities covering a large fraction of the surface, which were proven to be absent on the UHV points using surface spectroscopies. So we believe that the correlation in eqn (5) and Fig. 2b is more accurate than that in Fig. 1 and 2a.

Fig. 3 shows how the adhesion energy for a given metal varies from oxide to oxide. We have plotted these adhesion energies *versus* the heat of formation of the oxide support per mole of oxygen ($\Delta H_{f,OxSup}$), as suggested by earlier correlations of more limited data with this parameter.[96,115] The points for liquid Ni, Fe and Co are combined into a single line since, according to Fig. 1, these three metals should

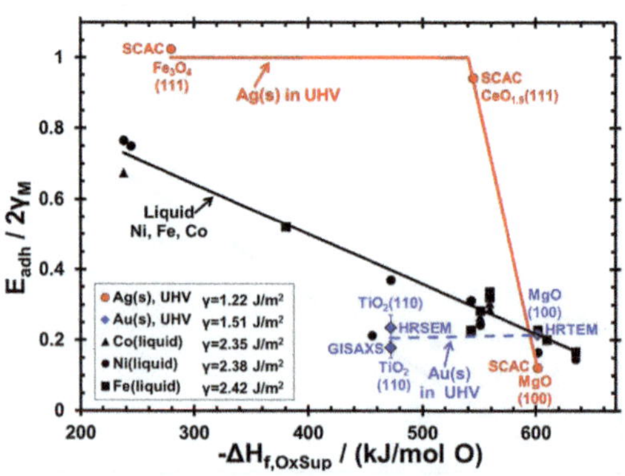

Fig. 3 Adhesion energies of the same metal (normalized to $2\gamma_M$) on different oxide supports plotted *versus* the heat of formation of the oxide support per mole of oxygen ($\Delta H_{f,OxSup}$). Different colors of points and lines are for different metals. Black points are for liquid Ni, Fe and Co droplets on oxide surfaces whose cleanliness was not proven, from Table 1. These metals are grouped together since they have very similar sublimation energies (and thus appear at nearly the same place on the *x*-axis of Fig. 1). Colored points are for solid metals collected on clean oxide surfaces in UHV, from Table 2. Lines through the subsets of data are the best linear fits.

have about the same adhesion energy to any given oxide, as indeed is observed here. As seen, E_{adh} for a given metal (or similar group of metals like Fe/Ni/Co) correlates almost linearly with $-\Delta H_{f,OxSup}$. As $-\Delta H_{f,OxSup}$ decreases (*i.e.*, for oxides where the oxygen in the oxide lattice is less stable), the adhesion energy increases. This linear increase continues only up to a point, since E_{adh} must saturate at $2\gamma_M$. This is its maximum possible value since complete wetting occurs when E_{adh} reaches $2\gamma_M$.[51]

To clarify the above discussion, we should point out that deciding which oxide of a given metal is more stable depends upon which perspective is being considered: the stability of the oxygen atoms or that of the metal atoms in that oxide. Consider the case of Cu oxides. The equilibrium

$$Cu_2O(s) \rightleftharpoons CuO(s) + Cu(s)$$

defines the stability and reactivity of O in these two oxides. If you have only a small amount of O, in which oxide form is it most stable? The answer is: the one with highest $-\Delta H_{f,ox}$ per mol O, which in this case is Cu_2O. (Actually, one should really compare free energies of formation, but we will neglect entropic differences here.) On the other hand, the equilibrium

$$Cu_2O(s) + 1/2\ O_2(g) \rightleftharpoons 2\ CuO(s)$$

defines the relative metal stability and reactivity of the metal in the oxides. If you have only a small amount of Cu in excess O_2, in which form does Cu prefer to be? In this case the answer is: the one with highest $-\Delta H_{f,ox}$ per mol Cu, which is CuO instead. Choosing the most stable oxide depends on which perspective you are taking: that of the O stability/reactivity (as, for example, desired in Fig. 3 where the x-axis approximates the reactivity of the O on the surfaces of different oxides), or that of the metal stability/reactivity (like Fig. 2b, where the x-axis approximates instead the reactivity of different metals toward the same oxide lattice).

To explain this increase in adhesion energy with the decreasing magnitude of the heat of formation (or stability) of the oxide seen in Fig. 3, we will again use the principle of chemical potential mentioned above. Note that the oxide substrates with the least negative heats of formation per mole of O (*i.e.*, those on the far left in Fig. 3), have the highest chemical potential of oxygen atoms in their oxide lattice (and in their surfaces). We assume that the dominant contribution to the bonding between the metal surface and the surface of the oxide lattice is from the bonds of the metal atoms to the oxygen anions of the oxide surface, and not to the lattice anions. This is based on the fact that DFT calculations generally prove that late transition metal adatoms on stoichiometric oxide terraces prefer to bind to oxygen anions rather than cations (see, for example,[51,119,120]). We also recognize that the higher the chemical potential of the surface oxygen atoms in an oxide lattice, the more strongly these atoms can bond to the metal atoms of a metal surface to which it attaches. That is, the more weakly a surface oxygen ion is bound to its neighboring cations within the oxide lattice, the more strongly it can bond to a metal atom or metal particle that approaches its surface.

We see very different slopes for the different metals (and simultaneously different methods) in Fig. 3. We have no explanation for this, but the data for solids and clean surfaces are still very limited.

If we look only at the data in Table 2, which are from UHV studies on surfaces that were verifiably clean, and compare E_{adh} on pairs of oxides for the same metal, we get the following order of increasing adhesion energy:

$$MgO(100) \approx TiO_2(110) \leq \alpha\text{-}Al_2O_3(0001) < CeO_{2-x}(111) \leq Fe_3O_4(111). \quad (6)$$

According to the general trend from liquid metals that E_{adh} should decrease with $-\Delta H_{f,OxSup}$ in Fig. 3, this order should be: $MgO < Al_2O_3 < CeO_2 < TiO_2 \leq SiO_2 <$

Fe_3O_4, which is the same as eqn (6) for those five oxides except for $TiO_2(110)$, which binds more weakly than expected. Thus, this general trend in Fig. 3 should not be used too rigorously, especially since E_{adh} should also depend sensitively on the crystal face of the oxide. This illustrates that there seem to be other factors besides $\Delta H_{f,OxSup}$ that determine the differences between E_{adh} for different oxide surfaces. Differences between oxides, and even between different faces of the same oxide, in the ability of surface oxygen atoms to stabilize the system by rearranging their bonding to nearby cations after the oxide surface is created, will also determine the chemical potential of the surface oxygen atoms. The x-axis in Fig. 3 does not take these differences into account.

Since oxygen vacancies are known to accumulate under late transition metal nanoparticles on reducible oxide surfaces,[15] oxygen vacancies appear to stabilize metal/oxide interfacial bonding. This is also supported by our calorimetric measurements which show that Ag adsorption and adhesion energies on $CeO_{2-x}(111)$ increase with the extent of surface reduction (x),[95,121] and DFT calculations which show that oxygen vacancies stabilize metal adsorption on various oxides.[15,122–125] Thus, the availability and mobility of near-surface oxygen vacancies may also play a role in determining adhesion energies. The increase in Ce^{3+} signal in XPS seen upon Ag adsorption on $CeO_{2-x}(111)$[125,126] which was not accompanied by any measurable loss of electron density from the Ag,[126] could be due to the Ag-induced segregation of oxygen vacancies to the surface. The well-known availability and fast transport of oxygen ions in ceria could help explain the high adhesion energies of metals to $CeO_{2-x}(111)$.

Surface impurities surely have a large effect on adhesion energies too, so one should also be careful using the adhesion energies from Table 1 where the degree of cleanliness was uncertain. We should point out that "clean" oxide surfaces prepared under UHV conditions can vary in atomic-level structure and even composition depending on the details of the recipe used to prepare them. Thus, one must be careful in interpreting the phrase "clean surface" as used here. This means that impurity elements were proven to be below the detection limit of some surface spectroscopy (generally 1–2% of a monolayer). One should refer to the literature to learn more details about the exact structures of these "clean surfaces".

Hydroxyls are often present on the surfaces of oxide supports. We found that surface hydroxyls increase the initial heat of adsorption for Cu measured on MgO(100) by 6–17%,[127] showing that they stabilize metal/oxide bonding when the metal clusters are smaller than ten atoms. However, this added stability does not persist to high metal coverage, so it appears from those data that the high-coverage integral heat of adsorption goes down with –OH addition. This means that the addition of surface –OH decreases E_{adh}. Other studies have found that hydroxyls stabilize small Au clusters on MgO(100).[85,86]

4. The chemical potential of metal atoms in nanoparticles increases quickly with decreasing particle size below 6 nm, but less quickly on "strong" supports

We have previously reported measurements by single crystal adsorption calorimetry (SCAC) of the heats of adsorption of several metal atoms (Ag, Cu, Ca, Li, and Pb) onto clean and ordered single crystal surfaces of several different oxides, including MgO(100), $CeO_2(111)$, $Fe_3O_4(111)$, and the p(2 × 1) oxide film on Mo(100),[94,95,104,105,107,111,113,115,121,127–132] as well as onto a clean but disordered oxide film on W(100).[133] Fig. 4 shows the heat *versus* coverage data for several of these systems of most interest here. Heats are converted here to standard enthalpies of adsorption, $\Delta H_{ad}{}^0$, at the stated temperature. ("Standard" means 1 bar gas pressure, with ΔH_{ad} independent of pressure below 1 bar.)

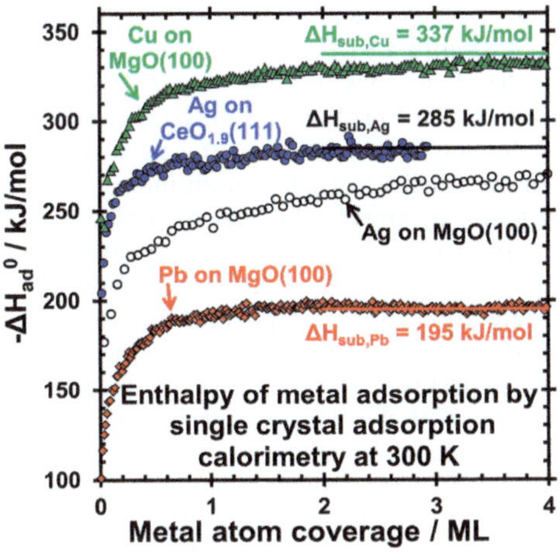

Fig. 4 Example heats of adsorption *versus* coverage data for metals on single-crystal oxides as measured by calorimetry at 300 K. Data are shown for: Pb, Ag and Cu on MgO(100) and Ag on $CeO_2(111)$ (with 5% oxygen vacancies in XPS probe depth). Data from refs. 104, 105, 111, 121. One ML equals the number of oxygen ions per unit area in the topmost atomic plane (1.12×10^{15} cm^{-2} for MgO(100) and 7.9×10^{14} cm^{-2} for $CeO_2(111)$). Horizontal lines mark the bulk heats of sublimation of the metal adsorbate.

For metals from the right side of the periodic table, the heat of adsorption starts low on these oxides and increases with coverage to eventually equal the bulk heat of sublimation, as seen for all the cases in Fig. 4. When such metals are vapor deposited onto single crystal surfaces of the types of oxides used as supports for metal catalysts, they typically grow as 3D clusters rather than wetting the surface.[51] The number of clusters per unit area grows initially with coverage, but usually saturates after a few percent of a monolayer (ML). With further increasing coverage, these clusters grow in size at nearly fixed number density until the particles grow together and coalesce.[51,134] For the systems in Fig. 4, this saturation cluster density is expected to have been reached already after the first pulse with the fluxes used (~0.02 ML pulse^{-1}). A variety of observations support the conclusion that isolated adatoms of metals like Ag, Cu and Pd on such oxides are highly mobile at room temperature, and diffuse rapidly from site to site across the oxide until finding a growing metal cluster.[118] The increase in heat with coverage in Fig. 4 is then dominated by the increasing number of metal–metal bonds that are formed upon adsorption as the cluster size grows: The fourth atom that adds to a trimer can make 3 metal–metal bonds at most, whereas a metal atom that adds to a kink site on an extended metal surface of an FCC(111) or HCP(0001) metal makes six metal–metal bonds, giving twice the heat of adsorption in the simplest pairwise bond additivity model whereby these metal–metal bonds have the same energy independent of size, and only nearest-neighbor bonding is significant. The final heat saturates at the bulk heat of sublimation of the metal, as expected for metal atoms adsorbing on an extended metal surface (or the surface of a very large metal nanoparticle).

However, there are subtle differences in heat of metal adsorption *versus* coverage between different supports that arise from the differences in the bond energies (adhesion energies) between a given metal and the different oxide surfaces. Indeed, these are exactly the types of heat of adsorption *versus* coverage data that were integrated to obtain the adhesion energies summarized in the SCAC entries in Table 2 above.

To separate out the effects of metal particle size from the effect of the support oxide on the heats of adsorption such as those in Fig. 4, it is better to plot these heats of adsorption *versus* the average metal particle size at each coverage instead of directly *versus* coverage. We have done this for a variety of systems, by estimating the average particle size *versus* coverage from various spectroscopic measurements (XPS, AES and LEIS), assuming particles have the shape of hemispherical caps and a fixed number density of particles per unit area.[94,95,107] The data for Ag are summarized in Fig. 5, where the heat of adsorption is plotted *versus* the effective particle diameter (a measure of the particle volume, assuming hemispherical shape). As seen, the heat of adsorption increases rapidly with Ag particle size below 6 nm, but then saturates at the heat of sublimation (285 kJ mol^{-1} [114] for Ag). This increase with size is dominated by the effect of particle size on the number of metal–metal bonds per atom, as noted above. However, for particles smaller than 6 nm, there are also very important differences between the supports: the heat of adsorption of Ag is much larger for the same particle size on the Fe_3O_4(111) and reduced CeO_2(111) surfaces than on MgO(100). This is due to the higher adhesion energy of Ag nanoparticles to Fe_3O_4(111) and CeO_{2-x}(111) than to MgO(100), as summarized in Table 2 and discussed already above. Part of the heat of Ag adsorption here comes from the metal/oxide adhesion energy, since the average contact area of the particles on the oxide support also increases with coverage or size. Since these metal particles nucleate mainly at steps,[51,95,125] the stronger bonding at defects also contributes. However, the steps here comprise only ~5% of the surface, so when the particles exceed ~1.5 nm diameter, most of the metal atoms at the metal/oxide interface are bonded to oxide terraces and *not* to ions at the oxide's steps.

Comparing the more reduced $CeO_{1.8}$(111) surface with $CeO_{1.9}$(111) in Fig. 5 also shows that Ag binds considerably more strongly to Ag clusters smaller than 2 nm when there are more oxygen vacancies present. This increase in stability at surface vacancies is consistent with DFT predictions.[122] Note too that these oxygen vacancies are expected to be mainly at the step edges on CeO_{2-x}(111).[121]

Neglecting entropy differences, the chemical potential of a metal atom in a particle of diameter D is higher than that in the bulk metal (infinite diameter, which we set to zero as the reference state) by an amount:

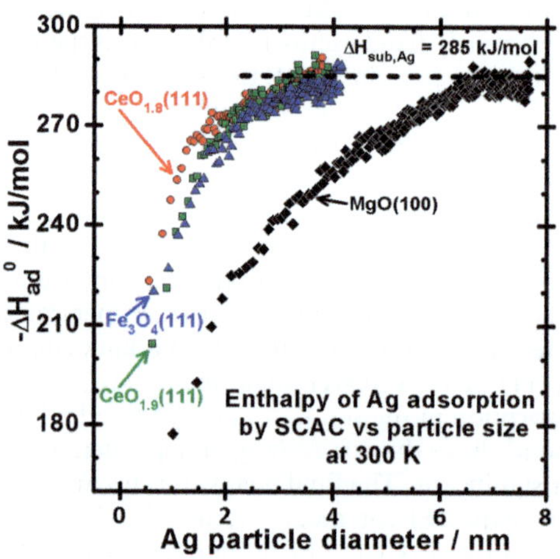

Fig. 5 The heat of Ag atom adsorption during experiments where Ag is vapor deposited onto oxide surfaces at 300 K where Ag atoms transiently adsorb on clean parts of the oxide surface but quickly diffuse across the surface and add to growing Ag particles on the surface. Plotted here is the measured heat of Ag atom adsorption *versus* the Ag particle diameter to which it adds (*i.e.*, the average Ag particle size at the Ag coverage corresponding to that heat value). Data are shown for four different surfaces: Fe_3O_4(111) thin film and two CeO_2(111) thin films with different extents of surface reduction (x = ~0.1 and 0.2 in CeO_{2-x}), all grown on Pt(111) to 4 or 5 nm thickness, and a 4 nm thick MgO(100) film grown on Mo(100). The data for Fe_3O_4(111) are from ref. 107 and those for CeO_2(111) and MgO(100) are from ref. 95. Adapted with permission from ref. 107. Copyright 2011 Royal Society of Chemistry.

$$\mu(D) - \mu(\infty) = \mu(D) = \Delta H_{ad}(D) + \Delta H_{sub,M}, \tag{7}$$

where $\Delta H_{ad}(D)$ is the differential enthalpy of adsorption at that particle diameter. Thus, the heats in Fig. 5, when added to $\Delta H_{sub,M}$ directly, equal this chemical potential $\mu(D)$. These data have been replotted in this format in Fig. 6, to more directly reflect $\mu(D)$.

Anchored metal nanoparticles frequently sinter or coarsen with time during use, starting from a collection of many small, highly dispersed nanoparticles and eventually converting to their thermodynamically-preferred state: fewer, larger particles.[51,135–140] The value $\mu(D)$ directly measures the thermodynamic driving force for such sintering. As seen in Fig. 6, it is large at small D and drops to zero as $-\Delta H_{ad}(D)$ approaches $\Delta H_{sub,M}$. This approach to $\Delta H_{sub,M}$ occurs at much smaller particle size on $Fe_3O_4(111)$ and reduced $CeO_2(111)$ surfaces than on MgO(100). Thus, the driving force for sintering disappears at much smaller particle size for Ag on $Fe_3O_4(111)$ and $CeO_{2-x}(111)$ surfaces than on MgO(100). This is entirely attributable to the much larger adhesion energies of Ag to these surfaces than to MgO(100), discussed above. Clearly, sintering should be slower for small Ag nanoparticles on CeO_{2-x} and $Fe_3O_4(111)$ surfaces than on MgO(100) if the rate accelerates with the thermodynamic driving force as expected. Indeed, rate equations derived based on microkinetic models of sintering mechanisms show that the quantity $\mu(D)$ enters the expression for the rate of sintering quite directly as a negative contribution to the apparent activation energy.[94,118,141–143] This is true whether sintering occurs mainly *via* the Ostwald ripening mechanism, or instead *via* particle diffusion/coalescence. Thus, sintering must be much slower on CeO_{2-x} and $Fe_3O_4(111)$ surfaces than on MgO(100). Indeed, it has been reported that ceria is a more sinter-resistant support for late transition metals than other oxides.[144,145] This is also consistent with our scanning tunneling microscopy (STM) and non-contact atomic force microscopy (nc-AFM) results which showed that Pd nanoparticles sinter far more rapidly on α-$Al_2O_3(0001)$[146] than on $CeO_{2-x}(111)$.[118]

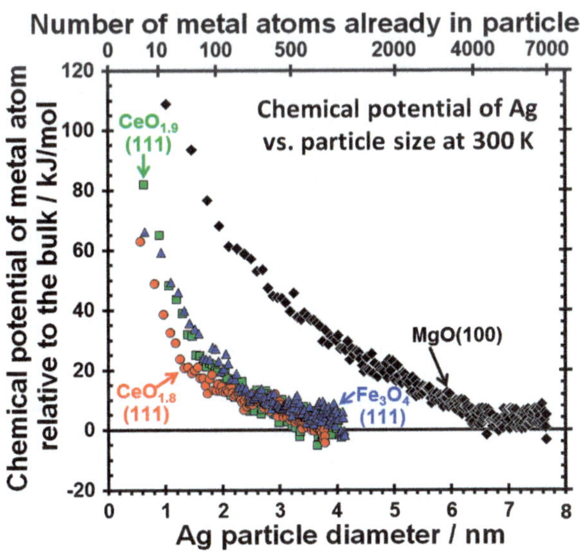

Fig. 6 The chemical potential of Ag atoms in Ag nanoparticles on different oxide surfaces *versus* the average Ag particle size. The chemical potential here is estimated by the difference in magnitude between the bulk heat of sublimation of the metal and its differential heat of adsorption on particles of that size (using the data in Fig. 5), and neglects entropy differences.

In summary, the chemical potential of metal atoms in a nanoparticle of effective diameter D, $\mu(D)$, increases very quickly with decreasing particle size below ~6 nm. Exactly how quickly $\mu(D)$ increases with decreasing D depends on the support. This occurs much more quickly for "weak" supports like MgO(100) (*i.e.*, ones that have a low adhesion energy to the metal particle) than for "strong" supports like CeO_{2-x}(111) or Fe_3O_4(111) (which have large adhesion energies to the particles).

The metal particle shape should also affect the metal chemical potential, since different facets have quite different surface energies,[102] and the product of these energies times the area of each facet adds to the total free energy of the particle and its derivative with size. In general, one would expect that comparing two particles of the same volume and area of attachment to the same support, the shape with the larger fraction of higher-energy facets will have the higher metal chemical potential. To our knowledge, there are no experimental measurements to support this prediction.

5. Increasing the chemical potential of surface metal atoms by decreasing particle size or anchoring them to a "weaker" support increases the strength with which they bond adsorbed catalytic intermediates

We next show that the reactivity of a metal atom in a supported metal nanoparticle correlates with the chemical potential of that metal atom. In general, the higher the chemical potential of an atom, the more strongly it can bond to a new neighbor. A common example of this is found in organic chemistry, where the strength with which C atoms bind to other C atoms is known to depend on the energetic stability of the C atoms involved: Relative to the elements in their standard states (graphite and $H_{2,gas}$), the standard enthalpy of formation per mole C of $-CH_x$ species increases in the order $-CH_3 < -CH_2 < -CH$.[147,148] Consequently, if two of each of these species is bonded together at their C atoms, the resulting C–C bond energy increases in this same order, in this case rather dramatically, from 380 to 730 to 970 kJ mol^{-1} as the number of H atoms on each of these two C atoms decreases from 3 (ethane) to 2 (ethene) to 1 (ethyne). Note that this occurs while the total bond order for each of the two C atoms remains fixed at 4. Similarly, a common trend in organometallic chemistry is that the reactivity of a metal center increases as its "degree of coordinative unsaturation" increases: the higher the degree of coordinative unsaturation of an atom, the higher its energy and the more strongly it can bond to other species. The simple bond-energy bond-order (BEBO) conservation model[149–151] quantified this qualitative trend and even proved accurate in estimating activation energies for gas-phase organic reactions.[149]

Extending this qualitative picture to metal atoms in catalytic surfaces leads to the following prediction: the more weakly a surface metal atom is bound to the surface of a catalytic material (*i.e.*, the higher its chemical potential), the more strongly that metal atom is expected to covalently chemisorb small molecules. This is consistent with the well-known trend that low-coordination metal atoms at steps on metal surfaces bind small molecules more strongly than do more stable metal atoms in higher-coordination environments such as in close-packed terraces. The same effect is commonly seen in bimetallic surfaces, but going in the opposite direction when the metal atoms have lower chemical potential on the bimetallic surface than on a pure metal surface of the same element. For example, when a metal atom resides in a monolayer supported on some other metal to which it binds more strongly than to itself (*i.e.*, when its chemical potential is lower than on the surface of the pure metal), as, for example, in Pd monolayers on Ta(110), Nb(110), Re(0001) and W(110), it binds small molecules like CO more weakly, often dramatically so.[152]

Since surface metal atoms on a given support are bound more weakly to smaller metal nanoparticles than to large metal particles in the size range below 4 nm (Fig. 5), and have lower coordination number there, they have higher chemical potential on the smaller particles. Thus, one expects metal atoms in the smaller metal particles to bind small adsorbed catalytic reaction intermediates more strongly than larger particles, for a given support. Consistent with this are numerous cases for which CO has been reported to bind more strongly to smaller particles of late transition metals in this size range, as we summarized previously,[51] including Pt on bulk α-Al_2O_3(0001), Pt on a highly ordered thin alumina film on NiAl(110), Cu on ZnO(000−1)-O, Pt on ZnO(000−1)-O, Pt on ZnO(0001)-Zn, and Pd on MgO(100). Similarly, adsorbed NO shows greater dissociation probability on smaller supported nanoparticles of Pt, Pd and Rh.[51] Also consistent with this model, the desorption activation energies for O adatoms on Au nanoparticles on TiO_2(110) to produce O_2 gas are ~50 kJ (mol O_2)$^{-1}$ larger from the smallest Au particles (which were only 1.3 layers thick, on average) compared to the largest particles (which resemble step sites on bulk Au(211)). This 50 kJ per mol O_2 difference implies that O adatoms are at least 25 kJ mol^{-1} more stable on the smallest Au particles than on bulk Au.[107,153]

However, the calorimetrically measured heats of adsorption we reported for CO on Pd nanoparticles on Fe_3O_4(111) go in the opposite direction with particle size. The initial heat of adsorption increased by almost 30 kJ mol^{-1} as the particle size increased from 2 to 4 nm, with the particle sizes characterized by STM.[154,155] In marked contrast, the initial heat of adsorption (desorption activation energy) decreased by 33 kJ mol^{-1} as Pd particle size increased from 2 to 5 nm on MgO(100).[156] The only way to explain these apparently contradictory results, whereby the strength of bonding to CO changes by ~25% with particle size in the same size range but in opposite directions, is to attribute it to differences in the way the Pd particles interact with the two different supports used here: Fe_3O_4(111) and MgO(100). Indeed, the chemical potential concept can again explain at least the qualitative difference between these supports. As shown in Fig. 3, metal particles have much higher adhesion energies to Fe_3O_4(111) (which is on the far left side of this figure) than on MgO(100) (which is on the far right). Thus, the chemical potential of a Pd atom in a 2 nm Pd particle on Fe_3O_4(111) will be much lower than for that same particle on MgO(100). As shown in Fig. 6, the chemical potential of Ag atoms is lower by ~50 kJ mol^{-1} for 2 nm Ag particles on Fe_3O_4(111) than for that same sized particle on MgO(100). Indeed, the chemical potential of Ag atoms in 2 nm Ag particles on Fe_3O_4(111) is so low it is already within 20 kJ mol^{-1} of the bulk metal, whereas on MgO(100) it is still 70 kJ mol^{-1} higher than bulk Ag(s). If this difference in chemical potential for 2 nm Ag also applies to 2 nm Pd, at least qualitatively as expected, it explains why the 2 nm Pd particle bonds CO much more strongly when on MgO(100) than on Fe_3O_4(111). This does not, however, explain why the trend with size goes in the opposite direction on Fe_3O_4(111) from what would be expected based on the size dependence summarized above, which should apply to any support. It was suggested during this Faraday Discussion that this might be connected to a weakening of the chemisorptive interaction due to the contraction of the lattice parameter of the Pd particle with decreasing size. Such lattice contractions, however, only partially reduce the effect of particle size on metal atom chemical potential (as obvious in Fig. 6), so that this effect is unlikely to be large enough to explain the complete *reversal* in the trend of CO adsorption energy with particle size observed. The size range where this change in CO stability was observed (2 to 4 nm) also seems to be too large for this type of effect.[157] Some other effect is likely at work here, which has yet to be identified. Perhaps the 2 nm Pd particles absorb a tiny amount of Fe or FeO_x, which could be too small to distinguish from the large signal for the Fe_3O_4(111) support in surface spectroscopies, but still in sufficient quantity to weaken the Pd bond strength to CO.

In any case, it seems that this concept of metal atom chemical potential will also prove to be useful for predicting how chemisorption bond strengths depend on the support material for a given particle size of late transition metal. That is, for a given particle size below 6 nm, metal atoms are much less stable on a "weak" support like

MgO(100) than on "strong" supports like CeO_2(111) and Fe_3O_4(111). Thus, they are predicted to bind small adsorbates more strongly. They should also reach the large-particle limit in this respect at smaller particle sizes when on "strong" supports than when on "weak" supports.

The discussion above is related to small adsorbates like O, CO and NO on later transition metal particles, where the chemical bonds are rather strong and dominated by their covalent character. When the adsorbate is only held to the surface by van der Waals interactions, the trend with particle size mentioned above may not hold. That is because larger metal particles are more polarizable, and thus are likely to make stronger van der Waals bonds to adsorbates. Indeed, methane has been shown to bind more strongly to larger Pd particles when supported on a highly ordered but few-atomic-layer-thick alumina film on NiAl(110), and this trend continues to rather large sizes (above 7 nm).[158]

Acknowledgements

The authors gratefully acknowledge financial support for this work by DOE-OBES Chemical Sciences Division under grant #DE-FG02-96ER14630, and Claude Henry for providing access to his book chapter before publication. CTC also thanks his present and former students and postdocs, whose hard work and superb data and publications made this paper possible.

References

1 D. W. Goodman, *Chem. Rev.*, 1995, **95**, 523–536.
2 K. Coulter, X. P. Xu and D. W. Goodman, *J. Phys. Chem.*, 1994, **98**, 1245–1249.
3 W. Eberhardt, *Surf. Sci.*, 2002, **500**, 242–270.
4 C. E. Gigola, H. R. Aduriz and P. Bodnariuk, *Appl. Catal.*, 1986, **27**, 133–144.
5 J. R. Engstrom, D. W. Goodman and W. H. Weinberg, *J. Am. Chem. Soc.*, 1986, **108**, 4653–4655.
6 P. L. J. Gunter, J. W. Niemantsverdriet, F. H. Ribeiro and G. A. Somorjai, *Catal. Rev. Sci. Eng.*, 1997, **39**, 77–168.
7 M. Haruta, *CATTECH*, 2002, **6**, 102–115.
8 M. Haruta, *Catal. Today*, 1997, **36**, 153–166.
9 A. Wolf and F. Schuth, *Appl. Catal., A*, 2002, **226**, 1–13.
10 A. L. de Oliveira, A. Wolf and F. Schuth, *Catal. Lett.*, 2001, **73**, 157–160.
11 W. J. Shen, Y. Ichihashi and Y. Matsumura, *Catal. Lett.*, 2002, **79**, 125–127.
12 W. J. Shen, A. Kobayashi, Y. Ichihashi, Y. Matsumura and M. Haruta, *Catal. Lett.*, 2001, **73**, 161–165.
13 Y. Matsumura, W. J. Shen, Y. Ichihashi and M. Okumura, *J. Catal.*, 2001, **197**, 267–272.
14 M. Vaarkamp, J. T. Miller, F. S. Modica and D. C. Koningsberger, *J. Catal.*, 1996, **163**, 294–305.
15 J. A. Rodriguez, G. Liu, T. Jirsak, J. Hrbek, Z. P. Chang, J. Dvorak and A. Maiti, *J. Am. Chem. Soc.*, 2002, **124**, 5242–5250.
16 S. Guerin, B. E. Hayden, C. E. Lee, C. Mormiche, J. R. Owen, A. E. Russell, B. Theobald and D. Thompsett, *J. Comb. Chem.*, 2004, **6**, 149–158.
17 B. E. Hayden, D. Pletcher, M. E. Rendall and J. P. Suchsland, *J. Phys. Chem. C*, 2007, **111**, 17044–17051.
18 B. E. Hayden, D. Pletcher, J. P. Suchsland and L. J. Williams, *Phys. Chem. Chem. Phys.*, 2009, **11**, 9141–9148.
19 M. Valden, X. Lai and D. W. Goodman, *Science*, 1998, **281**, 1647–1650.
20 M. Valden, S. Pak, X. Lai and D. W. Goodman, *Catal. Lett.*, 1998, **56**, 7–10.
21 U. Heiz and E. L. Bullock, *J. Mater. Chem.*, 2004, **14**, 564–577.
22 K. Judai, S. Abbet, A. S. Worz, U. Heiz and C. R. Henry, *J. Am. Chem. Soc.*, 2004, **126**, 2732–2737.
23 K. Judai, S. Abbet, A. S. Worz, A. M. Ferrari, L. Giordano, G. Pacchioni and U. Heiz, *J. Mol. Catal. A: Chem.*, 2003, **199**, 103–113.
24 S. S. Lee, C. Y. Fan, T. P. Wu and S. L. Anderson, *J. Am. Chem. Soc.*, 2004, **126**, 5682–5683.
25 P. Liu and J. A. Rodriguez, *J. Chem. Phys.*, 2007, **126**, 164705.

26 J. A. Rodriguez, P. Liu, J. Hrbek, J. Evans and M. Perez, *Angew. Chem., Int. Ed.*, 2007, **46**, 1329–1332.
27 J. L. Lu, H. J. Gao, S. Shaikhutdinov and H.-J. Freund, *Catal. Lett.*, 2007, **114**, 8–16.
28 J. A. Rodriguez, P. Liu, X. Wang, W. Wen, J. Hanson, J. Hrbek, M. Perez and J. Evans, *Catal. Today*, 2009, **143**, 45–50.
29 S. Mostafa, F. Behafarid, J. R. Croy, L. K. Ono, L. Li, J. C. Yang, A. I. Frenkel and B. Roldan Cuenya, *J. Am. Chem. Soc.*, 2010, **132**, 15714–15719.
30 A. Goguet, F. Meunier, J. P. Breen, R. Burch, M. I. Petch and A. F. Ghenciu, *J. Catal.*, 2004, **226**, 382–392.
31 D. Tibiletti, A. Goguet, F. C. Meunier, J. P. Breen and R. Burch, *Chem. Commun.*, 2004, 1636–1637.
32 G. A. Deluga, J. R. Salge, L. D. Schmidt and X. E. Verykios, *Science*, 2004, **303**, 993–997.
33 D. Andreeva, V. Idakiev, T. Tabakova, L. Ilieva, P. Falaras, A. Bourlinos and A. Travlos, *Catal. Today*, 2002, **72**, 51–57.
34 X. S. Liu, O. Korotkikh and R. Farrauto, *Appl. Catal., A*, 2002, **226**, 293–303.
35 Q. Fu, A. Weber and M. Flytzani-Stephanopoulos, *Catal. Lett.*, 2001, **77**, 87–95.
36 M. Flytzani-Stephanopoulos, *MRS Bull.*, 2001, **26**, 885–889.
37 S. Hilaire, X. Wang, T. Luo, R. J. Gorte and J. Wagner, *Appl. Catal., A*, 2001, **215**, 271–278.
38 T. Bunluesin, R. J. Gorte and G. W. Graham, *Appl. Catal., B*, 1998, **15**, 107–114.
39 M. L. Toebes, J. A. van Dillen and Y. P. de Jong, *J. Mol. Catal. A: Chem.*, 2001, **173**, 75–98.
40 F. Charreteur, F. Jaouen, S. Ruggeri and J. P. Dodelet, *Electrochim. Acta*, 2008, **53**, 2925–2938.
41 J. Graciani, J. J. Plata, J. F. Sanz, P. Liu and J. A. Rodriguez, *J. Chem. Phys.*, 2010, **132**, 104703.
42 J. A. Rodriguez, J. Graciani, J. Evans, J. B. Park, F. Yang, D. Stacchiola, S. D. Senanayake, S. G. Ma, M. Perez, P. Liu, J. F. Sanz and J. Hrbek, *Angew. Chem., Int. Ed.*, 2009, **48**, 8047–8050.
43 J. A. Rodriguez, X. Wang, P. Liu, W. Wen, J. C. Hanson, J. Hrbek, M. Perez and J. Evans, *Top. Catal.*, 2007, **44**, 73–81.
44 Z. Yan, S. Chinta, A. A. Mohamed, J. P. Fackler and D. W. Goodman, *J. Am. Chem. Soc.*, 2005, **127**, 1604–1605.
45 A. Vijay, G. Mills and H. Metiu, *J. Chem. Phys.*, 2003, **118**, 6536–6551.
46 E. Wahlstrom, N. Lopez, R. Schaub, P. Thostrup, A. Ronnau, C. Africh, E. Laegsgaard, J. K. Norskov and F. Besenbacher, *Phys. Rev. Lett.*, 2003, **90**, 026101.
47 A. Sanchez, S. Abbet, U. Heiz, W. D. Schneider, H. Hakkinen, R. N. Barnett and U. Landman, *J. Phys. Chem. A*, 1999, **103**, 9573–9578.
48 Y. H. Zhang, M. L. Toebes, A. van der Eerden, W. E. O'Grady, K. P. de Jong and D. C. Koningsberger, *J. Phys. Chem. B*, 2004, **108**, 18509–18519.
49 C. R. Henry, C. Chapon, S. Giorgio and C. Goyhenex, in *Chemisorption and Reactivity on Supported Clusters and Thin Films*, ed. R. M. Lambert and G. Pacchioni, Kluwer Acad. Publ., Amsterdam, 1997, p. 117.
50 C. R. Henry, *Surf. Sci. Rep.*, 1998, **31**, 235–325.
51 C. T. Campbell, *Surf. Sci. Rep.*, 1997, **27**, 1–111.
52 H.-J. Freund, *Angew. Chem., Int. Ed. Engl.*, 1997, **36**, 452.
53 M. Bäumer and H.-J. Freund, *Prog. Surf. Sci.*, 1999, **61**, 127–198.
54 H.-J. Freund, M. Bäumer and H. Kuhlenbeck, *Adv. Catal.*, 2000, **45**, 333–384.
55 R. Persaud and T. E. Madey, in *The Chemical Physics of Solid Surfaces and Heterogeneous Catalysis*, ed. D. A. King and D. P. Woodruff, Elsevier, Amsterdam, 1997, vol. 8.
56 D. R. Rainer and D. W. Goodman, *J. Mol. Catal. A: Chem.*, 1998, **131**, 259–283.
57 U. Diebold, J. M. Pan and T. E. Madey, *Surf. Sci.*, 1995, **331–333**, 845–854.
58 R. J. Lad, *Surf. Rev. Lett.*, 1995, **2**, 109–126.
59 V. Johanek, M. Laurin, J. Hoffmann, S. Schauermann, A. W. Grant, B. Kasemo, J. Libuda and H.-J. Freund, *Surf. Sci.*, 2004, **561**, L218–L224.
60 S. Bertarione, D. Scarano, A. Zecchina, V. Johanek, J. Hoffmann, S. Schauermann, J. Libuda, G. Rupprechter and H.-J. Freund, *J. Catal.*, 2004, **223**, 64–73.
61 A. M. Doyle, S. K. Shaikhutdinov and H.-J. Freund, *J. Catal.*, 2004, **223**, 444–453.
62 C. Lemire, R. Meyer, S. Shaikhutdinov and H.-J. Freund, *Angew. Chem., Int. Ed.*, 2004, **43**, 118–121.
63 R. Meyer, M. Bäumer, S. K. Shaikhutdinov and H.-J. Freund, *Surf. Sci.*, 2003, **546**, L813–L819.
64 A. M. Doyle, S. K. Shaikhutdinov, S. D. Jackson and H.-J. Freund, *Angew. Chem., Int. Ed.*, 2003, **42**, 5240–5243.
65 D. W. Goodman, *J. Catal.*, 2003, **216**, 213–222.

66 A. Kolmakov and D. W. Goodman, *Chem. Rec.*, 2002, **2**, 446–457.
67 A. K. Santra and D. W. Goodman, *J. Phys.: Condens. Matter*, 2003, **15**, R31–R62.
68 J. Oh, T. Kondo, D. Hatake, Y. Iwasaki, Y. Honma, Y. Suda, D. Sekiba, H. Kudo and J. Nakamura, *J. Phys. Chem. Lett.*, 2010, **1**, 463–466.
69 S. H. Lee, S. S. Han, J. K. Kang, J. H. Ryu and H. M. Lee, *Surf. Sci.*, 2008, **602**, 1433–1439.
70 J. P. Jalkanen, M. Halonen, D. Fernandez-Torre, K. Laasonen and L. Halonen, *J. Phys. Chem. A*, 2007, **111**, 12317–12326.
71 S. Nemsak, K. Masek and V. Matolin, *Surf. Sci.*, 2007, **601**, 4475–4478.
72 J. Akola and H. Hakkinen, *Phys. Rev. B: Condens. Matter Mater. Phys.*, 2006, **74**, 165404.
73 Z. Song, T. H. Cai, J. C. Hanson, J. A. Rodriguez and J. Hrbek, *J. Am. Chem. Soc.*, 2004, **126**, 8576–8584.
74 M. R. C. Hunt and R. E. Palmer, *Philos. Trans. R. Soc. London, Ser. A*, 1998, **356**, 231–247.
75 J. C. Parlebas, P. Kruger, M. Taguchi, C. Demangeat and A. Kotani, *Surf. Sci.*, 1997, **377–379**, 177–180.
76 H. Xu, H. Permana, Y. Lu and K. Y. S. Ng, *Surf. Sci.*, 1995, **325**, 285–293.
77 F. J. C. S. Aires, P. Sautet, J. L. Rousset, G. Fuchs and P. Melinon, *J. Vac. Sci. Technol., B*, 1994, **12**, 1776–1779.
78 G. W. Clark and L. L. Kesmodel, *J. Vac. Sci. Technol., B*, 1993, **11**, 131–136.
79 U. Muller, K. Sattler, J. Xhie, N. Venkateswaran and G. Raina, *J. Vac. Sci. Technol., B*, 1991, **9**, 829–832.
80 M. Bowker, P. Stone, R. Bennett and N. Perkins, *Surf. Sci.*, 2002, **511**, 435–448.
81 M. Bowker, P. Stone, P. Morrall, R. Smith, R. Bennett, N. Perkins, R. Kvon, C. Pang, E. Fourre and M. Hall, *J. Catal.*, 2005, **234**, 172–181.
82 M. Bowker, *Surf. Sci.*, 2009, **603**, 2359–2362.
83 M. Bowker, M. Broughton, A. Carley, P. Davies, D. Morgan, J. Crouch, G. Lalev, S. Dimov and D. T. Pham, *Langmuir*, 2010, **26**, 16261–16266.
84 J. B. Park, J. Graciani, J. Evans, D. Stacchiola, S. G. Ma, P. Liu, A. Nambu, J. F. Sanz, J. Hrbek and J. A. Rodriguez, *Proc. Natl. Acad. Sci. U. S. A.*, 2009, **106**, 4975–4980.
85 M. A. Brown, E. Carrasco, M. Sterrer and H.-J. Freund, *J. Am. Chem. Soc.*, 2010, **132**, 4064–4065.
86 M. A. Brown, Y. Fujimori, F. Ringleb, X. Shao, F. Stavale, N. Nilius, M. Sterrer and H.-J. Freund, *J. Am. Chem. Soc.*, 2011, **133**, 10668–10676.
87 S. M. McClure, M. Lundwall, Z. Zhou, F. Yang and D. W. Goodman, *Catal. Lett.*, 2009, **133**, 298–306.
88 S. M. McClure, M. J. Lundwall and D. W. Goodman, *Proc. Natl. Acad. Sci. U. S. A.*, 2010, **108**, 931–936.
89 S. M. McClure and D. W. Goodman, *Top. Catal.*, 2011, **54**, 349–362.
90 S. Shaikhutdinov and H.-J. Freund, *ChemPhysChem*, 2013, **14**, 71–77.
91 A. Bruix, J. A. Rodriguez, P. J. Ramirez, S. D. Senanayake, J. Evans, J. B. Park, D. Stacchiola, P. Liu, J. Hrbek and F. Illas, *J. Am. Chem. Soc.*, 2012, **134**, 8968–8974.
92 B. Roldan Cuenya, *Thin Solid Films*, 2010, **518**, 3127–3150.
93 F. Behafarid and B. Roldan Cuenya, *Surf. Sci.*, 2012, **606**, 908–918.
94 C. T. Campbell, S. C. Parker and D. E. Starr, *Science*, 2002, **298**, 811–814.
95 J. A. Farmer and C. T. Campbell, *Science*, 2010, **329**, 933–936.
96 J. G. Li, *J. Am. Ceram. Soc.*, 1992, **75**, 3118–3126.
97 D. Chatain, I. Rivollet and N. Eustathopoulos, *J. Chim. Phys. Phys.-Chim. Biol.*, 1986, **83**, 561–567.
98 D. Chatain, I. Rivollet and N. Eustathopoulos, *J. Chim. Phys. Phys.-Chim. Biol.*, 1987, **84**, 201–203.
99 D. Chatain, L. Coudurier and N. Eustathopoulos, *Rev. Phys. Appl.*, 1988, **23**, 1055–1064.
100 R. Sangiorgi, M. L. Muolo, D. Chatain and N. Eustathopoulos, *J. Am. Ceram. Soc.*, 1988, **71**, 742–748.
101 W. L. Winterbottom, *Acta Metall.*, 1967, **15**, 303–310.
102 L. Vitos, A. V. Ruban, H. L. Skriver and J. Kollar, *Surf. Sci.*, 1998, **411**, 186–202.
103 C. Bombis, A. Emundts, M. Nowicki and H. P. Bonzel, *Surf. Sci.*, 2002, **511**, 83–96.
104 D. E. Starr, D. J. Bald, J. E. Musgrove, J. T. Ranney and C. T. Campbell, *J. Chem. Phys.*, 2001, **114**, 3752–3764.
105 J. H. Larsen, J. T. Ranney, D. E. Starr, J. E. Musgrove and C. T. Campbell, *Phys. Rev. B: Condens. Matter*, 2001, **63**, 195410.
106 C. T. Campbell and J. R. V. Sellers, *Chem. Rev.*, 2013, **113**(6), 4106–4135.
107 C. T. Campbell, J. C. Sharp, Y. X. Yao, E. M. Karp and T. L. Silbaugh, *Faraday Discuss.*, 2011, **152**, 227–239.

108 S. Giorgio, C. Chapon, C. R. Henry, G. Nihoul and J. M. Penisson, *Philos. Mag. A*, 1991, **64**, 87–96.
109 H. Graoui, S. Giorgio and C. R. Henry, *Surf. Sci.*, 1998, **417**, 350–360.
110 C. Revenant, F. Leroy, R. Lazzari, G. Renaud and C. R. Henry, *Phys. Rev. B: Condens. Matter Mater. Phys.*, 2004, **69**, 035411.
111 J. T. Ranney, D. E. Starr, J. E. Musgrove, D. J. Bald and C. T. Campbell, *Faraday Discuss.*, 1999, **114**, 195–208.
112 J. Olander, R. Lazzari, J. Jupille, B. Mangili, J. Goniakowski and G. Renaud, *Phys. Rev. B: Condens. Matter Mater. Phys.*, 2007, **76**, 075409.
113 J. H. Larsen, D. E. Starr and C. T. Campbell, *J. Chem. Thermodyn.*, 2001, **33**, 333–345.
114 W. M. Haynes, ed., *CRC Handbook of Chemistry and Physics (Internet Version)*, CRC Press/Taylor and Francis, BocaRaton, FL, 2012.
115 C. T. Campbell and D. E. Starr, *J. Am. Chem. Soc.*, 2002, **124**, 9212–9218.
116 M. Hu, S. Noda and H. Komiyama, *Surf. Sci.*, 2002, **513**, 530–538.
117 F. Didier and J. Jupille, *Surf. Sci.*, 1994, **314**, 378–384.
118 C. T. Campbell, *Acc. Chem. Res.*, 2013, DOI: 10.1021/ar3003514.
119 L. J. Xu, G. Henkelman, C. T. Campbell and H. Jonsson, *Phys. Rev. Lett.*, 2005, **95**, 146103.
120 M. M. Branda, N. C. Hernandez, J. F. Sanz and F. Illas, *J. Phys. Chem. C*, 2010, **114**, 1934–1941.
121 J. A. Farmer, J. H. Baricuatro and C. T. Campbell, *J. Phys. Chem. C*, 2010, **114**, 17166–17172.
122 J. H. Wang, M. L. Liu and M. C. Lin, *Solid State Ionics*, 2006, **177**, 939–947.
123 C. J. Zhang, A. Michaelides and S. J. Jenkins, *Phys. Chem. Chem. Phys.*, 2011, **13**, 22–33.
124 G. Pacchioni, *Phys. Chem. Chem. Phys.*, 2013, **15**, 1737–1757.
125 P. Luches, F. Pagliuca, S. Valeri, F. Illas, G. Preda and G. Pacchioni, *J. Phys. Chem. C*, 2012, **116**, 1122–1132.
126 D. D. Kong, G. D. Wang, Y. H. Pan, S. W. Hu, J. B. Hou, H. B. Pan, C. T. Campbell and J. F. Zhu, *J. Phys. Chem. C*, 2011, **115**, 6715–6725.
127 D. E. Starr, S. F. Diaz, J. E. Musgrove, J. T. Ranney, D. J. Bald, L. Nelen, H. Ihm and C. T. Campbell, *Surf. Sci.*, 2002, **515**, 13–20.
128 C. T. Campbell and O. Lytken, *Surf. Sci.*, 2009, **603**, 1365–1372.
129 D. E. Starr and C. T. Campbell, *J. Phys. Chem. B*, 2001, **105**, 3776–3782.
130 J. Zhu, J. A. Farmer, N. Ruzycki, L. Xu, C. T. Campbell and G. Henkelman, *J. Am. Chem. Soc.*, 2008, **130**, 2314–2322.
131 J. A. Farmer, C. T. Campbell, L. Xu and G. Henkelman, *J. Am. Chem. Soc.*, 2009, **131**, 3098–3103.
132 J. A. Farmer, N. Ruzycki, J. F. Zhu and C. T. Campbell, *Phys. Rev. B: Condens. Matter Mater. Phys.*, 2009, **80**, 035418.
133 J. T. Stuckless, D. E. Starr, D. J. Bald and C. T. Campbell, *J. Chem. Phys.*, 1997, **107**, 5547–5553.
134 J. A. Venables, *Introduction to Surface and Thin Film Processes*, Cambridge University Press, Cambridge, 2000.
135 P. Wynblatt and N. A. Gjostein, in *Progess in Solid State Chemistry*, ed. J. O. McCaldin and G. A. Somorjai, Pergamon, Oxford, 1975, vol. 9, p. 21.
136 P. Wynblatt and N. A. Gjostein, *Acta Metall.*, 1976, **24**, 1165–1174.
137 E. Ruckenstein and D. B. Dadyburjor, *Rev. Chem. Eng.*, 1983, **1**, 251–282.
138 G. A. Fuentes and E. Salinas-Rodriguez, *Catalyst Deactivation 1997*, 1997, vol. 111, pp. 573–583.
139 C. H. Bartholomew, *Catalyst Deactivation 1997*, 1997, vol. 111, pp. 585–592.
140 J. Sehested, *J. Catal.*, 2003, **217**, 417–426.
141 S. C. Parker and C. T. Campbell, *Phys. Rev. B: Condens. Matter Mater. Phys.*, 2007, **75**, 035430.
142 S. C. Parker and C. T. Campbell, *Top. Catal.*, 2007, **44**, 3–13.
143 M. J. J. Jak, C. Konstapel, A. van Kreuningen, J. Chrost, J. Verhoeven and J. W. M. Frenken, *Surf. Sci.*, 2001, **474**, 28–36.
144 D. Kalakkad, A. K. Datye and H. Robota, *Appl. Catal., B*, 1992, **1**, 191–219.
145 J. M. Schwartz and L. D. Schmidt, *J. Catal.*, 1992, **138**, 283–293.
146 S. L. Tait, L. T. Ngo, Q. M. Yu, S. C. Fain and C. T. Campbell, *J. Chem. Phys.*, 2005, **122**, 064712.
147 M. W. Chase, Jr., *J. Phys. Chem. Ref. Data*, 1998, **Monograph 9**, 1–1951.
148 NIST Chemistry Webbook, NIST Standard Reference Database Number 69, http://webbook.nist.gov/chemistry/.
149 H. S. Johnston and C. Parr, *J. Am. Chem. Soc.*, 1963, **85**, 2544–2551.
150 W. H. Weinberg and R. P. Merrill, *Surf. Sci.*, 1972, **33**, 493–515.

151 W. H. Weinberg, *J. Vac. Sci. Technol.*, 1973, **10**, 89–94.
152 C. T. Campbell, in *Handbook of Heterogeneous Catalysis*, ed. G. Ertl, H. Knözinger and J. Weitkamp, VCH, Weinheim, Germany, 1997, ch. 4.5, pp. 814–826.
153 V. Bondzie, S. C. Parker and C. T. Campbell, *Catal. Lett.*, 2011, **141**, 1721–1721.
154 J.-H. Fischer-Wolfarth, J. A. Farmer, J. M. Flores-Camacho, A. Genest, I. V. Yudanov, N. Rosch, C. T. Campbell, S. Schauermann and H.-J. Freund, *Phys. Rev. B: Condens. Matter Mater. Phys.*, 2010, **81**, 241416.
155 J. M. Flores-Camacho, J.-H. Fischer-Wolfarth, M. Peter, C. T. Campbell, S. Schauermann and H.-J. Freund, *Phys. Chem. Chem. Phys.*, 2011, **13**, 16800–16810.
156 C. Duriez, C. R. Henry and C. Chapon, *Surf. Sci.*, 1991, **253**, 190–204.
157 L. Li, A. H. Larsen, N. A. Romero, V. A. Morozov, C. Glinsvad, F. Abild-Pedersen, J. Greeley, K. W. Jacobsen and J. K. Norskov, *J. Phys. Chem. Lett.*, 2013, **4**, 222–226.
158 K. Watanabe, Y. Matsumoto, M. Kampling, K. Al-Shamery and H.-J. Freund, *Angew. Chem., Int. Ed.*, 1999, **38**, 2192–2194.
159 R. Lazzari, G. Renaud, J. Jupille and F. Leroy, *Phys. Rev. B*, 2007, **76**, 125412.
160 L. Zhang, F. Cosandey, R. Persaud and T. E. Madey, *Surf. Sci.*, 1999, **439**, 73–85.

Exploring surface science and restructuring in reactive atmospheres of colloidally prepared bimetallic CuNi and CuCo nanoparticles on SiO_2 *in situ* using ambient pressure X-ray photoelectron spectroscopy†

Simon K. Beaumont,‡[a] Selim Alayoglu,[a] Vladimir V. Pushkarev,‡[a] Zhi Liu,[c] Norbert Kruse[d] and Gabor A. Somorjai*[ab]

Received 10th December 2012, Accepted 2nd January 2013
DOI: 10.1039/c2fd20145c

Bimetallic nanoparticles (~11 nm diameter) of CuNi and CuCo were prepared by a new synthetic route and the 1 : 1 atomic ratio of their constituent elements confirmed using STEM-EDS at a single particle level. These nanoparticles, supported on the native oxide layer of a silicon wafer, were studied *in situ* in a series of reactive gas atmospheres ($H_2 \rightarrow CO$ or $CO/H_2 \rightarrow O_2 \rightarrow H_2$) using ambient pressure X-ray photoelectron spectroscopy (AP-XPS). Despite the deliberate similarity of nickel and cobalt with respect to copper, their restructuring behaviour is different. CuNi nanoparticles were found to surface segregate nickel in H_2, but copper in O_2 reversibly, while CuCo nanoparticles were found to surface segregate copper irreversibly when exposed to O_2, such that the surface remains copper rich when re-exposed to H_2. Both systems also restructure in opposition to the behaviour predicted by the surface free energies and enthalpies of oxide formation of the elements from which they are comprised – factors that are seen to control restructuring and surface segregation in many similar systems.

Introduction

Bimetallic heterogeneous catalysts have long been known to exhibit interesting or useful properties not found in their individual constituent metals. For this reason, they find extensive use in many important chemical processes. Obvious examples include the use of bimetallic catalysts containing either Pt or Pd for isomerisation, olefin hydrogenation and alcohol synthesis, or PtSn in hydrocarbon reforming and dehydrogenation reactions.[1] Increasingly, there are numerous reports in the

[a]*Department of Chemistry, University of California, Berkeley, CA 94720, USA. E-mail: somorjai@berkeley.edu*
[b]*Chemical Sciences Division, Lawrence Berkeley National Laboratory, Berkeley, CA 94720, USA*
[c]*Advanced Light Source, Lawrence Berkeley National Laboratory, Berkeley, CA 94720, USA*
[d]*Universiteé Libre de Bruxelles, Campus de la Plaine, CPMPT–CP 243, 50 Av. F. D. Roosevelt, 1050 Bruxelles, Belgium*

† Electronic Supplementary Information (ESI) available: Additional experimental information and SEM characterisation; Additional XPS data. See DOI: 10.1039/c2fd20145c
‡ Current addresses: SKB: Durham University, Department of Chemistry, Science Site, South Road, Durham DH3 1LE, UK; VVP: Dow Corning Corp., 2200 Salzburg Road, Midland, 48686 MI, USA.

literature of new bimetallic combinations being used as heterogeneous catalysts for a large variety of chemical transformations. A current example of this is the pressure to develop or improve catalytic transformations to tackle dwindling fossil fuel reserves – this has led to a huge interest in syngas conversion chemistry. In this field many bimetallic systems have been considered,[2] but also some very systematic theoretical efforts have been employed to consider all plausible combinations of bimetallic surfaces.[3] Another merit of bimetallics has been the possibility of lowering cost while maintaining performance through the use of a precious metal in combination with a more abundant, cheaper metal, with the properties of the former still being retained. An example of this is the suggested use of Pt skin–Ni core nanoparticles as PEM fuel cell catalysts.[4]

Early research in the use of heterogeneous bimetallic catalysts began by developing the well known theories of 'ensemble' *vs.* 'electronic' effect,[5] often responsible for the enhanced properties conferred upon a catalyst by surface alloying. The advent of ultra high vacuum science in the latter half of the 20th century allowed surface systems to be studied in detail and much work was done to explore and understand the surface segregation of metals.[6] For bulk materials their behaviours in vacuum can be understood as a thermodynamic competition of entropy, cohesive and surface energies.[7] Lattice strain and relaxation effects were generally considered too small to be important.[6] In the present paper we focus on exploring the changes on going from a bulk material to bimetallic metal nanoparticles, of which heterogeneous catalysts are normally comprised. This changes a number of important factors by essentially diminishing the importance of bulk energy terms relative to surface energy terms on the thermodynamic equilibrium structure of the particle. These are likely to include:

• The greater importance of surface free energy. (It is also noteworthy that the surface energy itself will be particle size dependant,[8] as has been shown experimentally for PtRh nanoparticles.[9])

• Surface–adsorbate interactions becoming a significant energetic term when balanced by very little 'bulk'. Furthermore, surface oxide or hydride formation a few atomic layers deep may account for a large fraction of the particle (as there are many more surface atoms) and thus be a major energetic term.

• Diminished importance of kinetic limitations; in many bulk systems it is valid to assume only the top layer or layers of an alloy are restructured at moderate temperature,[6] whereas for nanoparticles complete restructuring of the material to reach thermodynamic equilibrium has been seen on the 15 min timescale of an ambient pressure XPS experiment.[10]

• Entropy of mixing is less important as partial surface segregation represents a much higher fraction of the possible configurations available to the atoms within the nanoparticle (in contrast to a bulk material where entropy strongly promotes mixing into the bulk).

• The low coordination of many surface sites means relaxation and strain effects can be expected to also differ from those in a bulk material, as seen for RuPt core–shell nanoparticles.[11]

The first two of these points concerning the increased importance of surface energy and adsorbate–surface interactions are particularly significant in catalysis, where very reactive, high pressure gases (or liquids) are usually present. In 2004 Lang *et al.* reported apparently anomalous results in the infrared characterisation of AuPt clusters on SiO_2, using CO as a probe molecule.[12] They observed the Au signal decrease and the Pt signal growth under exposure to CO at moderate temperature (140 °C). This was attributed to surface segregation resulting from the strong interaction of Pt with CO (while Au has the lower surface free energy in vacuum).

As indicated in Table 1, in the last decade the phenomena of bimetallic nanoparticle restructuring has been the focus of considerable work by our group and others, in particular utilising new *in situ* techniques, such as ambient pressure XPS. Indeed, bimetallic nanoparticle catalysts have been the focus of a variety of *in situ* studies

Table 1 Reported bimetallic nanoparticles systems studied under reactive gas conditions and the resulting structural effects observed

System studied	Technique used	Conditions and segregation effect		Factors controlling structure
AuCu[18]	*In situ* XANES	500 °C H_2 Cu migrates to core	3–400 °C O_2/CO CuO_x surface patches	N/a
AuPd[19]	*In situ* XAFS, PDF and DRIFTS	Vac; RT No surface enrichment	350 °C CO Pd rich surface	CO–Pd bond formation
AuPd[20]	AP-XPS	Vac; RT Pd rich surface	200 °C O_2/CO Au-rich particles, restructure to Au surfaces, others remain Pd rich	Surface/cohesive energy; adsorption enthalpies
AuPt[12]	Infrared of CO stretching	Vac; RT Au migrates to surface	140 °C CO Pt enriched on surface	Workfunction *vs.* M–CO bond
SnPt[21]	*In situ* XAFS/ QXAFS	400 °C O_2 SnO_2 on surface	400 °C H_2 returns to Pt_3Sn	Structural oxidation
PtRh[10]	AP-XPS	300 °C NO Rh surface segregation	300 °C H_2 Pt enriched surface	$\Delta H^{\ominus}{}_f$\{oxide\} *vs.* surface free energy
PdRh[22,10]	AP-XPS	300 °C NO Rh surface segregation	300 °C NO + CO[a] or H_2 Pd surface segregation	Rh oxide formation
PtPd[10,22]	AP-XPS	300 °C NO Pd remains on surface	300 °C NO + CO[a] or H_2 Pd remains on surface	$\Delta H^{\ominus}{}_f$\{oxide\} *vs.* surface free energy
PtCu[19]	*In situ* XAFS, PDF and DRIFTS	Vac; RT Cu rich surface	350 °C CO Pt rich surface	CO–Pt surface adsorption
PtCo[23]	Electrochemistry	200 °C CO Pt surface segregation		Segregation *vs.* surface mixing; Pt–CO *vs.* Co–CO
PtCo[24]	AP-XPS, *in situ* NEXAFS	247 °C O_2 Co enriched at surface	247 °C H_2 Pt segregates to surface	Kinetically controlled
PtCo[25]	AP-XPS and ETEM	Vac; RT Both metals on surface	250 °C in H_2 Pt forms surface shell	N/a
PtCo[26]	AP-XPS, *in situ* NEXAFS	125 °C H_2 Pt surface segregation	125 °C in CO/O_2 Co segregates to surface, not oxidised	Adsorbate induced restructuring
AgCu[27]	AP-XPS	247 °C O_2 and O_2/C_2H_4 Cu enriched at surface	247 °C H_2 less Cu at surface	Surface energies and CuO formation

[a] NO + CO is reducing.

and the details of these are well reviewed elsewhere.[13] The ambient pressure XPS technique is particularly powerful for addressing the question of how and when nanoparticles undergo restructuring. It enables element and oxidation state specific information to be collected in the 0.1–10 torr range – where there are already $\sim 10^6$ collisions per second per surface atom making mass transfer of adsorbates in the gas phase prior to adsorption cease to be rate limiting in most cases.[14] This technique, described elsewhere,[15] circumvents the usual limitation that emitted photoelectrons have too short a mean free path in ambient gas to be measured. This is achieved by bringing a small, differentially pumped aperture cone very close to the sample surface. This allows gas pressure at the sample to be maintained, while photoelectrons emitted from the sample quickly escape through the aperture to near vacuum conditions, enabling them to be collected and analysed. In combination with synchrotron radiation, AP-XPS is especially powerful, allowing highly surface sensitive measurements and depth profiling by varying the incident photon energy.[16] (This utilises the fact the incident photon energy can be tuned to limit the possible mean free path of photoelectrons produced from a given core level, thus controlling the depth from within the sample that the spectra obtained represent.)[17]

Table 1 illustrates many examples in which oxidation or formation of surface oxides is suggested to be the dominant factor in their restructuring. This has either been for systems such as PtRh where neither has a strong propensity to form oxides (it is their relative difference) or for systems such as PtCo where the size mismatch of the atoms is very large and may be contributory to restructuring effects. All these examples involve at least one heavy element (4d or 5d) and to our knowledge these types of studies haven't been conducted on any combinations of 3d metals. These have much higher propensities to form oxides and very similar sizes. Owing to their abundance, and thus lower cost, they are also a very common choice as possible elements to involve in forming new bimetallic catalysts. Accordingly, we have chosen to consider Cu in combination with Co and Ni. These three all have similar metallic radii (128, 125 and 125 pm respectively) and much greater enthalpies of oxide formation per metal atom than the precious metals studied previously.[28]

Here, CuCo and CuNi nanoparticles of 1 : 1 atomic composition of the two metals and of similar sizes (~11 nm) were prepared using a new colloidal synthesis. These have then been supported on a planar SiO_2 surface and characterised in oxidising and reducing environments using ambient pressure XPS. With this new data, we aim to inform a discussion of how such bimetallic nanoparticles behave under catalytic conditions and the importance of addressing this issue experimentally and theoretically in the preparation of new heterogeneous bimetallic catalysts.

Experimental methods

A Synthesis of $Cu_{50}Co_{50}$ and $Cu_{50}Ni_{50}$ nanoparticles

All chemicals were used as obtained from the supplier. $Ni(acac)_2$ was stored in darkness.

11 nm $Cu_{50}Co_{50}$ nanoparticles. The metal precursor salts were dissolved in a reducing, coordinating solvent (oleylamine) at 80 °C. Typically, $Cu(acac)_2$ (27.1 mg, Aldrich, ≥99.99% pure) and $Co(acac)_2$ (26.7 mg, Aldrich, ≥99.0% pure) were dissolved in oleylamine (5 mL) in a 50 mL round bottom flask. The flask was evacuated and subsequently flushed with Ar. The evacuation/purging cycle was repeated several times before the solution was purged by bubbling Ar through it for 15 min. Finally, the reaction flask was submerged in an oil bath preheated to 230 °C. The solution turned black, indicating the formation of colloidal particles after approximately 2 min. This colloidal suspension was then aged at 230 °C for 10 min prior to terminating the reaction by rapid cooling by removing it from the oil bath. After reaching room temperature the nanoparticles were precipitated from the colloidal solution with acetone, centrifuged and re-dispersed in toluene. As synthesized nanoparticles were then stored in toluene until further use.

11 nm $Cu_{50}Ni_{50}$ nanoparticles. The synthetic procedure was identical to that of the $Cu_{50}Co_{50}$ nanoparticles, except that $Ni(acac)_2$ (27.3 mg, Aldrich, 95.0% pure) was used in place of $Co(acac)_2$.

B Ambient pressure XPS

AP-XPS experiments were conducted at Beamline 9.3.2 at the Advanced Light Source, Lawrence Berkeley National Laboratory. Samples were prepared as dipcoat films of nanoparticles to yield 2D films on 275 μm thick Si wafers. The choice of Si wafers to anchor the nanoparticles to was selected to mimic a typical heterogeneous catalyst oxide support and avoid any risk of alloying with a gold surface, often used as the support when preparing 2D films of nanoparticles. Ambient Pressure XPS was conducted using an apparatus described elsewhere.[29] The instrument also takes advantage of the tunable incident photon energy of synchrotron-generated X-rays to enable acquisition of both very surface sensitive measurements as well as near-surface depth profiles of both oxidation state and elemental composition of the samples. Owing to the range of accessible energies with this instrument, spectra were collected using electrons excited from the Cu 3p, Co 3p and Ni 3p levels (rather than the 2p levels usually used in a conventional laboratory XPS when using an Al Kα source). The photoionisation cross section for these elements with a 380 eV source energy is comparable to that of the 2p levels when using an Al Kα source. All energies are corrected with respect to the Fermi edge (recorded immediately prior to or following the spectrum of interest, without changing the incident photon energy). Relative concentrations in the region of the particle probed by the XPS were calculated and plotted by fitting a Shirley background using XPSPeak and then integrating. The obtained results were also corrected for the relevant photoionisation cross sections[30] in the calculation of the relative composition of the two metals. Maximum temperatures for heating in each gas were selected based on SEM characterisation with *ex situ* thermal treatments to minimise the chance of any agglomeration occurring due to thermal activation. Probing depths are estimated using the method of Seah and Dench as described in the ESI.† The possibility of beam damage or heating causing agglomeration was examined using SEM before and after AP-XPS experiments for both samples and the results are provided in the ESI.†

C Additional characterization

TEM and HR-TEM images, and STEM/EDS phase spectra were obtained using Jeol 2100F (scanning) transmission electron microscope. Samples were prepared by drop casting the nanoparticles (in toluene) on to an ultra thin carbon-coated gold TEM grid. A Zeiss Ultra55 scanning electron microscope was then used for imaging and confirming the integrity of the dipcoated films of nanoparticles on Si wafers before and after the AP-XPS studies.

Results

CuNi bimetallic nanoparticles

Fig. 1a shows representative TEM images of the as prepared, unsupported 10.6 ± 1.3 nm CuNi nanoparticles, prepared using an oleylamine capping agent as described above. Their corresponding size distribution is indicated above the image. Fig. 1b shows a High Angle Annular Dark Field (HAADF) image of a nanoparticle in the same sample in which TEM (STEM) linescans using Energy Dispersive X-ray Spectra (EDS) were acquired for Cu K- and Ni K-edges as shown. These linescans confirm the presence of both metals within the individual particle and show that the as prepared particle contains a copper enriched core and a more nickel rich shell, when measured under the vacuum conditions of the TEM. (The Cu signal is strong only near the middle of the linescan, whereas the Ni is strongest at the edges, where a greater amount of the near surface region of the sphere is probed.) Additionally, this technique was used to confirm the relative metal content for 10 individual particles and it was found that they contained 53.7 ± 9.6 at. % Cu, which compares

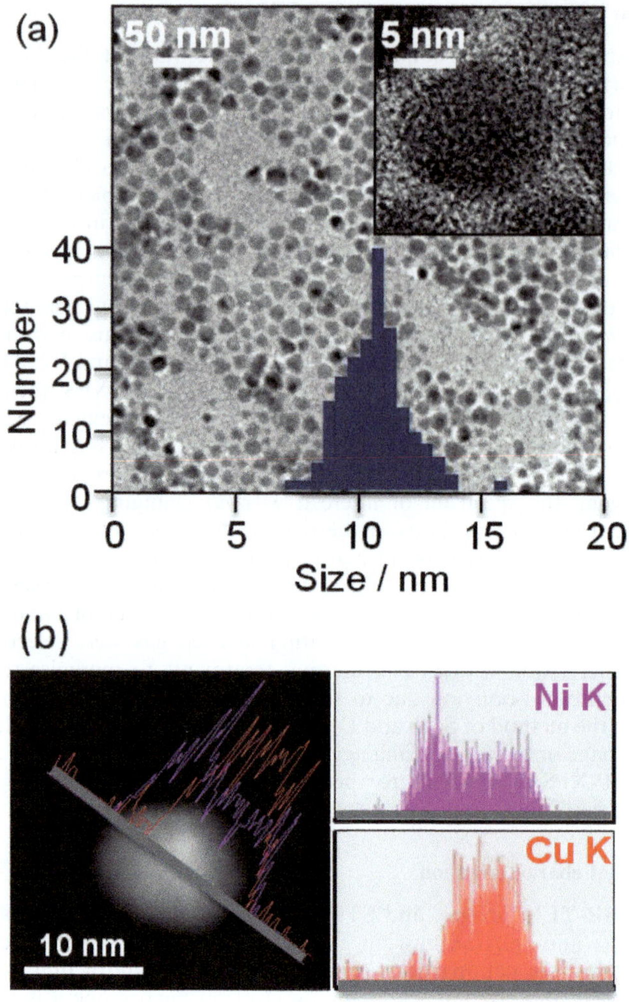

Fig. 1 (a) Representative HR-TEM images of the as prepared CuNi nanoparticles, with the sample size distribution obtained by counting a large number of nanoparticles from different regions of the TEM grid superimposed above the image. (b) High Angle Annular Dark Field (HAADF) image of a CuNi nanoparticle, the grey line indicating the line along which scanning TEM (STEM) linescans using Energy Dispersive X-ray Spectra (EDS) were acquired for Cu K- and Ni K-edges as shown on the right.

favourably to an average Cu content recorded over an area enclosing many particles of 57.2 at. % Cu. These values are also consistent with the 50 at.% Cu content targeted from the synthesis. This control over particle size and composition is crucial for enabling us to study reconstruction effects, because the spectral signals originate from a material in which all particles are near identical, rather being the average of many different systems, which would mask the effects we are seeking to probe.

The as prepared nanoparticles, supported on the silicon oxide surface of a silicon wafer, were then investigated using ambient pressure XPS. The start point of the XPS data is directly comparable to the TEM data above as they are both recorded on the as prepared sample, at room temperature and after evacuation in a vacuum chamber. Spectra were acquired in vacuum and during the heating steps and introduction of 100 mtorr of H_2. (See ESI – as these initial steps prior to heat and gas treatments are likely still dominated by the as synthesized structure and may not be the result of thermally activated rearrangements they will not be discussed in detail).† Heating in H_2 results in an enhancement of the Ni/Cu ratio in the near-surface region (~1 nm probing depth). The major changes of interest occur on switching from reducing (H_2 or CO) to oxidising atmospheres and back again as shown in Fig. 2. The spectra shown represent the top ~0.8 nm of the 10 nm diameter nanoparticles and so correspond mostly to the top 2–3 atomic layers. It is clear that nickel is dominant on the surface in H_2 (~60/40), but O_2 brings copper to the surface to produce a slightly copper rich surface. However, on re-exposure to H_2 this effect is immediately and fully reversed. In H_2 and O_2 within the accessible energy range, the sample was also probed using a higher incident photon energy sampling the top 4–5 atomic layers (~1.2 nm) to address the question of whether this effect occurs purely on the surface layer of the particles or whether it involves the top several layers of the particles. Comparison of these spectra (see ESI)† does not show any significant change between probing 2–3 layers and 4–5 layers deep, indicating this is not an effect dominated by a single surface layer.

As the presence of oxygen allows the formation of metal oxide it is also instructive to consider the O 1s spectra in Fig. 3, acquired during the same experiment as the Cu and Ni 3p spectra shown in Fig. 2 (steps are numbered in both figures). Since some Si 2p signal can be seen (suggesting a nanoparticle coverage of 95%), it should be remembered that some of the oxygen seen originates from the surface oxide layer of the silicon wafer and not from the nanoparticles themselves – and this is likely to be the origin of the peak at around 532.5 eV that remains even in the presence of H_2 at 450 °C. In the presence of O_2 or CO, gas phase peaks (marked in Fig. 3) are seen to the higher binding energy side of the 532.5 eV peak as a significant number of gas molecules will inherently be sampled (in contrast to an ultra high vacuum experiment).[31] However, the key feature to note is the sudden appearance in O_2 of the lower binding energy features centred at 531.0 and 528.4 eV, which is likely indicative of the formation of metal oxides.[32] Re-examination of the Cu and Ni 3p spectra in Fig. 2b indicate that only the Cu appears to have been significantly

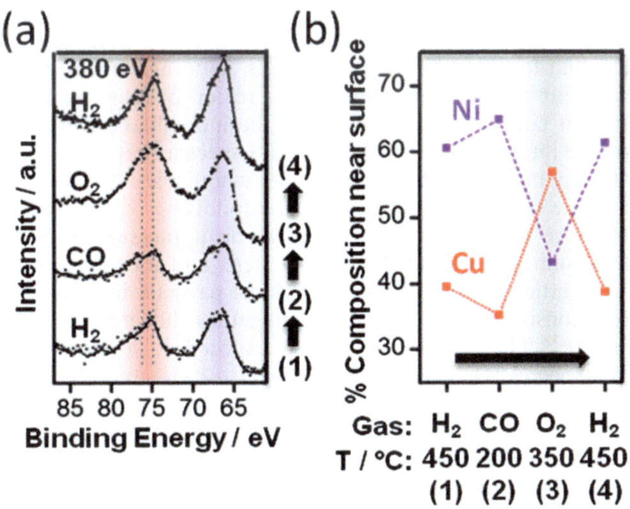

Fig. 2 (a) Ambient pressure Cu (red) and Ni (purple) 3p XP Spectra of CuNi nanoparticles supported on the native oxide layer of a silicon wafer during exposure to a series of gases as indicated (0.1 torr) using an incident photon energy of 380 eV, corresponding to a mean free path of escaping electrons of 0.8 nm. (b) Graph showing the corresponding % of Cu (red) and Ni (purple) in this top 0.8 nm surface layer of the nanoparticles for each condition in (a) (after correction for the relevant photon energy photoelectron cross sections and removal of a Shirley background).

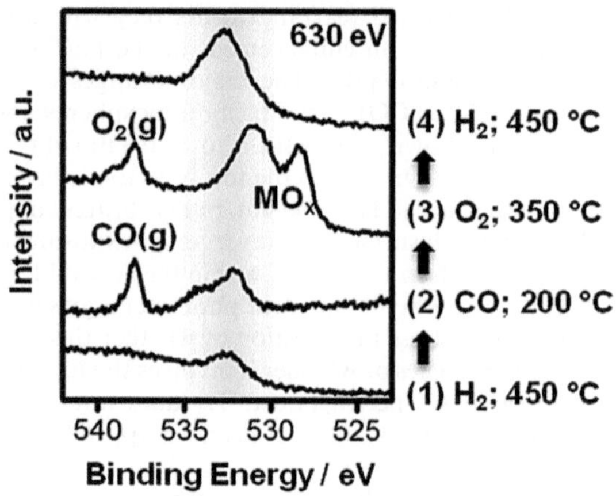

Fig. 3 Ambient pressure O 1s XP Spectra of CuNi nanoparticles supported on the native oxide layer of a silicon wafer during exposure to a series of gases as indicated (0.1 torr) using an incident photon energy of 630 eV The data was acquired in the same series of numbered experiments as shown in Fig. 2.

oxidised (inferred from a higher binding energy contribution to the Cu 3p) while the Ni 3p peak width appears unchanged.

CuCo bimetallic nanoparticles

Fig. 4a shows representative TEM images of the unsupported 11.2 ± 1.4 nm CuCo nanoparticles prepared in an analogous way to the CuNi particles discussed above, along with their corresponding size distribution. Again, STEM-EDS linescans as shown in Fig. 4b were also acquired to investigate the structure of as prepared particles (under the vacuum conditions of the microscope). This time the core–shell type segregation is visibly more marked than in the CuNi case, with the shell nearer the surface appearing to contain most of the cobalt. Overall the expected 1 : 1 Cu : Co atomic ratio is obtained – for 10 individual particles, on average each individual particle was found to comprise 55 ± 7% Cu, in good agreement with the average of 54% Cu content obtained sampling over a region of many nanoparticles. This is important as it confirms good control of the individual particle compositions, necessary for the interpretation of the AP-XPS spectra that follow – for instance, formation of stray monometallic nanoparticles is unlikely.

As before, these CuCo nanoparticles, after supporting on the oxide surface of a silicon wafer, were then investigated using ambient pressure XPS. Spectra were acquired in vacuum and during the heating steps and introduction of 100 mtorr of H_2 with very little change seen (see ESI);† the particles initially appear Co rich at the surface, consistent with the large segregation that was seen in the STEM-EDS experiment.[33] The samples were then exposed to a series of reactive gas mixtures (atmospheres ($H_2 \rightarrow 1:2$ $CO/H_2 \rightarrow O_2 \rightarrow H_2$) and the spectra obtained in the Cu and Co 3p region that correspond to a probing depth of 0.8 nm are shown in Fig. 5a, along with the calculated surface compositions for each set of conditions. Again, in the presence of oxygen copper is clearly brought to the surface, however the ratio of Cu : Co does not change reversibly back once re-exposed to H_2 and heated to 450 °C. The spectra obtained using different probing depths (Fig. 6) indicate that although initially the high surface concentration of Co is likely to be only the very surface layer, the segregation of Cu that occurs in oxygen is a number of layers deep (since spectra probing 1.2 nm and 0.8 nm produce very similar % Cu compositions after exposure to oxygen).

O 1s spectra corresponding to the H_2 and O_2 conditions in the experiment above are shown in Fig. 7. As before, there is the sudden appearance of a large low binding

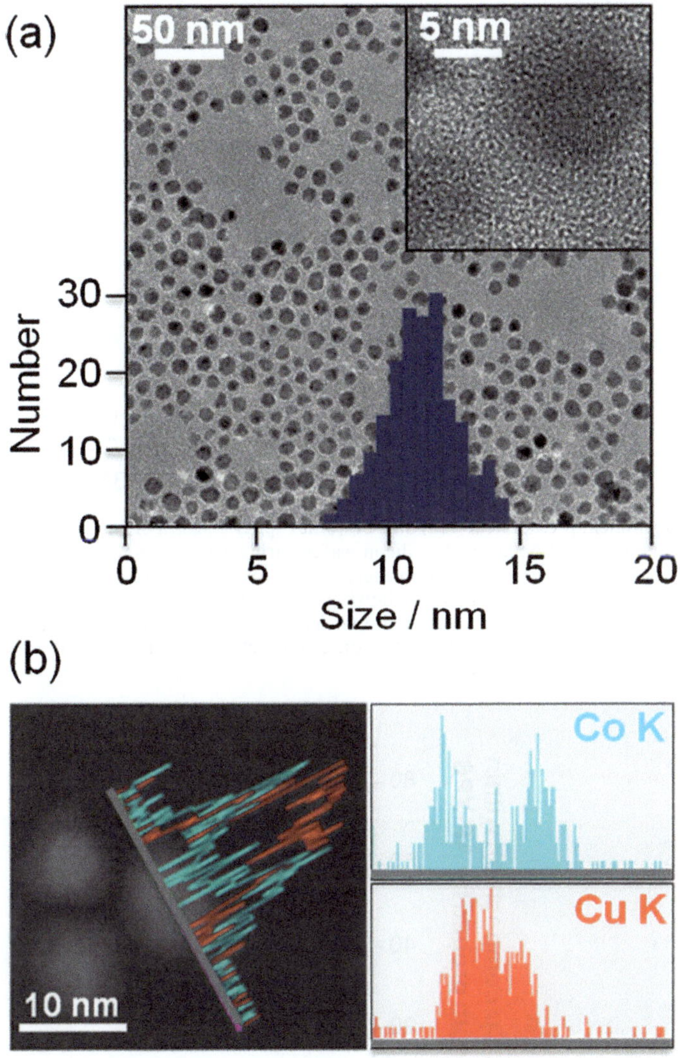

Fig. 4 (a) Representative HR-TEM images of the as prepared CuCo nanoparticles, with the sample size distribution obtained by counting a large number of nanoparticles from different regions of the TEM grid superimposed above the image. (b) High Angle Annular Dark Field (HAADF) image of CuCo nanoparticles, the grey line indicating the line along which scanning TEM (STEM) linescans using Energy Dispersive X-ray Spectra (EDS) were acquired for Cu K- and Co K-edges as shown on the right.

energy feature around 529 eV in the presence of oxygen, attributable to metal oxide,[32] which is almost totally removed again on re-exposure to H_2. This time, however, comparison to the spectra in Fig. 5 and 6 suggests that both metals are oxidised, as the peaks are shifted to higher binding energy by over 1 eV and have a rounder shape. This latter effect results from the convolution of more than one asymmetric $p_{1/2}$-$p_{3/2}$ doublet (expected intensity ratio of 1 : 2). For the quality of data acquired from these materials under ambient pressure conditions it is not appropriate to mathematically fit these by de-convolution, however, it is clear by comparison to the asymmetric $p_{1/2}$-$p_{3/2}$ doublet features upon re-exposure to H_2 that these have a significant, higher binding energy contribution – both metals are

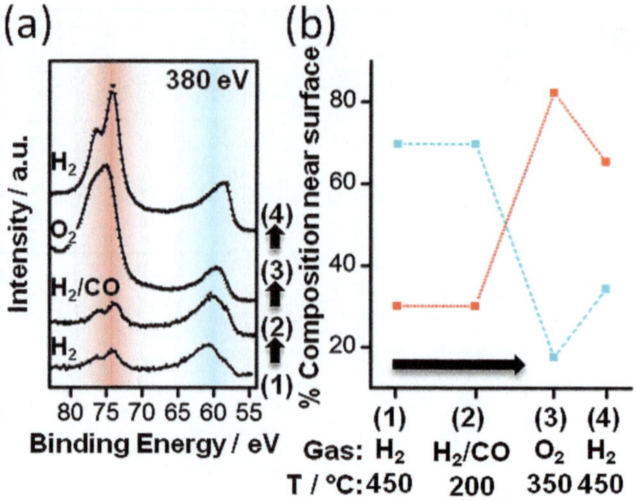

Fig. 5 (a) Ambient pressure Cu (red) and Co (blue) 3p XP Spectra of CuCo nanoparticles supported on the native oxide layer of a silicon wafer during exposure to a series of gases as indicated (0.1 torr) using an incident photon energy of 380 eV, corresponding to a mean free path of escaping electrons of 0.8 nm. (b) Graph showing the corresponding % of Cu (red) and Co (blue) in this top 0.8 nm surface layer of the nanoparticles for each condition in (a) (after correction for the relevant photon energy photoelectron cross sections and removal of a Shirley background).

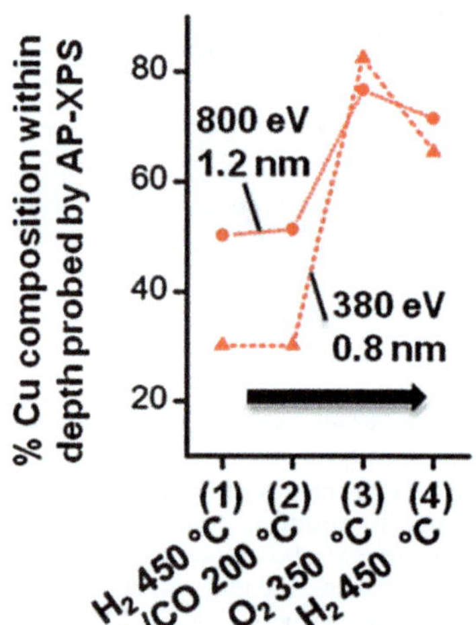

Fig. 6 Relative % of Cu found in surface layers corresponding to 1.2 nm (circles) and 0.8 nm (triangles) mean free paths of escaping electrons during the same series of experiments as shown in Fig. 5, obtained from Cu and Co 3p intensities using incident photon energies of 800 eV and 380 eV respectively (after accounting for the relevant photon energy photoelectron cross sections and the removal of a Shirley background).

at least partially oxidised. In contrast to the CuNi case, the CuCo already appears to have a small low binding energy contribution under H_2 before exposure to O_2. This may be the result of stable cobalt oxides being present, which cannot be efficiently reduced under these conditions. There is also a feature remaining on re-exposure to H_2, although from its position (~533 eV) it is likely this may be the result of irreversible further oxidation of the silicon support (in addition to the original native

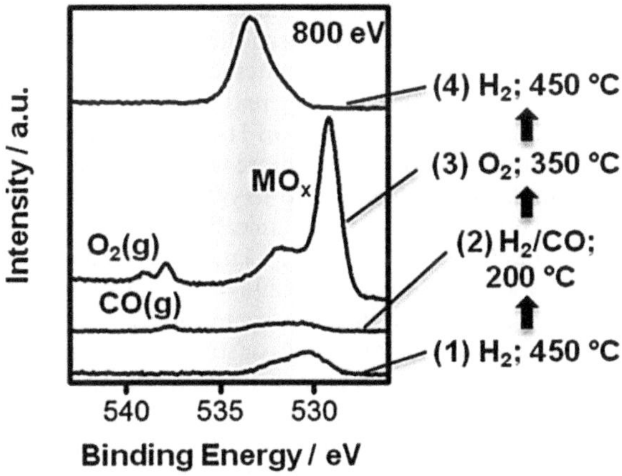

Fig. 7 Ambient pressure O 1s XP Spectra of CuCo nanoparticles supported on the native oxide layer of a silicon wafer during exposure to a series of gases as indicated (0.1 torr) using incident photon energy of 800 eV The data was acquired in the same series of numbered experiments as shown in Fig. 5.

oxide layer), rather than oxidation of the nanoparticles, since the Cu and Co signals appear to indicate limited oxide remains.

Discussion

In the previous section it has been identified that in both cases Cu is segregated to the surface on exposure to O_2. For CuNi the surface returns reversibly to being Ni rich upon exposure to H_2, whereas for CuCo the Cu remains at the surface. In both cases metal oxide formation is observed in oxygen and disappears again on heating in H_2. For CuNi it appears the oxidation occurs predominantly for Cu and not Ni, whereas for CuCo both metals are oxidised.

It is well known that for bulk materials Cu is expected to surface segregate (in the absence of reactive gases) and has been reported to do so in many experimental studies.[34] This is generally true for both CuCo and CuNi,[34] except at very high Ni contents not relevant to the present study.[35] In the case of bulk alloys (both experimental and theoretical studies have largely been performed on single crystal surfaces), the restructuring is understood for the CuNi case as a balance of cohesive energy (lattice bonding), entropy and surface free energy and surface layer relaxations.[36] Fewer studies have been performed on CuCo, perhaps because of the large miscibility gap in their phase diagram.[37] In our case copper is never seen to be significantly enriched above the overall bulk composition of 50% at the surface for the CuNi case, although is enriched further on exposure to oxygen in the case of CuCo. For CuCo it could be argued that heating allowed it to overcome kinetic barriers to diffusion or restructuring, however this occurs on lowering the temperature from 450 to 350 °C ruling out the possibility that the gas is not involved in this transformation. Accordingly, it appears neither case behaves as expected for the bulk materials.

The balance of cohesive energies favours Cu segregation[38] (the higher cohesive energy element being expected to be favourably located in the core, where more bonds with like atoms are possible), but this is not dependant on which gases are present. As outlined in the introduction, for many bimetallic nanoparticles it has been found that they restructure on exposure to reactive gases, such as O_2, as a result of competition between oxide formation enthalpy and surface free energy. However, in here copper has the lowest surface free energy (1.9 J m^{-2} *vs.* Co at 2.7 J m^{-2} and Ni at 2.4 J m^{-2}) and so in vacuum or H_2 would be expected to be on the surface. For CuCo, in which segregation is mostly confined only to the surface layer, this may explain why Cu remains on the surface after oxygen has caused the segregation to occur. For CuNi, however, it does not explain why nickel segregates to the surface in H_2. It is well known that H_2 can dissociate on nickel surfaces (whereas on copper this requires significant thermal activation). Efficient chemisorption of H_2 is therefore a plausible candidate for this nickel segregation effect, but much more work is needed to fully understand this phenomenon. Similarly, on exposure to oxygen the formation of metal oxides occurs, however this is not to the thermodynamically preferred oxide as seen in previously studied systems ($\Delta H^{\ominus}{}_f\{CuO\} = -157.3$ kJ mol^{-1} is significantly less than either $\Delta H^{\ominus}{}_f\{CoO\} = -237.9$ kJ mol^{-1} or $\Delta H^{\ominus}{}_f\{NiO\} = -244.4$ kJ mol^{-1}). This therefore cannot act as the driving force to surface segregate copper – but as it only occurs on exposure to oxygen it seems likely that kinetics or strain effects related to oxide formation must also participate. If kinetics on timescales longer than the ~40 min of repeat XPS measurements (that suggest stability on this timeframe) are important this could result from diffusion issues or the kinetics of oxide formation and reduction. (In relation to the former, it is interesting to note that interdiffusion in bulk materials is known to be at least 5 times slower for Cu into Ni or CuNi alloys compared to the reverse case of Ni into Cu or CuNi alloys at relevant temperatures.[39]) Given that the oxide formed is in contact with a layer of silicon oxide it is not clear if this interaction is strong enough to also influence the stability of the oxides; it would be very valuable to address this question in future work, but silica was picked for initial use as it is a good mimic for a heterogeneous metal–metal oxide catalyst. Another interesting possibility comes from the case of PtCu in CO gas, which restructures to give a Cu rich surface alloy on a single crystal both observed experimentally and on the basis of DFT calculations.[40] This is attributed to a uniquely stable surface structure that is preferentially formed in place of the expected strong Pt–CO bonding seen elsewhere, and such effects could also be at work with the copper present in both systems studied here. Another potential issue in addressing the question of why restructuring can occur in some cases would be the presence of defects or stacking faults altering the kinetics of restructuring – although this doesn't appear to have been reported in the above cases.

Although more detailed investigation is required to understand the nuances of these two systems, *in situ* ambient pressure XPS demonstrates unambiguously the importance of evaluating the surface structure under real reaction conditions in order to understand the surface composition of the actual catalytic surface at work. For instance, although in conditions not directly replicated here, CuNi catalysts for solid oxide fuel cells containing only 10% Ni were still found to cause significant hydrocarbon dissociation (cracking is typically associated with nickel not copper) at temperatures below 800 °C.[41] Although electronic effects have been invoked to explain this involving sub-surface nickel altering the properties of the copper, the present results offer an alternative explanation – nickel rather than copper may be segregated to the surface under the conditions of the reaction. CuNi has also recently been identified by theory and experiment to be an active and highly selective catalyst for methanol synthesis from syngas.[3] Given the strong likelihood of nickel segregation in such catalysts under reducing conditions, their optimisation and development for use as practical catalysts would considerably benefit from an improved understanding of this behaviour. These are just two examples, but given the prevalence of bimetallic catalytic systems, increasing our knowledge of these dramatic restructuring effects is undoubtedly important in understanding their surface chemistry and enabling the rational development of improved catalytic systems.

Conclusions

Bimetallic nanoparticles of 1 : 1 atomic composition Cu : Ni and Cu : Co (as confirmed by STEM-EDS) and ~11 nm diameter were prepared using a new synthetic method. This has provided two new bimetallic nanoparticle systems, substantially different to those studied previously in the context of surface segregation and restructuring. In vacuum, the as prepared nanoparticles have Ni and Co enriched shells, respectively, and a Cu rich core. Exposure of these nanoparticles supported on planar SiO_2 to reactive gases results in surface segregation determined *in situ* using ambient pressure XPS. Exposure to O_2 results in a relative increase in the percentage of Cu at the surface as compared to H_2 at 450 °C. For CuNi this change is reversible on re-exposure of the nanoparticles to H_2, while the CuCo remains Cu rich at the surface on re-exposure to H_2 at 450 °C. These observations are different to those theoretically and experimentally seen for bulk single crystals in vacuum and clearly not dominated by factors such as surface free energy and the enthalpy of oxide formation as seen in previous studies of bimetallic nanoparticles. Further studies of different sizes and compositions are likely to improve our understanding of the important balance of factors that control these bimetallic systems and others. However, the unexpected results of the present study serve to underline the importance of accounting for these as yet, not fully understood, restructuring effects. This is especially so for heterogeneous catalysis, where surface composition is crucial and high temperature and pressure reactive gases are usually present.

Acknowledgements

GAS acknowledges support from the Director, Office of Sciences, Office of Basic Energy Sciences, Division of Material Sciences of the US Department of Energy (DOE) under contract DE-AC02-05CH11231. NK gratefully acknowledges funding from Total, France. The user work at the Advanced Light Source and Molecular Foundry is supported by the Director, Office of Science, Office of Basic Energy Sciences, of the US Department of Energy under Contract No. DE-AC02-05CH11231. The authors are also grateful to Rui Chang at Beamline 9.3.2 for his practical assistance.

References

1 J. A. Rodriguez, in *Handbook of Heterogeneous Catalysis*, ed. G. Ertl, H. Knözinger, F. Schüth and J. Weitkamp, Wiley-VCH, Weinheim, 2nd edn, 2006, vol. 3, pp. 1298–1309.
2 V. R. Calderone, N. R. Shiju, D. Curulla Ferré and G. Rothenberg, *Green Chem.*, 2011, **13**, 1950.
3 F. Studt, F. Abild-Pedersen, Q. Wuc, A. D. Jensen, B. Temel, J.-D. Grunwaldt and J. K. Nørskov, *J. Catal.*, 2012, **293**, 51.
4 Y. Chen, F. Yang, Y. Dai, W. Wang and S. Chen, *J. Phys. Chem. C*, 2008, **112**, 1645.
5 W. M. H. Sachtler, *Faraday Discuss. Chem. Soc.*, 1981, **72**, 7.
6 A. Christensen, A. V. Ruban, P. Stoltze, K. W. Jacobsen, H. L. Skriver, J. K. Nörskov and F. Besenbacher, *Phys. Rev. B: Condens. Matter*, 1997, **56**, 5822 and references therein.
7 G. A. Somorjai and Y. Li, *Introduction to surface chemistry and catalysis*, Wiley, Hoboken, 2nd edn, 2010 pp. 296–302.
8 G. Ouyang, C. X. Wang and G. W. Yang, *Chem. Rev.*, 2009, **109**, 4221.
9 J. Y. Park, Y. Zhang, S. H. Joo, Y. Jung and G. A. Somorjai, *Catal. Today*, 2012, **181**, 133.
10 F. Tao, M. E. Grass, Y. Zhang, D. R. Butcher, F. Aksoy, S. Aloni, V. Altoe, S. Alayoglu, J. R. Renzas, C.-K. Tsung, Z. Zhu, Z. Liu, M. Salmeron and G. A. Somorjai, *J. Am. Chem. Soc.*, 2010, **132**, 8697.
11 S. Alayoglu, A. U. Nilekar, M. Mavrikakis and B. Eichhorn, *Nat. Mater.*, 2008, **7**, 333.
12 H. Lang, S. Maldonado, K. J. Stevenson and B. D. Chandler, *J. Am. Chem. Soc.*, 2004, **126**, 12949.
13 F. Tao, S. Zhang, L. Nguyen and X. Zhang, *Chem. Soc. Rev.*, 2012, **41**, 7980.
14 Assuming hydrogen to be an ideal gas colliding with a surface of 10^{15} atoms cm^{-2} at a pressure of 1 torr; D. P. Woodruff and T. A. Delchar, *Modern Techniques of Surface Science*, Cambridge University Press, Cambridge, 1994.
15 H. Siegbahn, L. Asplund, P. Kelfve, K. Hamrin, L. Karlsson and K. Siegbahn, *J. Electron Spectrosc. Relat. Phenom.*, 1974, **5**, 1059.

16 D. F. Ogletree, H. Bluhm, G. Lebedev, C. S. Fadley, Z. Hussain and M. Salmeron, *Rev. Sci. Instrum.*, 2002, **73**, 3872.
17 M. P. Seah and W. A. Dench, *Surf. Interface Anal.*, 1979, **1**, 2.
18 X. Liu, A. Wang, L. Li, T. Zhang, C.-Y. Mou and J.-F. Lee, *J. Catal.*, 2011, **278**, 288.
19 S. M. Oxford, P. L. Lee, P. J. Chupas, K. W. Chapman, M. C. Kung and H. H. Kung, *J. Phys. Chem. C*, 2010, **114**, 17085.
20 S. Alayoglu, F. Tao, V. Altoe, C. Specht, Z. Zhu, F. Aksoy, D. R. Butcher, R. J. Renzas, Z. Liu and G. A. Somorjai, *Catal. Lett.*, 2011, **141**, 633.
21 Y. Uemura, Y. Inada, K. K. Bando, T. Sasaki, N. Kamiuchi, K. Eguchi, A. Yagishita, M. Nomura, M. Tada and Y. Iwasawa, *J. Phys. Chem. C*, 2011, **115**, 5823.
22 F. Tao, M. E. Grass, Y. Zhang, D. R. Butcher, J. R. Renzas, Z. Liu, J. Y. Chung, B. S. Mun, M. Salmeron and G. A. Somorjai, *Science*, 2008, **322**, 932.
23 K. J. J. Mayrhofer, V. Juhart, K. Hartl, M. Hanzlik and M. Arenz, *Angew. Chem., Int. Ed.*, 2009, **48**, 3529.
24 V. Papaefthimiou, T. Dintzer, V. Dupuis, A. Tamion, F. Tournus, D. Teschner, M. Hævecker, A. Knop-Gericke, R. Schlögl and S. Zafeiratos, *J. Phys. Chem. Lett.*, 2011, **2**, 900.
25 S. Alayoglu, S. K. Beaumont, F. Zheng, V. V. Pushkarev, H. Zheng, V. Iablokov, Z. Liu, J. Guo, N. Kruse and G. A. Somorjai, *Top. Catal.*, 2011, **54**, 778.
26 F. Zheng, S. Alayoglu, V. V. Pushkarev, S. K. Beaumont, C. Specht, F. Aksoy, Z. Liu, J. Guo and G. A. Somorjai, *Catal. Today*, 2012, **182**, 54.
27 S. Piccinin, S. Zafeiratos, C. Stampfl, T. W. Hansen, M. Hævecker, D. Teschner, V. I. Bukhtiyarov, F. Girgsdies, A. Knop-Gericke, R. Schlögl and M. Scheffler, *Phys. Rev. Lett.*, 2010, **104**, 035503.
28 ($\Delta H^{\ominus}{}_f\{CuO\} = -157.3$ kJ mol^{-1}; $\Delta H^{\ominus}{}_f\{CoO\} = -237.9$ kJ mol^{-1}; $\Delta H^{\ominus}{}_f\{NiO\} = -244.4$ kJ mol^{-1}; $\Delta H^{\ominus}{}_f\{Rh_2O_3\} = -343$ kJ mol^{-1}, but -172 kJ mol^{-1} per Rh atom – taken from David Lide, *CRC Handbook of Chemistry and Physics*, Taylor and Francis, Raton, 87th edn, 2004).
29 M. E. Grass, P. G. Karlsson, F. Aksoy, M. Lundqvist, B. Wannberg, B. S. Mun, Z. Hussain and Z. Liu, *Rev. Sci. Instrum.*, 2010, **81**, 053106.
30 J. J. Yeah and I. Lindau, *Atomic Data And Nuclear Data Tables*, 1985, **32**, 55.
31 It should be remembered that it is likely that some nickel carbonyl forms under these conditions and could influence segregation, however $Ni(CO)_4$ is likely to be thermally desorbed if formed; V. Medvedev, R. Börner and N. Kruse, *Surf. Sci.*, 1998, **401**, L371.
32 NIST X-ray Photoelectron Spectroscopy Database, Version 4.1 (National Institute of Standards and Technology, Gaithersburg, 2012); http://srdata.nist.gov/xps/.
33 It should be remembered, as mentioned above for CuNi, that prior to heating and the exposure of the sample to reactive gases the cobalt surface enrichment observed may be the result of the preparative procedure/surface contaminants not yet thermally desorbed from the surface. For this reason the data prior to heating in H_2 at 450 °C will not be discussed further but is provided as supplementary information for completeness.
34 N. M. Galea, D. Knapp and T. Ziegler, *J. Catal.*, 2007, **247**, 20.
35 T. Sakurai, T. Hashizume, A. Jimbo, A. Sakai and S. Hyodo, PRL, 1985, **55**, 514.
36 H. Y. Wang, R. Najafabadi, D. J. Srolovitz and R. LeSar, *Phys. Rev. B: Condens. Matter*, 1992, **45**, 12028.
37 T. Nishizawa and K. Ishida, *Bull. Alloy Phase Diagrams*, 1984, **5**, 161.
38 Cohesive energies: 3.49 eV per atom of Cu compared to 4.44 eV and 4.39 eV per atom of Niand Co, respectively; C. Kittel, *Introduction to Solid State Physics*, John Wiley and Sons, Inc, Hoboken, 8th edn, 2005.
39 D. B. Butrymowicz, J. R. Manning and M. E. Read, *J. Phys. Chem. Ref. Data*, 1976, **5**, 103.
40 K. J. Andersson, F. Calle-Vallejo, J. Rossmeisl and L. Chorkendorff, *J. Am. Chem. Soc.*, 2009, **131**, 2404.
41 H. Kim, C. Lu, W. L. Worrell, J. M. Vohs and R. J. Gorte, *J. Electrochem. Soc.*, 2002, **149**, A247–A250.

The structure of AuPd nanoalloys anchored on spherical polyelectrolyte brushes determined by X-ray absorption spectroscopy

Julian Kaiser,[ab] Wojciech Szczerba,[c] Heinrich Riesemeier,[c] Uwe Reinholz,[c] Martin Radtke,[c] Martin Albrecht,[d] Yan Lu[ab] and Matthias Ballauff*[ab]

Received 27th November 2012, Accepted 25th January 2013
DOI: 10.1039/c3fd20132e

Well-defined and facetted bimetallic gold–palladium nanoalloys have been synthesized and anchored in spherical polyelectrolyte brushes (SPB) as composite particles (AuPd@SPB). These particles are better catalysts in aqueous phase than the pure metals. The atomistic arrangement of these nanoalloys has been analysed by extended X-ray absorption fine structure (EXAFS) spectroscopy at the Au–L_3 and the Pd–K absorption edge. The samples with high amounts of gold appear as almost statistically mixed random alloys. Alloy compositions with less gold show slight enrichment of Pd at the surface of the particle. In addition, signals of non-metallic palladium appear at the Pd–K edge which indicate the presence of the Pd^{2+} species in addition to metallic palladium. The relation of these structural features to the catalytic activity is discussed.

Introduction

Recently, we have demonstrated that metal nanoparticles can be generated and immobilized in spherical polyelectrolyte brushes (SPB).[1–7] Metallic nanoparticles, such as Au, Pd and Pt, with a size of 1–2 nm are generated directly in the brush system under mild conditions at ambient temperature and aqueous environment. The composite particles are robust and stable and can be used as catalysts in aqueous solution. The reduction of 4-nitrophenol (4-NiP) to 4-aminophenol by sodium borohydride has been used as a model reaction to test the catalytic activity of different metal nanoparticles.[8] The kinetics of this reaction can be analyzed with great precision. A Langmuir–Hinshelwood model was found to give a fully quantitative description of the kinetic data.[6,7] Hence, the entire catalytic cycle proceeds on the surface of the metal nanoparticles.

Bimetallic particles have received much attention[9–14] due to their enhanced activity in catalytic applications. Thus, Hutchings *et al.* demonstrated that AuPd nanoalloys exhibit a high catalytic activity for various reactions.[15–21] Recently, they found enhanced catalytic performances in the direct synthesis of hydrogen peroxide by

[a]*F-I2 Soft Matter and Functional Materials, Helmholtz-Zentrum Berlin, Hahn-Meitner-Platz 1, 14109 Berlin, Germany*
[b]*Department of Physics, Humboldt University Berlin, Newtonstr. 15, 12489 Berlin, Germany. E-mail: Matthias.Ballauff@helmholtz-berlin.de; Fax: +49 (0) 30 8062 43071; Tel: +49 (0) 30 8062 2308*
[c]*BAM Federal Institute for Materials Research and Testing, Div 1.3, Richard-Willstätter-Straße 11, 12489 Berlin, Germany; Fax: +49 30 8104 1137; Tel: +49 30 8104 1139*
[d]*Leibniz Institute for Crystal Growth Berlin, Max-Born-Straße, 12489 Berlin, Germany; Fax: +49 (0) 30 63923003; Tel: +49 (0) 30 63923094*

AuPd catalysts.[15–17,22–24] Moreover, AuPd bimetallic nanoparticles are known to show higher catalytic activities compared to their neat metal nanoparticles in hydrogenation reactions,[25–27] and degradation reactions.[18,28,29] An enhanced catalytic activity has been reported for oxidation reactions, *e.g.*, the oxidation of alcohols, toluene, propene and propanediol.[19–21,30–34] The reason therefore is supposed to lie in a reduction of the hydrocarbon decomposition with a simultaneous enhancement of the benzene desorption.[35–37] In addition, the arrangement of Pd atoms at a pristine surface as it would be the case in pure Pd nanoparticles, is changed by the Au atoms in the case of AuPd nanoalloys. This may have an impact on the catalytic enhancement of bimetallic AuPd nanoparticles compared to their pure metal nanoparticles.[37]

In our previous study, AuPd nanoalloys have been generated and anchored in SPB particles as shown in Fig. 1. Long chains of a polyelectrolyte are grafted densely to a solid core particle made of polystyrene.[38] It has been found that the nanoalloy systems have a much higher catalytic activity as compared to Au, Pd and Pt nanoparticles.[12] In particular, AuPd nanoalloy particles with a composition of 75 mol% gold and 25 mol% palladium has found to exhibit the highest catalytic activity. Structural analysis of the nanoparticles of alloys of Pd and Au[12] synthesized on the SPB pointed to a random alloy. This was mainly concluded from the linear dependence of the lattice distance on the composition. This distance was derived from high-resolution transmission electron microscopy (HR-TEM). The AuPd nanoalloys appear as facetted particles in HR-TEM and the face-centered cubic (fcc) structure can be seen clearly and no indication of a core–shell structure was found. However, the resolution of HR-TEM would not suffice to exclude a single layer of one metallic compound on the surface of the nanoparticles.

Extended X-ray absorption fine structure spectroscopy is the method of choice for the determination of the arrangement of atoms for these nanoparticles.[39] Moreover, this method gives information about the electronic state of the respective metal atoms. In a number of recent studies, EXAFS was applied to determine the fine structure of AuPd nanoalloys.[27,40–49] In most studies EXAFS is applied on AuPd nanoalloys in the size range of 1–7 nm[14,27,34,40,42–48,50] and many EXAFS measurements make use of both the Au–L_3 and the Pd–K edge.[14,34,40,42–47,49] Thus, by EXAFS measurements on both edges, Krumeich *et al.*[47] showed that partially oxidized palladium appears at the Pd edge for AuPd nanoalloys supported by TiO_2, but vanish for AuPd nanoalloys stabilized on a Al_2O_3 support. Additionally, an inhomogeneous Au–Pd distribution was found and indicated a slight Pd enrichment in the shell and an Au enhancement in the core of the nanoalloy.[47] Similar results have been

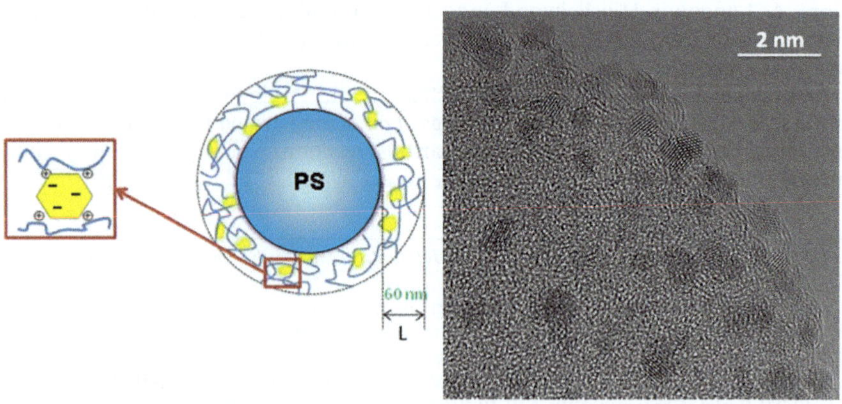

Fig. 1 Scheme of Au/Pd@SPB nanoalloys (left) and a representative TEM image (right) of $Au_{75}Pd_{25}$ nanoalloys.

obtained for AuPd nanoalloys containing Pd–O signals stabilized on various supports.[27,34,40,43–47] Furthermore, the Pd–O coordination numbers can be much higher than Au–Pd or Pd–Pd CNs.[34] However, some reported EXAFS studies of AuPd nanoalloys indicate the additional presence of Pd–O at the Pd absorption edge, resulting from an oxidation of surface Pd atoms. Furthermore, slight enrichments of Au atoms in the core are obtained.[27,34,40,47] All investigations performed by EXAFS so far demonstrate that this method is uniquely suited to study AuPd nanoalloys with a precision unsurpassed by any other method. Moreover, the results from these studies show that the structure and the electronic state of the nanoalloy depend on the conditions of synthesis and the support. Thus, there is a tendency of Pd to enrich at the surface of the nanoalloys. Since most catalytic reactions proceed on the surface of the nanoparticles, these details are of central importance.

Here we present EXAFS studies on nanoalloys of Au and Pd anchored on spherical polyelectrolyte brushes (SPB). A cationic brush system, containing poly(2-aminoethylmethacrylate hydrochloride) (PAEMH) as the polyelectrolyte chain has been used.[51] The catalytic activity of these nanoalloys has been the subject of detailed study recently.[12] Thus, in the present study we can correlate the fine structure of these nanoalloys with their catalytic activity.

Experimental

Preparation of AuPd Nanoalloys

The synthesis of gold–palladium nanoalloys was done as described in an earlier study.[12] 5 mL of dissolved $HAuCl_4$ solution was added drop wise to 100 mL of SPB latex suspension (solid content: 0.1 wt%).[12] Afterwards a certain amount of Na_2PdCl_4 was dissolved in 5 mL water and added drop wise. The total amount of metal salts is 4.5×10^{-5} mol. The molar ratio of the nanoalloy could be changed by the amount gold and palladium metal salt. Subsequently the mixture was stirred for 30 minutes under N_2 to remove oxygen from the liquid. Then the metal ions were reduced by a three-fold excess of $NaBH_4$. The reduction of $AuCl_4^-$ and $PdCl_4^{2-}$ ions could be followed easily by a change of color that the turbid latex solution turned from pale yellow to grey. After completion of the reaction the suspension was placed in an ultrafiltration cell and a five-fold excess of deionized water was passed through the cell.[12] The precise size was evaluated by counting the size of hundreds of nanoparticles by TEM measurements. The solvent of the composite particles suspension was removed by freeze-drying. The metal content for the composite particles was measured by TGA and evaluated to be approx. 5 wt%, which can be used for measurements directly after drying without further dilution.

EXAFS measurements

Extended X-ray absorption fine structure measurements were performed at DESY in Hamburg at the X beamline of the synchrotron laboratory Hasylab. The energy range was set from 11 400 to 12 719 eV. The edge step for the Au–L_3 edge has been obtained at 11 919 eV. Experiments were carried out on solid powder samples of 2 mm thickness at room temperature. The standard geometry of the sample stage allowed measuring sample and reference in one run using transmission and fluorescence detection for the sample. For the reference foil, the transmission signal has been collected. For this energy range, a Si(111) crystal has been used as a double crystal monochromator. The spot size of the incident beam was set to 2.5 mm $\times$ 1.0 mm (width $\times$ height). A multi(7)-element Si(Li) fluorescence detector (CANBERRA, Analytical Instruments) was adjusted at an angle of 90° to the X-ray beam and 45° to the sample to collect the fluorescence signal. Measurements at BESSY II in Berlin were performed at beamline KMC2 under similar conditions. For the fluorescence signal, an X-Flash Silicon drift detector (SDD) from Bruker was used. The transmission signal was detected by an Oxford Danfysik IC PLUS

50 ion chamber. The monochromator consists of a double Bragg(111) Si–Ge gradient concentration crystal with a germanium concentration of about 10% that varies along the crystal.[52,53]

In addition, the Pd XAFS measurements at the K-edge ($E_0 = 24\ 350$ eV) and additional Au–L_3 edge ($E_0 = 11\ 919$ eV) measurements were performed at the BAM line of BESSY II.[54] The beam was monochromatized using a Si(111) double-crystal monochromator (DCM) with an energy resolution of about 2×10^{-4}. The spectra were recorded in the pre-edge region ($E_0 - 100$ eV to $E_0 - 20$ eV) with an energy resolution of 5 eV. The X-ray absorption near edge structure (XANES) was scanned with an energy resolution of 1 eV, whereas the EXAFS region, above $E_0 + 50$ eV, was analysed with a constant step in the k-space of 0.05 Å^{-1}. The fluorescence radiation was measured at an angle of 90° with respect to the X-ray beam; the sample was tilted 45° with respect to both the X-ray beam and the direction of detection. The intensity of the emitted fluorescent radiation is proportional to the absorption, therefore the obtained excitation spectra are equivalent to conventional absorption spectra.[58] The fluorescence signal has been collected by an energy dispersive 100 mm^2 silicon drift detector (KETEK GmbH, Munich). The incident beam intensity was monitored by use of Oxford Danfysik ion chamber.

EXAFS data evaluation

For data analysis and evaluation, the open source software packages Athena and Artemis based on IFEFFIT was used.[55] For background subtraction the Rbkg value was set to 1.1. A Hanning-type Fourier window[56] for forward Fourier transform is set to a k-range of 3–12 Å^{-1} with a dk-value of 1. All samples were measured a minimum of ten times and merged in $\mu(E)$ to increase the signal to noise ratio. Fig. 2 displays EXAFS $\chi(k)$ plots[56] of AuPd nanoalloys after background subtraction at the Au–L_3 and Pd–K absorption edge.

For the fit of the data of the AuPd@SPB nanoalloys a crystal identification file (CIF) has been modified as follows: Starting from a CIF for bulk gold (for evaluation at the Au–L_3 absorption edge), Au atoms have been replaced by Pd atoms until the ratio of Au : Pd in the CIF is equal to the ratio of Au : Pd in the sample. For the interatomic distance between Au and Pd the average length (2.814 Å) between the interatomic distance of Pd–Pd and Au–Au has been chosen. At the Pd–K absorption edge a CIF for bulk palladium has been modified analogously. In addition, another CIF of PdO has been modified so that one Pd is the absorbing atom being surrounded by O atoms which represents a monolayer of Pd^{2+} partially covering the surface of the nanoalloy. These two models have been applied simultaneously at the Pd–K absorption edge to fit the Pd signal and the Pd^{2+} signal for AuPd@SPB nanoalloys and Pd@SPB nanoparticles.

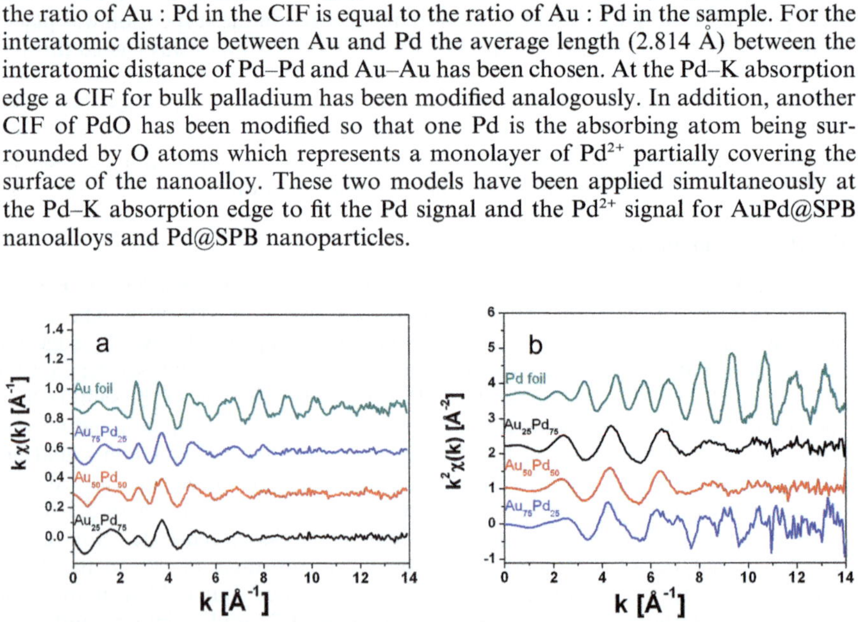

Fig. 2 $\chi(k)$ *vs.* k plots of EXAFS spectra of AuPd nanoalloys at Au–L_3 edge (a) and Pd–K edge (b).

Results and discussion

AuPd nanoalloys, Au nanoparticles (NP) and Pd NP were synthesized as described in previous work.[12] The particle diameters have been evaluated by TEM as 1 nm (Pd NP), 1.7 nm ($Au_{25}Pd_{75}$), 2.2 nm ($Au_{50}Pd_{50}$), 2.5 nm ($Au_{75}Pd_{25}$) and 2.7 nm (Au NP) with an uncertainty of approximately 0.2 nm, respectively. Samples containing Au : Pd ratio of 25 : 75 ($Au_{25}Pd_{75}$), 50 : 50 ($Au_{50}Pd_{50}$) and 75 : 25 ($Au_{75}Pd_{25}$) as well as neat Au and Pd NPs are studied by EXAFS. Fourier transformed (FT) spectra of the samples are presented in Fig. 3, 4, and 5.

Fig. 3 shows the imaginary part and magnitude FT EXAFS spectra of Au NP, $Au_{75}Pd_{25}$, $Au_{50}Pd_{50}$ and $Au_{25}Pd_{75}$ nanoalloys measured at the Au–L_3 absorption edge. It is found that FT $\chi(R)$ plots in R-space[56] show some shift of peaks to lower distances with the decrease of Au amount in alloy compositions and more gold atoms are surrounded by palladium. This is in accord with the Vegard's law that expresses linear dependence of lattice parameters on the ratio of the two alloy compounds. Additionally, the amplitude decreases with decreasing amount of gold. This is due to lower coordination numbers for small particles compared to gold foil on the one hand and due to fewer Au–Au bonds with decreasing amount of gold on the other hand. To determine the exact structure, the first shell of Au–Pd nanoalloys' EXAFS spectra are fitted. A model is created containing Au and Pd atoms with respect to the alloy composition. For a $Au_{50}Pd_{50}$ alloy, for example, 6 of 12 NN (nearest neighbor) atoms are defined as Au or Pd respectively. In order to interpret the data models, the FEFF code has been utilized using the Athena and Artemis software package.[55] The fits are shown in Fig. 4. The fitting parameters are reported in Table 1.

To determine the arrangement of the two different elements in the cluster, the coordination number (CN) for each element has to be estimated. It is obvious that the coordination number decreases with decreasing particle size due to fewer metal–metal bonds for surface atoms. In bulk material, the coordination number is 12 for gold and palladium. For particles in the range of 1.5 and 3 nm, it varies between approx. 7 and 9.[57] Liu *et al.*[27] studied AuPd nanoparticles at the Au–L_3 edge and reported rules to determine the alloy structures by comparing CN values of Au–Au and Au–Pd bonds. At the Au absorption edge, there are two coordination numbers. The CN for Au–Au describes the average amount of Au neighbor atoms, Au–Pd represents the number of Pd neighbor atoms next to the absorbing gold atom. In the case of random alloys, the ratio between Au–Au and Au–Pd coordination numbers has to be equivalent to the molar Au to Pd ratio of these elements in the cluster. Additionally, the sum of Au–Au and Au–Pd CNs should be less than 12 depending on the cluster size. If it is 12 or near, all gold atoms are surrounded completely by neighbor atoms. Therefore all gold atoms are located in the center of the particle. On the other hand, the smaller the sum of Au–Au and Au–Pd

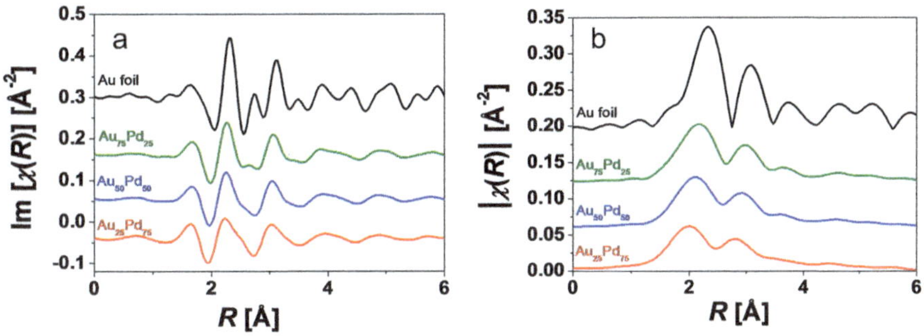

Fig. 3 Au–L_3 edge EXAFS spectra of AuPd nanoalloys and Au reference foil. Imaginary part (a) and magnitude of Fourier transform (b).

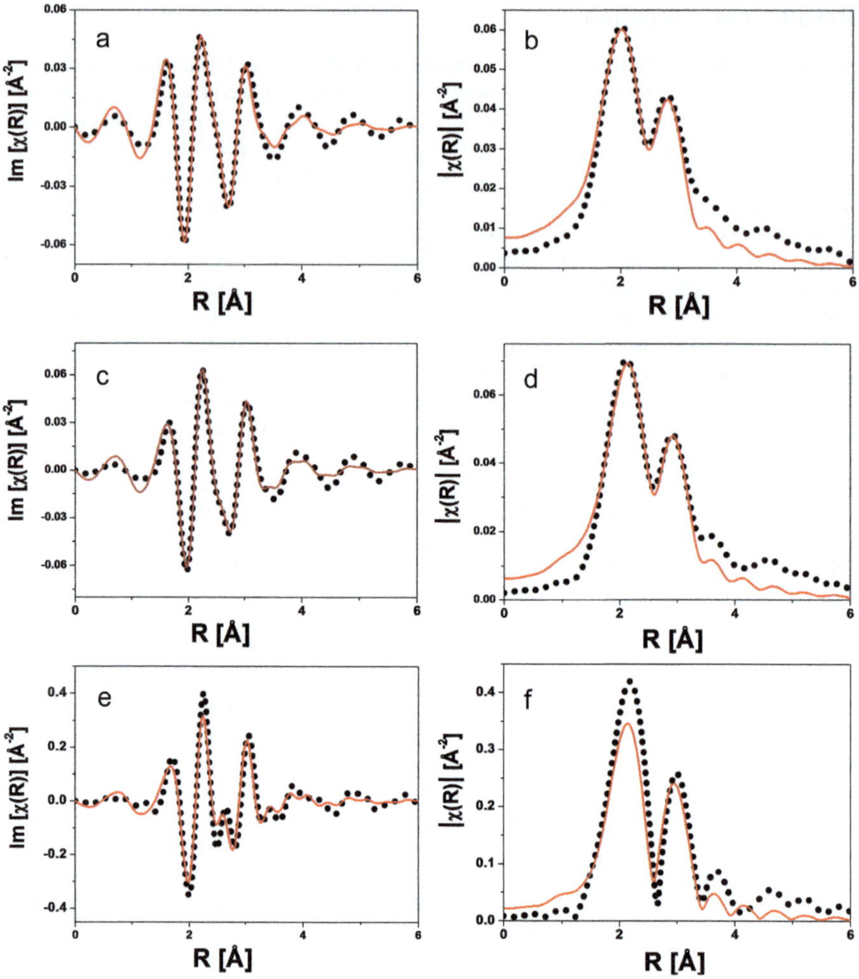

Fig. 4 Au–L_3 edge EXAFS single-shell fits of $Au_{25}Pd_{75}$ (a, b), $Au_{50}Pd_{50}$ (c, d) and $Au_{75}Pd_{25}$ (e, f) nanoalloys. Imaginary part (a, c, e) and FT magnitude (b, d, f). The dotted and solid lines represent experimental data and the fit, respectively.

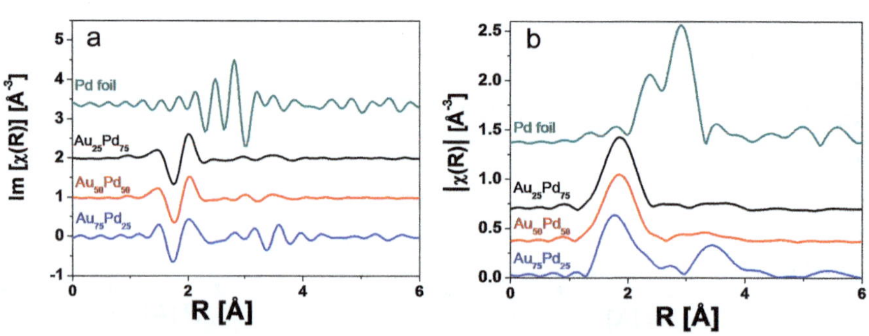

Fig. 5 Pd–K edge EXAFS spectra of AuPd nanoalloys and Pd reference foil. Imaginary part (a) and magnitude of Fourier transform (b).

Table 1 Results of the EXAFS fit parameters at the Au-L-3 edge. N = coordination number (CN), N_{total} = sum of N at one absorption edge, R = distance, σ^2 = Debye–Waller Factor, ΔE_0 = inner core correction

Sample	Edge	Pair	N	N_{total}	R [Å]	σ^2 [Å^{-2}]	ΔE_0 [eV]
Au–NP	Au–L_3	Au–Au	7.5 ± 0.3	7.5 ± 0.5	2.82 ± 0.01	0.012 ± 0.001	1.6 ± 0.3
Au75Pd25	Au–L_3	Au–Au	4.6 ± 0.3	5.7 ± 0.5	2.79 ± 0.01	0.008 ± 0.001	1.7 ± 0.5
	Au–L_3	Au–Pd	1.1 ± 0.2		2.71 ± 0.01	0.008 ± 0.001	1.7 ± 0.5
Au50Pd50	Au–L_3	Au–Au	3.9 ± 0.3	5.2 ± 0.5	2.77 ± 0.01	0.008 ± 0.001	0.6 ± 0.4
	Au–L_3	Au–Pd	1.3 ± 0.2		2.70 ± 0.01	0.008 ± 0.001	0.6 ± 0.4
Au25Pd75	Au–L_3	Au–Au	3.3 ± 0.2	4.6 ± 0.3	2.70 ± 0.01	0.008 ± 0.001	−3.4± 0.4
	Au–L_3	Au–Pd	1.3 ± 0.1		2.63 ± 0.01	0.008 ± 0.001	−3.4± 0.4

CNs the more Au atoms are located at the surface of the particle. Further discrimination between Au–Au and Au–Pd coordination numbers allows precise structural evaluations which is discussed in the following paragraph.[27,34,40,42,43,47,49]

The parameters of Table 1 are hence interpreted as follows: For homogeneous random alloys, the ratio between Au–Au bonds (N_{Au-Au}) and Au–Pd bonds (N_{Au-Pd}) has to be equal to the atomistic ratio in the alloy composition. As seen in Table 1, the experimental values deviate considerably from this prediction. In addition, the total coordination number N_{total} which is the sum of the average number of N_{Au-Au} and N_{Au-Pd}, is much less than 12. It would be in the range of 8–12 assuming all Au atoms are located in the center. Therefore a strict Au-core Pd-shell structure can be excluded. The number of Au–Au bonds for the $Au_{50}Pd_{50}$ and $Au_{25}Pd_{75}$ are 3.9 and 3.3, respectively, which is much higher than N_{Au-Au} of 2.6 and 1.2 respectively, which it is calculated from a random alloy. However, for the $Au_{75}Pd_{25}$ sample, N_{Au-Au} and N_{Au-Pd} is 4.6 and 1.1, which is close to the theoretical value of 4.3 and 1.4 of an $Au_{75}Pd_{25}$ random alloy. Therefore the $Au_{75}Pd_{25}$ sample indeed exhibits a random alloy character at the Au–L_3 absorption edge. For Au NP, an average coordination number of 7.5 is obtained. This value is reasonable with respect to higher number of surface atoms with much fewer neighbor atoms, compared to Au bulk. The interatomic distance (R) of Au–Pd is shorter than that of Au–Au due to a lower lattice parameter in Pd compared to Au. Additionally, the R values decrease with increasing amount of Pd and decreasing of particle size. The parameters σ^2 and ΔE_0 are in good agreement with reports in the literature[27,34,47] keeping in mind that bulk models are applied to nanoparticles.

The EXAFS spectra of the Pd–K edge display more complicated features. The spectra of AuPd alloys together with a Pd reference foil are shown in Fig. 5. First of all, the amplitude of AuPd alloys is much smaller than that of Pd foil. Comparing the AuPd samples, the amplitude decreases with decreasing amount of Pd which is to the lower Pd concentration. However, the main peak position does not differ much comparing the three different alloy compositions. Obviously, the main peak in the range of 1–2 Å does not point to metallic Pd signal by comparison to the Pd foil spectrum. Only a small peak in the range of 2–3.5 Å appears in all AuPd alloy spectra which can be attributed to the metallic Pd signal. Hence, the nonmetallic species generates a strong signal in the spectrum. Therefore only a qualitative evaluation can be achieved by applying theoretical models to experimental data by fits. The $Au_{75}Pd_{25}$ sample shows another feature in the range of 2.5–4 Å and differs from both other alloys.

For data evaluation at the Pd–K edge, a theoretical model of palladium(II)oxide and a second model consisting of fcc bulk AuPd random alloy is applied simultaneously (see experimental section). The experimental data together with fits are shown in Fig. 6. The fit parameters are reported in Table 2.

As shown in Table 2, for the Pd NP, the coordination number of Pd is very small, which is mainly due to the small size of these particles. Concerning a theoretical Pd cluster of 1 nm diameter, more than 70% of Pd atoms are supposed to be located at the surface. However, the average number of nonmetallic Pd bonds in the first coordination shell is 2, the average number of Pd–Pd bonds is 3.4. This indicates that a considerable part of Pd^{2+} in Pd nanoparticles is located at the surface of the particles.

In the case of $Au_{50}Pd_{50}$ and $Au_{25}Pd_{75}$, smaller values of N_{total} are obtained. This shows that a large amount of Pd atoms are located on the surface area of the particles. The N_{total} values at the Au–L_3 edge of both particles are much less than that for Au bulk metal and Au NP. Hence, a location of all Pd atoms at the surface is impossible. For both $Au_{25}Pd_{75}$ and $Au_{50}Pd_{50}$ particles, the N_{Pd2+}values are approx. 65% of N_{total}. This means that approximately 35% of Pd is metallic. By evaluation of CN, the metallic species is predicted to be mostly located inside the particle. This is in good agreement with the discussion about the Au–L_3 edge, where an Au gradient to the particle center is predicted. In addition, the ratio between Pd–Pd and Pd–Au bonds is not equal to the molar Au to Pd ratio in the sample which denotes that the structure of a statistically mixed random alloy has to be excluded in the

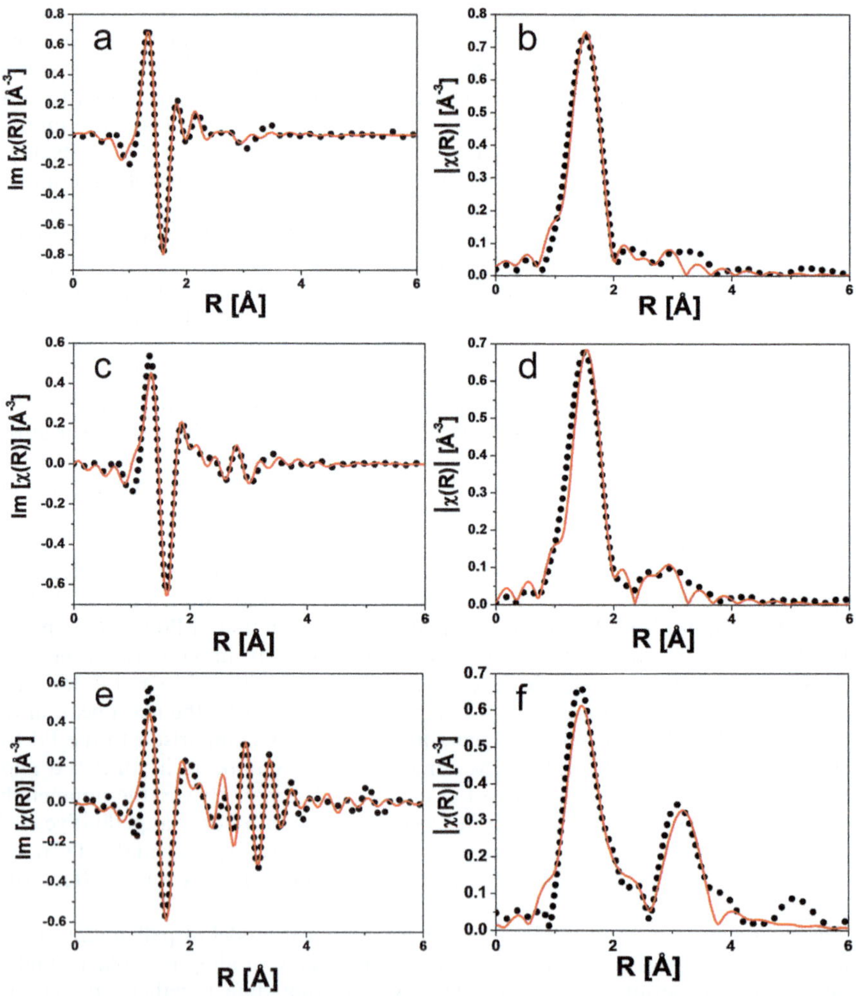

Fig. 6 Pd–K edge EXAFS single-shell fits of $Au_{25}Pd_{75}$ (a, b), $Au_{50}Pd_{50}$ (c, d) and $Au_{75}Pd_{25}$ (e, f) nanoalloys. Imaginary part (a, c, e) and FT magnitude (b, d, f). The dotted and solid lines represent experimental data and the fit, respectively.

Table 2 Results of the EXAFS fit parameters at the Pd–K edge. N = coordination number (CN), N_{total} = sum of N at one absorption edge, R = distance, σ^2 = Debye–Waller Factor, ΔE_0 = inner core correction

Sample	Edge	Pair	N	N_{total}	R [Å]	σ^2 [Å^{-2}]	ΔE_0 [eV]
Au75Pd25	Pd–K	Pd–Pd	2.3 ± 0.3	7.3 ± 0.8	2.76 ± 0.04	0.007 ± 0.002	−4.9 ± 1.0
	Pd–K	Pd–Au	3.3 ± 0.4		2.79 ± 0.01	0.007 ± 0.002	−4.9± 1.0
	Pd–K	Pd^{2+}	1.7 ± 0.2		2.02 ± 0.01	0.004 ± 0.001	0.0 ± 0.6
Au50Pd50	Pd–K	Pd–Pd	0.4 ± 0.1	2.5 ± 0.4	2.76 ± 0.04	0.007 ± 0.003	0.5 ± 0.5
	Pd–K	Pd–Au	0.5 ± 0.2		2.80 ± 0.04	0.007 ± 0.003	0.5 ± 0.5
	Pd–K	Pd^{2+}	1.6 ± 0.1		2.05 ± 0.01	0.001 ± 0.001	0.5 ± 0.5
Au25Pd75	Pd–K	Pd–Pd	0.3 ± 0.1	2.2 ± 0.3	2.77 ± 0.01	0.007 ± 0.003	3.0 ± 0.4
	Pd–K	Pd–Au	0.4 ± 0.1		2.76 ± 0.01	0.007 ± 0.003	3.0 ± 0.4
	Pd–K	Pd^{2+}	1.5 ± 0.1		2.05 ± 0.01	0.001 ± 0.001	3.0 ± 0.4
Pd–NP	Pd–K	Pd–Pd	3.4 ± 0.2	5.4 ± 0.3	2.72 ± 0.01	0.007 ± 0.003	1.0 ± 0.4
	Pd–K	Pd^{2+}	2.0 ± 0.1		2.05 ± 0.01	0.000 ± 0.001	1.0 ± 0.4

case of $Au_{25}Pd_{75}$ and $Au_{50}Pd_{50}$ nanoalloys. This confirms slight enrichment of Pd atoms at the surface area.

For the $Au_{75}Pd_{25}$ sample, the N_{total} value at the Pd–K edge is nearly as high as the one at the Au–L_3 edge. This indicates that approximately the same amount of Pd atoms as that of Au atoms are located inside the particle. Comparing $N_{Pd2+}$to N_{total}, 23% of palladium is supposed to be in a higher oxidation state. The higher amount of Pd–Pd bonds compared to Pd–Au bonds suggests a slight partial enrichment of palladium at the surface. These results are in good agreement with reports in the literature.[27,34,40,47]

Thus, the EXAFS data taken at the Au–L_3 edge and Pd–K edge demonstrate that the $Au_{50}Pd_{50}$ and $Au_{25}Pd_{75}$ alloys have slightly different structures to the $Au_{75}Pd_{25}$ alloy. For the latter alloy a structure close to theoretical random alloy is obtained while the two samples with a lower amount of gold show a slight increasing gradient of gold towards the particle center. At the Pd–K edge, non-metallic peaks for Pd^{2+} are found in the spectra in addition to Au–Pd and Pd–Pd signals.

These results obtained by EXAFS can now be compared to the catalytic activity of the same nanoparticles given recently.[12] Firstly, all nanoalloys are better catalysts than the neat metal nanoparticles. Thus, the non-metallic character of the Pd atoms on the surface of the $Au_{25}Pd_{75}$ and the $Au_{50}Pd_{50}$ nanoalloys seems to be of minor importance. However, the previous discussion of the electronic structure has come to the conclusion that an increasing number of Pd atoms in the nanoparticles leads to a HOMO that is mainly located at the surface.[12] This indicates that the Pd atoms located at the surface are likely to increase the catalytic activity. The maximum of catalytic activity, however, is not seen for alloys with maximal but minimal content of Pd. Moreover, the discussion of the electronic structure has revealed that a small change of the geometric structure leads to marked changes of the HOMO.[12] Hence, summing up the presently known facts one may conclude that the active sites on the surface consist of small groups of Pd atoms sitting at the surface, most probably at surface defects.

Conclusion

Well-defined and facetted gold–palladium nanoalloys have been synthesized and anchored in spherical polyelectrolyte brushes (SPB). Catalytic study using reduction of 4-NiP as model reaction demonstrates that Au–Pd nanoalloy systems, especially $Au_{75}Pd_{25}$, have a much higher catalytic activity compared to that of pure metal nanoparticles, which is highly related to their structures. EXAFS analysis of the alloy samples indicates that the $Au_{75}Pd_{25}$ alloy sample shows different structural behavior compared to $Au_{50}Pd_{50}$ and $Au_{25}Pd_{75}$ alloys. The $Au_{50}Pd_{50}$ and $Au_{25}Pd_{75}$ alloys exhibit a slight enrichment of Pd at the surface and Au in the core, while for the $Au_{75}Pd_{25}$ particle a structure which is closer to a random alloy is obtained. Thus more Pd atoms are located on the surface for $Au_{50}Pd_{50}$ and $Au_{25}Pd_{75}$ than for $Au_{75}Pd_{25}$. In addition, a partial coverage by Pd^{2+} at the surface is suggested by EXAFS evaluation for the Au–Pd alloy particles. These data together with the results presented recently[12] suggest that Pd atoms located at surface defects may constitute the active sites.

Acknowledgements

We gratefully acknowledge the assistance of Dr. M. Murphy at the HASYLAB/DESY synchrotron facility during EXAFS measurements. We acknowledge the colleagues from the Helmholtz-Zentrum Berlin – electron storage ring BESSY II for provision of synchrotron radiation at beamline KMC-2 and would like to thank Dr. V. Vidal, Dr. I. Zizak and Prof. Dr. A. Erko for support and assistance.

References

1 M. Schrinner, M. Ballauff, Y. Talmon, Y. Kauffmann, J. Thun, M. Moller and J. Breu, *Science*, 2009, **323**, 617–620.
2 Y. Lu, A. Wittemann and M. Ballauff, *Macromol. Rapid Commun.*, 2009, **30**, 806–815.
3 Y. Mei, Y. Lu, F. Polzer, M. Ballauff and M. Drechsler, *Chem. Mater.*, 2007, **19**, 1062–1069.
4 Y. Mei, G. Sharma, Y. Lu, M. Ballauff, M. Drechsler, T. Irrgang and R. Kempe, *Langmuir*, 2005, **21**, 12229–12234.
5 Y. Lu, Y. Mei, R. Walker, M. Ballauff and M. Drechsler, *Polymer*, 2006, **47**, 4985–4995.
6 S. Wunder, Y. Lu, M. Albrecht and M. Ballauff, *ACS Catal.*, 2011, **1**, 908–916.
7 S. Wunder, F. Polzer, Y. Lu, Y. Mei and M. Ballauff, *J. Phys. Chem. C*, 2010, **114**, 8814–8820.
8 P. Herves, M. Perez-Lorenzo, L. M. Liz-Marzan, J. Dzubiella, Y. Lu and M. Ballauff, *Chem. Soc. Rev.*, 2012, **41**, 5577–5587.
9 R. Ferrando, J. Jellinek and R. L. Johnston, *Chem. Rev.*, 2008, **108**(3), 845–910.
10 N. R. Shiju, *Appl. Catal., A*, 2009, **356**, 1–17.
11 G. J. Hutchings, *Dalton Trans.*, 2008, 5523–5536.
12 J. Kaiser, L. Leppert, H. Welz, F. Polzer, S. Wunder, N. Wanderka, M. Albrecht, T. Lunkenbein, J. Breu, S. Kümmel, Y. Lu and M. Ballauff, *Phys. Chem. Chem. Phys.*, 2012, **14**, 6487–6495.
13 M. Schrinner, S. Proch, Y. Mei, R. Kempe, N. Miyajima and M. Ballauff, *Adv. Mater.*, 2008, **20**, 1928–1933.
14 N. Toshima and T. Yonezawa, *New J. Chem.*, 1998, **22**, 1179–1201.
15 E. N. Ntainjua, M. Piccinini, S. J. Freakley, J. C. Pritchard, J. K. Edwards, A. F. Carley and G. J. Hutchings, *Green Chem.*, 2012, **14**, 170–181.
16 M. Piccinini, J. K. Edwards, J. A. Moulijn and G. J. Hutchings, *Catal. Sci. Technol.*, 2012, **2**, 1908–1913.
17 M. Sankar, Q. He, M. Morad, J. Pritchard, S. J. Freakley, J. K. Edwards, S. H. Taylor, D. J. Morgan, A. F. Carley, D. W. Knight, C. J. Kiely and G. J. Hutchings, *ACS Nano*, 2012, **6**, 6600–6613.
18 R. Su, R. Tiruvalam, Q. He, N. Dimitratos, L. Kesavan, C. Hammond, J. A. Lopez-Sanchez, R. Bechstein, C. J. Kiely, G. J. Hutchings and F. Besenbacher, *ACS Nano*, 2012, **6**, 6284–6292.
19 D. I. Enache, J. K. Edwards, P. Landon, B. Solsona-Espriu, A. F. Carley, A. A. Herzing, M. Watanabe, C. J. Kiely, D. W. Knight and G. J. Hutchings, *Science*, 2006, **311**, 362–365.
20 L. Kesavan, R. Tiruvalam, M. H. Ab Rahim, M. I. bin Saiman, D. I. Enache, R. L. Jenkins, N. Dimitratos, J. A. Lopez-Sanchez, S. H. Taylor, D. W. Knight, C. J. Kiely and G. J. Hutchings, *Science*, 2011, **331**, 195–199.
21 G. L. Brett, P. J. Miedziak, N. Dimitratos, J. A. Lopez-Sanchez, N. F. Dummer, R. Tiruvalam, C. J. Kiely, D. W. Knight, S. H. Taylor, D. J. Morgan, A. F. Carleya and G. J. Hutchings, *Catal. Sci. Technol.*, 2012, **2**, 97–104.
22 P. Landon, P. J. Collier, A. J. Papworth, C. J. Kiely and G. J. Hutchings, *Chem. Commun.*, 2002, 2058–2059.

23 A. Staykov, T. Kamachi, T. Ishihara and K. Yoshizawa, *J. Phys. Chem. C*, 2008, **112**, 19501–19505.
24 Y. F. Han, Z. Zhong, K. Ramesh, F. Chen, L. Chen, T. White, Q. Tay, S. N. Yaakub and Z. Wang, *J. Phys. Chem. C*, 2007, **111**, 8410–8413.
25 S. Remita, M. Mostafavi and M. O. Delcourt, *Radiat. Phys. Chem.*, 1996, **47**, 275–279.
26 H. Bonnemann, U. Endruschat, B. Tesche, A. Rufinska, C. W. Lehmann, F. E. Wagner, G. Filoti, V. Parvulescu and V. I. Parvulescu, *Eur. J. Inorg. Chem.*, 2000, 819–822.
27 F. Liu, D. Wechsler and P. Zhang, *Chem. Phys. Lett.*, 2008, **461**, 254–259.
28 X. Wei, X. F. Yang, A. Q. Wang, L. Li, X. Y. Liu, T. Zhang, C. Y. Mou and J. Li, *J. Phys. Chem. C*, 2012, **116**, 6222–6232.
29 M. W. Shao, H. Wang, M. L. Zhang, D. D. D. Ma and S. T. Lee, *Appl. Phys. Lett.*, 2008, **93**, 243110.
30 F. Ksar, L. Ramos, B. Keita, L. Nadjo, P. Beaunier and H. Remita, *Chem. Mater.*, 2009, **21**, 3677–3683.
31 Y. T. Chen, H. M. Lim, Q. H. Tang, Y. T. Gao, T. Sun, Q. Y. Yan and Y. H. Yang, *Appl. Catal., A*, 2010, **380**, 55–65.
32 D. Wang, A. Villa, P. Spontoni, D. S. Su and L. Prati, *Chem.–Eur. J.*, 2010, **16**, 10007–10013.
33 M. Hosseini, T. Barakat, R. Cousin, A. Aboukais, B. L. Su, G. De Weireld and S. Siffert, *Appl. Catal., B*, 2012, **111–112**, 218–224.
34 S. Marx and A. Baiker, *J. Phys. Chem. C*, 2009, **113**, 6191–6201.
35 C. J. Baddeley, R. M. Ormerod, A. W. Stephenson and R. M. Lambert, *J. Phys. Chem.*, 1995, **99**, 5146–5151.
36 C. J. Baddeley, M. Tikhov, C. Hardacre, J. R. Lomas and R. M. Lambert, *J. Phys. Chem.*, 1996, **100**, 2189–2194.
37 A. F. Lee, C. J. Baddeley, C. Hardacre, R. M. Ormerod, R. M. Lambert, G. Schmid and H. West, *J. Phys. Chem.*, 1995, **99**, 6096–6102.
38 M. Ballauff, *Prog. Polym. Sci.*, 2007, **32**, 1135–1151.
39 A. I. Frenkel, A. Yevick, C. Cooper, R. Vasic, Modeling the Structure and Composition of Nanoparticles by Extended X-Ray Absorption Fine-Structure Spectroscopy. In *Annual Review of Analytical Chemistry, Vol 4*, R. G. Cooks and E. S. Yeung, Eds. Annual Reviews, Palo Alto, 2011, Vol. 4, pp. 23–39.
40 C. Evangelisti, E. Schiavi, L. A. Aronica, A. M. Caporusso, G. Vitulli, L. Bertinetti, G. Martra, A. Balerna and S. Mobilio, *J. Catal.*, 2012, **286**, 224–236.
41 Y. L. Fang, J. T. Miller, N. Guo, K. N. Heck, P. J. J. Alvarez and M. S. Wong, *Catal. Today*, 2011, **160**, 96–102.
42 X. W. Teng, Q. Wang, P. Liu, W. Han, A. Frenkel, W. Wen, N. Marinkovic, J. C. Hanson and J. A. Rodriguez, *J. Am. Chem. Soc.*, 2008, **130**, 1093–1101.
43 M. R. Knecht, M. G. Weir, A. I. Frenkel and R. M. Crooks, *Chem. Mater.*, 2008, **20**, 1019–1028.
44 M. G. Weir, M. R. Knecht, A. I. Frenkel and R. M. Crooks, *Langmuir*, 2010, **26**, 1137–1146.
45 P. Dash, T. Bond, C. Fowler, W. Hou, N. Coombs and R. W. J. Scott, *J. Phys. Chem. C*, 2009, **113**, 12719–12730.
46 W. C. Ketchie, M. Murayama and R. J. Davis, *J. Catal.*, 2007, **250**, 264–273.
47 F. Krumeich, S. Marx, A. Baiker and R. Nesper, *Z. Anorg. Allg. Chem.*, 2011, **637**, 875–881.
48 F. Liu and P. Zhang, *Appl. Phys. Lett.*, 2010, **96**, 043105.
49 M. Harada, K. Asakura and N. Toshima, *J. Phys. Chem.*, 1993, **97**, 5103–5114.
50 J. W. Couves and P. Meehan, *Phys. B*, 1995, **208–209**, 665–667.
51 X. Guo, A. Weiss and M. Ballauff, *Macromolecules*, 1999, **32**, 6043–6046.
52 A. Erko, I. Packe, W. Gudat, N. Abrosimov and A. Firsov, A graded crystal monochromator at BESSY II. In *Advances in X-Ray Optics*, A. K. Freund, T. Ishikawa, A. M. Khounsary, D. C. Mancini, A. G. Michette and S. Oestreich, Eds. Spie-Int Soc Optical Engineering, Bellingham, 2001, Vol. 4145, pp. 122–128.
53 A. Erko, I. Packe, C. Hellwig, M. Fieber-Erdmann, O. Pawlizki, M. Veldkamp and W. Gudat, KMC-2: the new x-ray beamline at BESSY II. In *Synchrotron Radiation Instrumentation*, P. Pianetta, J. Arthur and S. Brennan, Eds. Amer Inst Physics, Melville, 2000, Vol. 521, pp. 415–418.
54 H. Riesemeier, K. Ecker, W. Gorner, B. R. Muller, M. Radtke and M. Krumrey, *X-Ray Spectrom.*, 2005, **34**, 160–163.
55 B. Ravel and M. Newville, *J. Synchrotron Radiat.*, 2005, **12**, 537–541.
56 S. D. Kelly and D. H. B. Ravel, *Analysis of soils and mineralsusing X-ray absorption spectroscopy.*, *Soil Sci. Soc. Am.*, Madison, 2008.
57 R. E. Benfield, *J. Chem. Soc., Faraday Trans.*, 1992, **88**, 1107–1110.
58 E. Strub, R. Plarre, M. Radtke, U. Reinholz, H. Riesemeier, U. Schoknecht, K. Urban and P. Jüngel, *Nucl. Instrum. Methods Phys. Res., Sect. B*, 2008, **266**, 2405–2407.

Semi-hydrogenation of alkynes at single crystal, nanoparticle and biogenic nanoparticle surfaces: the role of defects in Lindlar-type catalysts and the origin of their selectivity

G. A. Attard,*[a] J. A. Bennett,[b] I. Mikheenko,[c] P. Jenkins,[a] S. Guan,[a] L. E. Macaskie,[c] J. Wood[b] and A. J. Wain[d]

Received 22nd January 2013, Accepted 24th January 2013
DOI: 10.1039/c3fd00007a

For the first time, the method of shell-isolated nanoparticle Raman spectroscopy (SHINERS) is used in combination with cyclic voltammetry (CV) and reactivity studies to investigate the adsorption behaviour of a series of three alkynes undergoing hydrogenation on nanoparticle, single crystal and bacteria/graphite-supported platinum surfaces. It is found that a strong association of alkynes with defect sites to produce a long-lived di-sigma/pi-alkene surface complex allows for deep hydrogenation of this intermediate to the alkane product. In contrast, when platinum surface defect sites are blocked by either bismuth or polyvinylpyrrolidone (PVP) (and thus leaving behind only Pt{111} terrace adsorption sites), large increases in selectivity to the semi-hydrogenation product are observed for all three alkynes. This finding is consistent with SHINERS collected from both well-ordered and roughened Pt{111} electrodes which revealed that the di-sigma/pi-bonded surface intermediate is hardly formed at all on Pt{111} unless defect sites are introduced *via* electrochemical roughening. As a general method of producing selective catalysts, the elimination of toxic heavy metals from Lindlar-type catalyst, used commonly in organic chemistry, and their replacement by more benign, organic species adsorbed at defect sites is discussed.

Introduction

The improvement of catalyst selectivity is a research area of major interest and importance. Ligand design and optimisation in homogeneous catalysts is a common route for modifying and increasing selectivity in a given reaction, for example, the development of the triphenylphosphine ligand into a diverse range of more complex architectures, such as 2,2′-bis(diphenylphosphino)-1,1′-binapthyl or BINAP ligand and its derivatives.[1–4] Despite the excellent activities and selectivities afforded by such homogeneous catalysts, difficulties with their separation and recycling prevent their common usage in large scale industrial processes.

The use of surface-modifying adsorbates, such as metal salts, the cinchona alkaloids or polymers, is a well explored route for tuning selective heterogeneous catalysts.[5–7] Recent studies have explored the concept of intrinsic selectivity at the

[a]*School of Chemistry, Cardiff University, Park Place, Cardiff CF10 3AT, UK. E-mail: attard@cardiff.ac.uk; Fax: +44 (0) 2920 874030; Tel: +44 (0) 2920 874069*
[b]*School of Chemical Engineering, University of Birmingham, Birmingham B15 2TT, UK*
[c]*School of Biosciences, University of Birmingham, Birmingham B15 2TT, UK*
[d]*National Physical Laboratory, Teddington, TW11 0LW, UK*

surface of a metal particle.[8,9] It is now known that different crystallite topographies (*e.g.* terraces, corners, edges and adatoms), interacting differently with certain adsorbates, lead to differences in reactivity and selectivity. Thus, the shape and size of the metal particle itself can affect its properties as a catalyst, an effect which may be termed structure sensitivity. By fully understanding the surface chemistry of the metal particle and the causes of site-specificity, catalyst reactivity may be tuned by changing the shape of the metal particle itself, without the need for additional ligand architectures.

The idea of a structure sensitive reaction has been demonstrated by Crespo-Quesada *et al.*,[8] who compared the selectivity of cubic, octahedral and cuboctahedral particles of platinum in the hydrogenation of 2-methyl-3-butyn-2-ol. The different proportions of terraces and edges/corners present on the three shapes led to distinct selectivity in the hydrogenation. The results of Crespo-Quesada *et al.* indicated that two distinct active sites were present on the particles; semi-hydrogenation of the alkyne occurred preferentially on terraces whereas further hydrogenation to the alkane occurred over edges. A similar study was conducted by Schmidt *et al.*, who used cubic, octahedral and cuboctahedral platinum particles in ethyl pyruvate enantioselective hydrogenation.[9] The reaction was found to be shape selective, with increasing Pt{111}/Pt{100} ratio leading to an increase in both rate and enantiomeric excess.

Lee *et al.* demonstrated structure sensitivity in the *cis–trans* isomerisation of 2-butene, where Pt{111} single crystals favoured *trans*- to *cis*-isomerisation.[10] The opposite selectivity was observed over Pt{100} facets. They used colloidal techniques to prepare cubic and tetrahedral particles of platinum, which exposed only {100} and {111} facets respectively. The catalyst composed of {111} surfaces not only favoured the isomerisation of *trans*- to *cis*-alkene but also gave higher alkene hydrogenation selectivity.[11]

The above investigations required the synthesis of metal particles of a specific shape. An alternative route for exploring the reactivities of different surface types is to use a selective poison to block a particular group of heterogeneous sites on the catalyst surface. Adsorbates such as bismuth and sulphur adsorb preferentially upon specific sites, preventing or limiting any reactions which occur mainly upon them. This approach was used previously by Anderson *et al.*,[12] who utilised bismuth and lead as selective poisons for palladium catalysts in the hydrogenation of 1- and 2-hexyne. Bismuth blocked surface defect sites, suppressing the double bond-shift of terminal to internal alkenes, which occurs rapidly at these sites. Lead preferentially adsorbed on terraces and caused a reduction in the degree of *cis–trans* isomerisation. Since then, more recent work has demonstrated the importance of surface modification in the surface chemistry of alkynes.[13–17]

Previous work by Attard *et al.* has also investigated the effect of selective surface poisons on catalytic properties.[18,19] A study on 2-butyne-1,4-diol (butynediol) demonstrated that 2-butene-1,4-diol (butenediol) selectivity could be substantially improved, at the cost of reaction rate, by increasing surface coverage of bismuth on platinum.[20] Since Bi is known to preferentially adsorb at defect sites,[21] these sites were associated with unselective alkyne to alkane transformations whereas Pt{111} sites facilitated selective hydrogenation of the alkyne to alkene, in agreement with conclusions drawn by Crespo-Quesada and co-workers.[8] The rates of *cis*- to *trans*-isomerisation were not affected in these studies, indicating that these reactions took place over Pt terraces in agreement with Lee *et al.*[10] In addition, organic fragments left over after cleaning of Pt nanoparticles grown in the bacterium *E. coli* were shown to increase the selectivity towards semi-hydrogenation products using 2-butyne-1,4-diol as substrate.[20] The location of such fragments at defect sites was shown to be responsible for this type of behaviour. A striking feature of this work was the elimination of toxic heavy metals from the catalyst whilst maintaining selectivity.

In order to understand fully the changes in catalytic properties caused by modifying the surface of an active metal particle it is necessary to characterise the surface

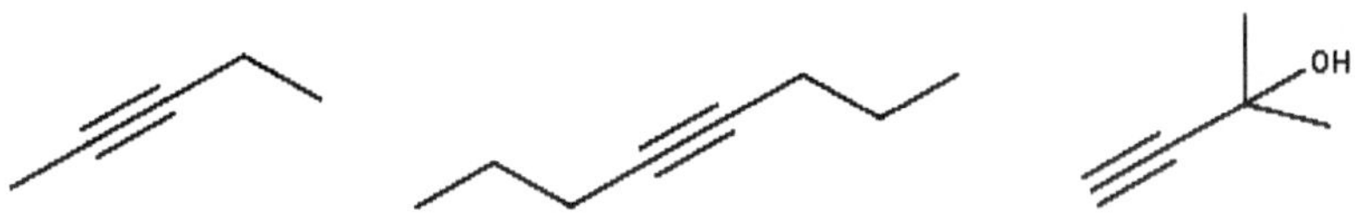

Scheme 1 Structural formulae of 2-pentyne, 4-octyne and 2-methyl-3-butyn-2-ol.

with a range of complementary techniques. Previously, Attard *et al.* have used cyclic voltammetry (CV) to probe the changes at the catalyst surface after various modifications.[19,20,22] This technique can be used to determine quantitatively the percentages of different surface sites available for adsorption[23] but lacks insight concerning the nature of the adsorbate or its bonding.

Surface Enhanced Raman Spectroscopy (SERS) is an excellent method for investigating molecular vibrations of adsorbed species.[24] SERS was first observed from molecules adsorbed at roughened silver electrode surfaces[25] and later at other roughened surfaces, particularly gold and copper. SERS is only weakly observed from metals such as platinum.[26] Later work however showed that the "roughening" effect could be reproduced by depositing metal nanoparticles of appropriate size onto a solid surface.[27] The selectivity and sensitivity of SERS extends to a wide variety of interfacial systems, including *in situ* and ambient analysis of electrochemical, catalytic, biological, and organic systems.[28–30] In the present study, two variants of surface Raman spectroscopy are introduced in order to obtain information concerning the nature of the surface intermediates forming when alkynes are undergoing hydrogenation at platinum surfaces. In the first of these, Au@Pt (meaning Pt shell on gold core) core–shell nanoparticles are deposited on polycrystalline platinum electrodes.[31] By limiting the gold core to a diameter of approximately 55 nm and the platinum shell to 2 nm, the nanoparticle reproduces surface-enhanced Raman properties similar to roughened gold but exhibits the surface chemistry of a normal platinum surface. Hence, SERS from a weakly SERS-active (but catalytically far more interesting) metal surface may readily be collected. In a more recent development, Tian and co-workers[32] demonstrated that Au@SiO_2 (meaning silica shell on gold core) nanoparticles (55 nm diameter gold core, 2 nm thick silica shell) deposited on a single crystal electrode could be used to observe SERS even from a well-defined single crystal surface. This breakthrough in the application of SERS was designated by these workers as shell-isolated nanoparticle enhanced Raman spectroscopy (SHINERS) and the method holds great potential for future *in situ* catalytic studies. Here, we exploit the technique to study, for the first time, molecular vibrations of adsorbed intermediates formed during the hydrogenation of a series of alkynes. Two internal alkynes, 2-pentyne and 4-octyne and a terminal alkyne with a polar functionality, 2-methyl-3-butyn-2-ol (see Scheme 1) were hydrogenated using a graphite supported 5% wt. platinum catalyst. The 5% wt. Pt/C catalyst was also modified with both Bi and PVP (polyvinylpyrrolidone). PVP was chosen since it is used as a capping agent for regulating particle size during growth and has previously been implicated in improved selectivity towards semi-hydrogenation of alkynes.[14,33] In addition, a biogenic analogue supported on bacterial biomass (bioPt) was tested for selective alkyne hydrogenation. The role of surface defects as arbiters of alkyne hydrogenation selectivity together with possible alternatives to toxic heavy metals in Lindlar-type catalysts will be explored.

Experimental

Hydrogenation reactions

Hydrogenation reactions were carried out in a 500 ml Baskerville stirred autoclave reactor. The catalyst (30 mg, 5% wt. Pt/C or 5% Pt on biomass) and isopropanol (150 ml) were added to the reactor, which was then flushed with nitrogen gas before being sealed. The mixture was heated to 40 °C, and stirred at 500 rpm under a flow of hydrogen for 10 minutes. The reactor was purged with nitrogen and the substrate added (4 ml). The reactor was then sealed, pressurised with hydrogen (2 bar) and stirred (1000 rpm, Rushton turbine) at 40 °C. Samples were taken periodically for analysis by gas chromatography to monitor the progress of the reaction.

Gas chromatography

Sample analysis was performed by gas chromatography, using a Varian CP-3380 with a flame ionisation detector and 30 m Supelco Gamma Dex 225 column. The oven temperature for each substrate was as follows: 2-methyl-3-butyn-2-ol, 65 °C; 4-octyne, 80 °C; 2-pentyne, with oven temperature initially at 30 °C, hold for 1 min, ramp to 80 °C at 15 °C min^{-1}. An injection volume of 0.5 μL was used.

Preparation of platinum nanoparticles supported on bacterial biomass

Cells of *Escherichia coli* were grown aerobically or under anaerobic respiratory conditions and harvested as described previously.[34] A known volume of the concentrated resting cell suspension was transferred under OFN (oxygen-free nitrogen) into 200 mL serum bottles containing an appropriate volume of degassed 2 mM Na_2PtCl_6 solution in 0.01 M HNO_3, pH 2.0 so that the final ratio (weight of Pt : dry weight of cells) was 1 : 19 to give a final loading of 5% Pt on biomass. Cells/Pt mixtures were left to stand for 60 min at 30 °C with occasional shaking to promote biosorption of Pt(IV) complexes[35] before H_2 was sparged through the suspension for 10 min (0.2 atm) to reduce cell surface-bound Pt(IV) into Pt (0). The Pt(0)-coated biomass (bioPt) was harvested by centrifugation (3000 *g*, 10 min, 4 °C) and washed three times in distilled water (dH_2O). Following a final acetone wash, the black precipitate was left to dry in air overnight. The resulting black powder containing bio-Pt(0) was finely ground in a mortar and tested for catalytic activity without any further processing (Table 1).

Electrochemical apparatus

A three-electrode electrochemical cell, described in detail previously,[36] was used and data was recorded *versus* a Pd/H reference electrode and a platinum counter electrode. Voltammetric data was recorded in degassed 0.5 M aqueous sulphuric acid (Aristar) at a sweep rate of 50 mV s^{-1} unless otherwise stated. The glassy carbon (GC) support electrode used for analysing the bioPt samples was cleaned and polished using alumina prior to use and a cyclic voltammogram (CV) was recorded to verify the cleanliness of the system.

Pt{*hkl*} single crystal electrodes were prepared using the Clavilier bead method.[37] In order to obtain *in situ* SERS data, the Pt single crystal electrode was modified with

Table 1 Description of catalysts tested

Catalyst	Description
Pt/C	5% wt. Pt on carbon reduced at 700 K under H_2/Ar
Pt/C + 10 ml PVP	5% wt. Pt on carbon reduced at 700 K under H_2/Ar
Pt/C + 12 ml PVP	5% wt. Pt on carbon reduced at 700 K under H_2/Ar
bioPt aero	5% wt. Pt supported on bacterial biomass grown aerobically; air dried
bioPt anaero	5% wt. Pt supported on bacterial biomass grown anaerobically; air dried

$Au@SiO_2$ NPs by allowing a small droplet of the nanoparticles suspended in water to dry out on the electrode surface to leave a film of nanoparticles behind (see below).

To create hydrogenation conditions, hydrogen gas was evolved at the working electrode by holding the potential at −0.03 V *versus* the Pd/H reference electrode. Hydrogen bubbles forming on the surface of the electrode, which might affect the detection of SERS signals, were removed by flowing the electrolyte through the cell using a peristaltic pump. To electrochemically clean the surface of the electrode without roughening the surface, potential cycling was performed between −0.10 and 0.85 V at 100 mV s^{-1} in 0.1 M H_2SO_4 (or $HClO_4$).[38]

Chemicals

Solutions of $HAuCl_4$ (Johnson Matthey, Assay 41.79%), sodium citrate dehydrate (Sigma-Aldrich, 99+%), (3-aminopropyl)trimethoxysilane (APTMS, Alfa Aesar, 97%), sodium silicate (Sigma-Aldrich, reagent grade), H_2SO_4 (BDH, Aristar grade), $HClO_4$ (Merck, Suprapur), 2-butyne-1,4-diol and 2-butene-1,4-diol (Sigma-Aldrich, 99%), bismuth nitrate pentahydrate (Sigma-Aldrich, 98+%), poly(vinylpyrrolidone) (PVP M_w = 10 000) were used as received and all aqueous solutions were prepared using ultrapure water (18.2 MΩ cm^{-1}). 2-Pentyne (Alfa Aesar, 99%), 4-octyne (Alfa Aesar, 99%) and 2-methyl-3-butyn-2-ol (Alfa Aesar, 98%) were used as supplied.

Nanoparticle synthesis

55 nm diameter Au seeds were prepared as follows:

100 mL of 0.01% (by mass) $HAuCl_4$ was refluxed at 112 °C. After 30 minutes, 600 μL of a tri-sodium citrate solution (38 mM) was added and the reaction mixture was then refluxed for another 30 min. To generate a platinum shell, the gold seeds were processed still further according to the method of Tian *et al.*[31] 1470 μL H_2PtCl_6 (1 mM) was added to 30 mL of the Au seed solution and the mixture refluxed at 80 °C. After 30 minutes, 1480 μL of ascorbic acid (10 mM) was then added into the mixture very slowly by using a peristaltic pump and the reflux continued for another 30 minutes to obtain Au@Pt NPs. $Au@SiO_2$ NPs were prepared by modification of the method of Tian *et al.*[32]

Briefly, 0.4 mL of a freshly prepared 1.0 mM aqueous solution of (3-aminopropyl)-trimethoxysilane (APTMS) was added to 30 mL of the Au seeds under vigorous stirring. 3.2 mL, 0.54 wt.%, sodium silicate solution, (pH of the silicate solution was adjusted to pH 10.28 using 0.1 M HCl) was added to the mixture. The reaction mixture was kept at 70 °C for 13 min to form $Au@SiO_2$ NPs with an ultrathin shell (2 nm).

Spherical Pt nanoparticles were prepared according to literature procedures.[39] Spherical platinum nanoparticles (4.5 ± 0.8 nm) were obtained by reduction of H_2PtCl_6 with hydrazine using water-in-oil micro-emulsions. In brief, this was achieved by mixing two micro-emulsions (one containing the metal precursor and the other containing the reducing agent) that have the same water to surfactant molar ratio and the same surfactant concentration, where polyethyleneglycol–dodecyl ether (BRIJ®30) was used as the surfactant.[39] For the platinum nanoparticles, 10 μL of the nanoparticle suspension was deposited on to a clean, polished gold bead electrode and dried under nitrogen. Electrochemical experiments were carried out in the same type of 3-electrode cell as used for the single crystal experiments. The supporting electrolyte, 0.1 M H_2SO_4, was degassed for 20 min prior to electrochemical measurements. In order to remove any leftover capping agent or any other weakly adsorbed contaminants, the nanoparticle modified gold electrodes were cycled between −0.3 V and 0.85 V.

Preparation of Pt/C catalyst selectively poisoned with PVP

For the Pt on graphite catalysts 1 g of 5% platinum/graphite was reduced at 700 K in a 5% H_2/N_2 gas stream for 4 hours. The catalyst was cooled overnight with the gas stream still flowing. The catalyst has been described fully previously.[20]

For selective poisoning with PVP between 0.3 and 0.5 g of 5% platinum/graphite was stirred with an aqueous 10^{-3} M PVP solution ($M_w = 10\,000$) for 30 minutes, filtered, rinsed thoroughly then air dried. Different volumes of the PVP solution were added in order to obtain the varying levels of catalyst loading (Table 1). CVs of the catalysts were obtained by pressing a small amount of sample onto a platinum mesh (flamed-annealed clean prior to use) which then formed the working electrode.

Raman spectroscopic measurements

Raman spectroscopic measurements were performed using a LabRam Raman Microscope (Horiba JobinYvon Ltd.) fitted with a HeNe laser ($\lambda = 633$ nm, output power 16 mW), with data recording and analysis performed using the proprietary LabSpec software (Horiba JobinYvon).[38]

Results and discussion

In situ Raman studies of alkyne adsorption

Fig. 1 shows SERS from a variety of platinum electrode surfaces undergoing the hydrogen evolution reaction (HER) in the presence of 0.1 M butynediol together with the Raman spectrum recorded using solid butynediol. The Raman peaks associated with bulk butynediol are readily identified at 2870, 2930 and 2955 cm^{-1} (C–H stretches), 2212 and 2283 cm^{-1} (C≡C triple bond stretches) and 1024 cm^{-1} (C–C stretches). For spherical platinum nanoparticles supported on a platinum electrode (Fig. 1b and 1c), a SER spectrum is obtained but the signal is rather weak, consistent with the known Raman behaviour of roughened platinum relative to the coinage metals copper, silver and gold.[26] Nevertheless, two new features are observed in the spectra from Pt nanoparticles under electrochemical conditions at 1580 cm^{-1} and 490 cm^{-1}. We ascribe the first of these peaks to an alkenic stretch associated

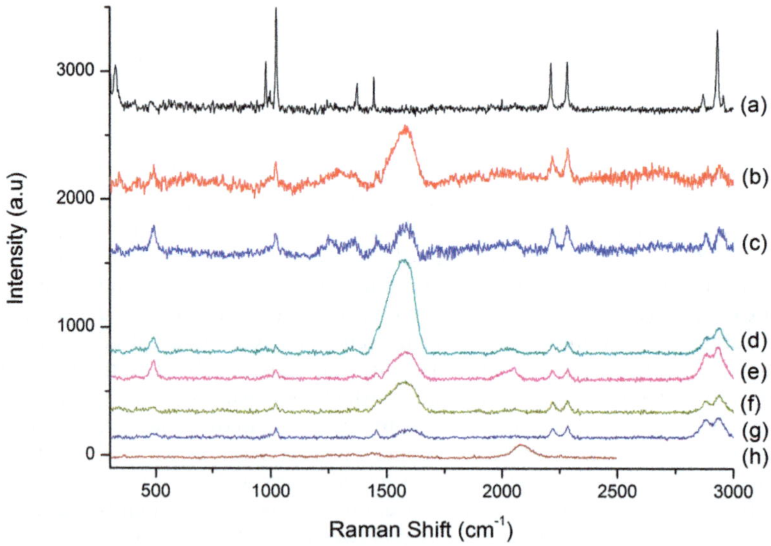

Fig. 1 Raman (a) and surface enhanced Raman spectra (b–g) of 2-butyne-1,4-diol. (h) shows hydrogen evolution at Au@Pt nanoparticles in the absence of 2-butyne-1,4-diol. (b) and (c) spherical Pt nanoparticles at double layer and HER potentials respectively. (d) and (e) Au@Pt nanoparticles at double layer and HER potentials respectively. (f) and (g) as for (d) and (e) except Pt defect sites blocked by bismuth.[20] Electrolyte consists of aqueous 0.1 M sulphuric acid together with 0.1 M 2-butyne-1,4-diol (b–g).

with a di-sigma/pi-bonded surface intermediate reported previously (Fig. 2).[40,41] The red shift in frequency of the 1580 cm^{-1} band compared to a normal non-conjugated alkene stretch (~1640 cm^{-1}) is due to a strong interaction of the double bond with the platinum surface.[42] In the absence of hydrogen (at electrode potentials in the double layer region) the 1580 cm^{-1} peak is more intense compared to when the electrode is evolving hydrogen. This would be consistent with a steady state situation whereby a relative decrease in the surface population of the di-sigma/pi-bonded surface complex in the presence of hydrogen is caused by hydrogenative desorption of the alkenic surface intermediate to the corresponding alkane. In the absence of hydrogen, one would therefore expect a greater surface population of the di-sigma/pi-bonded complex and a corresponding increase in band intensity. The peak at around 490 cm^{-1} would be consistent with a metal–carbon stretch vibration.

In order to increase the magnitude of the SERS signal over that obtainable from standard platinum substrates, experiments were next performed on Au@Pt core–shell nanoparticles, as described previously in the experimental section.[31] In Fig. 1d and 1e, typical data are shown using this approach for butynediol hydrogenation. Once again, peaks associated with solution phase and chemisorbed butynediol are observed in the spectra, this time with much better signal–noise ratio and with intensities an order of magnitude greater than those originating from the bulk electrolyte peaks. Under HER conditions, a broad series of peaks ranging from 1280 to 1575 cm^{-1} are immediately apparent. In addition, clear evidence of peaks in the range 2000–2050 cm^{-1} confirms the presence of adsorbed CO. Hence, the peak at around 490 cm^{-1} may be ascribed to the metal–carbon stretch of adsorbed CO[43] although weak metal–carbon vibrations from the sigma-bonded surface complex cannot be ruled out completely in this spectral region. It is deduced that a certain degree of stability may be ascribed to the di-sigma/pi-complex even under hydrogenation conditions although the magnitude of the 1574 cm^{-1} peak increases by a factor of three when no hydrogen is being evolved (Fig. 1d). In addition, the surface coverage of CO decreases in the absence of hydrogen gas as signified by the fall in intensity of both the 2000–2040 cm^{-1} and 490 cm^{-1} peaks. Interestingly, the intensity of the C–H stretch region also changes slightly when the electrode surface is undergoing hydrogen evolution giving rise to more intense peaks compared to their intensity at potentials away from the HER. We tentatively ascribe this behaviour to a change in the interaction of the (CH_2) groups with the surface, possibly due to a tilting towards the surface of the alkenic bond of the di-sigma/pi-complex under hydrogenation conditions as previously reported by Foucault and co-workers for the case of long chain surfactant molecules[44] (see Fig. 2).

In ref. 20 the importance of defect sites in promoting the hydrogenation of alkynes to alkanes was emphasised in that if defect sites were deliberately blocked by either bismuth or organic residues (from the host bacterium) a significant increase in selectivity towards semi-hydrogenation products was observed. Therefore, it seemed logical to re-examine spectroscopically and *in situ*, changes in the nature of the adsorbed intermediates derived from butynediol when such defect sites are blocked. For a surface in which 80% of defect sites on the Au@Pt nanoparticles are blocked by bismuth, Fig. 1f

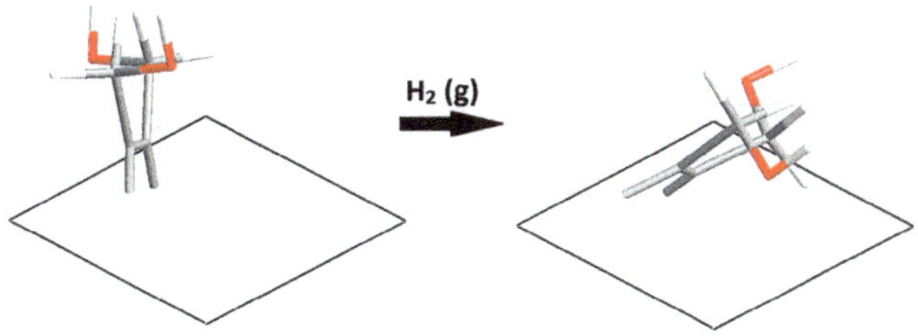

Fig. 2 Schematic diagram showing tilting of surface di-sigma/pi-bonded intermediate upon evolving hydrogen leading to greater interaction of methylene groups with surface.

and 1g reveal that substantial perturbations have indeed been promulgated relative to the SERS data obtained in the absence of bismuth. In particular, the intensity of the di-sigma/pi-bonded surface complex peak at 1574 cm^{-1} becomes very weak when the surface is undergoing HER (Fig. 1g) and only recovers to perhaps 25% of its original value (when adsorbed bismuth is absent, Fig. 1d) when data are collected at potentials positive of the HER. This suggests that the di-sigma/pi-bonded surface complex is strongly associated with defect sites. If prevented from forming at defect sites, large increases in semi-hydrogenation selectivity are known to result.[20] In addition, the intensity of peaks associated with adsorbed CO is also attenuated when bismuth is adsorbed at defect sites. This is consistent with previous electrochemical studies in which the suppression of CO formation, derived from organic molecules at platinum electrodes, was also noted when bismuth was adsorbed.[45]

In order to confirm that defect sites are intimately connected with a di-sigma/pi-surface complex formation from alkynes, electrochemical hydrogenation of butynediol from both pristine and defective single crystal Pt{111} electrodes using the SHINERS method was undertaken. In Fig. 3 are shown SHINERS data for butynediol hydrogenation at a Pt{111} electrode surface. Because 0.1 M perchloric acid was used as the electrolyte, a sharp peak at 934 cm^{-1} was always observed corresponding to the symmetric Cl–O stretch of the perchlorate anion of the electrochemical double layer. As seen with the Pt and Au@Pt core–shell nanoparticles, peaks ascribable to both adsorbed and electrolyte phase butynediol are observed in the SHINERS data. In contrast to previous results however, irrespective of whether or not the electrode was evolving hydrogen, the 1574 cm^{-1} peak hardly changed in size. We ascribe this behaviour to the absence of surface defect sites at the Pt{111} electrode surface. To substantiate this assertion further, the electrolyte was switched to one containing 0.1 M sulphuric acid and the electrode potential cycled to form/desorb an electrochemical oxide layer. This procedure is known to cause the formation of surface defects at the Pt{111} electrode surface which may be monitored by CV as (mainly) the appearance of {110} step sites at the expense of {111} terrace sites.[46] Fig. 4 shows the subsequent butynediol hydrogenation SHINERS data using the electrochemically roughened Pt{111} electrode in aqueous sulphuric acid. The data now reveal features more similar to those seen previously using the Au@Pt nanoparticles. For example, a substantial increase in the intensity of the 1580 cm^{-1} peak is made manifest when electrode potentials are positive of the HER region. In addition, the presence of adsorbed CO is now signified by enhanced

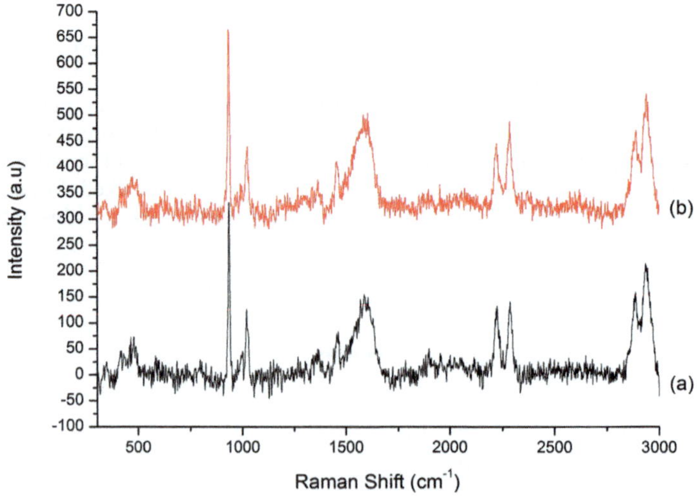

Fig. 3 SHINERS of 0.1 M 2-butyne-1,4-diol in 0.1 M perchloric acid adsorbed at Pt{111} at HER (a) and double layer (b) potentials.

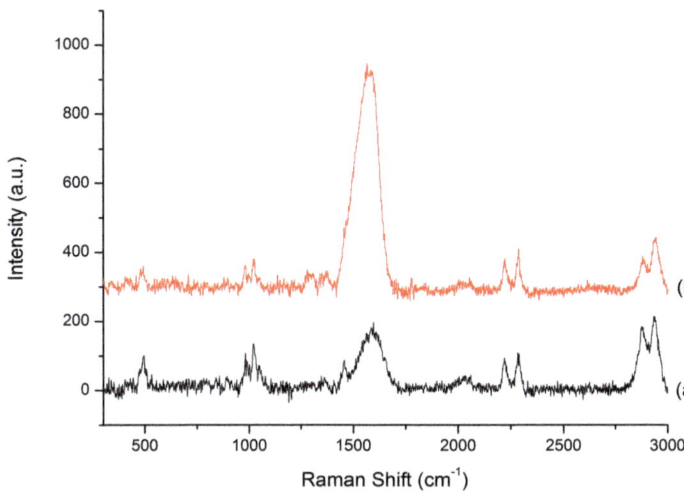

Fig. 4 Same as Fig. 3 except that Pt{111} surface has been electrochemically roughened and the perchloric acid electrolyte replaced with sulphuric acid.

peak intensity around 489 and 2032 cm^{-1}. Furthermore, a slight increase in the C–H stretch peak intensities at HER potentials, particularly the (CH_2) symmetric stretch at 2879 cm^{-1} relative to the asymmetric (CH_2) stretch at 2938 cm^{-1}, is seen.

Catalytic reactivity and selectivity measurements of the platinum catalysts

A series of Pt catalysts were tested for their efficacy in the selective hydrogenation of three alkynes. The substrates consisted of a symmetrical internal alkyne, 4-octyne, an unsymmetrical internal alkyne, 2-pentyne and a terminal alkyne with a polar functionality, 2-methyl-3-butyn-2-ol (MeBuOH from hereonin). The latter substrate was used in a study on shape selectivity by Crespo-Quesada *et al.*[33] where it was observed that increasing the particles edge/terrace ratio led to an increase in selectivity to alkene. This was due to a reduced rate of further hydrogenation to alkane, which occurred solely over edges.

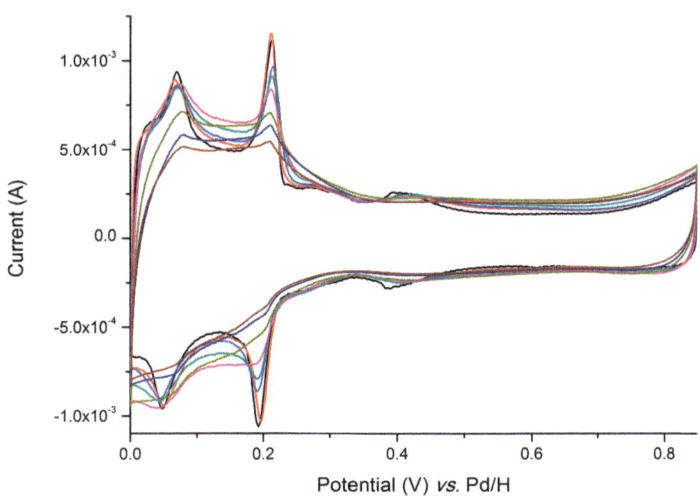

Fig. 5 Cyclic voltammetry data for increasing loadings of PVP on the reduced Pt on graphite catalyst in 0.5 M sulphuric acid. Sweep rate = 50 mV s^{-1}.

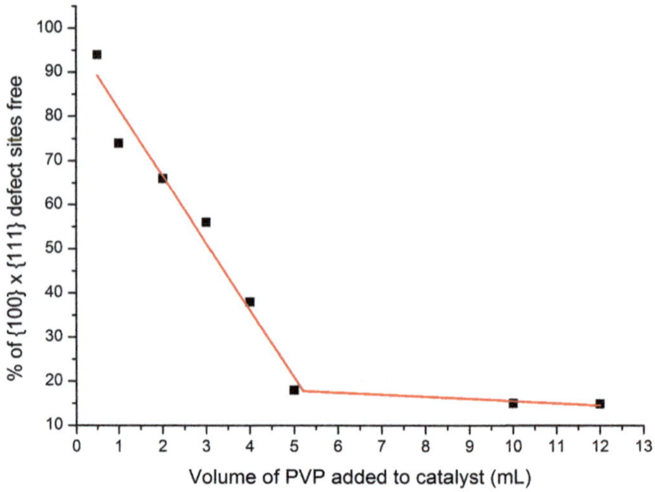

Fig. 6 Plot of {100} × {111} step peak intensity *versus* volume of PVP dosed. A critical blocking of the defect sites is achieved after 5 cm^3 of PVP is added.

The first catalyst to be studied was a standard 5 wt. % Pt/graphite catalyst reduced at 700 K and characterised fully in a previous study.[20] This catalyst in turn was modified with two different coverages of PVP. In ref. 47 it was demonstrated that PVP selectively blocks defect sites leaving behind Pt{111} terrace sites, in a similar fashion to bismuth. Fig. 5 depicts the changes in the CV response of the sintered catalyst upon adsorption of PVP. A key finding is the selective blocking of defect sites as signified by the attenuation in the intensity of the two peaks at approximately 0.05 ({110} sites) and 0.21 ({100} × {111} steps) V (Pd/H) with increasing PVP loading.[47] The attenuation in the broad peak centred at 0.40 V (Pd/H) associated with Pt{111} terraces is also seen for PVP adsorbed on Pt{111}[47] but in turn this is replaced by an increase in intensity around 0.6–0.7 V associated with OH rather than bi-sulphate adsorption at Pt{111} sites.[47] Hence, the supported Pt catalyst behaves in a similar way to the single crystal electrode surfaces so far as PVP adsorption is concerned.[47] It is difficult to assign a coverage to PVP since the number of Pt sites blocked per PVP polymer is not known. However, the residual hydrogen underpotential deposition (H UPD) peak area (charge) between 0 and 0.3 V relative to the un-dosed catalyst may be used to signify the number of "free" Pt sites after PVP adsorption. Unfortunately, even here there are ambiguities since after approximately 50% of H UPD sites are blocked, irrespective of how much PVP is added, no further attenuation in H UPD charge is observed. Rather, all electrosorption peaks broaden and the double layer capacitance increases. Hence, although a nominal coverage of PVP (in terms of H UPD sites blocked) may be determined up to 50% surface H UPD blocked, after this point a plateau is reached. Therefore, in Fig. 6 we plot the attenuation of the intensity of the {111} × {100} defect feature at 0.2 V (Pd/H) as a function of PVP amount added to the catalyst to at least highlight the extent of defect site blocking. A linear decrease in peak intensity is observed up to 5 cm^3 of PVP added. We therefore conclude that practically all defect sites are blocked after 5 cm^3 of PVP has been dosed onto a fixed mass of the catalyst. Hence, both the 10 cm^3 and 12 cm^3 dosed samples fall into the plateau region of this plot representing 2 and 2.4 times the critical amount of PVP associated with defect blocking. The catalysts that were tested are described in Table 1.

Fig. 7–9 show activity and selectivity data as a function of percentage conversion for the three substrates studied (MeBuOH, 2-pentyne and 4-octyne respectively) and Table 2 summarises activity and selectivity for all catalysts investigated. It was found that catalysts were active in the hydrogenation of all three substrates but there were a number of

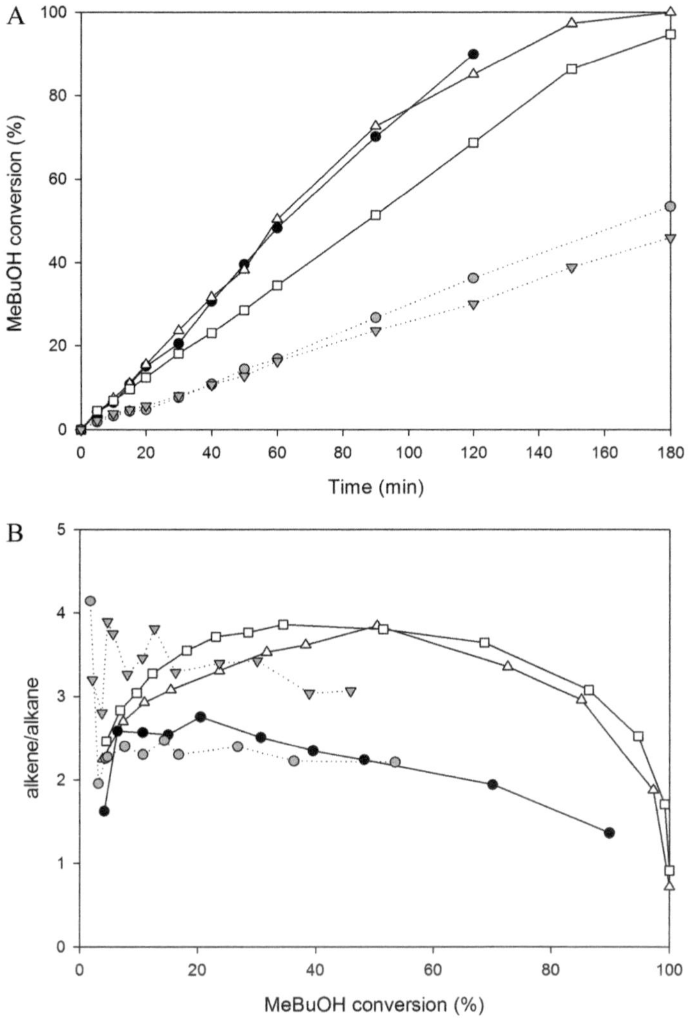

Fig. 7 Graphs of MeBuOH conversion *vs.* time (A) and alkene/alkane ratio *vs.* MeBuOH conversion (B) (-●- Pt/C, -△- Pt/C + 10ml PVP, -□- Pt/C + 12ml PVP, -●- aerobically grown bioPt, -▽- anaerobically grown bioPt).

significant differences in activity and selectivity. 100% alkyne conversion required between 30 to 300 minutes, depending on substrate and catalyst. For the sintered Pt/C and bioPt catalysts, activity towards hydrogenation was found to follow the trend:

$$\text{2-pentyne} > \text{4-octyne} > \text{MeBuOH}$$

Alkene/alkane selectivity was similar for both the sintered Pt/C and bioPt catalysts, irrespective of substrate (MeBuOH ~ 2–3, 2-pentyne ~ 6 and 4-octyne ~ 5). The biogenic catalysts gave lower rates than the sintered Pt/C catalysts (typically two to three times slower). It was difficult to assess the Pt surface area of the bioPt samples but a tentative suggestion for this behaviour (low activity) would be a lower active surface area being present compared to the graphite supported catalyst. In MeBuOH hydrogenation the bioPt gave much lower rates than all other catalysts, including the selectively poisoned derivatives.

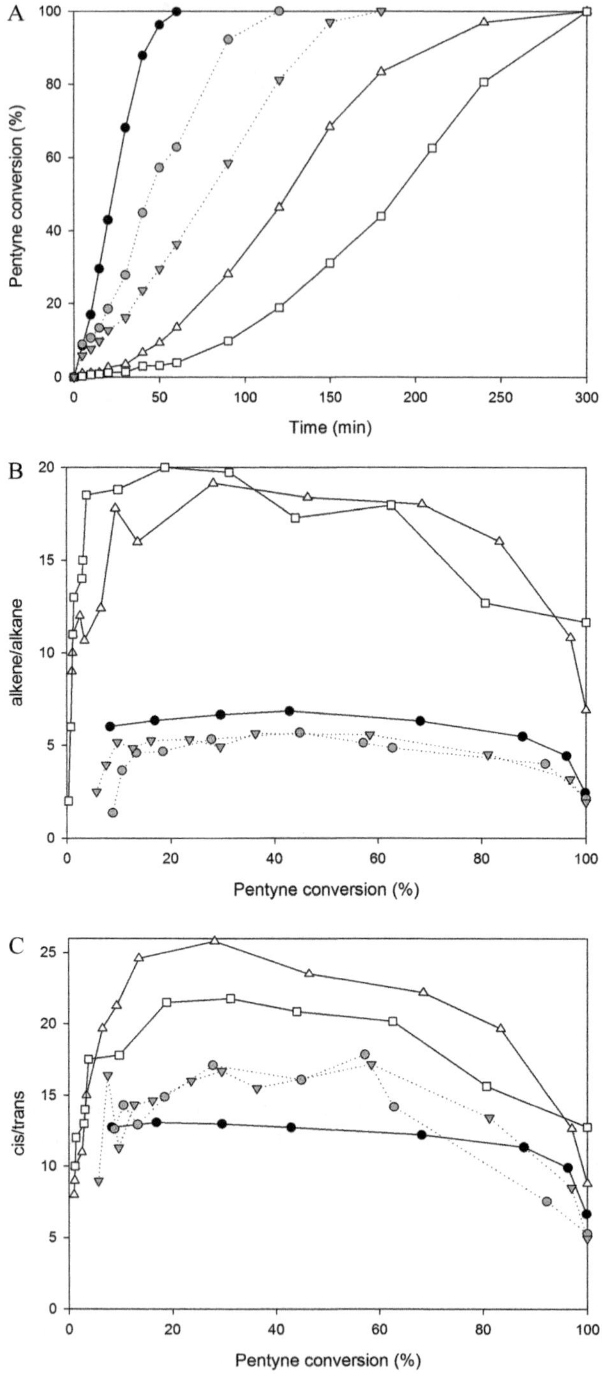

Fig. 8 Graphs of 2-pentyne conversion *vs.* time (A), alkene/alkane ratio *vs.* 2-pentyne conversion (B) and *cis*/*trans* *vs.* 2-pentyne conversion (C) (-●- Pt/C, -△- Pt/C + 10ml PVP, -□- Pt/C + 12ml PVP, -●- aerobically grown bioPt, -▼- anaerobically grown bioPt).

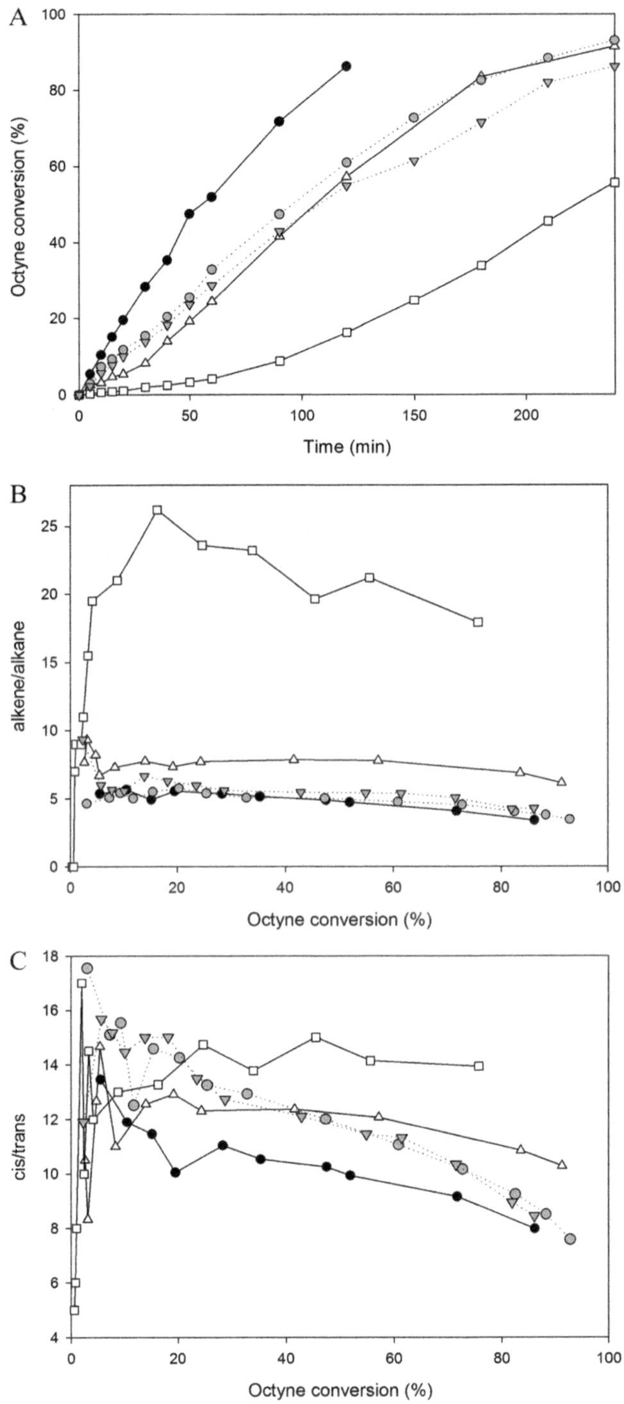

Fig. 9 Graphs of 4-octyne conversion *vs.* time (A), alkene/alkane ratio *vs.* 4-octyne conversion (B) and *cis*/*trans* *vs.* 4-octyne conversion (C) (-●- Pt/C, -△- Pt/C + 10ml PVP, -□- Pt/C + 12ml PVP, -●- aerobically grown bioPt, -▽- anaerobically grown bioPt).

Table 2 Activity and selectivity data for the range of catalysts tested. Values given are at 50% alkyne conversion for selectivity and for activity, the time in minutes for 50% conversion

		MeBuOH		2-pentyne			4-octyne		
Entry	Catalyst	Alkene/ alkane	Activity	Alkene/ alkane	*Cis*/ *trans*	Activity	Alkene/ alkane	*Cis*/ *trans*	Activity
1	Pt/C	2.2	60	6.6	12.4	17	4.7	10.0	51
2	Pt/C + 10 cm^3 of PVP	3.7	60	18.4	23.4	125	7.8	12.2	105
3	Pt/C + 12 cm^3 of PVP	3.6	88	17.3	20.1	185	20.2	14.6	225
4	bioPt aero	2.2	168	5.6	17.3	43	5.0	11.9	96
5	bioPt anaero	3.1	50% not reached	5.6	17.3	78	5.4	11.9	108

For the PVP poisoned catalyst series, increasing the surface coverage of PVP led to an expected decrease in reaction rate, due to a reduction in available active sites, but both PVP catalysts improved their selectivity towards alkene product over alkane. In the case of 2-pentyne and 4-octyne, this increase in selectivity was substantial (alkene/alkane ~18 for 2-pentyne and ~20 for 4-octyne compared to 5–6 for the unmodified catalyst). Inspection of the *cis-/trans*-selectivity entry for all substrates and catalysts reveals a correlation with alkene/alkane selectivity which would be consistent with both being related to the availability of Pt{111} terrace sites.[8,10]

In the light of the catalytic data reported above, with the conclusion that PVP enhances the reaction selectivity, electrochemical hydrogenation/SERS experiments were attempted using Au@Pt nanoparticles modified with PVP such that all defect sites were blocked (according to CV). Because of its ease of solubility in aqueous electrolytes compared to the other two alkyne hydrocarbons, MeBuOH was used as the substrate to undergo electrochemical hydrogenation. Fig. 10 shows SERS data corresponding to MeBuOH hydrogenation in the absence (a and b) and presence (c and d) of adsorbed PVP. In all spectra, the bulk bands of MeBuOH may be distinguished at 2119 cm^{-1} (C≡C triple bond stretch), 1447 cm^{-1} (asymmetric

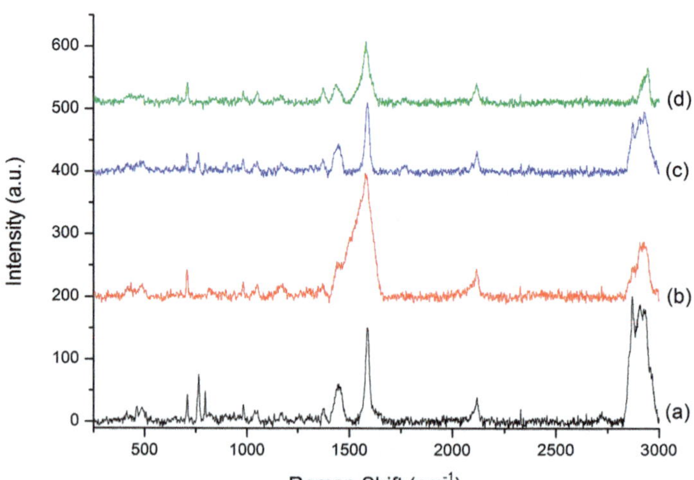

Fig. 10 SERS of 2-methyl-3-butyn-2-ol (a) at HER and (b) double layer potentials. (c) Same as (a) except surface dosed with PVP. (d) Same as (b) except surface dosed with PVP. Concentration of 2-methyl-3-butyn-2-ol was 0.1 M in 0.1 M aqueous sulphuric acid.

CH_3 bend), 1375 cm^{-1} (symmetric CH_3 bend) and 710 cm^{-1} (C–H primary alkyne bend). In Fig. 10a wherein the alkyne is undergoing electrochemical hydrogenation in the HER potential region, it is evident that new bands appear that cannot be ascribed to bulk dissolved species. In particular, an intense band appears at 1590 cm^{-1} consistent with the formation of a di-sigma/pi-bonded alkene surface complex as described previously. In addition a band at 775 cm^{-1} that could be associated with a C=C–H out of plane bend of the surface complex (although corresponding C=C–H stretches above 3000 cm^{-1} were sought extensively these were not observed). It may be that the surface selection rule plays some part in rendering C=C–H stretches rather weak.[48–50] When the potential is shifted away from the HER to the double layer region (Fig. 10b), a spectrum similar to those reported in Fig. 1 is observed whereby extra intensity accumulates in the range 1500–1600 cm^{-1} giving rise to a broad multiplicity of overlapping bands. This would be consistent with a greater coverage of surface complex forming in a variety of adsorption configurations as suggested earlier. In addition, some re-orientation of the adsorbed surface complexes may be deduced since, although surface coverage has increased, the C–H stretching peaks all decrease in intensity. Again, comparing the C–H stretch region in the absence and presence of hydrogen gas reveals a great increase in the intensity of the 2872 cm^{-1} band under hydrogenation conditions. Metal–carbon stretches around 490 cm^{-1} are consistent with adsorbed CO also being present but, if so, the corresponding C–O stretch expected around 2000 cm^{-1} is either very weak or not seen at all. In this case, it may be that the 490 cm^{-1} feature is due exclusively to bonding of the surface alkene complex to the platinum surface.

Repeating these SERS measurements using a PVP-modified surface instead allows for the effect of PVP to be investigated. Fig. 10c shows the PVP-modified surface undergoing hydrogen evolution in the presence of dissolved MeBuOH. The spectrum resembles very closely Fig. 10a except for a slight decrease in the intensity of all peaks. Under HER conditions, the 775 cm^{-1} feature is again noticeable as are the 1580 and 2872 cm^{-1} peaks. The bands from PVP are very weak in Fig. 10c although a weak feature is observed at 1760 cm^{-1} when PVP was adsorbed (significantly blue shifted compared to a normal cyclic amide stretch) in agreement with previous findings by El-Sayed *et al.* who also noted the same blue shift of the C=O stretch when PVP was adsorbed on silver.[51] We speculate that the amount of PVP adsorbed is actually just at the detection limit of our spectrometer although an observed stretching frequency of 1760 cm^{-1}, more commensurate with an isolated carbonyl substituent in a five-membered ring (rather than a cyclic amide) is still not understood at present. Future experiments are planned to clarify this finding.

When compared with Fig. 10b however, Fig. 10d reveals that a significant inhibition of di-sigma/pi-bonded surface complex arises following a modification of the surface with PVP. Hence, at potentials at which no hydrogen is being evolved, PVP acts to inhibit the uptake of alkyne. In fact a comparison between spectrum 10c and 10d shows very little difference except for the usual decease in C–H stretch intensity around 2900 cm^{-1} and the disappearance of the 775 cm^{-1} peak in the absence of hydrogen. We conclude that uptake of alkyne is inhibited by PVP and that, since the association of di-sigma/pi-bonded complex is associated with defects (see Fig. 1 and 4), PVP must be blocking these sites leading to an increase in selectivity and a decrease in activity. The nature of the adsorbed complex in the case of adsorbed MeBuOH is intriguing since it appears to be quite stable under HER conditions as signified by the intensity of the band at 1580 cm^{-1} (Fig. 10a and 10c). It is noteworthy also that the other two hydrocarbon alkynes tested gave rise to superior activity and selectivity compared to MeBuOH, with adsorbed PVP only making a moderate change to MeBuOH selective hydrogenation (see Table 2). It may be that this long-lived intermediate is responsible for this singular behaviour and future experiments are planned to investigate MeBuOH adsorption at Pt{*hkl*} surfaces in order to elucidate exactly how it interacts with various platinum adsorption sites.

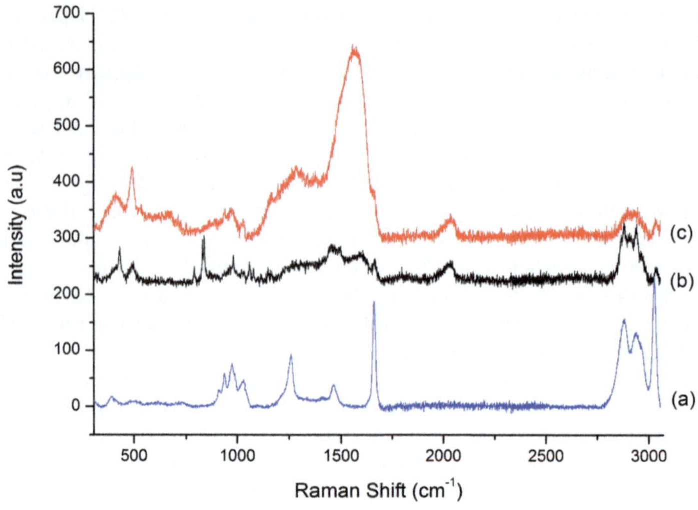

Fig. 11 Raman spectrum of bulk 2-butene-1,4-diol (a) and SERS of 0.5 M 2-butene-1,4-diol in 0.1 M sulphuric acid adsorbed on Au@Pt nanoparticles at HER (b) and double layer (c) potentials.

The final experiment we performed was to investigate the SERS of 2-butene-1,4-diol (butenediol) under HER conditions in order to compare with butynediol behaviour reported in Fig. 1. Fig. 11a shows the Raman spectrum of bulk butenediol. Intense bands at 3082, 2935 and 2876 cm^{-1} may be ascribed to C–H stretch vibrations of the C=C–H, CH_2 asymmetric and CH_2 symmetric stretches respectively. The C=C stretch is observed at 1660 cm^{-1} and the CH_2 distortion mode at 1464 cm^{-1}. The CH in-plane bend of the alkene is assigned to the 1259 cm^{-1} feature and the three bands around 1000 cm^{-1} are assigned to skeletal/C–C stretches. In Fig. 11b and 11c is shown data for butenediol adsorbed on Au@Pt nanoparticles at HER and double layer potentials respectively. It is striking that a similar intermediate band, as observed for butynediol adsorption, occurs at 1567 cm^{-1}. Other similarities would include evidence for CO formation (peaks at 2030 and 494 cm^{-1}) and a relative increase in intensity of all C–H stretches when hydrogen gas is being formed. These aspects strongly suggest a common surface di-sigma/pi-bonded surface intermediate being formed in both the alkyne and alkene surface reactions. Avery and Sheppard[41] who studied *cis*- and *trans*-but-2-ene and but-2-yne on Pt{111} reached exactly the same conclusion utilising thermal desorption spectroscopy and high resolution electron energy loss spectroscopy. They proposed the same common surface intermediate being formed at room temperature consisting of a di-sigma/pi-bonded alkene. It should be noted also that the bulk, solution phase butenediol peaks are very weak compared to the situation involving butynediol. This is still more remarkable when it is considered that the bulk concentration of butenediol in Fig. 11 was five times greater than the butynediol concentration in Fig. 1. This suggests that perhaps the much stronger interaction of alkynes compared to alkenes with metal surfaces causes a physisorbed layer of alkyne to form, in addition to the chemisorbed layer giving rise to stronger SERS signal for the "bulk" alkyne, especially when the decay in SER from the surface is reported to fall rapidly with distance.[52–54]

Discussion

A key finding of the present study from *in situ* spectroscopic measurements is that defect sites appear to be strongly associated with the formation of adsorbed di-sigma/pi-bonded alkenic intermediates following alkyne adsorption at Pt electrodes. Inspection of the surface Raman spectra also reveals that the C=C stretch band around 1580 cm^{-1} is actually rather broad and probably consists of several overlapping peaks in the range 1500–1600 cm^{-1}. Interestingly, even for Pt{111}, the C=C stretch band exhibited a substantial peak width suggesting a variety of bonding configurations associated with the surface intermediate. We suggest that variations in the angle of tilt between the surface and the plane of the alkene may be responsible for this behaviour.[40] It should also be noted that polymerisation of alkynes at surfaces has been reported[55] and it may be that interactions between pi-electrons on a number of adsorbed intermediates could be responsible for this peak-broadening effect. More work is required to elucidate the precise nature of the interaction giving rise to peak broadening. Nonetheless, there is a satisfactory confluence of reactivity, CV and SERS/SHINERS data pointing to the importance of defect sites in controlling hydrogenation selectivity.

In the case of PVP (as found for adsorbed bismuth on platinum[20]), selective blockage of defect sites resulting in only Pt{111} terraces being exposed to reactant also appears to be the *modus operandi* of the selective hydrogenation mechanism at metal surfaces. In fact, there are numerous examples in the literature already demonstrating improved alkyne hydrogenation selectivity as a consequence of dosing the catalyst with organic compounds, including PVP.[12,16,56–63] Crespo-Quesada *et al.* studied the selective hydrogenation of ethyne and looked at the behaviour of PVP-modified Pd catalysts and Pd catalysts in which PVP had been removed using ozone treatments.[33] Here, they found a marked increase in activity when PVP was removed, but a decrease in selectivity towards ethene product. These results are in accordance with the present study if defect sites blocked by PVP facilitate the best selectivity but only afford lower activities. The role of defects was also highlighted by these workers[14] based on catalytic measurements using shape-controlled nanoparticles. Our results are in accordance with these findings although our approach has been to take a pre-existing, standard graphite-supported catalyst and to monitor by CV the selective uptake of PVP (or bismuth[20]) by CV rather than starting with well-defined Pt nanoparticles. In a very recent paper by Kwon *et al.*,[15] the balance of interactions between the nanoparticle Pt catalyst surface and the capping agent/substrate in terms of differential adsorption energetics was shown to determine the overall selectivity of the hydrogenation of alkynes (including 4-octyne). In fact selectivity could be tuned using this approach.[15] Although no specific experiments carried out with shape-controlled nanoparticles were reported in ref. 15, it could be envisaged that one might expect particular substrates to interact with defect sites sufficiently strongly (or under conditions of a large excess of the substrate enabling displacement of PVP) such that the substrate could compete successfully with irreversibly adsorbed PVP for defect sites leading to a decrease in selectivity. This could also be true of course for inert site blockers such as bismuth. However, metal–metal surface bonds tend to be more stable than those between surface and capping agents such as PVP (compare for example thermal desorption measurements of Bi[64] and PVP[65]). Hence, this effect for bismuth and probably other inorganic additives would be small relative to surface modification by organic ligands. Thus, although the replacement of toxic metals by more benign adsorbates such as PVP is an important goal in heterogeneous selective hydrogenation by Lindlar-like catalysts, one has to ensure that binding is commensurate with the maintenance of good selectivity whilst achieving reasonable activity (high coverages of a strongly interacting capping agent would be detrimental to overall catalytic activity).

For the first time, two Pt catalysts on bacterial supports were compared in relation to selective alkyne hydrogenation at graphite-supported Pt and PVP-modified catalysts. Inspection of reactivity data reveals only marginal differences between aerobically and anaerobically produced bacterial samples. Based on the present findings, we deduce that little structural difference should be observed between both catalysts. In ref. 66, the BioPt catalyst produced anaerobically gave rise to little structural distinctiveness and resembled a polycrystalline surface. In terms of selectivity both BioPt samples were comparable to the reduced Pt/C catalyst and, therefore, it is deduced that the biomass plays no part in selectively blocking sites and, in fact, all BioPt adsorption sites are available to the alkyne. Hence, it is only after treatment to remove biomass that good selectivity is afforded[20] (resulting in partial blockage of defect sites by organic material). In the present study, PVP has proven to be a benign (relative to bismuth and lead) promoter of selective alkyne hydrogenation just like the biofragments reported in ref. 20.

Conclusions

The reactivity of a platinum surface was probed using a combination of shell-isolated nanoparticle Raman spectroscopy (SHINERS), cyclic voltammetry (CV) and hydrogenation studies. A range of particles, including nanoparticles, single crystal and supported particles, was studied. It is shown that alkyne adsorption on Pt defect sites yields a long-lived di-sigma/pi-alkene complex which may undergo further hydrogenation to produce alkane. Alkene selectivity can be increased by blocking defect sites, using PVP or Bi, leaving only Pt{111} terraces available for catalysis. SHINERS spectra of Pt{111} surfaces showed that the di-sigma/pi-alkene complex is only present in very small amounts, unless the surface is electrochemically roughened to create defect sites.

Acknowledgements

This work was in part financially supported by the UK National Measurement System. G.A.A., J.W. and L.E.M. also acknowledge the financial support of the EPSRC (grants EP/I007601/1 and EP/I007806/1 respectively).

References

1 L. Malatesta and C. Cariello, *J. Inorg. Nucl. Chem.*, 1958, **8**, 561–565.

2 A. Miyashita, H. Takaya, T. Souchi and R. Noyori, *Tetrahedron*, 1984, **40**(8), 1245–1253.

3 J. Fawcett, E. G. Hope, A. M. Stuart and A. J. West, *Polyhedron*, 2006, **25**(5), 1182–1186.

4 H. Altinel, G. Avsar, M. K. Yilmaz and B. Guzel, *J. Supercrit. Fluids*, 2009, **51**(2), 202–208.

5 J. Rajaram, A. P. S. Narula, H. P. S. Chawla and S. Dev, *Tetrahedron*, 1983, **39**(13), 2315–2322.

6 S. P. Griffiths, P. Johnston and P. B. Wells, *Appl. Catal., A*, 2000, **191**(1–2), 193–204.

7 M. Okamoto, T. Hirao and T. Yamaai, *J. Catal.*, 2010, **276**(2), 423–428.

8 M. Crespo-Quesada, A. Yarulin, M. Jin, Y. Xia and L. Kiwi-Minsker, *J. Am. Chem. Soc.*, 2011, **133**, 12787–12794.

9 E. Schmidt, A. Vargas, T. Mallat and A. Baiker, *J. Am. Chem. Soc.*, 2009, **131**, 12358–12367.

10 I. Lee, F. Delbecq, R. Morales, A. A. Manuel and F. Zaera, *Nat. Mater.*, 2009, **8**, 132–138.

11 I. Lee and F. Zaera, *J. Catal.*, 2010, **269**(2), 359–366.

12 J. A. Anderson, J. Mellor and R. P. K. Wells, *J. Catal.*, 2009, **261**(2), 208–216.

13 E. Gross and M. Asscher, *Langmuir*, 2010, **26**(21), 16226–16231.

14 M. Crespo-Quesada, F. Cárdenas-Lizana, A. Dessimoz and L. Kiwi-Minsker, *Catalysis*, 2012, **2**, 1773–1786.

15 S. G. Kwon, G. Krylova, A. Sumer, M. M. Schwartz, E. E. Bunel, C. L. Marshall, S. Chattopadhyay, B. Lee, J. Jellinek and E. V. Shevchenko, *Nano Lett.*, 2012, **12**, 5382–5388.

16 M. Crespo-Quesada, R. R. Dykeman, G. Laurenczy, P. J. Dyson and L. Kiwi-Minsker, *J. Catal.*, 2011, **279**, 66–74.

17 T. Mitsudome, Y. Takahashi, S. Ichikawa, T. Mizugaki, K. Jitsukawa and K. Kaneda, *Angew. Chem., Int. Ed.*, 2013, **52**, 1481–1485.

18 D. J. Jenkins, A. M. S. Alabdulrahman, G. A. Attard, K. G. Griffin, P. Johnstone and P. B. Wells, *J. Catal.*, 2005, **234**, 230–239.

19 G. A. Attard, K. G. Griffin, D. J. Jenkins, P. Johnston and P. B. Wells, *Catal. Today*, 2006, **114**(4), 346–352.

20 J. A. Bennett, G. A. Attard, K. Deplanche, M. Casadesus, S. E. Huxter, L. E. Macaskie and J. Wood, *ACS Catalysis*, 2012, **2**, 504–511.

21 E. Herrero, V. Climent and J. M. Feliu, *Electrochem. Commun.*, 2000, **2**(9), 636–640.

22 G. A. Attard, J. E. Gillies, C. A. Harris, D. J. Jenkins, P. Johnston, M. A. Price, D. J. Watson and P. B. Wells, *Appl. Catal., A*, 2001, **222**(1–2), 393–405.

23 P. Rodrıguez, J. Solla-Gullon, F. J. Vidal-Iglesias, E. Herrero, A. Aldaz and J. M. Feliu, *Anal. Chem.*, 2005, **77**, 5317–5323.
24 C. K. Klutse, A. Mayer, J. Wittkamper and B. M. Cullum, *J. Nanotechnol.*, 2012, 319038.
25 M. Fleischmann, P. J. Hendra and A. J. McQuillan, *Chem. Phys. Lett.*, 1974, **26**(2), 163–166.
26 J. Lasse, C. M. Aikens and G. C. Schatz, *Chem. Soc. Rev.*, 2008, **37**(5), 1061–1073.
27 J. J. Mock, *J. Chem. Phys.*, 2002, **116**(15), 6755–6760.
28 A. Campion and P. Kambhampati, *Chem. Soc. Rev.*, 1998, **27**, 241–250.
29 K. Kneipp, H. Kneipp, I. Itzkan, R. R. Dasar and M. S. Feld, *Chem. Rev.*, 1999, **99**, 2957–2975.
30 M. J. Weaver, A. Zou and H. Y. H. Chan, *Anal. Chem.*, 2000, **72**, 38A.
31 J. F. Li, Z. L. Yang, B. Ren, G. Liu, P. Fang, Y. Jiang, D. Y. Wu and Z. Q. Tian, *Langmuir*, 2006, **22**(25), 10372–10379.
32 J. F. Li, Y. F. Huang, Y. Ding, Z. L. Yang, S. B. Li, X. S. Zhou, F. R. Fan, W. Zhang, Z. Y. Zhou, D. Y. Wu, B. Ren, Z. L. Wang and Z. Q. Tian, *Nature*, 2010, **464**, 392–395.
33 M. Crespo-Quesada, J. M. Andanson, A. Yarulin, B. Lim, Y. Xia and L. Kiwi-Minsker, *Langmuir*, 2011, **27**, 7909–7916.
34 K. Deplanche, I. Caldelari, I. P. Mikheenko, F. Sargent and L. E. Macaskie, *Microbiology*, 2010, **156**, 2630–2640.
35 P. Yong, N. A. Rowson, P. G. Farr, I. R. Harris and L. E. Macaskie, *Biotechnol. Bioeng.*, 2002, **80**, 369–379.
36 R. W. Evans and G. A. Attard, *J. Electroanal. Chem.*, 1993, **345**, 337–350.
37 J. Clavilier, K. El-Achi and A. Rodes, *Chem. Phys.*, 1990, **141**, 1–14.
38 N. V. Rees, R. J. Taylor, Y. X. Jiang, I. R. Morgan, D. W. Knight and G. A. Attard, *J. Phys. Chem. C*, 2011, **115**, 1163–1170.
39 F. J. Vidal-Iglesias, R. M. Aran-Ais, J. Solla-Gullon, E. Herrero and J. M. Feliu, *Catalysis*, 2012, **2**(5), 901–910.
40 J. P. Camplin, J. K. Eve and E. M. McCash, *Phys. Chem. Chem. Phys.*, 2000, **2**, 4433–4440.
41 N. R. Avery and N. Sheppard, *Proc. R. Soc. London, Ser. A*, 1986, **405**, 1–25.
42 M. Bron and R. Holze, *Surf. Sci.*, 2000, **457**, 178–184.
43 P. Zhang, Y. X. Chen, J. Cai, S. Z. Liang, J. F. Li, A. Wang, B. Ren and Z. Q. Tian, *J. Phys. Chem. C*, 2009, **113**(40), 17518–17526.
44 R. Foucault, R. L. Birke and J. R. Lombardi, *Langmuir*, 2003, **19**, 8818–8827.
45 A. Saez, E. Exposito, J. Solla-Gullon, V. Montiel and A. Aldaz, *Electrochim. Acta*, 2012, **63**, 105–111.
46 A. Bjoerling and J. M. Feliu, *J. Electroanal. Chem.*, 2011, **662**(1), 17–24.
47 G. A. Attard, A. Brew and J.-Y. Ye, in preparation.
48 J. A. Creighton, *Surf. Sci.*, 1983, **124**, 209–219.
49 M. Moskovits and J. S. J. Suh, *J. Phys. Chem.*, 1984, **88**, 5526–5530.
50 M. A. Bryant and J. E. J. Pemberton, *J. Am. Chem. Soc.*, 1991, **113**, 3629–3637.
51 M. A. Mahmoud, C. E. Tabor and M. A. El-Sayed, *J. Phys. Chem. C*, 2009, **113**(14), 5493–5501.
52 K. A. Willets and R. P. Van Duyne, *Annu. Rev. Phys. Chem.*, 2007, **58**, 267–297.
53 Y. Fang, Y. Wei, C. Bai and L. Kan, *J. Phys. Chem.*, 1996, **100**, 17410–17413.
54 Z. Q. Tian, B. Ren, J. F. Li and Z. L. Yang, *Chem. Commun.*, 2007, 3514–3534.
55 B. Tardy and J. C. Bertolini, *J. Chim. Phys. Phys.-Chim. Biol.*, 1985, **82**, 407–414.
56 N. Semagina, A. Renken and L. Kiwi-Minsker, *J. Phys. Chem. C*, 2007, **111**, 13933–13937.
57 C. Evangelisti, N. Panziera, A. D'Alessio, L. Bertinetti, M. Botavina and G. Vitulli, *J. Catal.*, 2010, **272**, 246–252.
58 F. Klasovsky, P. Claus and D. Wolf, *Top. Catal.*, 2009, **52**, 412–423.
59 C. Lange, S. Storck, B. Tesche and W. F. Maier, *J. Catal.*, 1998, **175**, 280–293.
60 C. Lange, D. De Caro, A. Gamez, S. Storck, J. S. Bradley and W. F. Maier, *Langmuir*, 1999, **15**, 5333–5338.
61 N. Semagina, E. Joannet, S. Parra, E. Sulman, A. Renken and L. Kiwi-Minsker, *Appl. Catal., A*, 2005, **280**, 141–147.
62 M. G. Musolino, C. M. S. Cutrupi, A. Donato, D. Pietropaolo and R. Pietropaolo, *J. Mol. Catal. A: Chem.*, 2003, **195**, 147–157.
63 M. M. Telkar, C. V. Rode, R. V. Chaudhari, S. S. Joshi and A. M. Nalawade, *Appl. Catal., A*, 2004, **273**, 11–19.
64 M. T. Paffett, C. T. Campbell and T. N. Taylor, *J. Vac. Sci. Technol., A*, 1985, **3**(3, Pt. 1), 812–816.
65 Y. Borodko, S. E. Habas, M. Koebel, P. Yang, H. Frei and G. A. Somorjai, *J. Phys. Chem. B*, 2006, **110**, 23052–23059.
66 G. A. Attard, M. Casadesús, L. E. Macaskie and K. Deplanche, *Langmuir*, 2012, **28**, 5267–5274.

Pt-group bimetallic nanocrystals with high-index facets as high performance electrocatalysts

Na Tian,* Jing Xiao, Zhi-You Zhou, Hai-Xia Liu, Yu-Jia Deng, Long Huang, Bin-Bin Xu and Shi-Gang Sun*

Received 10th December 2012, Accepted 31st January 2013
DOI: 10.1039/c3fd20146e

Bimetallic nanocrystals bound by high-index facets are promising catalysts, as they have both electronic effects from alloying and surface structure effects from high-index facets. Herein, we mainly focused on electrochemical preparation of two new Pt–Rh nanocrystals with high-index facets: {830}-bound tetrahexahedron and {311}-bound trapezohedron, and their excellent electrocatalytic properties for ethanol oxidation, especially the ability to break C–C bonds. Combining previous results about surface-modified tetrahexahedral Pt nanocrystals, we discuss the correlation of alloy and surface effects of the bimetallic system.

1. Introduction

Pt-group metals are important catalyst materials in industry; Pt–Rh for petroleum reforming in the petrochemical industry, Pt–Pd–Rh for three-way convertors in the automobile industry, and Pt for oxygen reduction reaction (ORR) and direct oxidation of fuel molecules in fuel cells.[1] Due to their limited reserves and high price, and the increasing demand in industry, it becomes more and more a key issue to explore and design new types of catalysts with enhanced activity and stability.

The catalytic performance of nanocrystals (NCs) can be tuned either by their composition, which mediates electronic structure,[2,3] or by their shape, which determines the surface atomic arrangement and coordination.[4–7] As for shapes, nanocrystals bound by high-index facets are very promising, as they contain a high density of low-coordinated step atoms with high reactivity.[8,9] In 2007, we developed an electrochemical square-wave potential method, by which high surface energy of high-index facets was overcome through oxygen adsorption/desorption on Pt, and successfully synthesized tetrahexahedral Pt nanocrystals (THH Pt NCs) with {730} high-index facets.[10] In addition, high-index facets have more kinks and edges, which can promote the breaking of C–C bonds, and increase electrochemical conversion efficiency of fuel.[11,12] Inspired by these studies, the past few years have witnessed a wide variety of noble metal nanocrystals enclosed by high-index facets.[13–18]

Compared with monometallic nanocrystals, bimetallic ones are advantageous for a range of fundamental studies and applications, because the variations in compositions and spatial distributions provide additional handles for experimentally manipulating both the structures and properties.[19,20] For example, Wu *et al.* synthesized different Pt–M alloy (M = Co, Fe, Ni, Pd) nanocrystals with cubic and octahedral morphologies,[21] and found that octahedral Pt-alloy nanocrystals exhibit better ORR

State Key Laboratory of Physical Chemistry of Solid Surfaces, Department of Chemistry, College of Chemistry and Chemical Engineering, Xiamen University, Xiamen 361005, China. E-mail: tnsd@xmu.edu.cn; sgsun@xmu.edu.cn

electrocatalytic activity than cubic ones. These alloys are bound by low-index facets of {100} and {111}, when aiming for high-index faceted nanocrystals, the synthesis becomes challenging.

Xie *et al.* firstly reported the synthesis of concave hexoctahedral Au–Pd alloy nanocrystals enclosed by {431} high-index facets through chemical reduction.[22] Very recently, we also prepared THH Pd-based (*e.g.*, Pd–Pt) alloy nanocrystals bound by {10 3 0} high-index facets, whose catalytic activity is 22 times higher than that of commercial Pd black for formic acid oxidation at 0 V.[23] As for electrocatalysis, Pt-based alloys are more useful than Pd-based alloys, because a greater range of electrochemical reactions can be catalyzed by Pt alloy catalysts.[1] Studies confirmed that the addition of Rh to Pt can significantly enhance the cleavage of the C–C bond in the oxidation of ethanol, and such enhancement is attributed to the change of electronic properties caused by the addition of Rh.[24–26]

In this paper, we present the electrochemical synthesis of alloy Pt–Rh nanocrystals with high-index facets at high yields, and manipulate their shapes from tetrahexahedron to trapezohedron by changing the electrode potential. Correspondingly, the surfaces changed from {*hk*0} to {*hkk*} high-index facets. The as-prepared Pt–Rh alloy nanocrystals exhibit excellent performance towards ethanol electrooxidation, for breaking the C–C bond to form CO_2, due to the synergistic effect of alloy and high-index facets. We then combined some previous studies about surface modification of THH Pt NCs with foreign atoms, and discussed several effects involved in bimetallic nanocrystals with high-index facets, including bifunctional effect, third-body effect, ligand effect, strain effect, as well as surface structure (atomic arrangement) effect.

2. Electrochemical preparation of high-index faceted nanocrystals

Electrochemical preparation of tetrahexahedral and trapezohedral Pt–Rh alloy NCs: Alloy NCs were prepared in a standard three-electrode cell working with a 263A potentiostat (EG&G) controlled by a homemade software which generated a modified square-wave waveform as required. Saturated calomel electrode (SCE) was chosen as the reference electrode, and all potentials mentioned in the paper were quoted *versus* the SCE scale. The working electrode was a glassy carbon (GC, $\phi = 6$ mm, purchased from Takai Carbon Co., Ltd., Tokyo, Japan.) rod embedded into a Teflon holder. Before use, the GC electrode was mechanically polished using successively alumina powder of size 5, 1, and 0.3 μm. Typically, tetrahexahedral Pt–Rh (Pt–Rh THH) alloy NCs were directly electrodeposited onto the GC electrode in 400 μM K_2PtCl_6 + 20 μM $RhCl_3$ + 0.1 M H_2SO_4 solution by programmed potential stepping. In detail, the GC electrode was firstly subjected to a potential step from 1.20 to −0.30 V, and maintained for 140 ms to generate crystal nuclei. Then, square-wave potentials ($f = 100$ Hz) with lower (E_L) and upper (E_U) potential limits of 0.05 and 1.10 V were applied on the crystal nuclei for 20 min to grow large alloy THH NCs. The conditions of the preparation of trapezohedral Pt–Rh (Pt–Rh TPH) alloy NCs were the same as those of Pt–Rh THH NCs except that the E_L was 0.08 V.

For comparison, we also prepared pure Pt nanocrystals with THH and TPH shapes.[27] In this case, more concentrated Pt precursor solution (2 mM K_2PtCl_6 + 0.1 M H_2SO_4) was used. NCs were also directly electrodeposited onto the GC electrode by programmed potential stepping, similar to that used for the Pt–Rh alloy. The E_L and E_U potential limits were 0.05 and 1.10 V for Pt THH, and 0.07 and 1.10 V for Pt TPH, respectively, and the growth time was 20 min.

The morphology and structure of Pt–Rh NCs were characterized by scanning electron microscopy (SEM, Hitachi S-4800) and transmission electron microscopy (TEM, Tecnai-F30 at 300 kV and JEM-2100 at 200 kV).

3. Physical characterizations of alloy Pt–Rh nanocrystals with high-index facets

3.1 SEM characterizations

Fig. 1 and Fig. 2 show SEM images of THH and TPH Pt–Rh alloy NCs. The proportion of NCs with THH or TPH shape is over 90%. The average size of THH and TPH are about 61 and 66 nm, respectively. The shapes of THH and TPH can be finely discerned in the high magnification SEM images. The insets show corresponding geometric models of the shapes. THH and TPH are enclosed by 24 {*hk*0} and {*hkk*} high-index facets, respectively.[5] The THH shape can be seen as a cube with each face capped by a square pyramid. The difference of TPH from THH is that the straight line of cubic edge in THH is replaced by a fold line.[5]

Fig. 1c and Fig. 2c show the energy dispersive X-ray spectroscopy (EDS) of the THH and TPH Pt–Rh NCs, respectively. Both the samples contain minor amounts of Rh, *i.e.*, about 4%, close to the Rh content in the plating solution. High-index faceted Pt–Rh NCs with a higher percentage of Rh are rarely obtained, and the obtained NCs in the plating solution containing a higher ratio of Rh tend to show irregular surfaces. This may be due to the much higher surface energy of Rh than Pt,[28] and the relatively high lattice mismatch of Pt and Rh (3%). That is, the formation of high-index facets on Rh NCs is more challenging, and cannot be realized under current conditions.

3.2 TEM characterizations

The crystal Miller indices of THH can be determined by measuring the plane angles. Fig. 3a and 3b are the TEM image and corresponding SAED pattern of a THH Pt–Rh NC along the [001] orientation. Along this orientation, eight side {*hk*0} facets parallel to the [001] axis form an octagonal projection, as can be seen in the model of the same direction (shown in the inset). After carefully measuring the plane angles (*i.e.*, α and β), the Miller indices of the crystal are determined. For THH Pt–Rh NCs, the average values of angle α and β are 130.7° and 139.1°, respectively, which are very near to the theoretical values of 131.1° and 138.9° for {830} facets. This result indicates that the nanocrystals are mainly comprised of {830} facets. Fig. 3c shows the model of atomic arrangement of {830} facets, periodically composed of one {210} subfacet followed by two {310} subfacets. The density of step atoms on the {830} facets is as high as 4.56×10^{14} cm^{-2}, that is, 37.5% of the total number of atoms on the surface. Fig. 3d shows the STEM image and corresponding EDS elemental mapping of Pt and Rh in a THH Pt–Rh NC. The distribution of Pt and Rh is uniform, and the profiles are almost same, indicating the formation of an alloy.

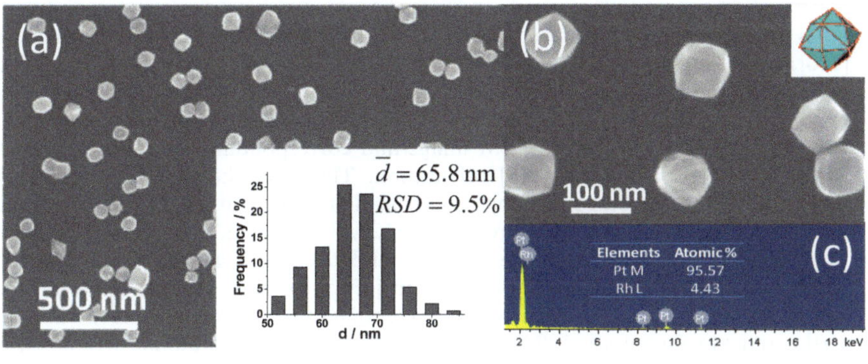

Fig. 1 Low (a) and high (b) magnification SEM images of tetrahexahedral (THH) Pt–Rh NCs. The inset in (a) shows size histogram, and the inset in (b) shows a model of a tetrahexahedron. (c) EDS of the THH Pt–Rh NCs.

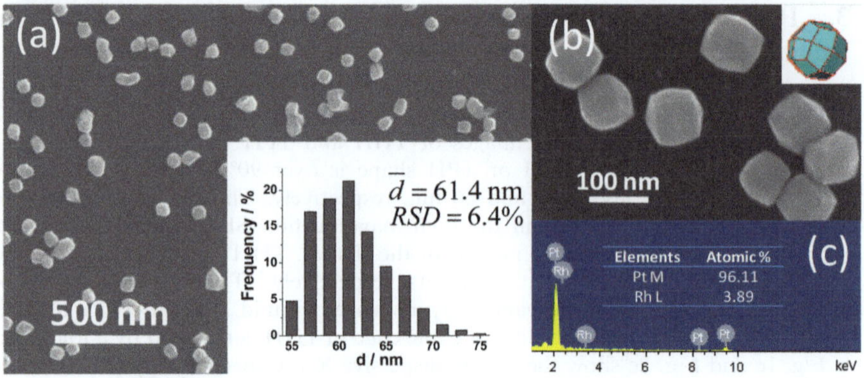

Fig. 2 Low (a) and high (b) magnification SEM images of trapezohedral (TPH) Pt–Rh NCs. The inset in (a) shows size histogram, and the inset in (b) shows a model of a trapezohedron. (c) EDS of the TPH Pt–Rh NCs.

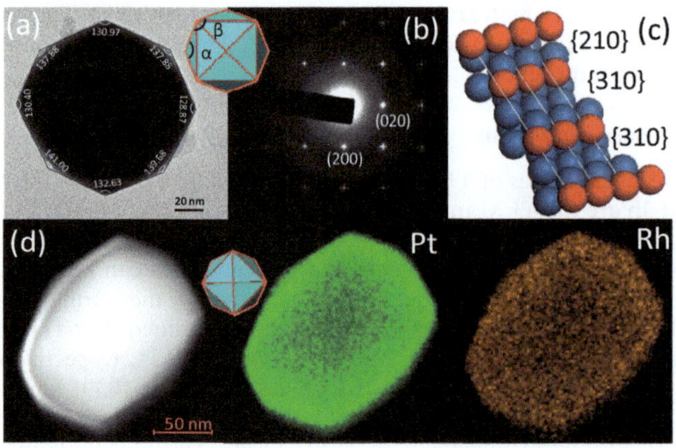

Fig. 3 TEM image (a) and SAED pattern (b) of a THH Pt–Rh NC recorded along the [001] crystal zone. The inset is a [001]-projected THH model. Miller indices were determined by measuring plane angles of α and β. (c) Atomic model of the {830} facet. (d) STEM image and EDS elemental mapping of Pt and Rh in a THH Pt–Rh NC, showing uniform distribution of Pt and Rh.

As for TEM characterization of TPH shape, the optimal crystal orientation is the [110] axis. Along the [001] axis, both TPH and THH shapes show a similar octagonal projection, and it is hard to distinguish them;[5] when along the [110] axis, TPH still exhibits an octagonal projection, while THH shows a hexagonal projection.[5,27] Fig. 4a and Fig. 4b show the TEM image and corresponding SAED pattern of a TPH Pt–Rh NC along the [110] orientation. The inset shows a [110]-projected TPH model, whose octagonal projection is similar to that in the TEM image. The average angles of α, β, and γ are 152.5°, 127.7°, and 132.7°, respectively. Note that along the [110] axis, the angles of α and β contain at least one projected edge, and only γ is formed through the projection of two adjacent {*hkk*} facets. The Miller indices of {311} can be easily identified through the measured value of γ (theoretical value of γ is 129.5°). Fig. 4c shows the model of atomic arrangement of {311} facets, which are composed of a terrace of two atomic widths of (111) symmetry, separated by a monatomic step of (100) symmetry. The density of stepped atoms with a coordination number of 7 on the (311) facets is as high as 7.83×10^{14}

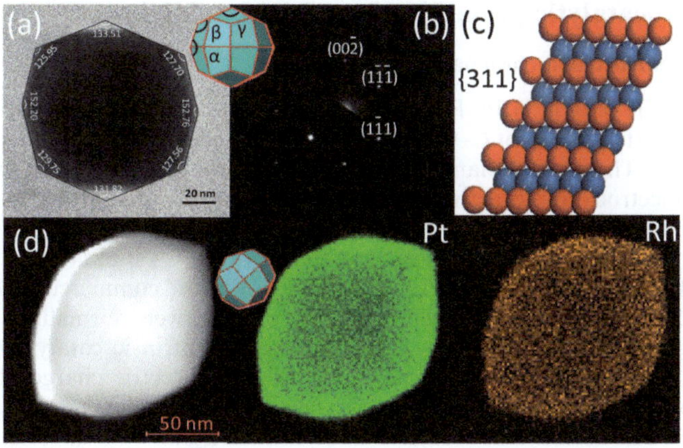

Fig. 4 TEM image (a) and SAED pattern (b) of a TPH Pt–Rh NC recorded along the [110] crystal zone. The inset is a [110]-projected TPH model. Miller indices were determined by measuring plane angles of γ. (c) Atomic model of the {311} facet. (d) STEM image and EDS elemental mapping of Pt and Rh in a TPH Pt–Rh NC, showing uniform distribution of Pt and Rh.

cm^{-2}, that is, 50% of the total number of atoms on the surface. The profiles of EDS elemental mapping of Pt and Rh in a TPH Pt–Rh NC (Fig. 4d), determined by STEM, are also uniform and show similar shapes, demonstrating the formation of a Pt–Rh alloy.

The formation of high-index facets on THH and TPH Pt–Rh NCs may also be the same as the growth mechanism of THH Pt NCs that we reported previously,[10] as the Rh content is rather low (about 4%) in the alloys. The repetitive oxygen adsorption/desorption induced by a square-wave potential only allows open-structure surfaces with low coordinated atoms to survive. We found that potential plays a vital role in tuning the shape/facets of the Pt–Rh alloy, as THH and TPH can be synthesized by just altering the lower potential of the square wave (0.05 V for THH and 0.08 V for TPH NCs). More intensive surface reconstruction from oxygen adsorption/desorption is introduced at higher E_L, as crystal growth rate decreases. The coordination numbers of stepped atoms on $\{hk0\}$ and $\{hkk\}$ facets are 6 and 7, respectively. Oxygen adsorption starts at lower potential on the $\{hk0\}$ facets than on the $\{hkk\}$ facets. Therefore, the formation of THH and TPH Pt–Rh NCs can be correlated with the intensity of surface reconstruction by oxygen adsorption/desorption, *i.e.*, more intensive oxygen adsorption/desorption will favor the formation of $\{hkk\}$ facets with low oxygen affinity. This result is different from our previous results of Pd nanorods.[29]

In the following section, we will discuss electrocatalytic performance of the synthesized Pt–Rh alloys. To reveal the role of the alloy more clearly, we also synthesized pure Pt NCs with THH and TPH shapes (see Fig. 5) as comparison catalysts.

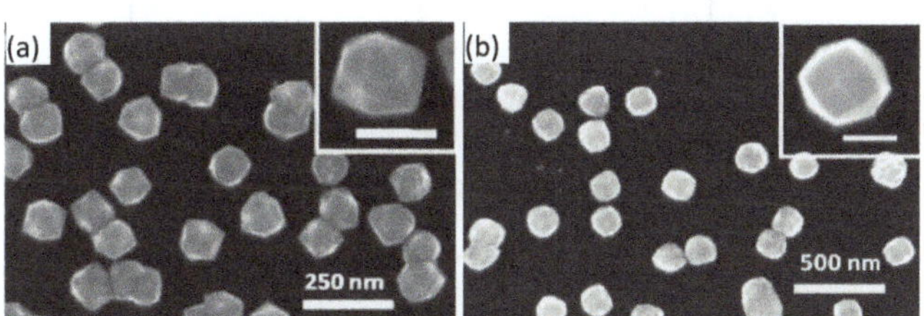

Fig. 5 SEM images of THH (a) and TPH (b) Pt nanocrystals. The insets show high-magnification SEM images (scale bar: 100 nm) to help identify shapes.

4. Electrocatalytic properties for ethanol oxidation

4.1 Voltammetric and chronoamperometric tests

The alloy THH and TPH Pt–Rh NCs are expected to have high catalytic activity, especially boosted ability of C–C breaking to produce more CO_2 for ethanol electrooxidation.[11,25] Therefore, ethanol electrooxidation was selected as the testing reaction. The electrocatalytic activity was tested in 0.1 M ethanol + 0.1 M $HClO_4$ solution at room temperature.

Fig. 6 shows the steady-state cyclic voltammograms (CVs) of ethanol electrooxidation on Pt–Rh THH, Pt–Rh TPH, Pt THH, Pt TPH and commercial Pt/C (~3 nm, Johnson Matthey Inc.) catalysts. The oxidation currents were normalized to the electrochemical surface area (ECSA) to investigate the intrinsic catalytic activity of different catalysts. The ECSA is calculated from the electric charge of hydrogen adsorption/desorption on Pt and Rh measured in 0.1 M H_2SO_4. As can be seen from Fig. 6, Pt–Rh alloys have higher catalytic activity than monometallic Pt catalysts, and Pt–Rh TPH shows the highest activity among all the catalysts. Based on the peak current density, the activity increases in the following order: Pt–Rh TPH (4.19 mA cm^{-2}) > Pt–Rh THH (3.27 mA cm^{-2}) > Pt TPH (2.90 mA cm^{-2}) > Pt THH (2.70 mA cm^{-2}) > Pt/C (0.66 mA cm^{-2}). The activity of Pt–Rh TPH is about 6.3 times that of the commercial Pt/C catalyst. In addition, there is a slight negative shift of peak potentials for alloy catalysts compared with monometallic Pt catalysts. The peak potentials of Pt–Rh TPH and THH are 0.535 V, lower than those of Pt THH and Pt TPH (both at 0.548 V) and commercial Pt/C (0.556 V). It should be pointed out that the commercial Pt/C has much smaller particle sizes (~3 nm) and so higher area, and in turn has a higher overall rate. That is, for catalytic activity per unit weight of Pt or Pt–Rh, the overall activity of these larger TPH and THH Pt–Rh NCs is less than that of the 3-nm commercial Pt nanoparticles (~25% as active). Future research will be needed to decrease the particle size. The catalysts are very stable during the electrochemical tests. Taking THH Pt–Rh NCs as an example, after 100 potential cycles, only ~5% loss of peak current density of ethanol electrooxidation was observed.

The above activity order can be attributed to both the electronic effects of the alloy and the surface structure effects of high-index facets. As for electronic effects, it has been reported that the addition of Rh to Pt will facilitate the electrocatalytic activity towards ethanol oxidation,[25,30,31] which may be ascribed to Rh electronic interaction (electron-withdrawing) with Pt, as well as lattice contraction effect (lattice constant: a_{Pt} = 0.392 nm; a_{Rh} = 0.380 nm).[25] The two electronic effects will result in the down-shift of d-band center, *i.e.*, weakening adsorption strength

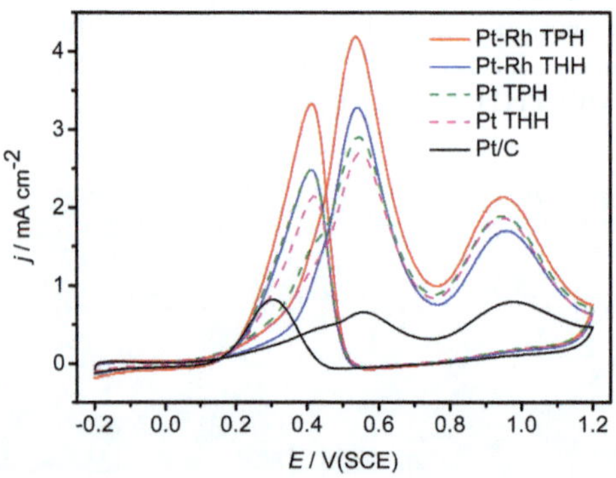

Fig. 6 Cyclic voltammograms of ethanol oxidation on Pt–Rh THH, Pt–Rh TPH, Pt THH, Pt TPH and commercial Pt/C catalysts at 50 mV s^{-1} in 0.1 M ethanol + 0.1 M $HClO_4$.

of adsorbates (*e.g.*, poisoning intermediates). As for the surface structure effect, {*hkk*} facets of Pt have high activity for ethanol oxidation.[18,32] For example, Huang *et al.* reported that octapod-shape Pt nanocrystals with {411} facets exhibit enhanced (2.3–5.6 times) electrocatalytic activity over commercial Pt catalysts towards ethanol oxidation.[18] In this study, the {311} facets on TPH Pt–Rh NCs have a higher density of step atoms than {830} facets on THH Pt–Rh NCs. Therefore, it is reasonable that {311}-bound TPH Pt–Rh NCs show the highest catalytic activity. Combination of electronic and surface structure effects leads to the above reasonable activity order: high-index faceted alloy Pt–Rh NCs > high-index faceted Pt NCs > Pt/C, and the highest activity was observed on TPH Pt–Rh NCs.

4.2 *In situ* FTIR study of ethanol electrooxidation

To evaluate the ability of breaking C–C bonds, we carried out electrochemical *in situ* FTIR spectroscopic studies of ethanol oxidation on the above catalysts. *In situ* FTIR experiments were conducted on a Nexus-870 FTIR spectrometer (Nicolet) with a liquid-N_2-cooled MCT-A detector. The experimental details were described previously.[33] A thin-layer IR cell with a CaF_2 disk window was employed. In this configuration, an un-polarized IR radiation was sequentially passed through a CaF_2 disc window and a thin-layer (~10 μm) solution, and then it was reflected by the electrode surface. So, both dissolved substances in the thin-layer solution and adsorbed species on the electrode surface can be detected under this external reflection mode.

Fig. 7a shows *in situ* FTIR spectra of ethanol oxidation at 0.50 V on Pt–Rh THH NCs, Pt–Rh TPH NCs, Pt THH, Pt TPH and commercial Pt/C catalysts. The band at 2343 cm^{-1} is attributed to CO_2, the final oxidation product through the cleavage of the C–C bond in ethanol. The band at about 2040 cm^{-1} is attributed to linearly-bonded CO (CO_L), a poisoning species formed during the oxidation of ethanol. The bands at 1715 and 1280 cm^{-1} are the stretching vibrations of C=O and C–O bonds in acetic acid (HAc). Usually, the latter is employed for quantitative analysis of acetic acid,[34] as the band at 1715 cm^{-1} is distorted by water absorption, and is contributed possibly by acetaldehyde (another product of ethanol electrooxidation). The ability to break C–C bonds is evaluated by the ratio of band intensities of CO_2 to that of acetic acid (I_{CO2}/I_{HAc}), namely the bands at 2343 cm^{-1} and 1280 cm^{-1}.

Fig. 7b shows the variation of I_{CO2}/I_{HAc} for ethanol oxidation on different catalysts at 0.50 V. The I_{CO2}/I_{HAc} declines in the same order as the catalytic activity, and Pt–Rh TPH NCs show the highest value, about 1.8 times that of the Pt/C. This result clearly indicates that high-index facets have greater ability to break the C–C bond, and the addition of Rh to Pt can further enhance this ability. As for fuel (ethanol) utilization, complete oxidation to CO_2 can provide 12 electrons per ethanol molecule, whereas only four electrons for incomplete oxidation to acetic acid. Clearly, the high selectivity to CO_2 on Pt–Rh TPH NCs will increase the efficiency of direct ethanol fuel cells. These results are consistent with previous studies: Pt/Rh alloy have higher CO_2 yields than pure Pt towards ethanol oxidation,[25,26] and Pt(*hkk*) single planes also have higher CO_2 yields than Pt(111).[35] The enhanced effect of Pt–Rh alloying was attributed to Rh promoting the dehydrogenation from ethanol, which facilitates the formation of adsorbed fragments with only one carbon and consequently the oxidation to CO_2.[36] Alternatively, Houtman and Barteau suggested that ethanol adsorbed on rhodium forms a cyclic adsorbate, which leads to a very stressed C–C bond based on UHV experiments.[37]

5. Discussions

5.1 Surface modification of high-index faceted Pt nanocrystals

Besides forming alloys, surface modification of the monolayer/sub-monolayer of foreign atoms on given nanocrystals with high-index facets is also a strategy to obtain bimetallic nanocrystals with high-index facets.

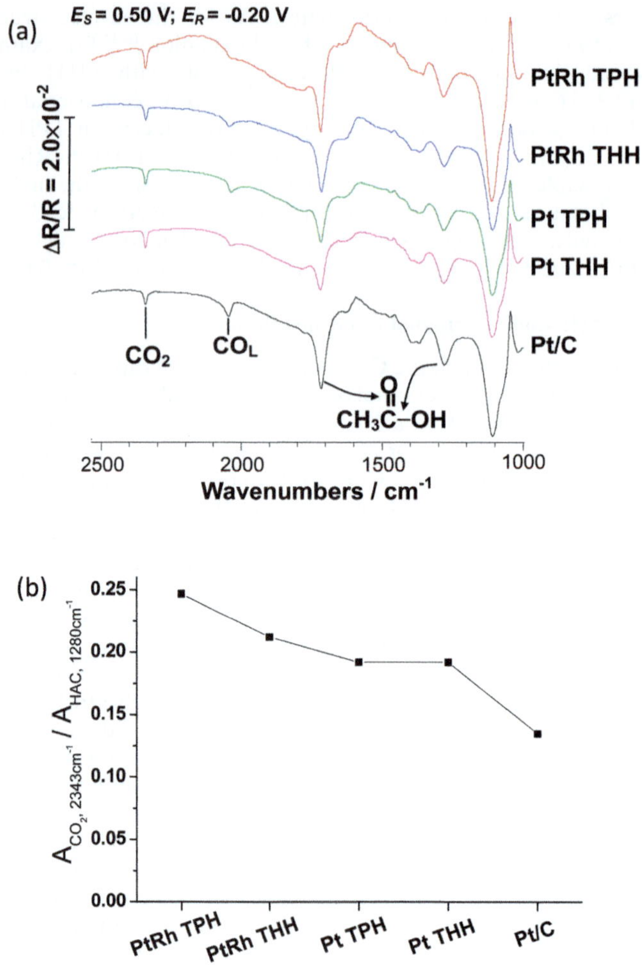

Fig. 7 (a) *In situ* FTIR spectra of ethanol oxidation on Pt–Rh TPH, Pt–Rh THH, Pt TPH, Pt THH and commercial Pt/C catalysts in 0.1 M ethanol + 0.1 M $HClO_4$. $E_S = 0.50$ V, $E_R = -0.20$ V. (b) Comparison of the ratio of band intensity of CO_2 to that of acetic acid on Pt–Rh TPH, Pt–Rh THH, Pt TPH, Pt THH and commercial Pt/C catalysts.

Previously, we used two pathways to modify THH Pt NCs through electrochemical methods, as shown in Fig. 8. One approach is direct electrochemical deposition of the target atoms on the surface of THH Pt NCs, usually done by repetitively potential cycling or holding the potential at a constant value in the solution containing very dilute target metal ions ($\sim 10^{-5}$ M). The coverage of foreign metal (θ), can be changed by simply changing the parameters like cycling number, metal ion concentration, and deposition time. The other approach is through under potential deposition (UPD) of copper followed by galvanic displacement, a well-known strategy developed by Adzic's group,[38,39] by which monolayer or sub-monolayer foreign atoms on the host metal surface can be easily obtained.

We have modified the {730}-faceted THH Pt NCs with Bi and Ru atoms through the electrodeposition approach,[40,41] and modified with Au through the Cu-UPD strategy.[42] By modifying the THH Pt NCs with Ru, the catalytic activity towards methanol oxidation is greatly improved at relatively low potential (0.2–0.5 V) compared with bare THH Pt NCs (Fig. 9a). The corresponding *in situ* FTIR studies

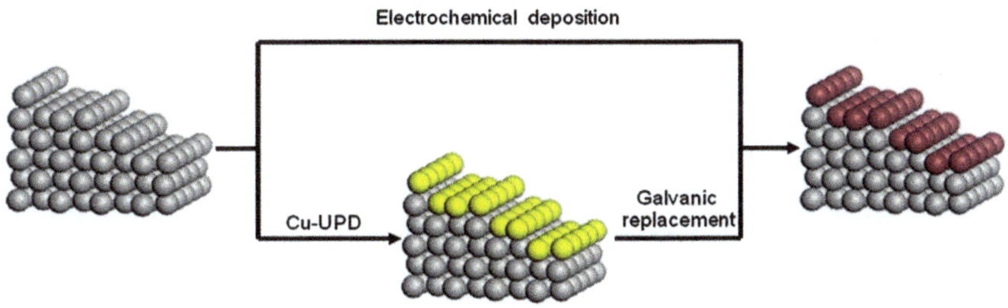

Fig. 8 Illustration of two electrochemical approaches for surface modification with foreign atoms.

and CO stripping (Fig. 9b) experiments confirm that the improved catalytic activity at low potential is mainly due to the promoting of oxidation of adsorbed CO (CO_{ad}) formed during the methanol electrooxidation. An optimum Ru coverage of around 0.5 coincides with the previously proposed bi-functional mechanism. According to this mechanism, Pt atoms act as active sites for methanol dehydrogenation to form reaction intermediates (*e.g.*, CO_{ad}); Ru (being less noble than Pt) activates H_2O and adsorbs OH species at lower potentials than are possible for Pt. The adsorbed CO at the Pt sites then reacts with the OH adsorbate at the Ru centers through a Langmuir–Hinshelwood mechanism, producing CO_2 and concomitantly liberating the Pt atoms for further adsorption. Moreover, it is interesting to observe that Ru-modified THH Pt NCs show higher catalytic activity than Ru-modified Pt black catalyst with similar θ_{Ru}. This may be due to that the high-index faceted THH Pt NCs have more steps than Pt black, and the modified Ru adatoms preferentially adsorb on Pt stepped sites, to reduce the poison on THH Pt NCs.

On the other hand, both Bi- and Au-decorated THH Pt NCs show greatly enhanced electrocatalytic activity towards formic acid oxidation (Fig. 10a). It has long been established that formic acid electrooxidation on Pt proceeds by two parallel pathways.[43] The direct path is a relatively facile process,[44] involving the formation of CO_2, by dehydrogenation through adsorbed HCOO as the reactive intermediate.[45] The much slower indirect route consists of the dehydration of HCOOH, through a surface blocking and poisoning intermediate of CO_{ad}.[45,46] Fortunately, the formation of CO_{ad} from HCOOH dehydration needs several (*e.g.* three) adjacent Pt surface sites, so the modification of Pt surfaces with foreign atoms can diminish the number of available Pt ensembles required for the dehydration of HCOOH to form CO_{ad}, so it can greatly suppresses the poison formation. This is the so-called third-body effect.[47] In the case of Au, the third-body effect plays a key role for the enhanced activity, since there is no promoting effect for CO_{ad} oxidation and the Au-coverage dependence of activity (Fig. 10b) is consistent with theoretical simulation (the optimum coverage is around 0.7).[47] As for Bi decoration, the catalytic activity is enhanced significantly, even 5 times higher than that of Au-modified THH Pt NCs. So, both third-body effect and electronic (ligand) effect contribute to the enhancement. Note that the catalytic activity monotonically increases with the increase of coverage of Bi on {730}-faceted THH Pt NCs, different from that of Bi-decorated bulk Pt(111) surfaces.[47] This result indicates that the enhanced mechanism is also correlated with the Pt surface structure and nano size effects.

5.2 Comparison between alloy and surface modification routes

When incorporating a foreign element into a host metal, two major electronic effects will be induced: strain effect and ligand effect.[48] The former is caused by size mismatch between the host and the foreign atoms, leading to a compression or expansion in the host metal lattice. As a result, the d-band center with respect to

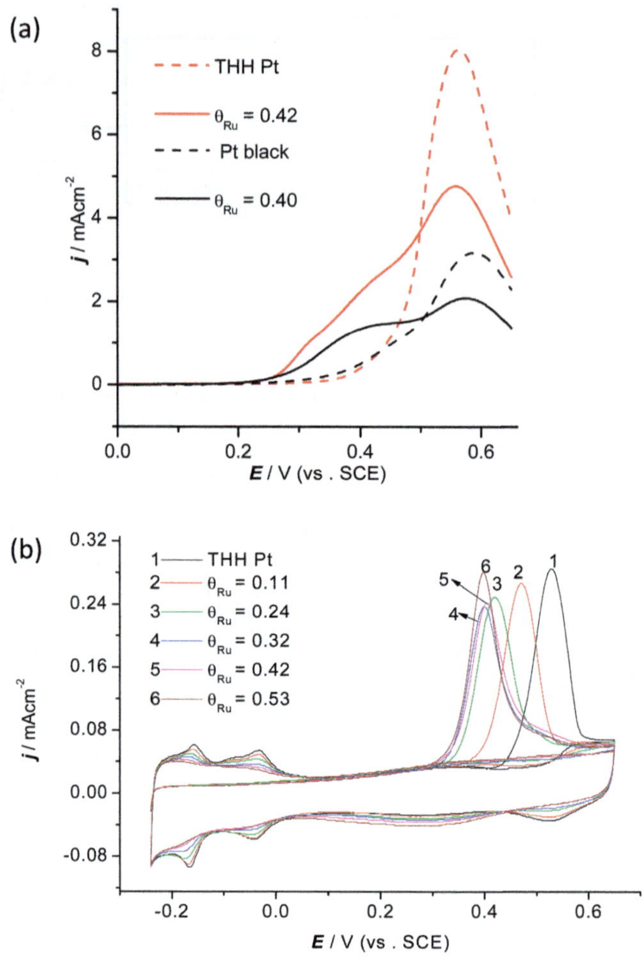

Fig. 9 (a) Electrocatalytic oxidation of methanol on Ru-modified THH Pt NCs and Pt black catalysts at 50 mV s^{-1} in 1.0 M CH_3OH + 0.1 M $HClO_4$. (b) CO oxidative stripping CVs for the bare and Ru modified THH Pt NC electrodes in 0.1 M H_2SO_4 at 50 mV s^{-1}.

the Fermi level will be shifted down or shifted up, which affects the binding energy of adsorbates;[3] while the latter refers to the electronic interactions between foreign and host atoms, *i.e.*, change in the chemical surrounding (ligand) of the host atom.

As mentioned above, there are two routes to build bimetallic nanocrystals with high-index facets. One is direct synthesis of (bulk) alloy nanocrystals with high-index facets through co-reduction or co-electrodeposition. The other route is through surface modification of foreign atoms on preformed nanocrystals with high-index facets.

In terms of synthesis, the surface-modification route is relatively easy, since there are a considerable number of monometallic nanocrystals with high-index facets synthesized,[13,14] and many foreign metals can be deposited with a wide range of coverage. Whereas the direct synthesis of alloy nanocrystals with high-index facets is quite challenging, as it involves lattice mismatch, and co-reduction of two metal ions usually with different equilibrium potentials or reduction kinetics. Although high-index faceted alloy nanocrystals are difficult in synthesis, their durability under reaction conditions is expected to be higher than that of their surface-modified counterparts, as monolayer/sub-monolayer modified atoms are easily leached out.

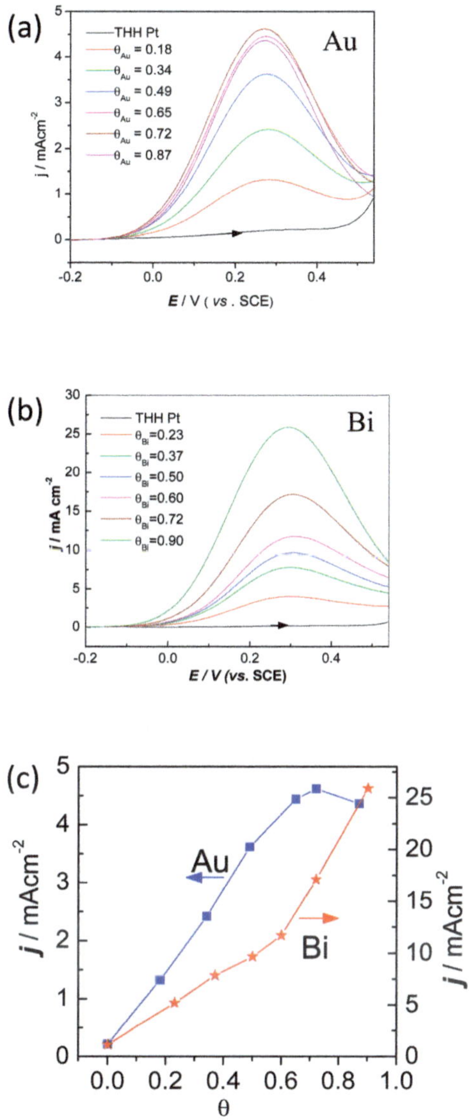

Fig. 10 Electrocatalytic oxidation of formic acid oxidation on Au-modified (a) and Bi-modified (b) THH Pt NCs at 50 mV s^{-1} in 0.25 M HCOOH + 0.5 M H_2SO_4. (c) Variation of peak current with the coverage of Au and Bi.

In terms of enhanced catalytic effects, surface-modified nanocrystals possess ligand effect, bifunctional effect, and third-body effect; while alloy nanocrystals also have strain effect besides the above-mentioned three effects. That is, there are more ways to manipulate both the structures and properties for alloy nanocrystals.

When it comes to surface structure, high-index facets have a unique advantage. Usually, foreign atoms preferentially occupy high-active step atoms on the host metal surface. So, the gain in catalytic activity by bimetallic effects will be partially offset by the loss of high-active step atoms.[49] This compromised effect is diminished for high-index faceted NCs, which have a high density of step atoms.[50] The higher activity of Ru-modified THH Pt NCs than Ru-modified Pt black justifies this hypothesis.

Overall, bimetallic nanocrystals with high-index facets provide a variety of structural parameters to tune their catalytic properties, which is of importance for both fundamental studies and practical applications.

Conclusions

Pt–Rh alloy nanocrystals with high-index facets have been successfully synthesized by an electrochemical method. The shapes of the nanocrystals can be tuned from tetrahexahedron (THH) to trapezohedron (TPH) by altering the electrode potential. Through the synergistic effects of high-index facets and electronic structure of the alloy, both THH and TPH Pt–Rh alloy nanocrystals exhibit excellent catalytic activity towards ethanol electrooxidation. This work gives light to the direction of tuning the composition, shape, as well as surface facets of metal nanocatalysts to improve their catalytic activity.

Acknowledgements

This study was supported by National Basic Research Program of China (973 Program) (2012CB215500), NSFC (21222310, 20933004, 21073152, and 21021002), Foundation for the Author of National Excellent Doctoral Dissertation of China (201126), and Program for New Century Excellent Talents in University (NCET-10-0715).

Notes and references

1 A. C. Chen and P. Holt-Hindle, *Chem. Rev.*, 2010, **110**, 3767.
2 V. R. Stamenkovic, B. Fowler, B. S. Mun, G. F. Wang, P. N. Ross, C. A. Lucas and N. M. Markovic, *Science*, 2007, **315**, 493.
3 J. K. Norskov, F. Abild-Pedersen, F. Studt and T. Bligaard, *Proc. Natl. Acad. Sci. U. S. A.*, 2011, **108**, 937.
4 Z. Y. Zhou, N. Tian, Z. Z. Huang, D. J. Chen and S. G. Sun, *Faraday Discuss.*, 2009, **140**, 81.
5 N. Tian, Z. Y. Zhou and S. G. Sun, *J. Phys. Chem. C*, 2008, **112**, 19801.
6 J. Solla-Gullon, F. J. Vidal-Iglesias and J. M. Feliu, *Annu. Rep. Prog. Chem., Sect. C*, 2011, **107**, 263.
7 Y. Xia, Y. J. Xiong, B. Lim and S. E. Skrabalak, *Angew. Chem., Int. Ed.*, 2009, **48**, 60.
8 N. D. Spencer, R. C. Schoonmaker and G. A. Somorjai, *Nature*, 1981, **294**, 643.
9 S. G. Sun, A. C. Chen, T. S. Huang, J. B. Li and Z. W. Tian, *J. Electroanal. Chem.*, 1992, **340**, 213.
10 N. Tian, Z. Y. Zhou, S. G. Sun, Y. Ding and Z. L. Wang, *Science*, 2007, **316**, 732.
11 Z. Y. Zhou, Z. Z. Huang, D. J. Chen, Q. Wang, N. Tian and S. G. Sun, *Angew. Chem., Int. Ed.*, 2010, **49**, 411.
12 G. Somorjai and D. Blakely, *Nature*, 1975, **258**, 580.
13 L. Zhang, W. Niu and G. Xu, *Nano Today*, 2012, **7**, 586.
14 Z. Y. Zhou, N. Tian, J. T. Li, I. Broadwell and S. G. Sun, *Chem. Soc. Rev.*, 2011, **40**, 4167.
15 T. Yu, D. Y. Kim, H. Zhang and Y. N. Xia, *Angew. Chem., Int. Ed.*, 2011, **50**, 2773.
16 C. L. Lu, K. S. Prasad, H. L. Wu, J. A. Ho and M. H. Huang, *J. Am. Chem. Soc.*, 2010, **132**, 14546.
17 Y. Yu, Q. B. Zhang, B. Liu and J. Y. Lee, *J. Am. Chem. Soc.*, 2010, **132**, 18258.
18 X. Q. Huang, Z. P. Zhao, J. M. Fan, Y. M. Tan and N. F. Zheng, *J. Am. Chem. Soc.*, 2011, **133**, 4718.
19 J. Zeng, C. Zhu, J. Tao, M. S. Jin, H. Zhang, Z. Y. Li, Y. M. Zhu and Y. N. Xia, *Angew. Chem., Int. Ed.*, 2012, **51**, 2354.
20 C. Zhu, J. Zeng, J. Tao, M. C. Johnson, I. Schmidt-Krey, L. Blubaugh, Y. M. Zhu, Z. Z. Gu and Y. N. Xia, *J. Am. Chem. Soc.*, 2012, **134**, 15822.
21 J. Wu, A. Gross and H. Yang, *Nano Lett.*, 2011, **11**, 798.
22 L. Zhang, J. Zhang, Q. Kuang, S. Xie, Z. Jiang, Z. Xie and L. Zheng, *J. Am. Chem. Soc.*, 2011, **133**, 17114.
23 Y. J. Deng, N. Tian, Z. Y. Zhou, R. Huang, Z. L. Liu, J. Xiao and S. G. Sun, *Chem. Sci.*, 2012, **3**, 1157.

24 J. De Souza, S. Queiroz, K. Bergamaski, E. Gonzalez and F. Nart, *J. Phys. Chem. B*, 2002, **106**, 9825.
25 F. Lima and E. Gonzalez, *Electrochim. Acta*, 2008, **53**, 2963.
26 Q. Yuan, Z. Y. Zhou, J. Zhuang and X. Wang, *Chem. Mater.*, 2010, **22**, 2395.
27 Y. Y. Li, Y. X. Jiang, M. H. Chen, H. G. Liao, R. Huang, Z. Y. Zhou, N. Tian, S. P. Chen and S. G. Sun, *Chem. Commun.*, 2012, **48**, 9531.
28 Y. N. Wen and H. M. Zhang, *Solid State Commun.*, 2007, **144**, 163.
29 N. Tian, Z. Y. Zhou and S. G. Sun, *Chem. Commun.*, 2009, 1502.
30 K. Bergamaski, E. R. Gonzalez and F. C. Nart, *Electrochim. Acta*, 2008, **53**, 4396.
31 J. P. I. de Souza, S. L. Queiroz, K. Bergamaski, E. R. Gonzalez and F. C. Nart, *J. Phys. Chem. B*, 2002, **106**, 9825.
32 L. Zhang, D. Q. Chen, Z. Y. Jiang, J. W. Zhang, S. F. Xie, Q. Kuang, Z. X. Xie and L. S. Zheng, *Nano Res.*, 2012, **5**, 181.
33 S. G. Sun and Y. Lin, *Electrochim. Acta*, 1998, **44**, 1153.
34 L. W. H. Leung and M. J. Weaver, *J. Phys. Chem.*, 1988, **92**, 4019.
35 D. J. Tarnowski and C. Korzeniewski, *J. Phys. Chem. B*, 1997, **101**, 253.
36 K. Bergamaski, J. F. Gomes, B. E. Goi and F. C. Nart, *Ecletica Quimica*, 2003, **28**, 87.
37 C. J. Houtman and M. A. Barteau, *J. Catal.*, 1991, **130**, 528.
38 J. Zhang, K. Sasaki, E. Sutter and R. R. Adzic, *Science*, 2007, **315**, 220.
39 S. R. Brankovic, J. X. Wang and R. R. Adzic, *Surf. Sci.*, 2001, **474**, L173.
40 Q. S. Chen, Z. Y. Zhou, F. J. Vidal-Iglesias, J. Solla-Gullon, J. M. Feliu and S. G. Sun, *J. Am. Chem. Soc.*, 2011, **133**, 12930.
41 H. X. Liu, N. Tian, M. P. Brandon, Z. Y. Zhou, J. L. Lin, C. Hardacre, W. F. Lin and S. G. Sun, *ACS Catal.*, 2012, **2**, 708.
42 H. X. Liu, N. Tian, M. P. Brandon, J. Pei, Z. C. Huangfu, C. Zhan, Z. Y. Zhou, C. Hardacre, W. F. Lin and S. G. Sun, *Phys. Chem. Chem. Phys.*, 2012, **14**, 16415.
43 A. Capon and R. Parsons, *J. Electroanal. Chem.*, 1973, **45**, 205.
44 T. Iwasita, X. Xia, E. Herrero and H.-D. Liess, *Langmuir*, 1996, **12**, 4260.
45 A. Miki, S. Ye and M. Osawa, *Chem. Commun.*, 2002, 1500.
46 S. Park, Y. Xie and M. J. Weaver, *Langmuir*, 2002, **18**, 5792.
47 E. Leiva, T. Iwasita, E. Herrero and J. M. Feliu, *Langmuir*, 1997, **13**, 6287.
48 L. Xiao, B. Huang, L. Zhuang and J. Lu, *RSC Adv.*, 2011, **1**, 1358.
49 V. Del Colle, A. Berna, G. Tremiliosi, E. Herrero and J. M. Feliu, *Phys. Chem. Chem. Phys.*, 2008, **10**, 3766.
50 V. Del Colle, J. Souza-Garcia, G. Tremiliosi, E. Herrero and J. M. Feliu, *Phys. Chem. Chem. Phys.*, 2011, **13**, 12163.

Shape-selected bimetallic nanoparticle electrocatalysts: evolution of their atomic-scale structure, chemical composition, and electrochemical reactivity under various chemical environments

Chunhua Cui,[a] Mahdi Ahmadi,[b] Farzad Behafarid,[b] Lin Gan,[a] Maximilian Neumann,[a] Marc Heggen,[c] Beatriz Roldan Cuenya*[b] and Peter Strasser*[a]

Received 23rd December 2012, Accepted 31st January 2013
DOI: 10.1039/c3fd20159g

Solid surfaces generally respond sensitively to their environment. Gas phase or liquid phase species may adsorb and react with individual surface atoms altering the solid-gas and solid-liquid electronic and chemical properties of the interface. A comprehensive understanding of chemical and electrochemical interfaces with respect to their responses to external stimuli is still missing. The evolution of the structure and composition of shape-selected octahedral PtNi nanoparticles (NPs) in response to chemical (gas-phase) and electrochemical (liquid-phase) environments was studied, and contrasted to that of pure Pt and spherical PtNi NPs. The NPs were exposed to thermal annealing in hydrogen, oxygen, and vacuum, and the resulting NP surface composition was analyzed using X-ray photoelectron spectroscopy (XPS). In gaseous environments, the presence of O_2 during annealing (300 °C) lead to a strong segregation of Ni species to the NP surface, the formation of NiO, and a Pt-rich NP core, while a similar treatment in H_2 lead to a more homogenous Pt-Ni alloy core, and a thinner NiO shell. Further, the initial presence of NiO species on the as-prepared samples was found to influence the atomic segregation trends upon low temperature annealing (300 °C). This is due to the fact that at this temperature nickel is only partially reduced, and NiO favors surface segregation. The effect of electrochemical cycling in acid and alkaline electrolytes on the structure and composition of the octahedral PtNi NPs was monitored using image-corrected high resolution transmission electron microscopy (TEM) and high-angle annular dark field scanning TEM (HAADF-STEM). Sample pretreatments in surface active oxygenates, such as oxygen and hydroxide anions, resulted in oxygen-enriched Ni surfaces (Ni oxides and/or hydroxides). Acid treatments were found to strongly reduce the content of Ni species on the NP surface, *via* its dissolution in the electrolyte, leading to a Pt-skeleton structure, with a thick Pt shell and a Pt-Ni core. The presence of Ni hydroxides on the NP surface was shown to improve the kinetics of the electrooxidation of CO and the electrocatalytic hydrogen

[a]*Department of Chemistry, Chemical Engineering Division, Technical University, Berlin, Germany. E-mail: pstrasser@tu-berlin.de*
[b]*Department of Physics, University of Central Florida, Orlando, FL 32816, USA. E-mail: roldan@ucf.edu*
[c]*Ernst Ruska-Centre for Microscopy and Spectroscopy with Electrons, Forschungszentrum Juelich, Germany*

evolution reactions. The affinity to water and the oxophilicity of Ni hydroxides are proposed as likely origin of the observed effects.

Metal alloy particles of nanometer dimensions typically show different surface catalytic reactivity as compared to their bulk alloy form. Since the surface-to-volume ratio increases with decreasing nanoparticle (NP) size, characteristics such as the surface atomic composition and geometrical distribution, alongside with their imperfections (defects), become increasingly relevant at controlling the catalytic behavior of individual metal NPs. This is the origin of size- and shape-dependent reactivities of metal NP catalysts.[1–4] While the exploration of NP size effects in catalysis has a long history,[5–10] our understanding of the role of the NP shape in heterogeneous catalysis and electrocatalysis is still in its infancy.[11–19] New wet-chemical colloidal synthesis methods have significantly improved our ability to control the size and shape of metal and oxide NPs.[18,20–22]

Other than the particles' geometry, the surface atomic composition of alloy NPs largely affects their catalytic behavior. Although the surface composition is initially determined by molecular processes taking place during the NP synthesis, it is a dynamic parameter. In fact, the surface composition of small NPs is highly sensitive to external stimuli, such as temperature or surface active components in gas- or liquid-phase environments during heterogeneous catalytic and electrocatalytic processes, respectively. In gas-phase catalysis as well as electrochemical catalysis, surface adsorption and reaction can induce surface restructuring[22,23] or selective surface segregation of an alloy component.[24–34] Owing to the presence of the segregated component, partially segregated surfaces may display bifunctional catalytic enhancement effects resulting in a new and unexpected catalytic behavior compared to the original surface. Along these lines, the catalytic activity of bimetallic surfaces could be turned on and off by suitable consecutive exposure to segregation-inducing environments.

Several previous experimental and theoretical studies have investigated temperature-[35–44] and adsorbate-driven[45–48] segregation process in Pt-Ni systems, including the effect of such segregation in their catalytic reactivity.[35,37,49–52] In general, segregation phenomena in bimetallic systems can be understood based on the surface energies and atomic radius of the two metals, with the element having a larger atomic radius and a lower surface energy segregating towards the free surface. In the case of platinum and nickel, although platinum has a slightly higher surface energy than Ni (2.475 J m^{-2} *vs.* 2.45 J m^{-2}),[53] due to the much larger atomic radius of Pt (1.39 Å) as compared to Ni (1.24 Å), the segregation of platinum toward the surface is expected, as it was reported for annealing studies in vacuum[35–47] and in H_2 environments.[47] However, for this material system there is experimental[37,38,47,54] and theoretical[35,40,42,47,50] evidence that Pt segregation would only result in the formation of the so called "Pt skin" or "sandwich structure", and not in a phase-separated core–shell structure. The Pt skin refers to a configuration in which the outermost layer is Pt-rich, while the second layer is Ni- rich. Remarkably, although adsorbates could change the segregation properties of the Pt-Ni system, the segregation phenomena appear to be limited to the top-most surface layers. For example, while some density functional theory (DFT) calculations did not predict any segregation due to chemisorbed oxygen,[48] others showed the formation of a Ni skin with the outermost layer enriched by Ni and the second layer by Pt.[47] In contrast, in cases where oxides might be readily formed, this behavior might be drastically modified, since oxygen might then not only be present at the surface or top-most subsurface region, but also in the bulk of the NPs. In particular, the formation of nickel oxide could contribute to the segregation of Ni to the NP surface, as discussed for various nickel alloys.[45]

In this contribution we investigate the evolution of the structure, composition, and catalytic reactivity of shape-selected octahedral Pt-Ni bimetallic NPs in response to various thermal, gas-phase, and electrochemical environments. We follow the changes of the near-surface composition of the PtNi(111) facets after exposure to oxygen and hydrogen gas environments using X-ray photoelectron spectroscopy (XPS) and contrast those to surface compositional transformations during electrochemical potential cycling pretreatments in alkaline and acidic environments. Acidic electrochemical environments dissolve Ni species from the surface leaving behind a Pt enriched interface. Oxygen and hydroxide (alkaline) environments promote the surface segregation of Ni(II) oxide species, which, owing to their oxophilicity and resulting affinity to water activation, show a significant promotion of the electrochemical CO oxidation and the electrocatalytic hydrogen evolution. A bifunctional synergetic effect between surface Pt atoms adjacent to Ni oxide domains is proposed.

Experimental

Catalyst synthesis

Spherical Pt NPs (Pt/C) supported on a high surface area carbon powder. Commercial Vulcan XC 72R supported Pt NPs with a 15% loading by weight with a BET value of 151 $m^2\ g^{-1}$ and mean size of about 5 nm was used as benchmark and standard catalyst[55,56,58] for CO electrooxidation and hydrogen evolution. The catalyst was employed as received.

Spherical PtNi NPs (s-PtNi/C) supported on a high surface area carbon powder. To synthesize spherical PtNi NPs, 0.3 mmol of $Ni(ac)_2/4H_2O$, 0.4 mmol of 1,2-tetradecandiol, 0.6 mL of oleic acid, and 0.6 mL of oleylamine were added into 40 mL of dioctylether in a three-neck flask. The solution temperature was then increased to 200 °C in an oil bath during 40 min. Subsequently, 0.3 mmol $Pt(acac)_2$ dissolved in 2 mL dichlorobenzene was injected into the reaction flask *via* a syringe. This solution was held at 200 °C for 1 h and then cooled down to room temperature (RT). After that, ethanol was added to precipitate the NPs, which were then collected after centrifugation (8000 rpm, 10 min). The NPs were further washed in ethanol and then redispersed in hexane. The latter solution was then mixed with Vulcan XC carbon by a ultrasonic treatment. To remove the residual surfactants, the obtained catalysts were heated at 180 °C in air for 1 h and further annealed at 400 °C in H_2 (4 vol% in Ar) for 4 h.[57]

Octahedral PtNi NPs (oct-PtNi/C) supported on a high surface area carbon powder. To synthesize octahedral PtNi NPs, 4 mM $Pt(acac)_2$ and 10 mM $Ni(acac)_2$ were mixed together in 100 mL DMF under ultrasonic treatment. The resulting homogeneous solution was transferred to a glass-lined stainless-steel autoclave and heated from RT to 120 °C within 10 min. The temperature was held for 42 h. Before washing the particles, the carbon support was added to the solution under ultrasonic treatment. Finally, the oct-PtNi/C was washed with ethanol/water several times.

For the XPS study, three identical PtNi samples containing octahedral NPs were prepared using the same NP solution. The NPs were drop-coated on atomically clean pyrolytic graphite (HOPG) substrates after sonicating the original NP solution for 10 min and subsequently annealed in O_2, H_2 and vacuum.

Catalyst film electrode preparation

A catalyst ink was prepared by mixing the as-prepared catalyst powder in 5 ml of an aqueous and isopropanol solution containing 5 wt% Nafion and ultrasonicated for 15 min. A 10 μl aliquot was dispensed onto the rotating disk electrode (RDE) resulting in a Pt loading of about 7μg cm^{-2} for s-PtNi/C, 8 μg cm^{-2} for oct-PtNi/C, and 9 μg cm^{-2} for Pt/C, respectively. Finally, the as-prepared RDE catalyst film was dried at 50 °C for 15 min.

Catalyst characterization

TEM and energy dispersive X-ray spectroscopy (EDX) measurements were performed in a FEI Tecnai G2 20 S-TWIN electron microscope (200 kV) equipped with a LaB_6 cathode. High-resolution TEM was performed in an image corrected FEI TITAN 80-300 electron microscope (300 kV). High-resolution STEM and electron energy loss spectroscopy (EELS) experiments were performed in a FEI TITAN 80-300 microscope (300 kV) equipped with a probe corrector and a HAADF detector. 'Z-Contrast' conditions were achieved using a probe semi-angle of 25 mrad and an inner collection angle of the detector of 70 mrad. EELS spectra were recorded with a Gatan image filter Tridiem 866ERS system. For the EELS line profile measurements, both Ni-$L_{2,3}$ and Pt-$M_{4,5}$ edges spectra were collected across individual NPs with the acquisition time of 2 s/spectrum. The relative intensities of the Ni-$L_{2,3}$ and Pt-$M_{4,5}$ edges were normalized with respect to their elemental scattering cross sections so that the intensity ratio can be related to the atomic ratio along the electron beam. Inductively coupled plasma (ICP) mass data were obtained by a 715-ES- ICP analysis system (Varian).

The morphology and dispersion of the PtNi NPs supported on HOPG was monitored *via* atomic force microscopy (AFM) (Digital Instruments, Nanoscope III). On these samples, thermal treatments in different environments were conducted in a high pressure (HP) cell (SPECS GmbH) that allows annealing up to 800 °C and pressures up to 20 bar. The HP cell was evacuated after each thermal treatment and the samples were directly transferred to the attached ultrahigh vacuum (UHV) system without exposure to air. XPS data were acquired *in situ* in a UHV chamber equipped with a monochromatic X-ray source (Al-kα, 1486.6 eV, 350 W) and an hemispherical electron analyzer (Phoibos 100, SPECS GmbH). For the XPS study, three identical PtNi samples containing octahedral NPs were annealed in 1 bar O_2, 1 bar H_2, and vacuum environments within the HP cell for 20 min. The temperature ramp was set to 1 °C s^{-1}. All XPS spectra were measured at RT. The high resolution XPS data were acquired using a pass energy, E_{pass}, of 18 eV. The XPS spectra were analyzed using the CASA XPS software.[61] All spectra were aligned using the C-1s peak from the graphite support at 284.3 eV as reference. The Ni/Pt atomic ratios were obtained from the area under the Ni-2p and Pt-4f peaks after normalization by the corresponding atomic sensitivity factors of 26.08 and 14.99, respectively.[61]

Electrochemical pretreatments

All electrochemical measurements were carried out in a conventional three electrode cell, employing a Pt gauze as counter electrode and a commercial reversible hydrogen electrode (RHE).[59,60] All potentials here are referenced to this reference electrode. Electrochemical catalyst pretreatments are illustrated in Fig. 1. Pt/C, s-PtNi/C and oct-PtNi/C catalysts were cycled between 0.06 and 0.6 V_{RHE} at 100mV s^{-1} in deaereated 0.1 M KOH alkaline or 0.1 M $HClO_4$ acidic solutions. Catalysts that underwent 1 and 20 cycles in alkaline are denoted as "1-alkaline" and "20-alkaline", respectively. s-PtNi/C and oct-PtNi/C catalysts that were cycled 100 times in 0.1 M $HClO_4$ under otherwise identical conditions were denoted "100-acid", see Fig. 1.

CO electrooxidation measurements

The electrochemically pretreated Pt/C, s-PtNi/C and oct-PtNi/C catalysts were tested for their electrooxidation (stripping) of adsorbed CO at saturation coverage. The CO stripping of "1-alkaline" and "20-alkaline" was performed in 0.1 M KOH alkaline solutions only, while the "100- acid" acid-pretreated samples were measured in both alkaline 0.1 M KOH and 0.1 M $HClO_4$ acid solutions, Fig. 1. CO adsorption was carried out at a constant electrode potential of +0.05 V *versus* RHE by flowing CO gas through the electrolyte for 15 min. Then, the gas was switched to N_2 for

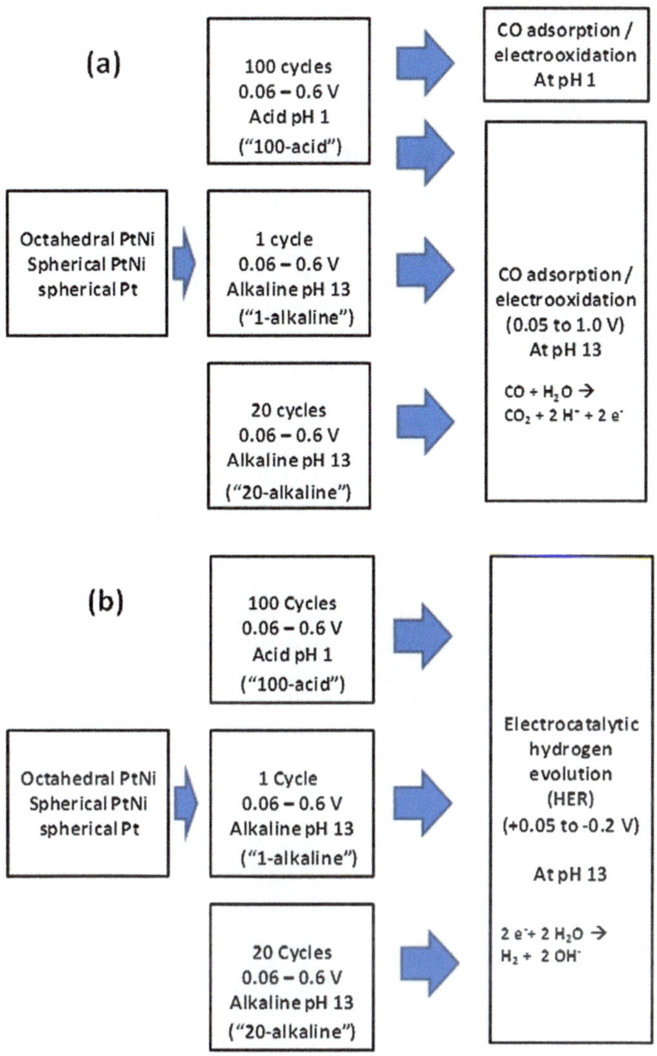

Fig. 1 (a) Nanoparticle catalyst pretreatments and electrocatalytic CO oxidation in acid and alkaline conditions, (b) pretreatments (acid and alkaline) and electrocatalytic hydrogen evolution reaction testing (alkaline) of Pt and PtNi NPs.

15 min to remove excess non-adsorbed CO gas in the electrolyte solution while keeping the potential constant. After that, the potential was scanned from +0.05 to +1.0 V *versus* RHE to record the CO stripping. One additional potential cycle was conducted in order to record the H_{upd} charge value as well as the electrochemical response of the surface at 50 mV s^{-1}.

Hydrogen evolution reaction measurements

After the electrochemical pretreatments, the RDE was transferred into an electrochemical cell containing a H_2-saturated 0.1 M KOH (pH 13) solution. In each experiment, the electrode was immersed in the solution at +0.05 V. Then, the HER polarization curves were recorded between −0.2 V and +0.05 V *versus* RHE at a scan rate of 1mV s^{-1} as shown in Fig. 1. The HER activities were normalized to the catalytically active real surface area (ECSA), which was evaluated by CO stripping.

Results and discussion

Structural and morphological characterization (TEM, AFM)

As described in the experimental section, octahedral PtNi bimetallic NPs with a bulk Pt : Ni ratio of 1 : 1 were prepared using wet-chemical methods (solvothermal process at elevated pressures),[62] with DMF serving as solvent and reducing agent.[63,64] The spherical NPs were prepared using an ambient pressure hot-injection technique[65] at 200 °C involving a high-boiling point solvent, surfactants and reducing agents as separate components.[66]

Fig. 2 shows TEM and HAADF STEM micrographs of as-prepared unsupported octahedral PtNi (oct-PtNi) NPs. The particles exhibited a narrow size distribution with a mean diameter of about 9 nm. EELS data[64] exhibited the composition distribution and STEM image in Fig. 2d showed the octahedral structure in the PtNi nanoparticles. Fig. 3 shows TEM and HAADF STEM images of the unsupported and carbon-supported spherical PtNi (s-PtNi) NPs. Particle sizes were narrowly distributed around a mean of about 6 nm. X-Ray diffraction patterns (not shown) and EELS intensity line scans for Pt and Ni, Fig. 3(d), for selected individual NPs showed that the catalyst consisted of a single face-centered-cubic bimetallic alloy phase with a fairly homogeneous distribution of Pt and Ni atoms at the atomic scale. Pt to Ni intensity ratios across an individual NP are close to 1, which is consistent with the PtNi bulk composition. The particle center appears to be slightly enriched in Ni, indicating the formation of Ni-rich alloy seeds during the reduction process.

Fig. 4 shows an AFM image of the as prepared $Pt_{50}Ni_{50}$/HOPG sample. The average NP height obtained is 3.7 ± 0.8 nm. The atomic steps of the HOPG surface can be observed, and NP decoration of step edges is evident.

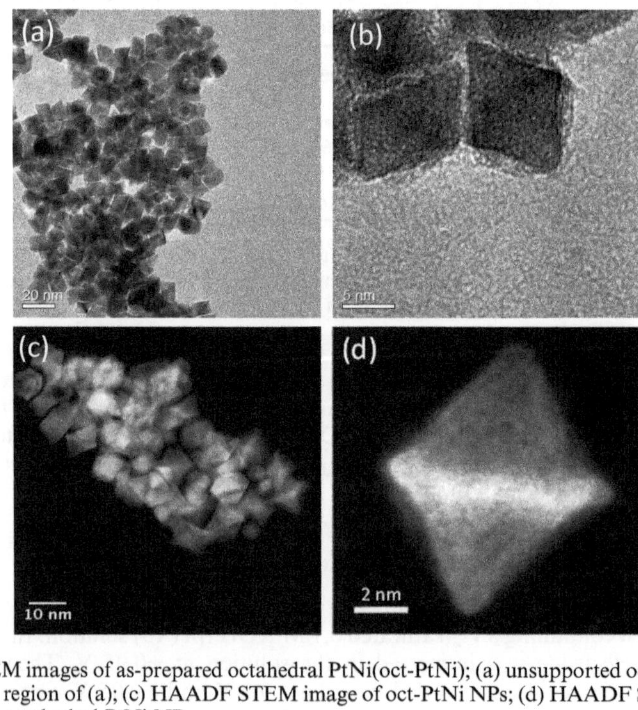

Fig. 2 TEM images of as-prepared octahedral PtNi(oct-PtNi); (a) unsupported oct-PtNi; (b) a zoomed-in region of (a); (c) HAADF STEM image of oct-PtNi NPs; (d) HAADF STEM image of a single octahedral PtNi NP.

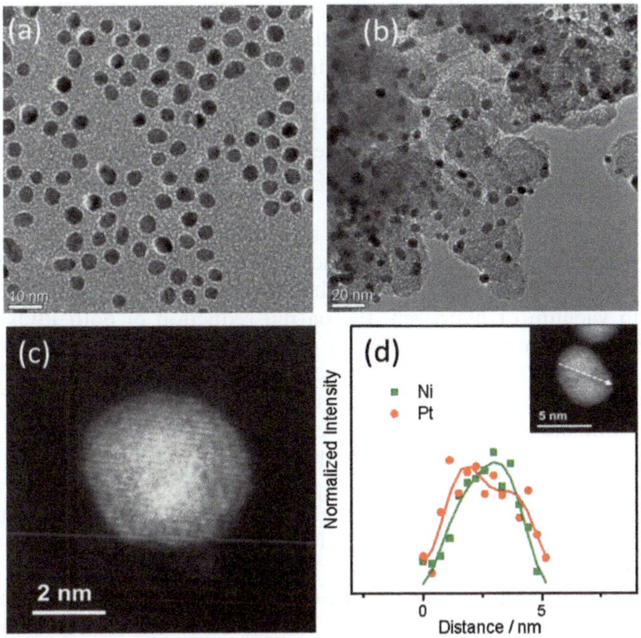

Fig. 3 TEM images of as-prepared spherical PtNi(s-PtNi) NPs: (a) unsupported s-PtNi, (b) carbon-supported s-PtNi, (c) HAADF-TEM image of an individual PtNi NP, (d) STEM EELS line scan (along the arrow in the inset) of a single PtNi NP showing the spatial distribution of Ni and Pt.

Evolution of surface composition under chemically reactive gas and liquid environments

Gas-phase pretreatments. Fig. 5 shows XPS data from the Pt-4f and Ni-2p core level regions of one of the as-prepared samples together with spectra acquired on identical samples after *in situ* thermal treatments at 300 °C for 40 min in different environments, namely in UHV, in 1 bar of O_2, and in 1 bar of H_2. Higher annealing

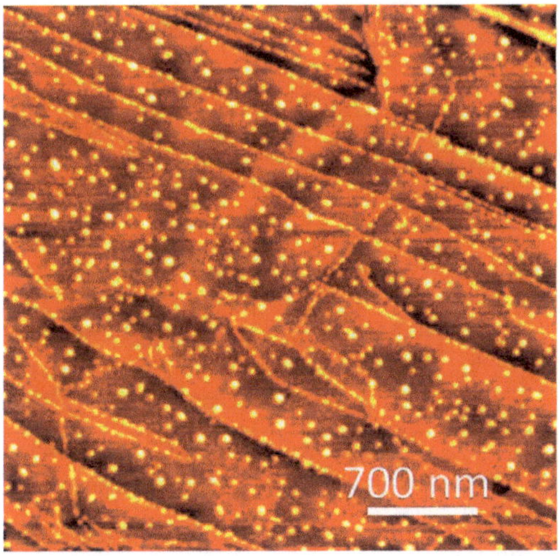

Fig. 4 AFM image of an as-prepared PtNi NP sample acquired at RT in air.

temperatures were avoided due to concerns regarding possible NP shape transformations. However, Monte Carlo simulations of PtNi NPs annealed in vacuum at 330 °C demonstrated no drastic change in the shape of octahedral NPs.[42]

The Pt-4f spectra in Fig. 5(a) were fitted with three doublets ($4f_{7/2}$ and $4f_{5/2}$) corresponding to Pt^0, Pt^{2+} and Pt^{4+} with binding energies in agreement with those previously reported.[67] The Ni-2p core level region was fitted with two doublets corresponding to Ni and Ni^{2+} ($2p_{3/2}$ and $2p_{1/2}$) and a third one accounting for satellite peaks.[68,69]

The content of the different Pt and Ni species as well as the Ni/Pt ratio after the different treatments is shown in Fig. 6. The as prepared samples contain metallic and oxidic Pt species (PtO and PtO_2), while the Ni is mainly in the form of NiO. Upon annealing in vacuum or hydrogen, the partial reduction of both PtOx and NiOx species is evident from the increase in the content of the respective metallic components, Fig. 5 and 6. While Pt is completely reduced in H_2 at 300 °C, a relatively large NiO content is still observed, as expected due to the higher affinity of nickel for oxygen as compared to platinum. Nevertheless, the most significant effect was observed upon annealing in oxygen, with not only the total oxidation of Ni, but also a drastic increase in the intensity of the Ni signal. On the other hand, for the Pt component, only an increase in the Pt^{4+} content is detected, with metallic Pt species being still present. These results suggest that under those conditions, a Pt-rich NP core might be protected by a NiO shell.

Fig. 6(c) shows the Ni/Pt ratio for the as prepared sample as well as that obtained after annealing in different environments. The Ni : Pt ratio increases in all cases after the thermal treatments, with the $(Ni : Pt)_{O_2} > (Ni : Pt)_{Vacuum} > (Ni : Pt)_{H_2}$. As it was mentioned in the introduction, the segregation of Pt induced by annealing in vacuum or H_2 environments would be hard to detect using XPS since it only affects the composition of the outermost layers of the NPs. The increase observed here is most likely due to initial presence of the oxide. To the best of our knowledge, there is no theoretical study predicting the segregation behavior of Pt and Ni in the presence of such oxide species. However, our XPS measurements revealed that as long as the Ni oxide is present in the NPs, a shell of NiO and a core of Pt will form upon heating, regardless of the chemical environment. Interestingly, the final content of Ni at the NP surface appears to be proportional to the extent of the oxidation of the Ni component after a given treatment. This can be seen by comparing the Ni^{2+} contributions in Fig. 6(b) and the Ni/Pt ratios in Fig. 6(c) after the different

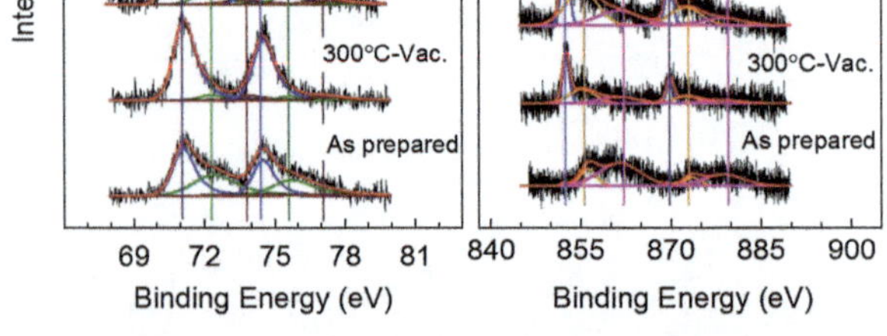

Fig. 5 XPS spectra from the (a) Pt-4f and (b) Ni-2p core level regions of PtNi NPs supported on HOPG acquired at RT before (as prepared) and after annealing in vacuum and in 1 bar of O_2 and H_2 at 300 °C for 20 min.

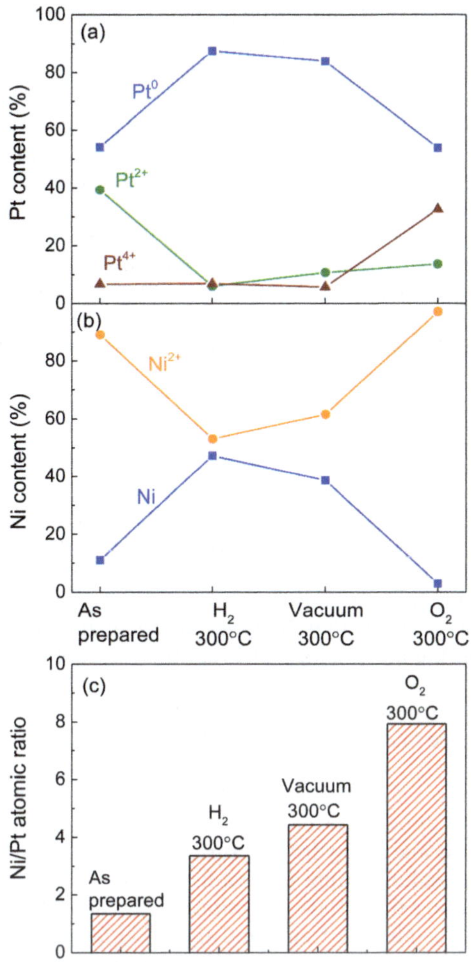

Fig. 6 Relative content of (a) Pt and (b) Ni species extracted from the XPS measurements in Fig. 5. (c) Ni/Pt atomic ratio obtained for the as prepared sample and analogous fresh samples after the indicated sample treatments.

treatments. For both parameters, the lowest content is observed after the thermal treatment in H_2, followed by the vacuum annealing, and the maximum after O_2 exposure at the same temperature (300 °C).

As was mentioned in the introduction, the formation of nickel oxide could contribute to the segregation of Ni to the NP surface. By reducing the NiO *via* annealing in vacuum or hydrogen environments, Ni can diffuse back into the Pt core and form a uniform Pt–Ni alloy. The onset temperature for Ni diffusion into Pt(111) single crystals and alloy formation has been experimentally reported to vary between 180 °C–380 °C, depending on the Ni coverage (0.8–3ML).[44] Fig. 7 displays a model of the distribution of Pt, Ni, PtOx and NiOx species on the octahedral NPs after the different *in situ* thermal treatments.

Electrochemical potential cycling pretreatments. Having explored the surface redox and segregation phenomena of oct-PtNi NPs in gas-phase chemical environments, we performed a similar study in electrochemical environments and tested the effect of the changes in the surface structure and composition on the catalytic reactivity.

To achieve that, we subjected samples containing carbon-supported octahedral (oct-PtNi) and spherical (s-PtNi) alloy NPs to three different potential cycling protocols, two in alkaline (0.1 m KOH) and one in acidic (0.1 m $HClO_4$) electrochemical environments. We chose a potential range from +0.05 to +0.6 V/RHE for each cycle such that the evolution of the surface composition would not be influenced by the formation of Pt oxides, while Ni and NiO surface species could form, segregate, or dissolve. The cycling pretreatments included: i) a single cycle in an alkaline solution representing a standard measurement close to the pristine state of the catalyst surface (1-alkaline), ii) 20 potential cycles in an alkaline solution, because this treatment yielded a stable cyclic voltammogram (not shown) and showed a maximum in catalytic HER performance (see below), and iii) 100 potential cycles in an acidic electrolyte, after which the cyclic voltammogram was stable with time and resembles in shape that of a pure Pt surface, indicating the complete loss of Ni or NiO from the top layer.

The effect of the pretreatments in the three electrochemical environments on the morphology and chemical composition of the oct-PtNi NPs is displayed in Fig. 8 and 9, where Z-contrast TEM scans along the <110> and <100> directions are shown.

In these images, the Z-contrast scales with nearly the power of 2 of the atomic number of the elements along the scanned electron beam direction, and hence carries compositional information along with thickness information. Fig. 8(a) and 8(d) represent the near-pristine state of the octahedral NPs.

Fig. 8(b) and 8(e) evidence that 20 cycles in the alkaline solution did not have a significant effect on the morphology or Z-contrast, except perhaps a slightly darkened contrast near the bottom vertex of the image in Fig. 8(b) (see arrow). Darkened regions can be explained by reduced specimen thickness or enrichment in the lighter element at comparable thickness. Since the alkaline environment precludes the dissolution of either Pt or Ni, an enrichment in Ni or Ni oxides is a likely explanation. Fig. 8(c) and 8(f) show a clearly increased Z-contrast between the bright central octahedral edges and the opposite facets and vertex regions. Since acidic conditions are known to dissolve Ni and Ni hydroxide surface species, a thinning of the specimen by selective metal dissolution appears as a likely explanation for the changes in the Z-contrast. Also, along the <100> direction in Fig. 8(f), the enhanced Z-contrast between the central octahedral frame and the (111) facet region is consistent with a reduced thickness of the facet region. To support the latter conclusion we performed additional STEM and high resolution TEM studies on acid-leached oct-PtNi. The high resolution TEM image in Fig. 9(b) confirms that the acid-cycled octahedra suffered from selective surface dissolution of Ni and Ni hydroxide species near the facet centers, resulting in a concave facet morphology. Fig. 9(c) illustrates schematically the morphology of the resulting Pt-rich concave octahedra NP formed during the treatment of oct-PtNi in acidic environments.

In order to gain further insight into the surface and near-surface composition of the electrochemically-treated samples, we have measured the octahedral and

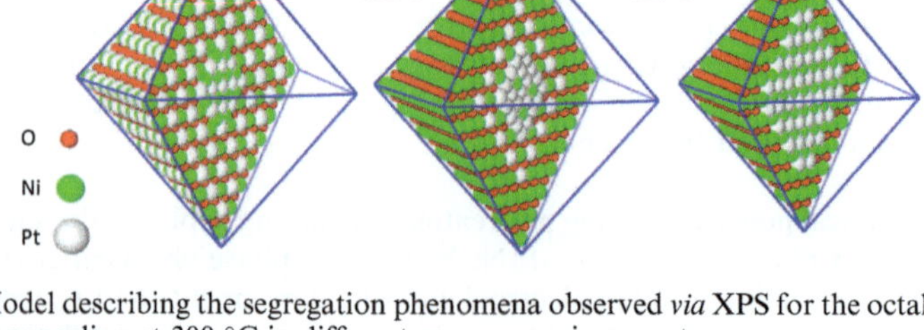

Fig. 7 Model describing the segregation phenomena observed *via* XPS for the octahedral PtNi NPs after annealing at 300 °C in different gaseous environments.

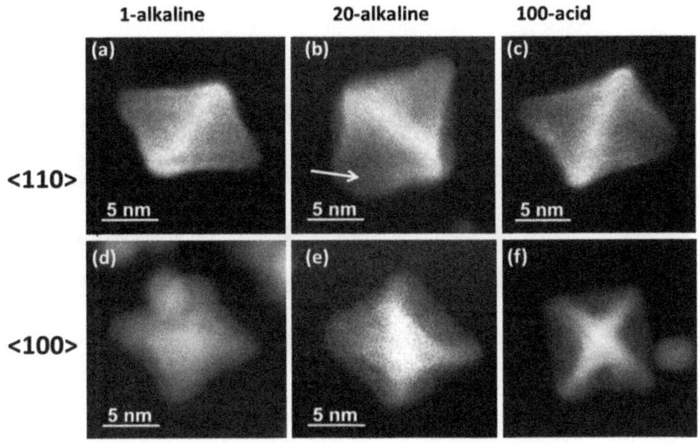

Fig. 8 (a) HAADF STEM images along the (a–c) <110> and (d–f) <100> zone axes of the three differently pretreated octahedral PtNi NPs (oct-PtNi/C): (a,d) "1-alkaline", (b,e) "20-alkaline", (c,f) "100-acid".

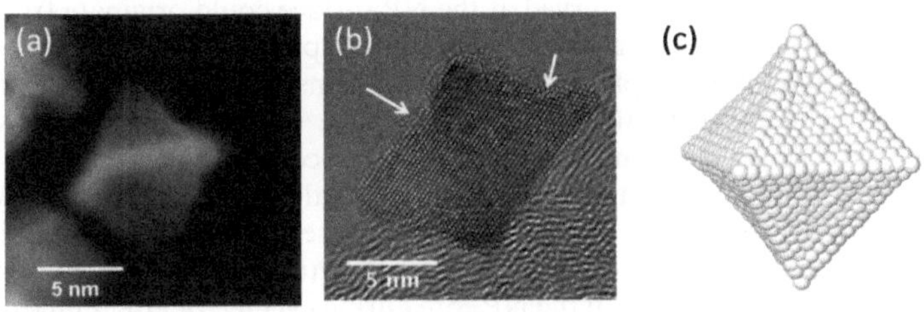

Fig. 9 (a) HAADF STEM image, (b) high resolution TEM image of an acid-leached concave octahedron, and (c) schematic concave octahedral structure of oct-PtNi/C after potential cycling in an acidic 0.1 m $HClO_4$ electrolyte. The model NP displays a Pt-rich structure after $Ni/Ni(OH)_2$ dissolution.

spherical PtNi NPs supported on flat adhesive graphite tabs after 1 and 20 cycles in the alkaline solution, and after 100 cycles in the acid solution *via ex situ* XPS. Fig. 10 shows Pt-4f and Ni-2p XPS data of oct-PtNi and s-PtNi NPs after the different electrochemical treatments. After one cycle from 0.05 V to 0.6 V and back to 0.05 V in an alkaline electrolyte, platinum is reduced to the metallic state, while Ni is partially oxidized in the s-PtNi NPs and completely oxidized in the oct-PtNi NPs, Fig. 11(a),(b). The initial Ni : Pt ratios in the electrochemical environment were significantly higher than those under UHV gas-phase conditions, suggesting the formation of a thick Ni-rich surface layer in the liquid environment. Additional potential cycling in the alkaline solution (20 and 100 cycles) lead to a slight decrease in the Ni : Pt ratio, signalling a Pt-richer surface/near surface NP region, Fig. 11(c). According to their Pourbaix behavior, Ni hydroxides cannot chemically dissolve under $pH = 13$ conditions. Thus, the decrease Ni/Pt in the ratio observed either originated from an electrochemical surface segregation of Pt species, or else, from a partial surface Ni hydroxides reduction at the low electrode potential limit (0.05 V) and subsequent Ni inward diffusion and Pt-Ni alloy formation inside the NP core. The appearance of a small Pt^{2+} component in the XPS data of the octahedral NPs might correlate with the facile electrochemical oxidation of low-coordinated surface Pt atoms near the corners and edges of the octahedral NP surface. It should also be noted that these samples were all measured after transfer in air to the UHV XPS system, and therefore, it cannot be ruled out that a fraction of

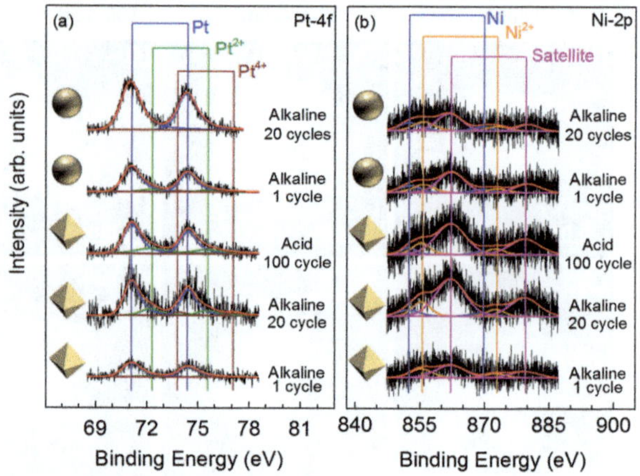

Fig. 10 XPS spectra from the (a) Pt-4f and (b) Ni-2p core level regions of octahedral and spherical PtNi NPs supported on a high surface area carbon black acquired after different voltammetry cycles in alkaline and acid solutions.

the oxidic Pt and Ni species observed at the NP surface could originate from the air exposure. Nevertheless, as mentioned before, PtOx species were not observed on the 1-alkaline oct-PtNi NP sample. Similar observations were made for the spherical PtNi NPs after identical treatments in the alkaline solution, Fig. 11. However, a larger decrease in the Ni content at the NP surface was obtained for s-PtNi as compared to oct-PtNi after the first 20 cycles in the alkaline solution. Whether that is exclusively attributable to a shape effect or whether a size effect is co-involved remains speculative given our data. Subsurface Ni atoms may prefer to move away from the surface and alloy within the NP core during cycling; or else, Pt may be more mobile due to its lower average surface coordination and hence preferably surface segregate. Also, in contrast to the octahedral NPs, the dominant Pt component in the spherical NPs remained metallic Pt.

After 20 and 100 voltammetry cycles in the acidic solution, a strong decrease in the Ni/Pt ratio is observed, Fig. 11(c). In light of the TEM measurements, such observation can be attributed to the selective surface Ni dissolution in the acid electrolytes. Interestingly, in this case platinum appears to be partially oxidized (Pt^{2+}), while nickel is partially reduced, Fig. 11(a),(b). Here, the partial oxidation of Pt might be again related to the low Pt coordination or due to the removal of the protective Ni shell into the electrolyte during the acidic potential cycling.

Electrocatalytic activity

CO oxidation. To study the impact of the electrochemical pretreatments and resulting surface structure of the oct-PtNi/C and s-PtNi/C NPs on their electrocatalytic properties, the pretreated "20-alkaline" and "100-acid" oct-PtNi/C samples were used for the electrocatalytic oxidation of a pre-adsorbed monolayer of CO during a cyclic voltammetric sweep. Conceptually, this electrochemical reactivity measurement is equivalent to a temperature ramp during a gas-phase catalytic temperature-programmed-oxidation (TPO) experiment. First, a study of the oxidative CO stripping of the "100-acid" oct-PtNi/C and s-PtNi/C samples was performed in acid and alkaline electrolytes, and compared to a pure Pt/C NP catalyst. The black traces marked "1." in Fig. 12(a),(b), and (c) show the first cyclic voltammograms (CVs) in acid. Between 0.05 V and 0.6 V all initial anodic voltammograms remained flat at vanishing current densities, consistent with a fully CO-covered, and hence poisoned and inactive Pt surface. CO electrooxidation on Pt started with a small

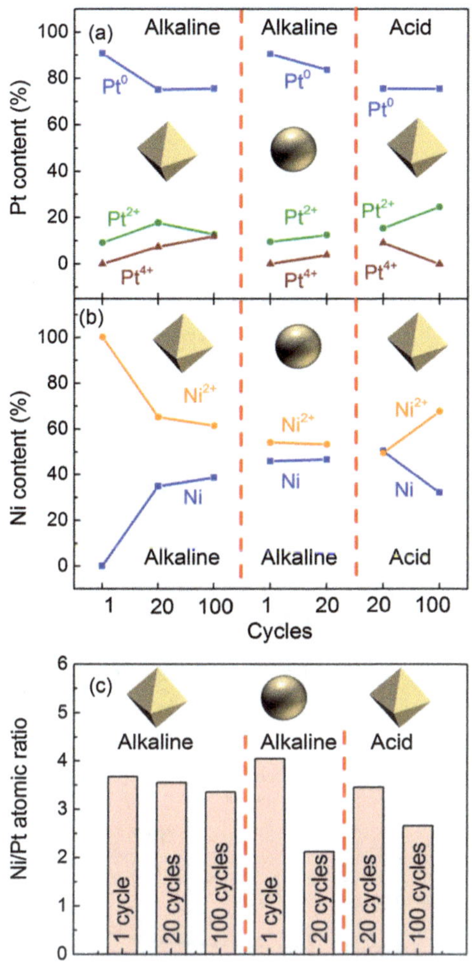

Fig. 11 Relative content of (a) Pt and (b) Ni species extracted from the XPS measurements in Fig. 10. (c) Ni/Pt atomic ratio obtained after a different number of voltammetry cycles in alkaline and acid solutions.

pre-peak feature around 0.72 V, followed by the main peak around 0.85 V, Fig. 12(a). The shape of the cyclic voltammogram is fully consistent with previous CO studies of Pt NPs of comparable size.[70,71] In contrast, the "100-acid" oct-PtNi/C and s-PtNi/C samples showed a 130 mV cathodic shift of the major CO oxidation peak. Following earlier arguments this can be attributed to a weaker interaction of surface Pt atoms with adsorbed CO. Alternatively, low-coordinated Pt atoms formed during the acidic leaching process could serve as active adsorption sites for water activation, resulting in adsorbed OH species that remove CO. The small differences between the acid-treated octahedral and the spherical NP catalysts suggest a fairly limited impact of the NP shape on the CO stripping process. Instead, the electrocatalytic properties appear to be controlled by the surface and near-surface composition.

In alkaline environments the voltammetric CO oxidation peaks showed a significant cathodic potential shift. The Pt/C NPs showed a CO stripping onset potential of about 0.6 V, Fig. 12(a), followed by a gradual current increase peaking at 0.7 V (marked as "Pt"). The potential shift can be explained by the presence of reactive adsorbed hydroxide anions, which cause CO oxidation at lower electrode potentials.

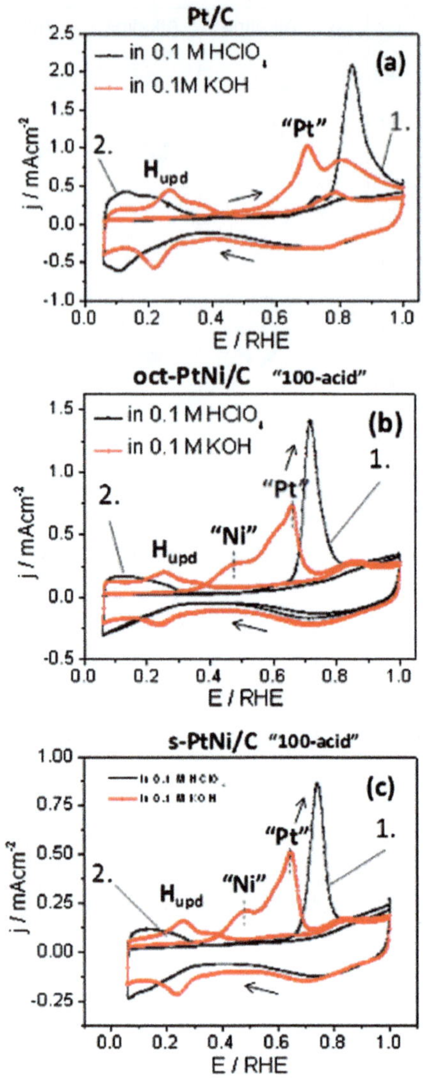

Fig. 12 Comparison of the electrocatalytic CO oxidation in alkaline (red) and acid solution (black) of: (a) Pt/C, (b) "100-acid" pretreated oct-PtNi/C, and (c) "100-acid" pretreated s-PtNi/C. Scan numbers are shown. Arrows indicate scan directions. "Pt" and "Ni" mark major CO oxidation peaks.

A severe cathodic shift in the CO oxidation onset potential was observed for the PtNi/C NPs. Both, octahedral and spherical "100-acid" PtNi NPs exhibited a new CO oxidation peak at around 0.5 V (marked "Ni"). Being shape-independent and absent for the pure Pt NPs, we conclude that this feature is associated with the presence of Ni or Ni-hydroxides in or near the catalyst surface. It is reasonable to assume that the Ni species effectively catalyze the activation of water molecules and serve as supply for active oxygenates to oxidize surface CO.

The second voltammetric cycle of the NPs in acid and alkaline environments (marked "2" in Fig. 12) showed the characteristic peak features associated with the underpotential deposition and removal of atomic hydrogen (H_{upd}) in acid and alkaline electrolytes.

A separate comparative study of the CO oxidation behavior of alkaline-pretreated octahedral and spherical NPs ("1-alkaline" and "20-alkaline" NPs) and acid-pretreated NPs ("100-acid") was carried out and the results are displayed in Fig. 13. The CO stripping curve of both octahedral and spherical "1-alkaline" NPs showed an unexpectedly intense CO removal peak (marked "Ni") in Fig. 13(a) just below 0.5 V, and hence very close to the position of the "Ni" peak of the "100-acid" NPs. More anodic, both "1-alkaline" samples showed a broad shoulder near the potential where the "100-acid" sample showed its most intense current density peak associated with CO stripping [marked "Pt" in Fig. 13(a)]. This suggests that Ni, Ni hydroxide species and maybe defects in or near the surface are responsible for the early CO oxidation onset.

Both, the octahedral and spherical PtNi/C "20-alkaline" samples exhibited a very similar CO oxidation voltammogram as compared to their "1-alkaline" sample. In light of the XPS results of Fig. 11, the "20-alkaline" CVs appear fairly insensitive to moderate changes in the surface Ni/Pt ratio (lower for the s-PtNi/C sample) or the metal oxidation state (higher content of cationic Pt species for oct-PtNi/C).

Table 1 provides a synopsis of the electrochemically active surface areas (CO-ECSA) evaluated from the various voltammetric stripping curves. The Pt/C NPs showed consistent ECSA values, in good agreement with earlier reports.[56,58,72,73] The ratio of CO-based ECSA and H_{upd}-based ECSA for Pt/C was 1.08, confirming consistency between the two surface area evaluation methods. CO-ECSA values of the PtNi/C NPs were also comparable between the acid and alkaline measurements (see Table 1). The somewhat larger oct-PtNi/C NPs showed higher ECSA values than the smaller spherical NPs, contrary to intuition. The reason for this could lie in the surface roughness of the concave octahedral NPs offering an increased number of CO adsorption sites. Furthermore, CO-ECSA values of the "1-alkaline" and "20-alkaline" samples were consistent, suggesting a negligible reduction in CO adsorption during prolonged potential cycling. It is noteworthy that the ratio of CO-ECSA and H_{upd}-ECSA was about 1.5 for both the spherical and the octahedral "100-acid" NPs. This area ratio has been linked to weakened hydrogen surface binding as compared to standard Pt. A value of the ratio larger than one has recently even been considered diagnostic for the formation of a Pt alloy skin surface, consisting of a single segregated Pt monolayer with the second layer being non-noble metal enriched. Unfortunately, due to the inelastic mean free path of electrons or XPS probe-depth, the XPS results cannot provide unambiguous evidence for the formation of a skin surface structure. The observed voltammetric charge ratio value of 1.5 means that the Pt skin-like hydrogen adsorption properties[70] can also appear in acid-dealloyed core–shell surfaces when treated in alkaline electrolytes, even in the absence of any thermal pretreatment. It is conceivable that the CO catalysis in

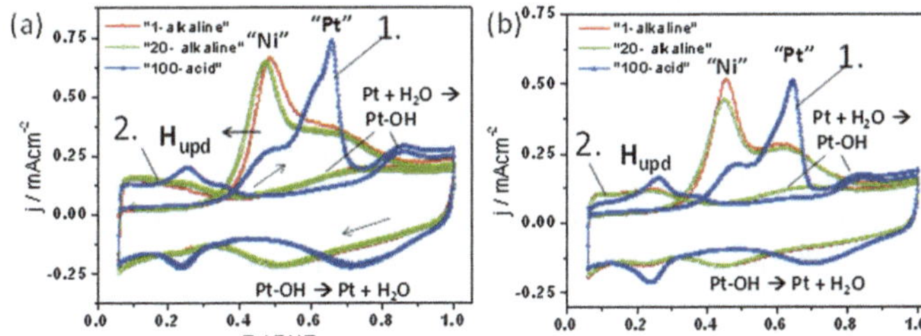

Fig. 13 First and second CO electrooxidation voltammograms of (a) oct-PtNi/C and (b) s-PtNi/C catalysts in alkaline solutions (pH 13) after different pretreatments. Arrows indicate scan directions, scan numbers are marked. H_{upd} denotes voltammetric features associated with underpotential deposition and stripping of atomic hydrogen. Voltammetric peaks associated with the formation and reduction of Pt-OH are indicated.

Table 1 CO stripping-based electrochemically active surface areas (CO-ECSA) performed in alkaline and acid environments

Catalyst pretreatment	Catalyst	CO-ECSA (alkaline) $m^2\ g^{-1}{}_{Pt}$	CO-ECSA (acid) $m^2\ g^{-1}{}_{Pt}$
	Pt	85	84.6[a]
"1-alkaline"	s-PtNi	53	N/A
	oct-PtNi	61	N/A
"20-alkaline"	s-PtNi	50	N/A
	oct-PtNi	60	N/A
"100-acid"	s-PtNi	37	38[a]
	oct-PtNi	50	48[a]

[a] CO-ECSA/H_{upd}-ECSA ratios at pH 1 : Pt/C : 1.08; s-PtNi/C : 1.53; oct-PtNi/C : 1.50.

alkaline electrolytes of the "100-acid" sample smoothed the surface by "electrochemical CO annealing" leading to the healing of surface defects,[24] while aiding Ni to segregate toward the surface and thus created a near-surface alloy closely resembling a skin structure. The latter hypothesis is supported by the fact that the CO stripping voltammograms of "100-acid" oct-PtNi/C and s-PtNi/C NPs closely resemble that of a skin-type PtNi.[70]

Overall, the CO electrooxidation study demonstrates how a simple variation in the pretreatment environment can drastically alter the catalytic behavior of alloy electrocatalysts through environment-induced segregation and the presence of distinct active species in or near the catalyst surface.

Hydrogen evolution in alkaline conditions. In addition to the electrocatalytic CO oxidation reaction, we chose the electrocatalytic reduction of water to molecular hydrogen (hydrogen evolution reaction, HER) in an alkaline electrolyte as a model reaction to test the impact of acid and alkaline surface pretreatments. This reaction is of great importance for neutral and alkaline water splitting for chemical energy storage applications. While alkaline water electrolysis is generally preferred based on catalyst cost, an efficient HER catalysts for alkaline media is missing. Pt is the best known HER electrocatalyst in acidic conditions. However, it displays severe overpotentials (efficiency losses) in alkaline solutions. Fig. 14(a) presents the cyclic voltammetric behavior of the "1-alkaline" pretreated Pt/C, oct-PtNi/C, and s-PtNi/C catalysts in an alkaline electrolyte over a wide potential range from 0.05 to 1.0 V in order to show the H_{upd} as well as the water activation/PtOH formation regions. The pure Pt NPs show sharper H_{upd} as well as PtOH features (0.3 V and 0.8 V), while the alloy NPs exhibit much broader potential waves in either region. In agreement with earlier studies on PtNi,[74] the alloy NPs exhibit a reduced H_{upd} charge due to the reduced interaction of atomic hydrogen with Ni-containing surface and subsurface layers. The onset of water activation occurs fairly early, namely at around +0.5 V for both catalysts. No obvious difference can be detected between the spherical and the octahedral NPs. Fig. 14(b) presents the hydrogen evolution reactivity in the alkaline electrolyte (blue, green, red) of Pt/C after the acid and alkaline pretreatments, and compares those with the activity of the hydrogen evolution on Pt in acidic electrolytes. The cathodic shift of the alkaline HER curves evidences that Pt is a much better electrocatalyst in acid than in alkaline solutions. The HER reactivity of Pt in alkaline solutions appears largely independent of the surface pretreatment, Fig. 14(b). In contrast to Pt, PtNi/C NPs showed a significantly improved HER behavior, approaching that of pure Pt in acidic conditions. We assume that the presence of Ni in the NPs is responsible for the general increase in the HER activity of the alloys. The pretreatments were found to induce noticeable differences in the HER activity. The "100-acid"-pretreated sample catalyzes the HER less efficiently as compared to the "20-alkaline" sample. The redistribution

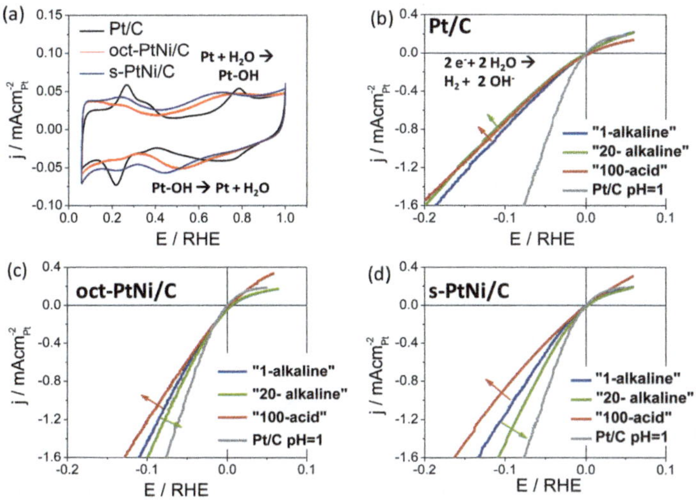

Fig. 14 (a) Cyclic voltammetry of spherical Pt/C, oct-PtNi/C, and s-PtNi/C NPs in 0.1 M KOH, (b)–(d) electrocatalytic HER activity in an alkaline 0.1 M KOH electrolyte (pH 13) of (b) Pt/C, (c) oct-PtNi/C, and (d) s-PtNi/C after pretreatments in alkaline and acid solutions.

of surface Pt and Ni hydroxide species coupled with the emergence of lower, more favorable surface Ni/Pt ratios for HER could explain the additional activity gain of the "20-alkaline" catalyst. On the other hand, acid cycling associated with a strong depletion in Ni hydroxide on the NP surface could be responsible for the observed decrease in HER activity. As the data of spherical s-PtNi/C NPs in Fig. 14(d) illustrate, the influence of the NP shape on the catalytic HER activity is relatively small.

Bifunctional electrocatalysts through environment-induced segregation. The CO oxidation and HER catalysis served as model systems to demonstrate the drastic effect that pretreatments in electrochemical environments can have on surface structure and composition as well as on catalytic activity. Fig. 15 illustrates our hypotheses, based on the TEM, XPS and reactivity results, as to the surface chemical transformation during the electrochemical measurements. After synthesis, the (111) surfaces of oct-PtNi/C NPs consisted of a mix of Pt surface atoms and surface $Ni(OH)_2$. After an initial alkaline pretreatment cycle ("1-alkaline") the surface formed a thick Ni hydroxide-rich layer, while the low electrode potentials reduced surface Pt oxides to metallic Pt. Ni was likely present in its α-Ni hydroxide phase (green balls), which is the most stable Ni surface compound in the electrode potential range considered.[75,76] Previous studies have shown that potential cycling can heal defects, and help remove low-coordinated adatoms analogous to thermal annealing in the gas phase.[77] Along these lines, the present cycling in alkaline caused surface compositional changes and geometric rearrangements in the surface to attain the energetically most favorable state of the alloy surface. The existence of hydroxide anions in the electrolyte generally favors the presence of Ni oxides in the top layer, ["20-alkaline" in Fig. 15]. However, XPS also evidenced that potential cycling reduced the surface Ni/Pt ratio. This could be due to either Pt surface segregation because of the Ni reduction, or to Ni inward diffusion and alloy formation in the NP core driven by the heat of alloying of Pt and Ni.

Cycling pretreatment in acid environments selectively leached a significant portion of the surface Ni hydroxide species from the surface and subsurface region of the octahedral catalysts,[66] resulting in an essentially Ni-free surface enclosing a subsurface PtNi alloy core (core–shell structural arrangement). When tested in acid

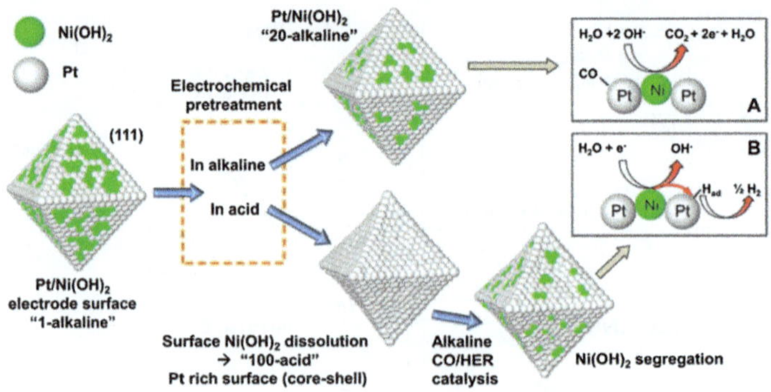

Fig. 15 Electrochemical surface dissolution/segregation processes in various electrochemical environments. CO oxidation and HER are enhanced due to bifunctional effects at Pt atoms adjacent to oxophilic $Ni(OH)_x$ islands.

conditions, the subsurface Ni of the acid-leached PtNi NPs caused only a slight CO oxidation activity enhancement as evidenced by the 130 mV shift of the major oxidation peak. When tested in alkaline solutions, however, ("alkaline CO/HER electrolysis" in Fig. 15) the hydroxide anions appeared to have caused a partial re-segregation of Ni species to the surface, evidenced by the emergence of the Ni-related CO oxidation peak around 0.5 V (marked "Ni"). The renewed presence of Ni surface species in the alkaline tested "100-acid" NPs was responsible for the activity enhancement in HER relative to pure Pt NPs.

Owing to their Ni hydroxide-rich surface, the alkaline-pretreated NPs showed a drastically increased CO oxidation activity around 0.5 V and displayed a catalytic HER activity approaching that of pure Pt in acidic conditions. Interestingly, the effect of the NP shape on surface chemistry and electrocatalysis was smaller than anticipated, and appeared negligible as compared to the drastic compositional segregation effects.

Mechanistically, the CO oxidation reaction proceeds *via* a Langmuir Hinshelwood reaction mechanism[71] according to:

$$CO + * \rightarrow CO_{ad} \tag{1}$$

$$OH^- + * \rightarrow OH_{ad} + e^- \tag{2}$$

$$CO_{ad} + OH_{ad} + OH^- \rightarrow CO_2 + H_2O + e^- + 2\,* \tag{3}$$

On CO-covered Pt and Pt alloy surfaces at potentials below 0.5 V, the CO oxidation cannot proceed due to missing active oxygenated surface species, symbolized by OH_{ad}. Hence, the adsorption of hydroxide anions and the resulting supply of active surface oxygenates represents the rate limiting step for the overall reaction. Pt NPs show an onset of OH formation more positive than 0.6 V [see Fig. 12(a)]. In the presence of Ni hydroxide islands or clusters at the NP surface, the supply of activated water or hydroxide according to reaction eqn (2) commences earlier, resulting in the observed cathodic shift of the CO oxidation [peak "Ni" in Fig. 12(b,c) and 13(a,b)]. This notion is illustrated in inset A of Fig. 15, where Ni surface species are shown to serve as active site for water and anion activation.

The HER is probably the most intensely studied reaction in electrochemistry.[78,79] Its mechanism in alkaline solution is treated as a combination of the Volmer reaction step (4) involving the dissociation of water to form a weakly adsorbed reactive H_{ad} intermediate, followed by either the Tafel recombination (5) or the Heyrovsky step (6) according to:

$$2\ H_2O + * + 2\ e^- \rightarrow 2\ H_{ad} + 2\ OH^- \quad (4)$$

$$2\ H_{ad} \rightarrow H_2 + 2\ * \quad (5)$$

$$H_{ad} + e^- + H_2O \rightarrow H_2 + OH^- + * \quad (6)$$

The initial formation of the active adsorbed hydrogen (4) from a water molecule is generally considered a slow reaction step, making the activation of water a critical process to enhance the HER catalysis in alkaline solutions. The HER in acidic solutions, in contrast, involves the initial discharge of hydronium ions, which proceeds much faster associated with high HER activity. The oxophilicity and affinity to water adsorption of Ni-hydroxides has been well established in the past.[74,80–83] Based on that, we attribute the increased catalytic HER activity of the segregated PtNi alloy NPs, in particular, of the alkaline-pretreated NPs, to the improved water dissociation at Ni hydroxide patches (clusters) on the NP surface as compared to Pt surface atoms (see inset B in Fig. 15). Pt surface atoms adjacent to the Ni hydroxide accommodate the discharged reactive atomic hydrogen (H_{ad}) and subsequently catalyze reactions (5) and/or (6). Combined, the $Pt/Ni(OH)_2$ system represents a bifunctional electrocatalyst with considerably enhanced HER activity. Furthermore, the study of the "100 acid" PtNi/C NPs demonstrated that efficient bifunctional Ni hydroxide cluster formation can originate from subsurface Ni species, and thus does not need to be carried out by deposition or coating of Ni clusters.[74]

Conclusions

Gas-phase chemical and liquid electrochemical environments can induce significant compositional and structural changes in solid alloy surfaces, especially those in the nm-size regime. Chemical and electrochemical environments form distinctly different types of interfaces to solids. For example, due to the absence or presence of an electrified double layer and mobile ions, distinctly different molecular interactions with surfaces are possible. However, our study demonstrates that gas–solid and liquid–solid interfaces also display commonalities, indicating certain unifying basic physico-chemical concepts.

Under various gaseous environments, octahedral PtNi NPs showed Ni : Pt ratios > 1 and followed the trend: $(Ni/Pt)_{O_2} > (Ni/Pt)_{Vacuum} > (Ni/Pt)_{H_2}$. Furthermore, the latter trend directly correlates with the relative Ni(II) species in our samples. Interestingly, the initial presence of NiOx species on the as-prepared samples played a pivotal role in the atomic segregation observed before the onset temperature for drastic NiOx reduction (*i.e.*, below 300 °C). In particular, when a large content of NiOx species was present on the as-prepared samples, preferential Ni segregation to the NP surface was observed, even in reducing environments.

In analogy to the oxygen gas-phase pretreatments, electrochemical hydroxide anions stabilized thick Ni hydroxide-rich surfaces during the initial potential cycle. Subsequent cycles established a surface with more balanced Pt : Ni hydroxide ratios with novel synergetic properties. Acidic electrochemical environments caused the selective removal of surface Ni species, a reactive pathway generally less common in gas-phase environments. The alkaline pretreated catalyst NPs showed significantly increased CO electrooxidation and electrocatalytic hydrogen evolution activity for which the water activation ability of surface-segregated Ni hydroxides is held responsible. Coupled with the formation of suitable adjacent Pt surface atomic ensembles, the Ni hydroxide islands represents the self-organized formation of efficient bifunctional catalysts.

We expect that similar electrochemical segregation phenomena could occur in other Pt bimetallic alloy NPs or surface reactions, especially where surface active electrolytes or reactants are involved. As an example, during the electrocatalytic reduction of molecular oxygen, the catalyst surface may transform under *operando* conditions and change its catalytic properties due to emerging synergistic bi- or multifunctionalities.

Acknowledgements

The work at TUB was supported by the U.S. Department of Energy, DOE-EERE award DE-EE0000458 *via* subcontract through General Motors. P.S. acknowledges support from the cluster of excellence in catalysis (UniCat). The work at UCF was funded by the U.S. National Science Foundation (NSF CHE-1213182).

References

1 D. L. Fedlheim and C. A. Foss, *Metal Nanoparticles: Synthesis Characterization & Applications*, CRC - Taylor Francis Group, London, 2001.
2 G. Schmid, *Nanoparticles: From Theory to Applications*, Wiley-VCH, New York, 2004.
3 D. Astruc, *Nanoparticles and Catalysis*, Wiley-VCH, New York, 2007.
4 M. Oezaslan, M. Heggen and P. Strasser, *J. Am. Chem. Soc.*, 2012, **134**, 514–524.
5 B. Coq and F. Figueras, *Coord. Chem. Rev.*, 1998, **178–180**, 1753–1783.
6 G. C. Bond, *Surf. Sci.*, 1985, **156**, 966–981.
7 M. Arai, *J. Chem. Eng. Jpn.*, 1997, **30**, 1123–1125.
8 H. Gentsch, N. Guillen, V. Hartel and M. Kopp, *Ber. Bunsen-Ges. Phys. Chem*, 1973, **77**, 1023–1023.
9 S. Mukerjee, *J. Appl. Electrochem.*, 1990, **20**, 537–548.
10 S. Eppell, G. S. Chottiner and D. A. Scherson, *Abstr. Pap. Am. Chem. S.*, 1989, **197**, 103–COLL.
11 I. Lee, R. Morales, M. A. Albiter and F. Zaera, *Proc. Natl. Acad. Sci. U. S. A.*, 2008, **105**, 15241–15246.
12 S. Linic and P. Christopher, *ChemCatChem*, 2010, **2**, 1061–1063.
13 S. Kundu and H. Liang, *J. Colloid Interface Sci.*, 2011, **354**, 597–606.
14 S. Mostafa, F. Behafarid, J. R. Croy, L. K. Ono, L. Li, J. C. Yang, A. I. Frenkel and B. Roldan Cuenya, *J. Am. Chem. Soc.*, 2010, **132**, 15714–15719.
15 B. Roldan Cuenya, A. I. Frenkel, S. Mostafa, F. Behafarid, J. R. Croy, L. K. Ono and Q. Wang, *Phys. Rev. B: Condens. Matter Mater. Phys.*, 2010, **82**, 155450.
16 L. K. Ono, D. Sudfeld and B. Roldan Cuenya, *Surf. Sci.*, 2006, **600**, 5041–5050.
17 M. T. M. Koper, *Nanoscale*, 2011, **3**, 2054–2073.
18 C. Coutanceau, P. Urchaga, S. Brimaud and S. Baranton, *Electrocatalysis*, 2012, **3**, 75–87.
19 Q.-S. Chen, F. J. Vidal-Iglesias, J. Solla-Gullón, S.-G. Sun and J. M. Feliu, *Chem. Sci.*, 2012, **3**, 136.
20 K. An and G. A. Somorjai, *ChemCatChem*, 2012, **4**, 1512–1524.
21 S. S. Cheong, J. D. Watt and R. D. Tilley, *Nanoscale*, 2010, **2**, 2045–2053.
22 M. Chen, B. H. Wu, J. Yang and N. F. Zheng, *Adv. Mater.*, 2012, **24**, 862–879.
23 F. Tao, M. E. Grass, Y. Zhang, D. R. Butcher, J. R. Renzas, Z. Liu, J. Y. Chung, B. S. Mun, M. Salmeron and G. Somorjai, *Science*, 2008, **322**, 932–934.
24 K. J. J. Mayrhofer, V. Juhart, K. Hartl, M. Hanzlik and M. Arenz, *Angew. Chem., Int. Ed.*, 2009, **48**, 3529–3531.
25 K. J. Andersson, F. Calle-Vallejo, J. Rossmeisl and L. Chorkendorff, *J. Am. Chem. Soc.*, 2009, **131**, 2404–2407.
26 J. Nerlov and I. Chorkendorff, *Catal. Lett.*, 1998, **54**, 171–176.
27 J. Nerlov and I. Chorkendorff, *J. Catal.*, 1999, **181**, 271–279.
28 R. T. Mu, Q. A. Fu, H. Xu, H. I. Zhang, Y. Y. Huang, Z. Jiang, S. O. Zhang, D. L. Tan and X. H. Bao, *J. Am. Chem. Soc.*, 2011, **133**, 1978–1986.
29 R. T. Mu, X. G. Guo, Q. Fu and X. H. Bao, *J. Phys. Chem. C*, 2011, **115**, 20590–20595.
30 K. Liu, A. Q. Wang and T. Zhang, *ACS Catal.*, 2012, **2**, 1165–1178.

31 D. Wang, H. L. Xin, R. Hovden, H. Wang, Y. Yu, D. A. Muller, F. J. DiSalvo and H. D. Abruña, *Nature Mater.*, 2012.
32 B. N. Wanjala, J. Luo, R. Loukrakpam, B. Fang, D. Mott, P. N. Njoki, M. Engelhard, H. R. Naslund, J. K. Wu, L. Wang, O. Malis and C.-J. Zhong, *Chem. Mater.*, 2010, **22**, 4282–4294.
33 S. Alayoglu, A. U. Nilekar, M. Mavrikakis and B. Eichhorn, *Nat. Mater.*, 2008, **7**, 333–338.
34 F. Tao, M. E. Grass, Y. Zhang, D. R. Butcher, J. R. Renzas, Z. Liu, J. Y. Chung, B. S. Mun, M. Salmeron and G. A. Somorjai, *Science*, 2008, **322**, 932–934.
35 Y. Ma and P. B. Balbuena, *Surf. Sci.*, 2008, **602**, 107–113.
36 G. F. Wang, M. A. Van Hove, P. N. Ross and M. I. Baskes, *Prog. Surf. Sci.*, 2005, **79**, 28–45.
37 V. Stamenković, T. J. Schmidt, P. N. Ross and N. M. Marković, *J. Electroanal. Chem.*, 2003, **554–555**, 191–199.
38 S. Deckers, F. Habraken, W. van der Weg, A. Denier van der Gon, B. Pluis, J. van der Veen and R. Baudoing, *Phys. Rev. B*, 1990, **42**, 3253–3259.
39 J. Eymery and J. C. Joud, *Surf. Sci.*, 1990, **231**, 419–426.
40 R. Najafabadi and D. J. Srolovitz, *Surf. Sci.*, 1993, **286**, 104–115.
41 A. V. Ruban, H. L. Skriver and J. K. Norskov, *Phys. Rev. B*, 1999, **59**, 15990–16000.
42 G. Wang, M. A. Van Hove, P. N. Ross and M. I. Baskes, *J. Chem. Phys.*, 2005, **122**, 024706.
43 A. U. Nilekar, A. V. Ruban and M. Mavrikakis, *Surf. Sci.*, 2009, **603**, 91–96.
44 C. W. Su, H. Y. Ho, C. S. Shern and R. H. Chen, *Surf. Sci.*, 2002, **499**, 103–108.
45 R. Mu, Q. Fu, H. Liu, D. Tan, R. Zhai and X. Bao, *Appl. Surf. Sci.*, 2009, **255**, 7296–7301.
46 T. Jacob, B. V. Merinov and W. A. Goddard, *Chem. Phys. Lett.*, 2004, **385**, 374–377.
47 C. A. Menning and J. G. Chen, *J. Power Sources*, 2010, **195**, 3140–3144.
48 Y. Ma and P. B. Balbuena, *Surf. Sci.*, 2009, **603**, 349–353.
49 J. R. Kitchin, J. K. Nørskov, M. A. Barteau and J. G. Chen, *J. Chem. Phys.*, 2004, **120**, 10240.
50 V. R. Stamenkovic, B. Fowler, B. S. Mun, G. Wang, P. N. Ross, C. A. Lucas and N. M. Markovic, *Science*, 2007, **315**, 493–497.
51 Y. Wu, S. Cai, D. Wang, W. He and Y. Li, *J. Am. Chem. Soc.*, 2012, **134**, 8975–8981.
52 P. Strasser, *Rev. Chem. Eng.*, 2009, **25**, 255–295.
53 F. R. Boer, R. Boom, W. Mattens, A. Miedema and A. Niessen, *Cohesion in metals: transition metal alloys*, Vol. 1, Elsevier Science Publishers B.V, 1988.
54 V. Stamenković, T. J. Schmidt, P. N. Ross and N. M. Marković, *J. Phys. Chem. B*, 2002, **106**, 11970–11979.
55 F. Hasche, M. Oezaslan and P. Strasser, *J. Electrochem. Soc.*, 2012, **159**, B25–B34.
56 F. Hasche, M. Oezaslan and P. Strasser, *ChemPhysChem*, 2012, **13**, 828–834.
57 M. Oezaslan, F. Hasché and P. Strasser, *Chem. Mater.*, 2011, **23**, 2159–2165.
58 F. Hasché, M. Oezaslan and P. Strasser, *ChemCatChem*, 2011, **3**, 1805–1813.
59 R. Alessio, D. B. Dell Amico, F. Calderazzo, U. Englert, A. Guarini, L. Labella and P. Strasser, *Helv. Chim. Acta*, 1998, **81**, 219–230.
60 Z. Liu, C. Yu, I. Rusakova, D. Huang and P. Strasser, *Top. Catal.*, 2008, **49**, 241–250.
61 http://www.casaxps.com.
62 S. G. Kwon and T. Hyeon, *Small*, 2011, **7**, 2685–2702.
63 M. K. Carpenter, T. E. Moylan, R. S. Kukreja, M. H. Atwan and M. M. Tessema, *J. Am. Chem. Soc.*, 2012, **134**, 8535–8542.
64 C. Cui, L. Gan, H.-H. Li, S.-H. Yu, M. Heggen and P. Strasser, *Nano Lett.*, 2012, **12**, 5885–5889.
65 S. H. Sun, C. B. Murray, D. Weller, L. Folks and A. Moser, *Science*, 2000, **287**, 1989–1992.
66 L. Gan, M. Heggen, S. Rudi and P. Strasser, *Nano Lett.*, 2012, **12**, 5423–5430.
67 L. K. Ono, J. R. Croy, H. Heinrich and B. Roldan Cuenya, *J. Phys. Chem. C*, 2011, **115**, 16856–16866.
68 A. P. Grosvenor, M. C. Biesinger, R. S. Smart and N. S. McIntyre, *Surf. Sci.*, 2006, **600**, 1771–1779.
69 M. C. Biesinger, B. P. Payne, L. W. M. Lau, A. Gerson and R. S. C. Smart, *Surf. Interface Anal.*, 2009, **41**, 324–332.
70 D. F. van der Vliet, C. Wang, D. G. Li, A. P. Paulikas, J. Greeley, R. B. Rankin, D. Strmcnik, D. Tripkovic, N. M. Markovic and V. R. Stamenkovic, *Angew. Chem., Int. Ed.*, 2012, **51**, 3139–3142.
71 M. Arenz, K. J. J. Mayrhofer, V. Stamenkovic, B. B. Blizanac, T. Tomoyuki, P. N. Ross and N. M. Markovic, *J. Am. Chem. Soc.*, 2005, **127**, 6819–6829.
72 S. Rudi, X. Tuaev and P. Strasser, *Electrocatal.*, 2012, 1–9.
73 F. Hasche, M. Oezaslan and P. Strasser, *Phys. Chem. Chem. Phys.*, 2010, **12**, 15251–15258.
74 R. Subbaraman, D. Tripkovic, D. Strmcnik, K. C. Chang, M. Uchimura, A. P. Paulikas, V. Stamenkovic and N. M. Stamenkovic, *Science*, 2011, **334**, 1256–1260.

75 M. Jayalakshmi, M. M. Rao and K. B. Kim, *Int. J. Electrochem. Sci.*, 2006, **1**, 324–333.
76 M. P. Zach and R. M. Penner, *Adv. Mater.*, 2000, **12**, 878–883.
77 D. S. Strmcnik, D. V. Tripkovic, D. van der Vliet, K. C. Chang, V. Komanicky, H. You, G. Karapetrov, J. Greeley, V. R. Stamenkovic and N. M. Markovic, *J. Am. Chem. Soc.*, 2008, **130**, 15332–15339.
78 H. Gerischer, *Z. Phys. Chem.*, 1956, **8**, 137–153.
79 B. E. Conway and J. O. M. Bockris, *J. Chem. Phys.*, 1957, **26**, 532–541.
80 Q. Han, K. R. Liu, J. S. Chen and X. J. Wei, *Int. J. Hydrogen Energy*, 2003, **28**, 1345–1352.
81 Q. Han, K. R. Liu, J. S. Chen and X. J. Wei, *Int. J. Hydrogen Energy*, 2003, **28**, 1207–1212.
82 J. Greeley, L. Kibler, A. M. El-Aziz, D. M. Kolb and J. K. Nørskov, *ChemPhysChem*, 2006, **7**, 1032–1035.
83 L. Birry and A. Lasia, *J. Appl. Electrochem.*, 2004, **34**, 735–749.

Helium droplets: a new route to nanoparticles

Adrian Boatwright,[a] Cheng Feng,[a] Daniel Spence,[a] Elspeth Latimer,[a] Chris Binns,[b] Andrew M. Ellis*[a] and Shengfu Yang*[a]

Received 30th November 2012, Accepted 2nd January 2013
DOI: 10.1039/c2fd20136d

Helium droplets are large helium clusters that are capable of picking up individual atoms and molecules and show promise as nano-reactors for the synthesis of unique nanoparticles. In particular, the sequential addition of materials of different types offers opportunities for the fabrication of novel core–shell nanoparticles that cannot be synthesised by other methods. To exploit this potential, here we have carried out a mass spectrometry investigation on metal clusters in order to establish how to control the doping conditions for the fabrication of nanoparticles in superfluid helium droplets, and in particular to develop a recipe to control core and shell ratios in the case of core–shell nanoparticles. Several types of metal nanoparticles, including pure Ag, Au and Ni nanoparticles, and Ag/Au and Ni/Au core–shell systems, have been synthesised and then removed from the helium droplets by deposition on substrates for *ex situ* investigations using high-resolution transmission electron microscopy (TEM) and X-ray photoelectron spectroscopy (XPS). The TEM imaging has been used to estimate the sizes of nanoparticles, which show a bi-model distribution under the conditions employed. We also present the first evidence that crystalline metal nanoparticles are formed by self-assembly of metal atoms in helium droplets. The XPS investigation of Ni/Au core–shell nanoparticles shows an absence of any Au 4f core-level shift that would occur on alloying of Au and Ni, which provides the first direct evidence for the successful formation of core–shell nanoparticles using superfluid helium droplets.

Introduction

Helium droplets are large helium clusters typically composed of 10^3–10^6 helium atoms,[1] although larger droplets can be made.[2] Cooled by evaporative loss of helium atoms, the droplets quickly reach a steady-state temperature of *ca.* 0.4 K for ^{4}He, which puts them firmly into the superfluid phase. When atoms or molecules collide with helium droplets they are captured with a near unity pick-up probability.[3] Given the huge thermal conductivity of superfluid helium and the speed with which energy can be dissipated by helium evaporation, any dopant added to a helium droplet will be cooled rapidly to the ambient temperature of 0.4 K. It is generally accepted that most atoms and all molecules penetrate the surface and locate themselves near the centre of helium droplets after being picked up; the only known exceptions are certain metallic atoms with diffuse electron distributions, such as the alkali metals and some alkaline earth metals, which tend to reside at the surface of helium droplets due to a strong Pauli repulsion between these metal atoms and helium atoms.[1,4]

When more than one atom or molecule is picked up by a helium droplet, agglomeration of dopants can occur leading to the formation of clusters. The agglomeration

[a]*Department of Chemistry, University of Leicester, Leicester, LE1 7RH, UK. E-mail: sfy1@le.ac.uk; ame2@le.ac.uk; Fax: +44 (0)116 252 3789; Tel: +44 (0)116 252 2127*
[b]*Department of Physics and Astronomy, University of Leicester, Leicester, LE1 7RH, UK*

of dopants is mainly driven by the intermolecular forces between the foreign species, and the frictionless migration of dopants in superfluid helium aids the aggregation process. The pick-up of dopants follows Poisson statistics and through control of the size of the helium droplets and the partial pressure of dopants in the pick-up region, the mean size of the dopant clusters, $<n>$, can be controlled across a wide range. For example, Slipchenko and co-workers have formed $(NH_3)_n$ clusters in helium droplets where the mean cluster size, $<n>$, varied across the range $2–10^4$,[5] and propyne (C_3H_4) clusters with $<n> = 10–10^4$.[6] One can easily imagine replacing these molecules with metal atoms and therefore forming large metallic clusters in the nanoscale size regime.

Only in a few cases have sizeable metal clusters in helium droplets been studied, such as Ag_n ($n\sim40$), Eu_n clusters ($n\sim20$),[7] Na_n[8] and K_n ($n\sim100$).[9] Larger metallic clusters composed of over 1000 atoms, such as Cd_n[10,11] and Mg_n,[11] have also been formed and have been confirmed by mass spectrometry. More recently, very large silver clusters composed of over 6000 Ag atoms have been synthesised.[12,13] These Ag_n nanoparticles were deposited on a solid surface by letting the doped helium droplet collide with a substrate, with the expectation that the helium will evaporate rapidly. Confirmation of successful formation of nanoparticles was provided by transmission electron microscopy (TEM). In an earlier study the same research group attempted to produce core–shell nanoparticles by the sequential addition of different types of metal atoms to the helium droplets[6] but no direct evidence for the successful formation of core–shell structures was provided.

In a recent study that was not aimed at nanoparticles, our research group has provided evidence for core–shell cluster formation in helium nanodroplets. This work employed mass spectrometry of binary clusters, such as N_2/H_2O and CO_2/H_2O.[14] The binary clusters were formed by sequential addition of the two dopants to helium droplets using two pick-up cells. By reversing the order in which the dopants were added to the helium droplets, changes in the mass spectra were observed which were explicable by core–shell formation.

Given their unique properties, helium droplets offer extraordinary possibilities for the growth of nanoparticles.[15] In particular the sequential addition of materials to helium droplets provides a potential route to core–shell nanoparticles with an almost limitless variety of compositions. Volatile materials can be added readily to helium droplets while non-volatile species, such as metal atoms, can be added by oven evaporation. Thus, for example, one could prepare novel core–shell nanoparticles with core and shell combinations that cannot be realised by any 'wet' chemical routes. Even more unusual systems, such as core–multiple-shell systems, can equally well be envisaged. Another predictable advantage for the formation of core–shell nanoparticles using helium droplets over high-temperature techniques is the minimized inter-diffusion between layers of distinct materials because the highest temperature that the nanoparticles might experience is the temperature of the deposition substrate. Other advantages could include soft-landing[12] since helium droplets act as a cushion and dissipate the kinetic energy when the doped droplets hit the deposition target. In addition the core size and shell thickness can be varied independently by controlling the partial pressures of materials in the pick-up region.

Here we present some of our initial results aimed at forming core–shell nanoparticles using superfluid helium droplets. To this end, we have constructed a purpose-built Ultra-High Vacuum (UHV) helium droplet system equipped with multiple pick-up cells in order to allow the routine fabrication of nanoparticles. After describing the experimental apparatus, we first show some mass spectrometric data for metal clusters and bimetallic systems, which we use to learn how to control the metal atom doping. The production of nanoparticles is then described, including Ag, Ni, Au, Ag/Au and Ni/Au systems, and specifically their deposition on substrates. The deposited nanoparticles are then analysed using transmission electron microscopy (TEM) and X-ray photoelectron spectroscopy (XPS). The aim of this work is to establish a systematic procedure for the controlled growth of nanoparticles inside superfluid helium droplets and to provide direct evidence for the formation of core–shell nanoparticles.

Experimental

1. Droplet formation and metal pick-up

An UHV helium droplet system has recently been constructed and is shown schematically in Fig. 1. The UHV capability is essential for the synthesis of high-purity nanoparticles without contamination from background water and other impurities, especially when large helium droplets are employed. To avoid oil contamination, the vacuum system is evacuated by Maglev turbomolecular pumps which are backed by scroll pumps.

A closed-cycle cryostat (Sumitomo RDK-415D) is used to pre-cool high-purity helium gas (99.9999%), an essential first step in forming helium droplets. Helium droplets are then formed by supersonic expansion of the helium through a pinhole nozzle with an aperture of 5 μm diameter. Typical operating stagnation pressures employed in this work were 15–20 bars and the temperature of the nozzle could be varied from 3 to 20 K. After the droplets are formed they are collimated into a helium droplet beam by passing through a skimmer with an aperture of 0.5 mm diameter. The cryostat is mounted on an XYZ manipulator, which allows precise alignment of the helium droplet beam relative to the skimmer. Beyond the skimmer are two oven evaporators (with the ability to add more) for evaporating metals into the path of the continuously flowing helium droplet beam. The ovens are made from alumina and are resistively heated by tantalum wires, which can reach a temperature up to 2500 °C, enabling the evaporation of a wide range of materials.

2. Nanoparticle deposition and analysis

Any nanoparticles formed within the helium droplets are allowed to collide with a solid substrate placed downstream between the oven evaporators and mass

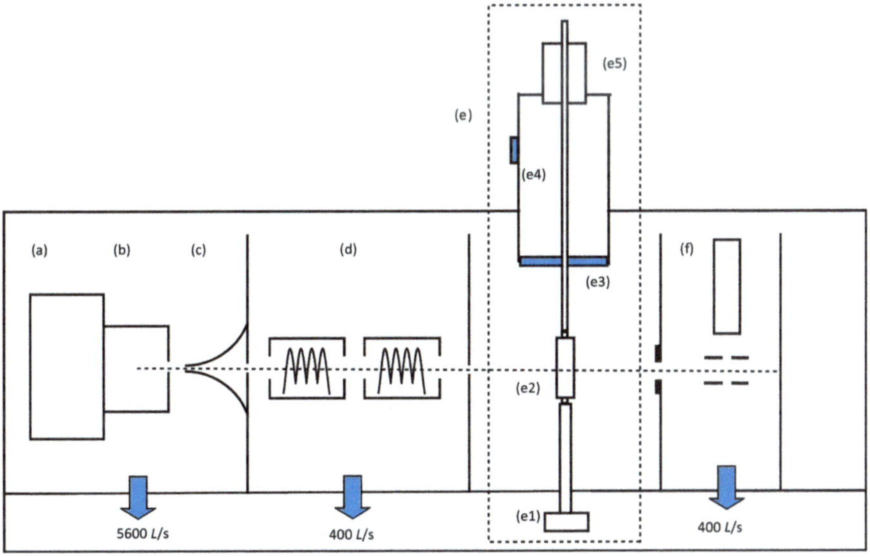

Fig. 1 Schematic of the UHV helium droplet source for the synthesis of core–shell nanoparticles. (a) XYZ manipulator; (b) cryostat; (c) skimmer; (d) oven evaporators; (e) deposition station composed of (e1) XYZ manipulator, (e2) sample holder, (e3) gate valve, (e4) load-lock chamber and (e5) magnetically coupled sample translator; (f) quadrupole mass spectrometer.

spectrometer (see Fig. 1), where a home-made deposition station is employed to manipulate the deposition targets. The deposition station is designed to load and unload the deposition targets without breaking the high vacuum. It is composed of a load-lock chamber, a magnetically coupled sample translator, a gate valve, a substrate holder and an XYZ manipulator. The sample translator is located in the load-lock chamber, while the XYZ manipulator is mounted on the UHV side opposite to the sample translator. The load-lock chamber is pumped by a small turbomolecular pump (50 L s^{-1}) and can be opened to air when the gate valve is shut. The substrate holder has a 3 $\times$ 3 array for loading multiple deposition targets. There is also a hole in the centre of the sample holder and by aligning this hole with the helium droplet beam, which can be done by monitoring the helium flow downstream using a mass spectrometer (see below), a reference position for the sample holder is established. The sample holder can then be translated to specific position by the XYZ manipulator so that the droplet beam can illuminate a selected substrate. Up to eight substrates can be loaded at a time and various types of substrates are used, depending on the intended post-deposition analysis technique. The deposition substrate holder can also be moved completely out of the path of the helium droplet beam to allow the unimpeded droplet beam to enter a quadrupole mass spectrometer (Extrel MAX 4000). The mass spectrometer is equipped with an electron impact ionization source and is used for preliminary diagnostics of the helium droplet source and to provide information on the pick-up of metal atoms.

For TEM imaging a JEOL 2100F transmission electron microscope, located at the University of Nottingham and equipped with a field-emission electron gun (FEG), was employed. The point resolution of this instrument is 0.19 nm. Both non-porous 5 nm thickness silicon grids (SiMPore, Inc.) and 3 nm ultra-thin carbon substrates on 300 mesh copper grids (Ted Pella, Inc.) were used for the deposition of Ag and Ag/Au nanoparticles. The typical deposition time for TEM imaging ranged from 30 s to a few minutes, depending on the conditions chosen for the helium droplet source and the oven evaporation of metals.

For XPS it is desirable to have a substrate that contributes little to the XPS spectrum, so highly ordered pyrolytic graphite (HOPG) is chosen. XPS measurements were performed using a Kratos AXIS ULTRA with a monochromatic Al Kα source (1487 eV) operated at a 10 mA emission current and 10 kV anode potential. The ULTRA XPS is used in fixed analyzer transmission (FAT) mode with a pass energy of 80 eV for wide scans and 20 eV for high resolution scans. The take-off angle of the photoelectron analyzer is 90° and the acceptance angle is 30° (in magnetic lens mode).

The XPS measurements focused on Ni and Ni/Au nanoparticles. These nanoparticles were produced with Ni in the first oven evaporator and Au in the second (see Fig. 1) and were collected on a deposition target for about 60 min to allow sufficient surface coverage for XPS analysis. In the outermost 10 nm of the surface the XPS can identify all elements present at concentration >0.05 atomic%, except H and He. All samples are mounted sequentially on the XPS standard bar using double-sided low tack silicone tape, which gives rise to trace contamination of silicon in the XPS spectrum. As the samples are held in place by tape they were treated as non-conducting because they were not grounded. A charge neutralizer filament above the sample surface generates low energy electrons, providing uniform charge neutralization.

Both wide survey scans and high-resolution scans were performed on each sample. The high resolution scans were charge-corrected to the main C 1s peak at 285 eV and were then quantified by normalizing the amounts of each element present using the Kratos sensitivity factors. The XPS peaks were assumed to be Gaussian, which then allows overlapping peaks to be resolved into individual components after proper subtraction of the baseline signal using the Shirley background subtraction method.[16]

XPS survey scans were performed for 10 min per sample and for each sample we selected three different sites on the substrate for XPS investigation, which are then averaged to quantify the signals. For high resolution scans data were accumulated for 5 min for each element (Ni 2p, Au 4f, O 1s, C 1s), where the energy range chosen is a function of the core electron orbital for each element type. All high-resolution scans were recorded at three locations per sample and the average is used for quantification purposes.

Results and discussion

1. Mass spectrometry of metal clusters

We start with mass spectrometry of metal clusters, from which we can learn about the performance of the helium droplet source and the temperature control of the ovens, and thus establish some of the basic requirements for the controlled pick-up of the core and shell materials.

Fig. 2 shows mass spectra of nickel clusters recorded at various oven temperatures and fixed helium droplet conditions. With a nozzle temperature of 11.0 K and a helium stagnation pressure of 15 bar behind the nozzle, an average helium droplet will consist of 10^4 helium atoms according to measurements of droplet sizes under similar conditions by Gomez *et al.*[17] At an oven temperature of 835 K the metal ion signal in the mass spectrum is dominated by Ni^+, but as the oven temperature is increased Ni_n^+ cluster cations are clearly observed in the mass spectra. Ni_n^+ cluster ions with n up to 8 can be seen in the mass spectrum when the oven temperature is increased to 932 K. A further increase of oven temperature leads to significant destruction of the helium droplets and at 966 K no helium cluster ions were observed in the mass spectrum, presumably because the addition of Ni to helium droplets has completely evaporated all of the helium. The oven temperature at which largest nickel clusters are observed in the mass spectrum, *i.e.*, at 932 K, is assumed to be close to the optimal doping conditions for a droplet composed of 10^4 helium atoms.

We now explore how to control the growth of clusters by varying the size of the helium droplets. Fig. 3 shows mass spectra from Ni-doped helium droplets recorded

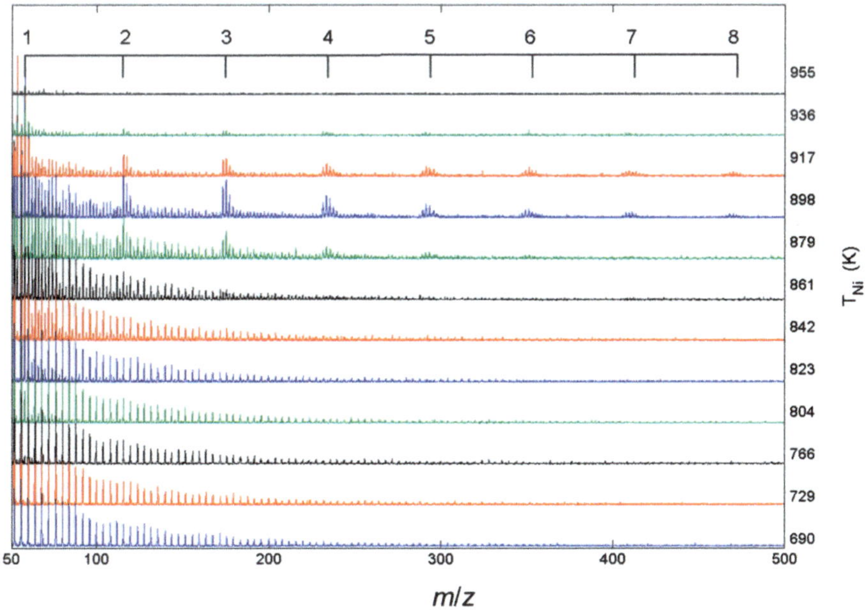

Fig. 2 Mass spectra from Ni-doped helium droplets at different oven temperatures (shown on the vertical scale on the right hand side). The Ni_n^+ cluster ions are identified by the value of n, which is shown above the tick marks at the top of the figure.

at various nozzle temperatures and a constant oven temperature of 950 K. To vary the size of helium droplets, we kept the stagnation pressure at 15 bar and the nozzle temperature, T_0, was varied. At $T_0 = 16$ K, only Ni^+ is observable in the mass spectrum, with the dimer cation becoming significant at 14 K. The maximum pick-up of Ni atoms, as deduced from the mass spectrum, is achieved at a source temperature of ~11 K. At even lower source temperatures, Ni_n^+ clusters fade away in the mass spectrum. Below 10 K, only Ni^+ is seen in the mass spectra, which is presumably due to the free effusion of Ni atoms into the mass spectrometer rather than any pick-up by helium droplets.

The disappearance of Ni_n^+ cluster ions at low source temperature can be attributed to the rapid increase of helium droplet sizes as the temperature is lowered.[17] Electron impact first ionizes a helium atom and the resulting He^+ ion can then undergo resonant hopping inside the helium droplet.[18] As the helium droplets become larger the probability of ionizing the dopant through a resonance charge transfer mechanism decreases. Theoretical modelling has predicted a mean free path of He^+ in helium droplets of 34 Å and for helium droplets composed of over 5×10^5 helium atoms ($T_0 \approx 9.5$ K), the charge transfer probability is expected to be negligibly small.[19] Hence at source temperatures below 10 K only He_n^+ clusters are expected to be seen in the mass spectra.

In another set of experiments we have formed Ag–Au binary clusters at a source temperature of 11.2 K and a stagnation pressure of 20 bar, corresponding to an expected mean helium droplet size of 1.6×10^4 helium atoms. Ag was added to the first pick-up cell and Au in the second. While maintaining a constant oven temperature of 825 K for Ag, we fine tuned the temperature of the Au-containing oven in order to optimize Ag–Au clusters in the mass spectrum (see Fig. 4). The largest Ag–Au cluster ions observed by mass spectrometry were composed of 4 Ag atoms and 7 Au atoms at an oven temperature of 873 K for Au, and so the Ag/Au ratio is roughly 1 : 1.75. A further increase of the oven temperature of Au led to destruction of helium droplets, as in the case of Ni clusters (see Fig. 2). The ratio can be fine tuned by variation of the oven temperatures, potentially allowing any desirable ratios between the quantities of core and shell materials.

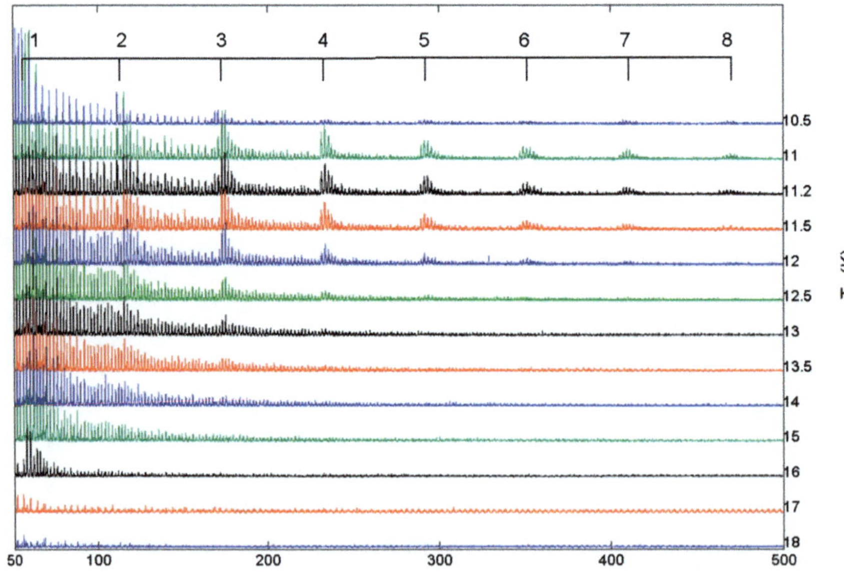

Fig. 3 Mass spectra of Ni clusters recorded at different nozzle temperatures.

2. Nanoparticles

The mass spectrometry results in the previous section are presented to show that we can control the relative quantities of materials and that mass spectrometry provides some useful diagnostic information about the pick-up process. However, the primary aim of this work is to generate and isolate nanoparticles and to judge the success or failure of this endeavour we need to move beyond mass spectrometry. Consequently, nanoparticles have been removed from the helium droplets by deposition onto TEM substrates for subsequent *ex situ* investigations.

Silver nanoparticles were initially deposited on lacey carbon substrates for TEM investigation and Fig. 5 shows a representative image obtained at a resolution of 20 nm. The silver nanoparticles were formed with an oven temperature of 850 K, a nozzle temperature of 9 K and a helium stagnation pressure of 15 bar, giving an average helium droplet size of $\sim 10^6$ helium atoms. Under these conditions we expect a supercritical helium expansion which should produce an exponential droplet size distribution rather than a lognormal distribution obtained for more gentle expansion conditions.[20] The TEM image shows a clear bi-model size distribution, *i.e.*, larger particles with an average diameter of 4 nm are accompanied by relatively smaller ones with diameters < 2 nm, with a dearth of particles between these two sizes. The reason for this bi-model distribution is not fully understood, but we can rule out the coalescence of nanoparticles for two reasons. First, the particle shapes are reasonably close to spherical, which would not be expected for particle agglomeration. Secondly, if agglomeration was the source of the larger particles then we would expect to see evidence of a grain boundary for the agglomerated particles. However, what we actually see from TEM, as discussed in more detail later, is a single crystal image (see Fig. 6c). We therefore suspect that the bi-model distribution is related to multi-centre growth of nanoparticles in the helium droplets.[13] When the droplets are sufficiently large, nucleation may take place at distinct sites within a droplet. Although speculative, we suggest that the larger particles are the product

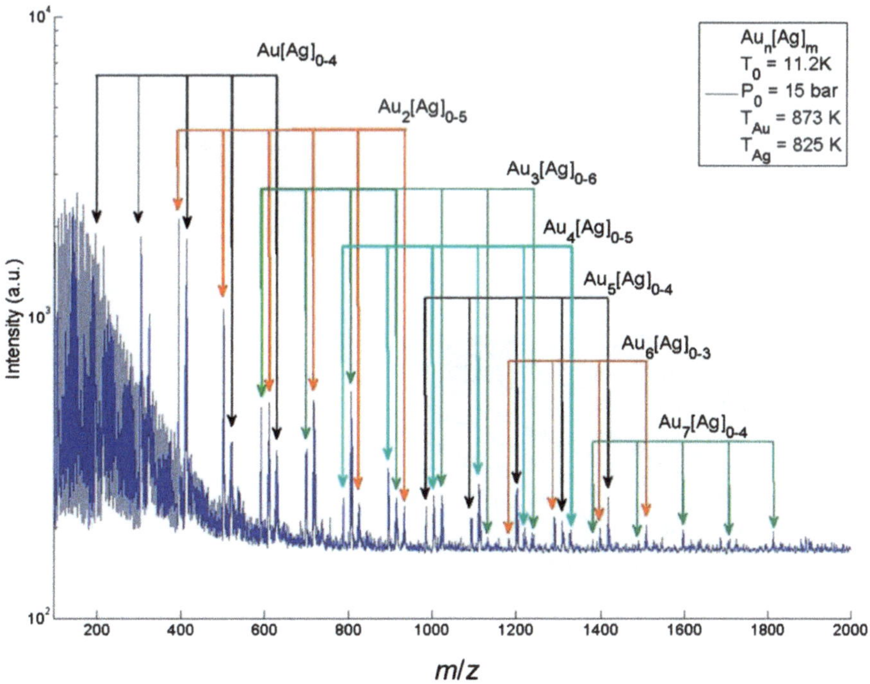

Fig. 4 Mass spectrum derived from Ag–Au binary clusters in helium droplets.

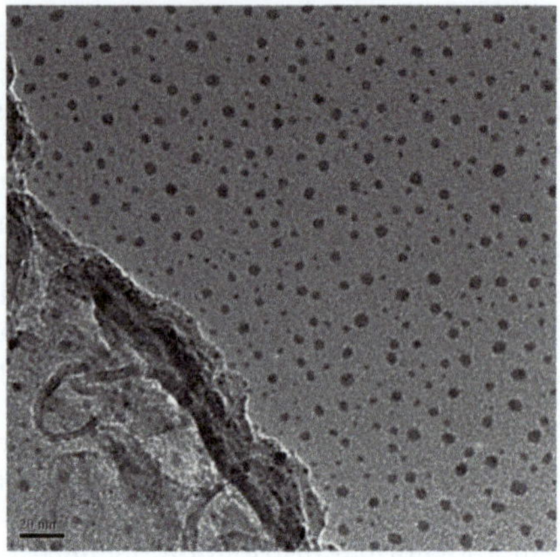

Fig. 5 TEM images of silver nanoparticles at a resolution of 20 nm.

of single-centre growth, whereas the smaller ones are the result of multi-centre growth in larger helium droplets present in the droplet beam. With this model it is necessary to assume that not all of the separate particles from multi-centre growth coalesce before deposition.

Following the mass spectrometry investigations on the Ag–Au binary clusters, we attempted to form bi-metallic nanoparticles with silver in the first oven and gold in the second. For illustration, we focus on nanoparticles grown using oven temperatures of 820 and 850 K for Ag and Au, respectively. The bi-metallic nanoparticles were produced by cooling the helium droplet source to 9 K and a stagnation pressure of 20 bar, which is expected to produce large droplets with a mean size of 2×10^6 helium atoms.[17] As the

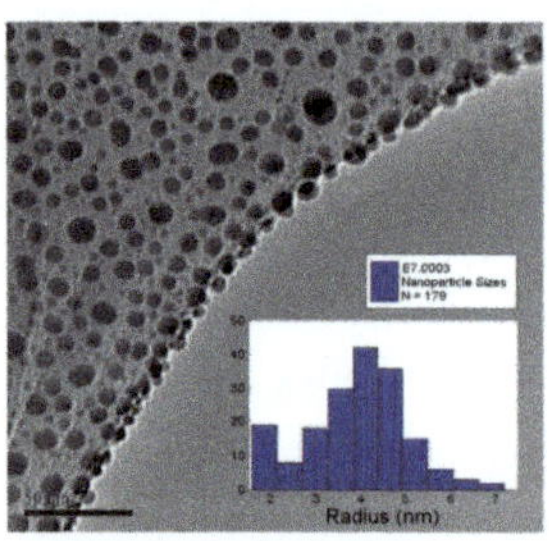

(a)

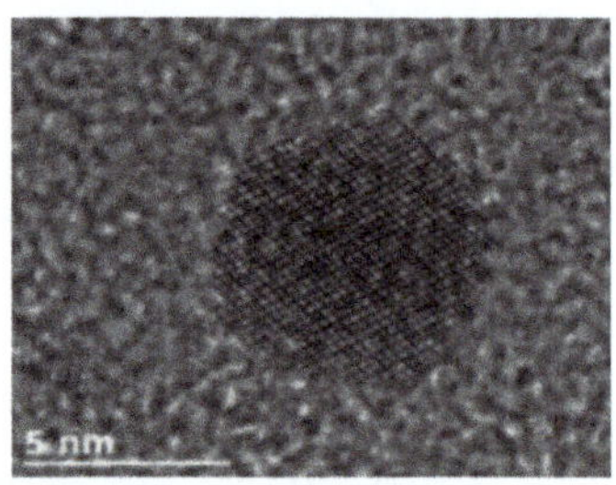

(b)

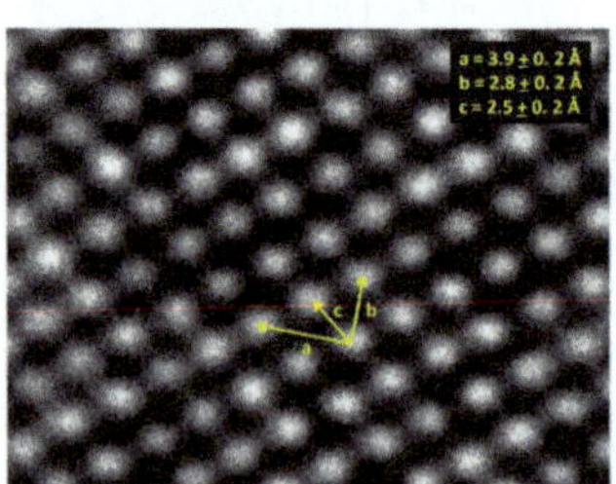

(c)

Fig. 6 Images of Ag/Au bi-metallic nanoparticles. (a) TEM image at a resolution of 50 nm; (b) high-resolution TEM image of a Ag–Au nanoparticle; (c) measurement of lattice constant. The distance $c = a/\sqrt{2}$, where a and b are the lattice constants.

TEM image in Fig. 6a shows, the Ag–Au nanoparticles produced under these conditions have a bi-model size distribution, with radii peaked at 1 nm and 4.5 nm.

From the TEM image in Fig. 6a it is unclear whether a core–shell structure has been formed by the sequential addition of silver and gold to helium droplets. Fig. 6b shows a high resolution TEM image of a specific nanoparticle, with clear evidence of a regular arrangement of atoms into a crystalline structure (see also Fig. 6c). This is the first time that the crystalline structure of a metal cluster grown in a helium droplet has been established, thus showing that metal atoms can self-assemble into a regular structure inside superfluid helium. Similar crystalline structures have been observed in high resolution images of other Ag/Au nanoparticles. Bulk gold and silver crystals adopt face-centred cubic (FCC) structures and their lattice constants are nearly identical, *i.e.*, 4.070 Å and 4.080 Å,[21] respectively. The distances *a* and *b* in Fig. 6c are the lattice constants and they are identical in a FCC structure. However, in the TEM image distance *a* is measured as 3.9 Å, which is quite close to the value in bulk gold and silver; but *b* is much smaller. The mismatch of *a* and *b* therefore indicates a tilted plane. This is consistent with the shapes of the particles seen in the TEM images, which look to be broadly spherical. The TEM images do not establish any core–shell structure, and nor would we expect them to given the very similar lattice constants of Ag and Au.

Finally, we present the synthesis of Ni/Au core–shell nanoparticles with Ni as the core and Au as the shell. Here we have employed XPS to characterise the resulting nanoparticles. In XPS spectra each element will give rise to a characteristic set of peaks in the photoelectron spectrum at kinetic energies determined by the photon energy and the respective core orbital binding energies. The presence of peaks at particular energies therefore indicates the presence of a specific element in the sample under study and their intensity is related to the concentration of the element within the sampled region. Furthermore any shift of the peak energy with respect to the bulk value indicates charge transfer or other changes in electron distribution associated with bonding. In particular, it has been shown in the case of Au layers grown on a stepped Ni(111) surface, which is a reasonably close analogue to a Ni nanoparticle surface, that alloying at the interface produces a 0.2 eV shift of the Au 4f peak to lower binding energy due to charge transfer.[22]

To investigate the core–shell structure, we first prepared pure Ni nanoparticles with Ni in the first oven evaporator and then deposited these nanoparticles on a graphene substrate. In a second experiment the gold oven was then switched on to produce Au vapour and NiAu particles were formed. In the end the Ni oven was switched off so that pure Au nanoparticles were produced. There are two possibilities for the binary nanoparticle structure, a core–shell configuration or an alloy. Although Au and Ni do not alloy at room temperature in the bulk, it cannot be ruled out in the case of nanoparticles after they have landed and warmed to ambient temperature and the first monolayer of Au on a stepped Ni surface does indeed from a surface alloy.[22] The Ni 2p and Au 4f core levels measured by XPS are shown in Fig. 7a and 7b for pure Ni and NiAu nanoparticles (Fig. 7a) and pure Au and NiAu nanoparticles (Fig. 7b)), respectively. The Ni 2p XPS spectrum is consistent with that of the recently reported XPS spectrum of pure Ni nanoparticles.[23] However, we cannot rule out some contribution from NiO because of the small shoulder on the low energy side of the Ni $2p_{3/2}$ peak and the intense satellite structure presented in the spectrum (860–865 eV). As expected, the Au 4f peak of Au and Ni/Au particles is shifted to higher binding energy when compared with the bulk value, *i.e.*, by ~0.3 eV. However, in Fig. 7b it is clear that there is no shift of the Au 4f peak commensurate with formation of a Au/Ni alloy when compared with the 4f peak of pure Au nanoparticles. Although we cannot rule out a degree of alloying at the interface, the XPS findings point to an atomic arrangement which is primarily a core–shell structure. This provides the *first* direct experimental evidence that a core–shell structure is formed and maintained when making nanoparticles in helium droplets and depositing them on a solid surface.

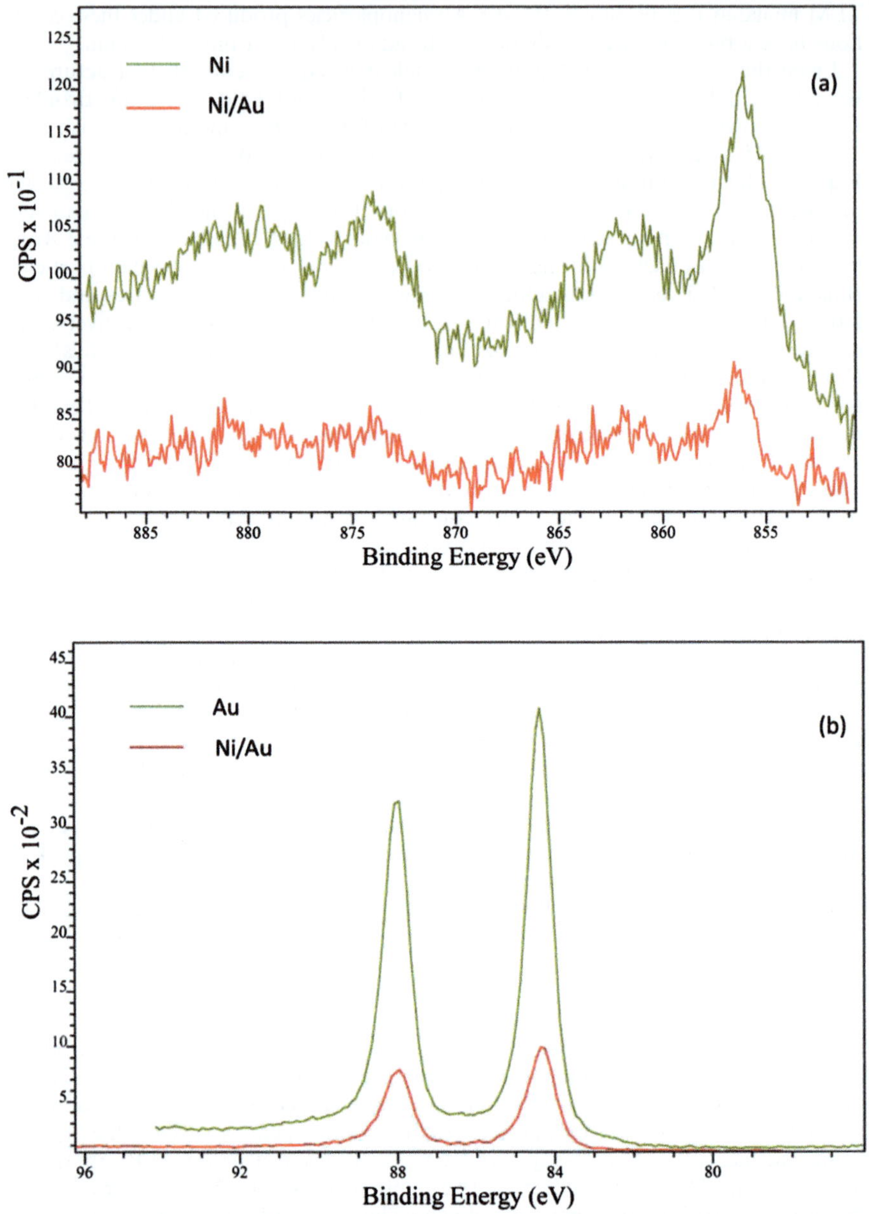

Fig. 7 XPS spectra of (a) Ni 2p core level in Ni particles and Ni/Au particles; (b) Au 4f core level in Au particles and Ni/Au particles.

Another technique that can be applied to directly measure the core–shell structure is very high-resolution TEM imaging and such experiments are planned for our nanoparticle samples using the EPSRC National Facility for Aberration-Corrected Scanning Transmission Electron Microscopy (SuperSTEM).

Conclusions

In this work we have presented our recent progress in nanoparticle synthesis using superfluid helium nanodroplets as nano-reactors. To establish a recipe for synthesising nanoparticles in a controllable way, we began with the mass spectrometry of metal-doped helium droplets, investigating three metals, both alone and in combination, *i.e.* Ag, Au, and Ni. This work has provided useful information on the doping process, particularly on the oven temperature needed to achieve maximum pick-up without significant destruction of helium droplets. The mass spectrum also indicates how the core and shell materials can be controlled in order to achieve the desired ratio of core and shell materials.

Nanoparticles, including Ag, Ni, Au, Ag/Au and Ni/Au systems were synthesised using large helium droplets, and were deposited on substrates for TEM imaging and XPS investigations. High-resolution TEM imaging has shown a crystalline structure in Ag–Au bi-metallic nanoparticles. This is an important result that provides the first clear experimental evidence that metal atoms can self-assemble into crystalline structures in helium nanodroplets. However, given the very similar lattice constants of bulk Au and Ag, the current TEM data do not provide any evidence for a core–shell structure of the Ag/Au nanoparticles, although we expect that our nanoparticles do have such a structure.

To obtain evidence for core–shell formation, we turned to XPS measurements of a different system; Ni/Au. The lack of any Au 4f shift that would result from the Ni and Au forming an alloy is taken as evidence for the successful formation and deposition of a core–shell system formed in helium nanodroplets.

Acknowledgements

SY, CB and AME wish to thank the UK EPSRC and the Leverhulme Trust for grant support, and the Advanced Microscopy Centre at the University of Leicester for providing TEM facilities. We also thank Emily Smith at the University of Nottingham for assisting XPS measurements and Michael Fay for providing TEM services.

References

1 J. P. Toennies and A. F. Vilesov, *Angew. Chem., Int. Ed.*, 2004, **43**, 2622.
2 R. E. Grisenti and J. P. Toennies, *Phys. Rev. Lett.*, 2003, **90**, 234501.
3 M. Lewerenz, B. Schilling and J. P. Toennies, *J. Chem. Phys.*, 1995, **102**, 8191.
4 O. Bünermann, G. Droppelmann, A. Hernando, R. Mayo and F. Stienkemeier, *J. Phys. Chem. A*, 2007, **111**, 12684.
5 M. N. Slipchenko, B. G. Sartakov and A. F. Vilesov, *J. Chem. Phys.*, 2008, **128**, 134509.
6 V. Mozhayskiy, M. N. Slipchenko, V. K. Adamchuk and A. F. Vilesov, *J. Chem. Phys.*, 2007, **127**, 094701.
7 J. D. Close, F. Federmann, K. Hoffmann and N. Quaas, *J. Low Temp. Phys.*, 1998, **111**, 661.
8 L. An der Lan, P. Bartl, C. Leidlmair, H. Schöbel, R. Jochum, S. Denifl, T. D. Märk, A. M. Ellis and P. Scheier, *J. Chem. Phys.*, 2011, **135**, 044309.
9 L. An der Lan, P. Bartl, C. Leidlmair, H. Schöbel, R. Jochum, S. Denifl, T. D. Märk, A. M. Ellis and P. Scheier, *Phys. Rev. B: Condens. Matter Mater. Phys.*, 2012, **85**, 115414.
10 J. Tiggesbäumker and F. Stienkemeier, *Phys. Chem. Chem. Phys.*, 2007, **9**, 4748.
11 T. Diederich, T. Döppner, T. Fennel, J. Tiggesbäumker and K.-H. Meiwes-Broer, *Phys. Rev. A: At., Mol., Opt. Phys.*, 2005, **72**, 23203.
12 E. Loginov, L. F. Gomez and A. F. Vilesov, *J. Phys. Chem. A*, 2011, **115**, 7199.
13 E. Loginov, L. F. Gomez, N. Chiang, A. Halder, N. Guggemos, V. V. Kresin and A. F. Vilesov, *Phys. Rev. Lett.*, 2011, **106**, 233401.
14 J. Liu, B. Shepperson, A. M. Ellis and S. Yang, *Phys. Chem. Chem. Phys.*, 2011, **13**, 13920.
15 S. Yang and A. M. Ellis, *Chem. Soc. Rev.*, 2013, **42**, 472.
16 J. E. Castle and A. M. Salvi, *J. Vac. Sci. Technol., A*, 2001, **19**, 1170.
17 L. F. Gomez, E. Loginov, R. Sliter and A. F. Vilesov, *J. Chem. Phys.*, 2011, **135**, 154201.
18 W. K. Lewis, M. Lindsay, R. J. Bemish and R. E. Miller, *J. Am. Chem. Soc.*, 2005, **127**, 7235.

19 A. M. Ellis and S. Yang, *Phys. Rev. A: At., Mol., Opt. Phys.*, 2007, **76**, 032714.
20 E. L. Knuth and U. Henne, *J. Chem. Phys.*, 1999, **110**, 2664.
21 A. Putnis, *Introduction to Mineral Sciences*, Cambridge University Press, 22 October, 1992.
22 S. Zafeiratos and S. Kennou, *Appl. Surf. Sci.*, 2001, **173**, 69.
23 P. Prieto, V. Nistor, K. Nouneh, M. Oyama, M. Abd-Lefdil and R. Díaz, *Appl. Surf. Sci.*, 2012, **258**, 8807.

General discussion

Professor Freund opened the discussion of the introductory lecture by Professor Campbell: For the interpretation of TPD data from nanoparticles on oxide surfaces and thin films, in particular, one has to take into account the changes in morphology and size due to migration of metal atoms across the surface and in the case of thin films into and out of the films.

Professor Campbell agreed: In many of the TPD experiments of Goodman that we re-analyzed in our *Chem. Rev.* paper,[1] he was able to do quantitative removal of both the metal nanoparticles and the oxide film by TPD. Having known Goodman, I am guessing that he would have looked for evidence for this in that TPD and in subsequent surface analysis of the remaining metal substrate. However, we cannot be sure, so this is a very good suggestion of caution with respect to interpretation. Thank you.

1 C. T. Campbell and J. R. V. Sellers, *Chem. Rev.*, 2013, DOI: 10.1021/cr300329s.

Professor Bowker questioned: In terms of your measurements of the heat of adsorption of metal atoms on growing nanoparticles at the very small limit, would we not expect oscillations in the strength of adsorption as shells of structure are completed, with a sudden decrease as very low coordination atoms are added to the particle (on a facet) and an increase as the newly growing shell has increasing average surface coordination sites available?

Professor Campbell responded: Certainly that is true, but our experiments surely have a broad particle size distribution so that our ability to observe such oscillations is poor. One might be able to see this by dosing size-selected metal clusters such as done in the later paper by Professor Anderson.

Professor Anderson asked Professor Campbell: Comparing the rate at which metal atom adsorption energies converge to the bulk, it looks like the convergence is faster on the more weakly binding oxides. Why is this?

Professor Campbell replied: The heat of metal adsorption converges to the large size limit more slowly with particle size but more rapidly with metal coverage. This latter effect is due to the fact that the more weakly binding oxides have fewer particles per unit area, so that the metal particles are much larger for a given metal coverage.

Dr Philip Davies enquired: Metal nanoparticle deposition onto catalyst supports usually involves deposition of a metal complex and reduction. Does this two step process result in different types of interaction between the particle and the surface?

Professor Campbell answered: It could lead to differences, but not if the final chemical composition of the oxide surface, the metal particle, and the oxide–metal interface are the same. Then, I would expect that metal migration within one particle is fast enough to ensure the same type of particle shape and interfacial geometry. However, getting all three chemical compositions to be the same after wet-chemical deposition as compared to preparation in the ultra-clean conditions of our UHV experiments is unlikely.

Professor Behm commented: You concluded from the particle size dependent adsorption energy of molecules on nanoparticles on the catalytic activity, assuming a Langmuir–Hinshelwood mechanism with the reacting molecules being adsorbed on the metal nanoparticles. I would like to comment on the mechanistic interpretation of the adsorption energy data, and I would like to illustrate that for CO oxidation on Au/TiO_2 or Au/CeO_2 catalysts, we think (and have provided ample evidence) that the active oxygen species in this case is surface lattice oxygen at the perimeter of the Au nanoparticle, which can be removed by reaction with CO or H_2, and the resulting O vacancy can be replenished by reaction with, *e.g.*, O_2 *via* a Au-assisted Mars–van Krevelen mechanism. In this case, one would also expect the Au-induced destabilization of the perimeter O atoms (the process does not occur in the absence of Au nanoparticles) to depend sensitively on the (particle size dependent) interaction of the O lattice oxygen at perimeter sites with the Au nanoparticles. Hence, one would expect a qualitatively comparable particle size dependence as for a reaction following a Langmuir–Hinshelwood mechanism. This example demonstrates that different mechanisms can be explained by the same adsorption data, asking for caution in the such kind of mechanistic interpretation.

Professor Campbell thanked Professor Behm for the comment and responded: I am aware of your beautiful transient kinetic data showing the important mechanistic role of this oxide lattice O at the perimeter of the Au islands. Jose A. Rodriguez has also shown evidence for the importance of perimeter sites. I also agree that the effect of Au particle size on the stability of O adatoms on the Au particles, which we observed[1] could also affect the rate of CO oxidation by this perimeter lattice O. Even if the stability of this perimeter lattice O is barely affected by the Au particle size, if it is populated by O_2 dissociation on the edges of the Au particles as you suggest, then that rate of O_2 activation should get faster with decreasing Au particle size below 6 nm. This is due to the stronger bonding of the product O adatoms to the Au, just as we proposed earlier for O_2 dissociation on any of the Au particle surfaces based on our observation of stronger O adatom bonding.[2]

1 C. T. Campbell, *Acc. Chem. Res.*, 2013, DOI: 10.1021/ar3003514.
2 V. A. Bondzie, S. C. Parker and C. T. Campbell, *Catal. Lett.*, 1999, **63**, 143–151.

Professor Hutchings commented: You have described the deposition of single metals onto supports and have determined the adsorption isotherms. Have you considered trying the deposit two metals at the same time? It may be possible to make supported alloys.

Professor Campbell responded: We have not tried to do this, but that is a good idea. Certainly it is possible to make such measurements. These could reveal if the heat of alloy formation depends on particle size and the choice of support, or how the adhesion and surface energies depend on the concentration of the 2nd metal.

Professor Hutchings then asked: In your presentation you discussed the interaction between the nanoparticles and the support. We have found that the support can have a marked effect on the way in which the nanoparticles wet the support when prepared using sol-immobilisation. We found that AuPd nanoparticles did not interact strongly with carbon surfaces but did with basic oxides such as MgO and ZnO. Perhaps this is influenced by the method of preparation, but can you comment on this?

Professor Campbell remarked: That is interesting. You allude to a good point which I failed to mention: the equilibrium particle shape can be analyzed to provide a quantitative estimate of the metal (or alloy)–support adhesion energy. The group of Claude R. Henry from Marseilles has some beautiful work in this area. The beauty of our ultra-high vacuum approach to particle deposition is that we can verify surface cleanliness and control surface composition. It may be that with wet-chemical approaches to particle deposition, unknown species decorate the support surface, or even concentrate only at the particle–support interface, which affect the adhesion energy between the particle and the support. Not knowing what these are is a disadvantage with respect to getting a deep basic understanding, but on the other hand, such effects do offer potentially powerful means to control particle stability if they can be understood at least well enough to enable good ideas about how to get the proper dopants to that interface.

Dr Philip Davies opened the discussion of the paper by Dr Beaumont: Fig. 3 of your paper shows O1s peaks at *ca.* 534 eV and 528.4 eV. The former is close to the binding energy expected for a nickel carbonyl[1] but the only examples of O1s binding energies of *ca.* 528 eV I know of are a species we identified on Cu(111) at low temperatures[2] and an "ionic" oxygen species observed on silver after high pressure exposures to $CO–O_2$ mixtures.[3] Can you comment on the assignment of this peak?

1 M. Barber, J. A. Connor, M. F. Guest, M. B. Hall, I. N. Hillier and W. N. E. Meredith, *Faraday Discuss. Chem. Soc.*, 1972, **54**, 219–226.
2 P. R. Davies, N. Shukla and D. J. Vincent, *J. Chem. Soc., Faraday Trans.*, 1995, **91** 2885–2890.
3 V. I. Bukhtiyarov, A. I. Boronin, I. P. Prosvirin and V. I. Savchenko, *J. Catal.*, 1994, **150**, 268–273.

Dr Beaumont responded: Thank you for the very helpful comments and references relating to the identity of the oxygen species seen in the XP spectra in Fig. 3. Certainly, some form of ionic oxygen (or OH) as a surface oxide is plausible here and the references seem to support that. Without further information/evidence specifically relating to these systems we can only speculate as to the exact species, but precise identification of this entity, while of interest, doesn't influence our primary message concerning the restructuring effect in these systems, although further work on understanding this could possibly provide a better understanding of how this species drives copper segregation to occur.

Dr Philip Davies then asked: In both Fig. 3 and 7 the peaks at 532.5 eV, assigned to oxygen from the silica substrate, show substantial changes in intensity during the experiment. In Fig. 3 for example, the peak disappears altogether after treatment in oxygen but reappears with greater intensity after treatment in hydrogen. Can you account for these changes in intensity? Could they be due to changes in the morphology of the bimetallic nanoparticles with perhaps an increased wetting of the silica under oxygen leading to a higher surface coverage?

Dr Beaumont replied: The greater intensity of peak (4) *vs.* (1) (before and after exposure to oxygen, but both recorded in H_2) can likely be explained by oxygen burn-off of any carbon remaining on the silica following solvent deposition of nanoparticles. The disappearance of this peak in oxygen is more complex, however, and could be attributed to one of several possibilities – the wetting suggested is certainly a possibility although I'm not aware of this phenomena occurring at 200 °C in other systems. Strong interactions of nanoparticles with mesoporous silica have very recently been seen to cause significant wetting for PVP-capped nanoparticles using XANES and SAXS in combination,[1] so changing the silica surface by oxidation could perhaps cause changes in wetting. It should be noted however that any wetting effect which occurs is not great enough to cause significant lasting agglomeration based on SEM measurements of the sample following XPS (Fig. S1). The absence of a clear silica O 1s signal in spectrum (3) could also be in part due to either: (a) the tightly packed films of nanoparticles undergoing a slight volume expansion on surface oxidation and thus obscuring more of the silica surface, or (b) the attenuation of the underlying silica substrate by physisorbed oxygen.

1 L. J. Giovanetti, J. M. Ramallo-López, M. Foxe, L. C. Jones, M. M. Koebel, G. A. Somorjai, A. F. Craievich, M. B. Salmeron and F. G. Requejo, *Small*, 2012, **8**, 468–473.

Professor Thornton noted: It may be that you generated hollow nanoparticles by your reaction sequence. Did you look for this with post-reaction electron microscopy/spectroscopy? If you have generated hollow nanoparticles does this explain the irreversibility of some processes, such as the O_2-induced Cu surface segregation in the case of the CuCo nanoparticle?

Dr Beaumont answered: We have regularly observed hollow nanoparticles during attempts to develop a reliable synthesis of spherical bimetallic CuCo nanoparticles – especially using certain solvents (mostly containing chlorine) and on long term exposure to air. The materials used in this work were specifically prepared *via* a route that mitigates this problem and does not generate hollow nanoparticles. This is confirmed by SEM recorded following the AP-XPS experiments and shown in the supporting information (Fig. S1).

Dr Yang said: In Fig. 1(b) and Fig. 4(b) there are pictures for the EDX spectra of CuNi and CuCo respectively. The CuCo shows a symmetric structure with a Cu-rich core but in case of CuNi, the Cu-rich core is clearly off-centre. Can you please comment on this?

Dr Beaumont responded: This small difference is most probably an artifact resulting from the way the measurement is obtained. The two elements are scanned sequentially and thus under the influence of the electron beam the sample heating makes it challenging to keep the material perfectly stationary and a small amount of drift in the position of the sample between measurements is to be expected.

Professor Anderson asked: You see surface segregation in your initially prepared particles that is opposite to what might be expected from the thermodynamics of the bulk metals. To what extent might this be due to the oleyl amine capping agent? Could there be different binding energies to copper, compared to the other metals, that might explain the initial structure?

Dr Beaumont answered: The initial structure could indeed be influenced by the capping agent, but likely also by the synthesis conditions which inherently are kinetically controlled (the most thermodynamically stable form is of course bulk material not nanoparticles). The fact we see segregation in CuCo is not entirely unexpected based upon their bulk miscibility, which is known to be poor. As to why Co is at the surface and Cu is not, it could be down to many synthetic parameters that govern the particle formation kinetics, including the interactions with oleylamine during the synthesis.

Professor Anderson then enquired: Do you know the fate of the capping agent – can you tell whether it all desorbs prior to your first AP-XPS experiments, or is there still a significant coverage on the particles at the start of analysis?

Dr Beaumont responded: Fig. S4 (initial heating from vacuum room temperature to 450 °C in H_2) illustrates this a little – much of the capping is definitely removed by annealing to 450 °C in H_2, which significantly enhances the intensities of Cu and Co 3p signals observed by XPS. In the absence of CO, the C 1s signal is also observed to decrease to a negligible amount and this could of course be on the metal or the silica – so we do not think there is a significant coverage – but traces can sadly never be ruled out (although the other (oxygen) species are present in overwhelmingly dominant quantities).

Professor Anderson further commented: Professor Attard's paper discusses selective binding of contaminants to reactive sites. Is it possible that even though most of the oleyl amine desorbs there might be some carbonaceous fragment that remains, perhaps in edge or other reactive sites, and that this might affect the particle reorganization? Do you see any carbon or nitrogen remaining on the samples after they have been heated in UHV?

Dr Beaumont replied: We do not see N 1s signals, we can see traces of carbon from C 1s, but some adventitious carbon will inevitably be present for any air-handled samples studied by XPS. Very trace contaminants could certainly effect reaction chemistry when used as catalysts (as per Prof. Attard's paper they can inhibit a specific site or as we observed with phosphine capping on Co they can have a longer range electronic effect[1]). We endeavour to use multiple capping agents to exclude this possibility in so far as is feasible when doing catalytic chemistry. For restructuring of nanoparticles in this way, however, we have not found any evidence of capping agents having a marked effect. For example, similar studies with PtCo used both oleic acid and PVP as (quite different) capping agents and the effects observed were similar in both cases.[2] Therefore, while it is possible it is not something we have any evidence for and the quantities of residual material are very low relative to the quantity of nanoparticle atoms undergoing restructuring.

1 V. Iablokov, S. K. Beaumont, S. Alayoglu, V. V. Pushkarev, C. Specht, J. Gao, A. P. Alivisatos, N. Kruse and G. A. Somorjai, *Nano Lett.*, 2012, **12**, 3091–3096.
2 S. Alayoglu, S. K. Beaumont, F. Zheng, V. V. Pushkarev, H. Zheng, V. Iablokov, Z. Liu, J. Guo, N. Kruse, and G. A. Somorjai, *Top. Catal.*, 2011, **54**, 778–785.

Professor Asakura remarked: Ni has larger affinity than Cu and Ni should be segregated after the O_2 treatment. I think nanoparticle may be oxidized after the treatment of O_2 at 350 °C. However what about the binding energies of CuO and NiO? Is there any possibility that the XPS of NiO overlaps that of Cu?

Dr Beaumont answered: The enthalpy of formation of NiO is indeed more favorable than CuO (−244 *vs.* −157 kJ mol^{-1}), but, as discussed in the manuscript, it is copper that appears to segregate to the surface. Therefore, the availability of kinetically accessible structures as well as pure thermodynamics must play a role in determining the surface structure. The same observation was made for Co *vs.* Cu, the enthalpy of formation of CoO also being greater than for CuO. With respect to potential overlap of binding energies between Ni and Cu, oxidation of all 3d transition metals typically results in small changes of the order of 1–2 eV, a change of 8 eV is not likely to be feasible/there is no possibility of the changes resulting from an overlap of NiO and Cu 3p levels as they are too far apart.

Professor Hutchings commented: Your preparation method is able to make CoCu nanoparticles with very defined compositions and particle sizes. At present the particle size is 11 nm. Supported Co nanoparticles were used as Fischer–Tropsch catalysts, and early catalysts did use Cu as a catalyst component. However, the latest studies indicate that Co nanoparticles need to be *ca.* 6 nm in diameter. Can you make smaller nanoparticles around 6 nm in size and have you considered testing your materials for Fischer–Tropsch synthesis and other hydrogenation reactions?

Dr Beaumont replied: Yes – CuCo has for some time held considerable promise as a catalyst for syngas conversion (*e.g.* US Patent No: 6,680,272) – especially if it is desirable to obtain higher alcohols, since many people anticipate it would combine the cobalt chemistry of Fischer–Tropsch and the "methanol synthesis" chemistry of copper. Experiments to try and use these materials to investigate this are a long term goal. Better synthetic control in our hands has been obtained for smaller (~2 nm) and larger (~10 nm) particles, however I believe the size effect around 6 nm you are referring to is for mass activity – plotted as TOF the chemistry is just less mass active and no different (in TOF or selectivity) for particles above 6 nm, so an investigation to understand the mechanistic questions with 11 nm particles should be just as useful. A larger problem with using these materials is the unknown effect of traces of amine based on our previous discovery of traces of phosphorous poisoning CO_2 hydrogenation reactions[1] and suggestions that NH_3 and HCN act as catalyst poisons and selectivity promoters in Co-catalyzed Fischer–Tropsch.[2] It may be possible to see meaningful changes if both Co and CoCu are prepared with oleylamine capping agents, however it would clearly be preferable to develop a synthetic method excluding N and P altogether, which is something we are continuing to work on.

1 V. Iablokov, S. K. Beaumont, S. Alayoglu, V. V. Pushkarev, C. Specht, J. Gao, A. P. Alivisatos, N. Kruse and G. A. Somorjai, *Nano Lett.*, 2012, **12**, 3091–3096.

2 S. C. Leviness, C. J. Mart, W. C. Behrmann, S. J. Hsia and D. R. Neskora, 1998, WO 98/50487.

Dr Boscoboinik communicated: This work is really interesting and of extremely high importance. There are just a couple of questions that came up when reading it that are not addressed in the paper. In Fig. 3 (XPS of the O1s region for the CuNi nanoparticle on the native oxide of the Si wafer), for the first spectrum from bottom to top there is already a peak which presumably comes from the O in the support. After the cycle of exposing the system to CO, O_2 and H_2 again this O1s peak grows significantly. While the reason for this is not discussed in the paper the author mentioned during the oral discussion that it could be related to having carbon on the surface originally which was eliminated later during the O_2 exposure, thus increasing the O1s peak.

The amount of carbon that would produce such an attenuation in the first spectrum would have to be quite significant. Do you have XPS spectra of carbon corresponding to these O spectra? Can you discard that the increase in the intensity of the O peak is related to sintering of the particles? If the change of the intensity of the O peak was due to burning the carbon contamination, wouldn't the large O peak be present in spectrum (3) of Fig. 3 already after O_2 exposure? (It shows up only after exposure to H_2 at 450 °C, spectrum (4) of Fig. 3).

The same question goes for Fig. 7, for the CuCo particles.

Again, this is a fascinating piece of work and of high relevance and a more detailed description of the spectra addressing these questions will be very useful.

Dr Beaumont communicated in reply: While it is agreed that the change due to carbon would require a significant amount, there may well be a layer of carbon present on the underlying silica following the deposition from solvent – and this is less likely to be removed in H_2 than from the metal due to a lack of hydrogen activation. Additionally, it must be remembered that here the O 1s spectra were obtained with an incident photon energy of 800 eV (rather than a conventional laboratory source), so the layer would not need to be very thick to get a significant attenuation because of the much smaller IMFP due to the lower kinetic energy of the photoelectrons.

We can exclude sintering fairly confidently on the basis of SEM experiments performed following the AP-XPS measurements that show the particle size and shape are preserved (Fig. S1 – both CuNi and CuCo).

The absence of this feature in the 3rd spectrum (the one during oxygen exposure) must be attributable to either wetting as suggested by Dr Philip Davies or possibly is in part due to attenuation as the surface layer expands on oxidation (discussed in more detail in the response to the question by Dr Philip Davies).

Professor Asakura opened the discussion of the paper by Dr Lu: I believe the analysis of EXAFS on bimetallic systems is still challenging. Could you explain why Pd–Au distances determined from the Pd K-edge and Au L3-edge were so different in Table 1 and 2? Those should be the same.

Dr Lu responded: First of all, it is evident that EXAFS takes the average over an ensemble of very small particles having an average size only in the range of 1–2 nm. Thus, the surface layer of atoms will have a marked influence on this average that includes the bulk structure as well as the surface structure of the particles. Atoms residing on the surface may have different distances to their neighbors. So we fully agree with the statement that the EXAFS analysis is quite challenging and other methods, *e.g.* TEM, are necessary to supplement these data. We hence did not argue on the atom–atom distances but rather on the coordination number of a given atom in an alloy.

In addition, it is worth noting that the error given in the tables for the Pd–Au distances R is only the uncertainty of the fit value. The uncertainty of atom–atom distances in both tables is $(\sigma^2)^{(1/2)}$. Moreover, there is a slight enhancement of the Pd atoms at the surface together with a concomitant enrichment of the Au atoms in the shell. Hence, EXAFS done at the Pd-edge probes the situation more on the surface while experiments at the Au-edge put more weight at the core of the particles. This may lead to slightly different atom–atom distances. In a truly random alloy, on the other hand, both values should agree. Indeed, in the $Au_{75}Pd_{25}$ alloy which is close to a random alloy structure, the Pd–Au distance in Table 1 (2.71 ± 0.089 nm) and in Table 2 (2.79 ± 0.083 nm) agrees within the limits of error. In case of the samples $Au_{50}Pd_{50}$ and $Au_{25}Pd_{75}$, which both exhibit an enrichment of Pd atoms at the surface, the deviation is beyond the margin of experimental error.

Professor Asakura then remarked: You explained the contraction of AuPd nanoparticles based on Vergard's law. But the Au–Au distance was shorter than the distance derived from Pd lattice constants. Is there any possibility for bond distance contraction due to the nano size effect?

Dr Lu answered: We measured the lattice constant of our Au/Pd nanoalloys by high-resolution TEM (HR-TEM) which we believe leads to the most precise values. Since the particles are of comparable magnitude, the lattice contraction due to the small size should shift all values in the same directions. Our EXAFS measurements on the neat Au and Pd nanoparticles point indeed to a contraction of *ca.* 1.5% for the size of 2 nm and of 0.8% for a size of 4 nm. This is in good agreement with data taken from the literature. In order to elucidate this point further, the lattice contraction of

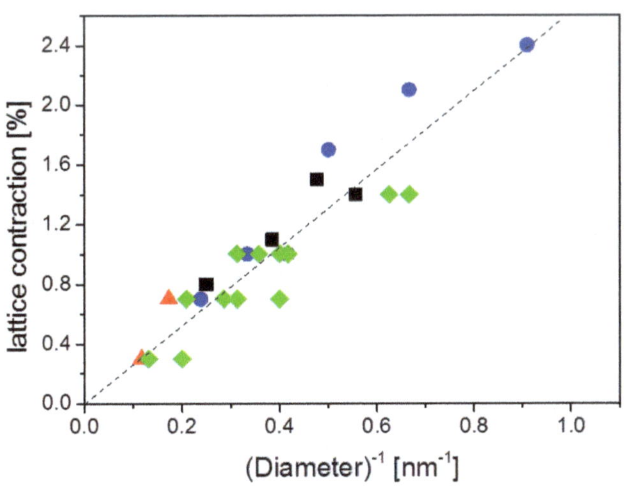

Fig. 1 Comparison of Au nanoparticles with different sizes and the lattice contraction from EXAFS experiments: Filled triangles (ref. 3), spheres (ref. 1), and squares (our study) show results of polymer-stabilized particles. Filled diamonds (ref. 4) represent nanoparticles supported by SiO_2, Al_2O_3, TiO_2, CeO_2, Nb_2O_5 and ZrO_2. The dashed line shows a mean square fit by a surface tension of 3.5 N m^{-1} (ref. 1).

our gold nanoparticles has been plotted against the reciprocal size of the particles in Fig. 1 according to Balerna *et al.*[1] Our own data are in good agreement with the data taken from the literature.[2–4]

1 A. Balerna, E. Bernieri, P. Picozzi, A. Reale, S. Santucci, E. Burattini and S. Mobilio, *Surf. Sci.*, 1985, **156**, 206–213.
2 H. Tsunoyama, H. Sakurai, N. Ichikuni, Y. Negishi and T. Tsukuda, *Langmuir*, 2004, **20**, 11293–11296.
3 W. Szczerba, H. Riesemeier and A. F. Thünemann, *Anal. Bioanal. Chem.*, 2010, **398**, 1967–1972.
4 J. T. Miller, A. J. Kropf, Y. Zha, J. R. Regalbuto, L. Delannoy, C. Louis, E. Bus and J. A. van Bokhoven, *J. Catal.*, 2006, **240**, 222–234.

Professor Ertl said: You mentioned that the Pd is non-metallic. So, what is the actual chemical state of your surface? Is it oxidic?

Dr Lu responded: Non-metallic Pd atoms on the surface of Au/Pd nanoalloys are a general phenomenon seen for particles differing widely in size. For our EXAFS measurements we found that nearly 70% of Pd in the system is non-metallic, which is mainly located on the surface of nanoalloy particles. Up to now, we have no clear indication on the surface groups bound to these atoms. These non-metallic Pd species might be PdO as the surface of Pd can be easily oxidized for samples stored at ambient conditions, especially for the particles smaller than 2 nm.[1–3] However, we tried to exclude oxygen during the synthesis of the particles as much as possible but still found the non-metallic Pd species. Hence, at the present moment we have no clear indication which surface group might be bound to the Pd atoms. Further work is necessary here in order to elucidate this important question.

1 Y.-S. Ho, C.-B. Wang and C.-T. Yeh, *J. Mol. Catal. A: Chem.*, 1996, **112**, 287.
2 M. Ohashi, K. D. Beard, S. Ma, D. A. Blom, J. St-Pierre, J. W. Van Zee and J. R. Monnier, *Electrochim. Acta*, 2010, **55**, 7376.
3 T. Wu, W. E. Kaden, W. A. Kunkel and S. L. Anderson, *Surf. Sci.*, 2009, **603**, 2764.

Professor Somorjai commented: Catalysis that occurs at the solid–liquid interface like in electrochemistry has a 1000-fold increase in density when compared with the same reaction at the catalyst–gas interface. Have you tried to carry out the same reaction at the catalyst–liquid and catalyst–gas interfaces?

Dr Lu replied: This is indeed a central question and we know of no work in which catalytic studies on metal nanoparticles are done both on the interface to the liquid as well as to the gas phase. Catalytic reactions on the surface of metal nanoparticles seem to be much better understood in the gas phase than in the liquid phase. This seems to be due to a lack of model reactions that work in the liquid phase. We believe the catalytic reduction of *p*-nitrophenol by borohydride has exactly filled in this gap. The kinetics of this reaction can be analysed by spectroscopic methods with high accuracy and there are a huge number of papers reporting kinetic constants obtained for various nanoparticles and alloys (see our recent review[1]). Up to now a consensus has emerged that the reduction of *p*-nitrophenol is a surface reaction.[2] No consensus, however, has been reached on the delay time that has been observed by many groups.[1] Following work done by Zhou *et al.* on Au-nanoparticles[3] we have presented clear evidence that this induction time is due to surface restructuring.[4] However, no direct proof by surface-sensitive methods has been given so far and further research along this line is necessary.

1 P. Herves, M. Perez-Lorenzo, L. M. Liz-Marzan, J. Dzubiella, Y. Lu and M. Ballauff, *Chem. Soc. Rev.*, 2012, **41**, 5577–5587.
2 S. Wunder, F. Polzer, Y. Lu, Y. Mei and M. Ballauff, *J. Phys. Chem. C*, 2010, **114**, 8814–8820.

3 X. Zhou, W. Xu, G. Liu, D. Panda and P. Chen, *J. Am. Chem. Soc.*, 2010, **132**, 138–146.
4 S. Wunder, Y. Lu, M. Albrecht and M. Ballauff, *ACS Catal.*, 2011, **1**, 908–916.

Professor Behm commented: The EXAFS-based data on the surface composition could be easily complemented by IR spectroscopy using adsorption of probe molecules such as CO, which often gains valuable information on the surface composition.

Dr Lu thanked Professor Behm for the suggestion and noted: The only concern is that in our study spherical polyelectrolyte brushes have been used as carrier system for the immobilization of AuPd nanoalloy particles. The presence of a polymeric carrier may disturb the adsorption for the IR spectroscopy measurement. But it is definitely worth trying this method and combining it with the EXAFS data.

Professor Hutchings said: If you consider the micrograph in Fig. 1 in the manuscript it appears that there is a range of particle sizes prepared by your method. I would suggest that aberration-corrected microscopy would be best to observe if very small clusters are also present with your preparation method. Have you investigated the composition of these individual nanoparticles as it is most likely there will be a size-dependent variation in composition? We have observed this for AuPd nanoparticles made by sol-immobilisation.[1] In this paper and in the previous paper[2] you can only refer to average compositions as the methods used average over many particles. It would be interesting to determine if there is a size-dependent variation in composition with your preparation method.

1 J. Kaiser, W. Szczerba, H. Riesemeier, U. Reinholz, M. Radtke, M. Albrecht, Y. Lu and M. Ballauff, *Faraday Discuss.*, 2013, DOI: 10.1039/c3fd20132e.
2 J. Kaiser, L. Leppert, H. Welz, F. Polzer, S. Wunder, N. Wanderka, M. Albrecht, T. Lunkenbein, J. Breu, S. Kümmel, Y. Lu and M. Ballauff, *Phys. Chem. Chem. Phys.*, 2012, **14**, 6487–6495.

Dr Lu replied: We agree that a change in composition that depends on size presents a serious issue. However, the size distribution of our Au/Pd particles is rather narrow and symmetric.[1] No indication of Ostwald ripening or aggregation is seen in these systems. In order to stress this point further we display in Fig. 2 the size distribution of the three Au/Pd nanoalloys under consideration in ref. 1. These findings are in full agreement with our recent studies on Au/Pt nanoalloys[2,3] which are excellent

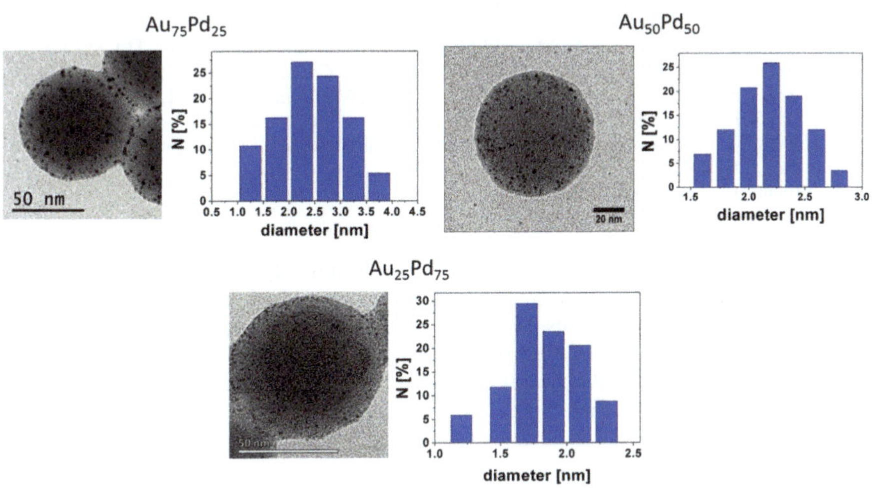

Fig. 2 TEM images for $Au_{75}Pd_{25}$, $Au_{50}Pd_{50}$, $Au_{25}Pd_{75}$ composite particles, and particle size distribution histograms of metal nanoparticles evaluated from the corresponding TEM images (ref. 1).

oxidation catalysts. We point out that we use a special system to generate and immobilize the nanoparticles. We used spherical polyelectrolyte brushes (SPB) that consist of a colloidal PS core onto which long linear polyelectrolyte chains are densely grafted. Metal ions are strongly localized within the polyelectrolyte brush layer and reduction with $NaBH_4$ leads to the formation of nanoalloy particles. The firm binding of the nanoparticles to the SPB prevents any further coagulation or coarsening of the nanoparticles. Also, the surface layer of the SPB in which the metal ions are bound has a thickness of only 50–60 nm.[2] Hence, diffusion of the reducing agent BH_4^- is required only over this distance leading to a fast nucleation.

1 J. Kaiser, L. Leppert, H. Welz, F. Polzer, S. Wunder, N. Wanderka, M. Albrecht, T. Lunkenbein, J. Breu, S. Kümmel, Y. Lu and M. Ballauff, *Phys. Chem. Chem. Phys.*, 2012, **14**, 6487–6495.
2 M. Schrinner, S. Proch, Y. Mei, R. Kempe, N. Miyajima and M. Ballauff, *Adv. Mater.*, 2008, **20**, 1928–1933.
3 M. Schrinner, M. Ballauff, Y. Talmon, Y. Kauffmann, J. Thun, M. Moller and J. Breu, *Science*, 2009, **323**, 617–620.

Mr Zhao queried: What is the reason for your choice of $NaBH_4$ as a hydrogen resource, as you mentioned in your presentation? Is there any influence of boron species on catalytic performance? Have you consider using other resources like formic acid or hydrogen gas?

Dr Lu responded: In all mechanistic studies[1–4] conducted by us on the catalytic activity of nanoparticles, the reduction of *p*-nitrophenol by borohydride has been chosen as a test. This reaction proceeds in water at ambient conditions and can be analyzed by spectroscopic methods with highest accuracy. In the meantime, there is a huge bulk of kinetic data available for this reaction. Recently, we argued that this reaction is a true model reaction that allows us to compare the catalytic activity of different nanoparticles in aqueous environment.[5] The interaction of borohydride ions with metallic nanoparticles is well studied[6] and the apparent rate constants can be obtained with unsurpassed precision. Moreover, we have presented clear evidence that surface restructuring is necessary to render Au nanoparticles catalytically active. A survey of the literature on this reaction may be found in our review.[5] Up to now, we have not tried with other resources like formic acid or hydrogen gas. It is certainly the next step to go when considering nanoalloys as catalysts in more practical applications.

1 Y. Mei, G. Sharma, Y. Lu, M. Ballauff, M. Drechsler, T. Irrgang and R. Kempe, *Langmuir*, 2005, **21**, 12229–12234.
2 S. Wunder, F. Polzer, Y. Lu, Y. Mei and M. Ballauff, *J. Phys. Chem. C*, 2010, **114**, 8814–8820.
3 S. Wunder, Y. Lu, M. Albrecht and M. Ballauff, *ACS Catal.*, 2011, **1**, 908–916.
4 S. Wu, J. Dzubiella, J. Kaiser, M. Drechsler, X. Guo, M. Ballauff and Y. Lu, *Angew. Chem., Int. Ed.*, 2012, **51**, 2229–2233.
5 P. Herves, M. Perez-Lorenzo, L. M. Liz-Marzan, J. Dzubiella, Y. Lu and M. Ballauff, *Chem. Soc. Rev.*, 2012, **41**, 5577–5587.
6 G. Guella, C. Zanchetta, B. Patton, and A. Miotello, *J. Phys. Chem. B*, 2006, **110**, 17024–17033.

(33:[31]31) **Professor Bowker** said: You suggest that very small alloy nanoparticles are random mixes as suggested by theory – but presumably this is on the basis of bulk thermodynamic considerations. Here we have very small nanoparticles (~2 nm) and much of the particle is surface, so surely such considerations do not apply. Also you indicate that much of the Pd is oxidised. If that is the case and there is such an intimate mix between Au and Pd, then there should be Au–O bonds too – do you see this in the EXAFS?

Dr Lu answered: As discussed in our answer to the first question raised by Professor Asakura, our arguments are mainly based on precise measurements by

HR-TEM and not on thermodynamic considerations. Theoretical evidence shows indeed that the thermodynamics of such nanoalloys may be more complicated than in macroscopic bulk phases. Thus, from classic molecular dynamics (MD) simulation,[1,2] stable ordered structures for Au–Pd bimetallic particles might be the Pd-core/Au-shell, random solid-solution and eutectic-like structure. Moreover, our analysis of the particles by EXAFS leaves no doubt that we are dealing with alloys.

EXAFS presents a sensitive tool to study the oxidation state of metal alloys. In our study, we found that the $Au_{50}Pd_{50}$ (size: ~2.2 nm) and $Au_{25}Pd_{75}$ (size: ~1.7 nm) alloys exhibit a slight enrichment of Pd at the surface and Au in the core, while for the $Au_{75}Pd_{25}$ particle (size: ~2.5 nm) a structure which is closer to a random alloy is obtained. This finding is in agreement with recent work done on Au/Pd nanoalloys prepared in widely different ways. The respective references are given in our article. In addition, some EXAFS studies of AuPd nanoalloys indicate the non-metallic Pd at the Pd absorption edge.[3–6] This finding is usually assigned to an oxidation of the Pd surface atoms. Slight enrichments of Au atoms in the core are obtained. Hence, non-metallic Pd atoms localized on the surface of the nanoparticles are a general phenomenon in these systems. However, we do not find the non-metallic gold at the Au L3-edge. We iterate that the non-metallic Pd species are mainly located on the surface of the particles, the metallic Pd atoms are localized within the core of the particles. Hence, we think that the finding of non-metallic Pd atoms on the surface of Au/Pd nanoalloys is a surface effect that merits further investigation.

1 H. B. Liu, U. Pal, A. Medina, C. Maldonado and J. A. Ascencio, *Phys. Rev. B*, 2005, **71**, 075403.
2 S. J. Mejía-Rosales, C. Fernández-Navarro, E. Pérez-Tijerina, D. A. Blom, L. F. Allard and M. José-Yacamán, *J. Phys. Chem. C*, 2007, **111**, 1256–1260.
3 F. Liu, D. Wechsler and P. Zhang, *Chem. Phys. Lett.*, 2008, **461**, 254–259.
4 S. Marx and A. Baiker, *J. Phys. Chem. C*, 2009, **113**, 6191–6201.
5 C. Evangelisti, E. Schiavi, L. A. Aronica, A. M. Caporusso, G. Vitulli, L. Bertinetti, G. Martra, A. Balerna and S. Mobilio, *J. Catal.*, 2012, **286**, 224–236.
6 F. Krumeich, S. Marx, A. Baiker and R. Nesper, *Z. Anorg. Allg. Chem.*, 2011, **637**, 875–881.

Professor Freund said: I was wondering about how uniquely one can model EXAFS spectra of bimetallic particles, given the small energy differences between different internal distributions of constituents.

Professor Ballauff replied on behalf of Dr Lu: We have discussed the various aspects of EXAFS in our answers to questions raised by Professors Asakura, Ertl and Bowker. Quite evidently, EXAFS is just one method to study nanoparticles and the results must be supplemented by other methods as *e.g.* high-resolution TEM. In general, there is a lot of literature devoted to EXAFS studies of nanoparticles and nanoalloys. In particular, the Au/Pd system has been studied quite frequently in the recent years because these investigations can be done at the Au-edge as well as at the Pd-edge conveniently (see the list of references in our paper). Having the information of two edges reduces the possible degrees of freedom when trying to fit a structural model to EXAFS data. Summarizing the results obtained by literature and by us it can be stated that the enrichment of Au in the cores of the particles and of Pd in the shell is a fact seen by many studies so far. Moreover, the non-metallic character of the Pd atoms localized at the surface of the particles is another established fact. Hence, EXAFS furnishes unique and valuable information on nanoalloys that is not easily available by other methods.

Dr Yang opened the discussion of the paper by Professor Attard: The authors have applied SERS technique in the *in situ* study of catalysis which is very interesting. The scattering cross-sections of Au nanoparticles are large for Au nanoparticles of diameter > 50 nm but for small nanoparticles, the scattering cross-sections are very small, which is clearly the reason that nanoparticles with a gold core of 55 nm have been applied in the paper. As small particles with a diameter no more than 5 nm is often the best for catalysis, I just wonder are there some other means that apply the SERS method to the catalytically active small particles?

Professor Attard replied: I am not sure is the honest answer! It is certainly true that for Pt (not normally considered a good SERS substrate) that nanoparticles of around 5 nm do give rise to reasonable SERS intensity data[1] although the signal is rather small. There is also work in the literature demonstrating that 5 nm diameter Au nanoparticles can actually give rise to a SERS signal.[2] We are presently modifying our silica-coated 55 nm gold core shell nanoparticles by depositing "small" gold nanoparticles onto the outer oxide surface. In this way, it may indeed be possible to monitor by SERS the surface chemistry of small metal nanoparticles that are attached to the oxide support of interest since the underlying Au core will act as the plasmonic material enhancing the vibrations of molecules adsorbed on the small gold particles attached to the oxide.

1 F. J. Vidal-Iglesias, J. Solla-Gullón, J. M. Orts, A. Rodes, J. M. Pérez and J. M. Feliu, *J. Phys. Chem. C*, 2012, **116**, 10781–10789.
2 K. Mougin, S. Darwich, L. Vidal, and H. Haidara, *Adv. Phys. Chem.*, 2010, DOI: 10.1155/2010/289371.

Dr Lai commented: Fig. 6 of the paper shows that adding more than 5 ml of PVP does not result in a further decrease in the amount of {100} × {111} defect sites. Based on that, one would expect the catalyst treated with 10 ml PVP to behave similarly to the catalyst treated with 12 ml PVP. Fig. 9b, however, shows that there is still a significant difference in selectivity between these two catalysts, while in Fig. 8b they behave similarly. My interpretation of this is that even though the total amount of defect sites is the same for the two catalysts, the nature of the remaining defect sites is still different, which can impact different reactions differently. Given that, could that be an avenue to further fine-tune the selectivity for different reactions, maybe in conjugation with varying the chain lengths of the PVP employed?

Professor Attard replied: It may well be! We have not tried different PVP chain lengths in our catalytic studies although it does not seem to make much of a difference so far as adsorption on platinum (as measured by CV) is concerned. Another point to consider though might be competition for metal sites by the substrate and the PVP. A recent paper[1] demonstrates that such a balance can lead to marked changes in selectivity for octyne hydrogenation in competition with primary alkylamine capping ligands.

1 S. G. Kwon, G. Krylova, A. Sumer, M. M. Schwartz, E. E. Bunel, C. L. Marshall, S. Chattopadhyay, B. Lee, J. Jellinek and E. V. Shevchenko, *Nano Lett.*, 2012, **12**, 5382–5388.

Professor Behm said: I very much like the SHINERS concept. I wonder, however, why this concept has exclusively been used to raise the sensitivity in Raman spectroscopy, and never in IR spectroscopy, *e.g.*, as an alternative to SEIRAS, allowing single crystal measurements also in ATR configurations. Is there any reason why this concept could not be transferred to IR spectroscopy?

Professor Attard responded: I see no reason at all and I think that it is a very good idea.

Professor Dr Suzer remarked: Although I find the paper quite interesting and novel, I have some comments about the origin of the broad peaks around 1580 cm^{-1}, and to a lesser extent around 1300 cm^{-1}, in almost all of the presented Raman spectra.

The lasers used, for realizing the SERS effect and enhancing the peak intensities by focusing the EM on the analytes, can also impart significant energy onto the nanoparticles to heat and cause thermal and/or photodegradation. As a result, the so-called D and G bands around the regions mentioned above are easily created during analyses of carbon materials. These bands are usually broad and must be distinguished from the real SERS bands. Moreover, the authors attach some effects to the position/shift of the 1580 cm^{-1} band, such as; "...strong interaction of the double bond with the Pt surface, *etc.*"

I suggest that these bands should be indicated by asterisks in all the figures, the nature of their existence, and measures taken for identifying and distinguishing them from the others should be thoroughly discussed. In addition, statements such as; "...strong interaction of the double bond with the Pt surface, *etc.*" should be completely removed from the entire text.

Professor Attard responded: This is a good remark since laser heating can indeed create the so-called G and D bands associated with carbonaceous/graphitic carbon from decomposition of organics. The first point I would make about the bands reported here however is that they are completely reversible in that if we change the potential from double layer to hydrogen evolution regions, the bands appear/disappear. In our experience, cracking of organics by the laser beam (which we always maintained at low power output (10%)) is that they form irreversibly and cannot be removed once formed. Second, although most of the Raman bands are broad, in Fig. 10, it is evident that the peak width of the band at 1580 cm^{-1} is very narrow, again unlike what is normally observed when laser-induced formation of carbon on metal surfaces is being observed. Finally, there was no time dependence of peak intensity in any of our measurements (the peaks did not grow with greater and greater exposure to the laser beam) and so although one can never completely rule out some carbon contamination, the behaviour observed and the weakness of D (or complete absence of G) bands in all our spectra suggests that this is not a particular problem here. So far as removing from the text references to "...strong interaction of the double bond with the Pt surface, *etc.*" we have to disagree since the bands we observe and their frequencies are completely consistent with this statement according to refs. 40 and 41 in our paper.

Professor Dr Suzer then asked: One way of distinguishing the artifact peaks, as a result of photodamage, is to follow them as a function of exposure time. If they grow in time, then they are surely artifacts.

Professor Attard responded: I agree entirely and I can confirm that there was no time dependence of their intensity with exposure.

Mr Zhao said: You impregnated PVP on prepared Pt/C catalyst. Can you determine the location of PVP in the catalyst? Was it on Pt or C?

Professor Attard replied: We can say that PVP definitely adsorbed on the Pt because we observe the attenuation in the hydrogen electrosorption peaks in CV. We cannot determine, however, whether or not PVP also adsorbed on the graphite support.

Professor Bowker remarked: I'm impressed by the apparently high activity of the bioPt. However, I guess it is not so easy to compare with the 5% Pt/C, since the bioPt is presumably 100% Pt. Do you know the surface area of the Pt available for the reaction, and the average particle sizes in each case?

Professor Attard responded: The average particle size of the bioPt is 2–5 nm (ref. 66 in our paper). The bioPt is actually only 5% by mass of the catalyst used.

The surface area of the 5% Pt/C is 2.1 m^2 g^{-1} with an average particle size of 14 nm but we are not sure of the metal surface area of the bioPt. We have found in the past that a reasonable proxy for total metal surface area present is the hydrogenation rate. From Table 2, the hydrogenation rate for the bioPt is always between 2–3 times slower than for the 5% Pt/C.

Professor Bowker then commented: The data appear to show very high activity for the bioPt, but it's not clear how many sites we have here – are there more or less surface Pt sites than for the Pt/C? If less then they are even more active than is apparent from the data.

Professor Attard replied: Based on my previous response this is difficult to answer without either a direct measurement of bioPt surface area or making assumptions concerning the relationship between surface area and rate.

Professor Bowker further asked: In the preparation of the bioPt in the Experimental part of the paper you mention grinding of the sample – is this to break up very big particles or does it fuse nanoparticles together?

Professor Attard answered: We have examined this point quite carefully. Extensive electron microscopy studies showed that grinding does not break the bacteria which support Pt nanoparticles and grinding has no effect on nanoparticle agglomeration or shape.

Professor Hutchings enquired: Do the nanoparticles sizes change on reaction, as very high activity is observed it may be that there is some restructuring occurring?

Professor Attard responded: We have no evidence for particle reconstruction after reaction nor changes in particle size but we can test this with TEM. There is certainly no change in surface morphology after re-cleaning of particles according to CV.

Professor Freund opened a general discussion of Dr Beaumont's, Dr Lu's and Professor Attard's papers: One of the important breakthroughs in surface science was the implementation of surface crystallography using LEED by the Somorjai and Ertl groups in the late sixties and seventies, so that we knew "where the atoms are" and what the contaminants are on the surface – I believe this is a goal we should achieve for supported nanoparticles with appropriate techniques.

Professor Somorjai responded: I agree that monitoring atomic positions in nanoparticles as they restructure or adsorb contaminants and molecules would be most important. We found that imaging NPs with STM is very difficult. Perhaps using EXAFS from a synchrotron to detect coordination numbers and changes under adsorption and reaction conditions would provide structural information and NEXAFS could determine changes of oxidation states.

Professor Bowker added: I would also point out that, in many way, most sensitive property of the surface is its chemistry. After all that is the basis of catalysis. Professor Campbell already reported measurements of the top layer by TPD and molecular beam methods. TPD very sensitive, though we need to use the right adsorbates to probe the surface chemistry, maybe molecular adsorbates, which are single site probes, on top adsorbates whose chemistry is different for different atoms. An example of this is work we have recently carried out using ethanol as a probe of the surface chemistry of Cu–Pd alloys. Here the alloy surface (50% of each in an ordered structure) adsorbs ethanol on both elemental components, and TPD shows the characteristic chemistry of the two elements, notwithstanding apparently different electronic structure from either individual metal. The Pd decarbonylates the ethanol giving methane and CO in TPD, whereas the Cu gives the dehydrogenation products of H_2 and acetaldehyde.

I'd also like to add a comment regarding the nomenclature of core–shell particles, which annoys me enormously, since it's quite incorrect. Au@Pt, should mean a shell of Au on Pt, whereas it misleadingly means the opposite. Perhaps Pt/Au would be a better way of describing it.

Professor Freund remarked: Only going this route from the simple to the more complex do we have a chance to understand the complexity of any system.

Professor Bowker noted: For probing surface sites it may be that you need more than one adsorbate which probe different sites – *e.g.* bidentate binders like carboxylates, or monodentate binders like alkoxides.

Professor Attard added: I would suggest that a powerful method of probing metallic catalytic surfaces that addresses both questions of surface morphology and the presence or absence of capping agents (contamination) would be cyclic voltammetry. It is particularly sensitive to local surface electronic structure associated with steps, terraces, kinks and also the surface composition in relation to bimetallic surfaces. It is unforgiving so far as contamination is concerned and for particular voltammetric "spikes", sensitivities down to <1% of a monolayer are readily realised.[1–5]

1 J. Solla-Gullón, P. Rodríguez, E. Herrero, A. Aldaza and J. M. Feliu, *Phys. Chem. Chem. Phys.*, 2008, **10**, 1359–1373.
2 F. J. Vidal-Iglesias, R. M. Arán-Ais, J. Solla-Gullón, E. Herrero and J. M. Feliu, *ACS Catal.*, 2012, **2**, 901–910.
3 F. J. Vidal-Iglesias, A. Al-Akl, D. J. Watson and G. A. Attard, *Electrochem. Commun.*, 2006, **8**, 1147–1150.
4 A. Al-Akl, G. Attard, R. Price and B. Timothy, *Phys. Chem. Chem. Phys.*, 2001, **3**, 3261–3268.
5 G. García and M. T. M. Koper, *Phys. Chem. Chem. Phys.*, 2008, **10**, 3802–3811.

Professor Ballauff responded to Professor Attard: This is a very good suggestion and application of this method may shed light on the central question which surface groups are located on the nanoparticles. However, we prepared our nanoparticles on the surface of spherical polyelectrolyte brushes and it is not yet clear whether the nanoalloy particles could be removed from this carrier system and placed on a suitable electrode without irreversible damage. I guess that the systems introduced in papers by Dr Tian and Professor Dr Strasser in this Discussion could be studied much easier by this method. Further work is necessary to clarify this question.

Professor Bowker also commented on the suggestion by Professor Attard: In comparing results with CV for catalysis or surface comparisons we must remember that CV is carried out under very different conditions, that is, under a water environment, and in the presence of net current flow.

Professor Attard replied: This is also an important point because if such *in situ* measurements are to be applicable in a general sense, the relationship between adsorption at different types of interface must be clearly understood. In our case, we can also adsorb hydrogen and substrate from the vapour phase of course without any water or other electrical double layer components being present. We note that in all of the systems that we have studied so far, little perturbation of the observed surface intermediates has been found[1] irrespective of whether or not the reaction was carried out on a hydrogen evolving electrode or when hydrogen is dosed directly from the gas phase.

1 R. J. Taylor, Y. X. Jiang, N. V. Rees, G. A. Attard, E. L. Jeffery and D. J. Willock, *J. Phys. Chem. C*, 2011, **115**, 21363–21372.

Professor Hermann commented: I seem to see a disagreement in this discussion: We agreed that understanding catalytic phenomena means understanding at a local *i.e.* atomic level. Now Prof. Bowker argues that chemistry (chemical analysis) itself can lead to a complete understanding. However, chemical analysis starts from quantities averaged over large regions and, thus, can yield understanding at an atomic level only in rare cases.

Professor Bowker replied: I certainly did not say "chemical analysis can lead to a complete understanding". I stated that adsorption/desorption measurements are the most sensitive and selective surface probes. However, we need a range of other techniques, including theoretical approaches, to define the nature of that interaction clearly and to fully understand the whole catalytic cycle.

Professor Freund made a comment: One way of making contact between studies on model systems and "real" systems is to use aberration corrected TEM which allows us to see individual atoms in and on the nanoparticles in wet-prepared systems. The results may then be compared directly with STM investigations gained under UHV conditions.

Professor Somorjai remarked: The major trend in catalysis science is to simplify the catalysts by eliminating active sites until we could identify the site as the chemistry that it was responsible for is removed. With this research approach we could never explore enzyme catalysis which demands a complex catalyst architecture for carrying out sequential reactions. The more complex the end product, the slower the turnover. We should consider exploring catalytic architectures where the reaction intermediates move from one active site to another toward forming the complex end product at the catalyst liquid (water) interface.

Professor Ertl closed the general discussion, noting: The discussion now touches the question about the best approach to the solution of a scientific problem, either top-down or bottom-up. My general experience tells me that the bottom-up approach, *i.e.* starting from understanding the elementary steps with simple model systems, is more successful.

Professor Anderson opened the discussion of the paper by Dr Tian: As you note in your paper, the mass activity of your catalysts is quite low, because the particle size is large. How small can you make these particles and still maintain the high index facets?

Dr Tian replied: The smallest perfect THH Pt NCs we have made is about 13 nm. Smaller ($<$10 nm) nanoparticles prepared by square wave potential usually show imperfect THH or not even well-defined shapes, but they still contain more step atoms than normal Pt nanoparticles and a higher activity, as evidenced by electrochemical tests.

Professor Anderson then asked: If you make small particles with high index facets, are they stable under repeated electrochemical cycling?

Dr Tian replied: The stability of Pt group metal nanocrystals with high index facets under repeated electrochemical cycling is very high, because they just grow under this condition. Square wave potential can induce repetitive oxygen adsorption/desorption (oxidation/reduction), which is one type of the repeated potential cycling. Under such conditions, high index facets are more stable than low index facets. So, Pt group metal nanocrystals with high index facets have high tolerance for repeated electrochemical cycling.

Professor Campbell noted: Fig. 10c showed a nice relationship between electrochemical rate and the concentration of gold and Bi on the catalyst which you interpreted in terms of a geometric factor for both gold and Bi but with an additional electronic factor for Bi to explain its much stronger promotional effect. I suggest that both might be due to the same geometric effect, but that Bi is much stronger simply because Bi atoms spread out uniformly across the Pt surface (due to slightly repulsive Bi–Bi interactions) whereas Au–Au interactions are attractive and so Au clusters into 2D islands (see ref. 1). Thus, Bi is much more effective than Au at creating a high density of small ensembles of available Pt sites.

1 C. T. Campbell, *Annu. Rev. Phys. Chem.*, 1990, **41**, 775–837.

Dr Tian responded: Yes, the Au adatoms tend to migrate and aggregate into clusters, which we have observed in the TEM measurement. The enhanced activity of Au modification can be attributed to the "third body effect". However, in the case of Bi modification, it is unreasonable to attribute the enhanced activity merely to the geometric effect (third body effect). As for formic acid electrooxidation on Pt, it is free from CO poisoning in the negative potential scan (CO has been stripped at the high potential in the positive scan). If only third body effect works, the oxidation current in the negative scan obtained on the modified surface should be smaller than that obtained on pure Pt, since some surface Pt sites are occupied by modified atoms merely as spectators. Our results showed that the oxidation current of formic acid in the negative scan is as high as 25 mA cm^{-2} on Bi-modified Pt THH NCs, while only 12 mA cm^{-2} on pure Pt THH NCs under the same conditions. So we conclude that there should also be electronic effect besides the third body effect for the case of Bi modification.

Dr Vincent enquired: For FTIR measurements in a thin-film electrochemical configuration there will be a build-up of product and depletion of reactant during the measurements. At what timepoint were your FTIR measurements recorded after stepping the potential to 0.5 V to initiate electrocatalysis? How realistic are your ratios of CO_2 to acetic acid given that species are likely to remain trapped in the film during measurements?

Dr Tian responded: Electrochemical *in situ* FTIR spectroscopy with a thin-layer configuration can really result in a build-up of product and depletion of reactant during the measurements. Under this condition, the middle products are more easily further oxidized in comparison with normal electrochemical tests with free mass transfer. So, during FTIR experiments, we kept nearly the same electrochemical surface areas for all samples and the same spectral collection procedures. That is, the spectra were collected just after the potential was stepped to 0.50 V, and the spectra collected within 300 s were averaged. The products in the thin layer really can dissolve slowly into the bulk solution, resulting in a decreased signal. The typical loss of spectral signal is about 20% after 100 s, if no new product formed during the test (such as in the case of the electrooxidation of adsorbed CO in the CO-free solution). As for the current study, ethanol can also diffuse into the thin layer from bulk solution, which can reduce the effect of product diffusion. Another issue is different diffusion rate. The diffusion coefficient of CO_2 and acetic acid are 1.92×10^{-5} and 1.21×10^{-5} cm^2 s^{-1}, respectively. That is, CO_2 can diffuse more quickly than acetic acid. So, the determined ratio of CO_2 to acetic acid should be slightly smaller than the real value.

Professor Hermann remarked: As you know, the (730) and (830) indexed surfaces are described as stepped/kinked surface structures with variable terrace widths. They are mathematically well defined. However, their experimental verification over large surface areas may be very difficult. How well can these surfaces be prepared experimentally?

Dr Tian replied: Yes, these complex facets are mathematically well defined, *e.g.*, (730) facets are composed by two (210) and one (310) subfacets, or be further decomposed into three (100) subterraces of widths 3, 3, 2 atom rows separated by (110) steps. It is difficult to prepare the bulk single crystal planes of (730) and (830) due to the difficulty of their orientation. In the THH Pt NPs, these facets could be prepared. We determined the Miller indices of these facets mainly by measurement of the crystal plane angles. Although the real surfaces may be not as well ordered as the models, the determined facets can still be seen as statistically averaged results. In addition, we also directly observed the (210) and (310) facets by HRTEM on the THH Pt NCs with {730} facets (Fig. 3).

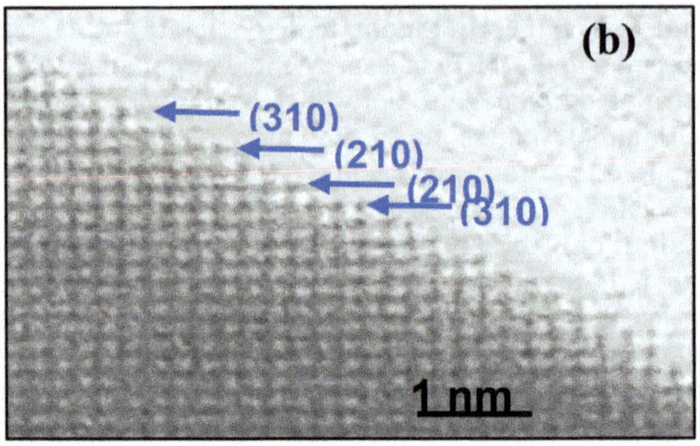

Fig. 3 HRTEM image of THH Pt NC. *Science*, 2007, **316**, 732.

Professor Behm asked: What is the reflectivity of glassy carbon in the IR range? This is just a technical question.

Dr Tian answered: The reflectivity of glassy carbon at 2000 cm^{-1} is about 0.20 for p-polarized light at a incidence angle of 60°. Under thin-layer configuration of electrochemical *in situ* FTIRS, its reflective capability is about 1/3 of smooth metal (*e.g.*, Pt) electrodes.

Professor Neurock questioned: How does the performance of your PtRh ethanol electrooxidation catalysts compare with that of the $PtRhSnO_2$ electrocatalysts reported by Adzic and co-workers?[1]

1 A. Kowal, M. Li, M. Shao, K. Sasaki, M. B. Vukmirovic, J. Zhang, N. S. Marinkovic, P. Liu, A. I. Frenkel and R. R. Adzic, *Nat. Mater.*, 2009, **8**, 325.

Dr Tian replied: The activity of high-index facets PtRh alloys is inferior to that of Adzic's $PtRhSnO_2$ electrocatalysts, but considerably higher than that of Pt/Rh(111) (same paper).

Professor Attard enquired: Do you have plans to use DEMS to analyse how much carbon dioxide product was produced rather than just relying on the infrared signature of CO_2 alone?

Dr Tian answered: Yes, it will be good to use DEMS to monitor the electrooxidation process of ethanol. Note that although DEMS can easily detect CO_2, it is insensitive to acetic acid with low volatility. So, it is hard to use DEMS to evaluate the ratio of CO_2 to acetic acid during ethanol oxidation. We are now trying to establish the combination of the DEMS method with electrochemical *in situ* FTIR spectroscopy. By this combined technology, we may get more insight into the reaction process of ethanol electrooxidation.

Professor Neurock addressed Dr Tian and Professor Dr Strasser: In catalysis as well as electrocatalysis, it is important to know what is on the surface under working conditions as this can significantly influence the catalytic behaviour. In both papers, do you know what the composition of the surface is? In Professor Dr Strasser's paper, are the active hydroxyl intermediates for CO oxidation on the NiO_x or do they form on Pt at the NiO interface?

Dr Tian responded: We have tried to use XPS to probe the surface composition. As shown in Fig. 4, the Rh 3d5/2 peak (principal line) is nearly overwhelmed by the Pt 4d5/2 peak. Through peak deconvolution and the correction of sensitivity factor, the Rh content on the near surface was determined to be about 4.6%, which is fairly consistent with the bulk Rh composition detected by the EDS (4.4%, Fig. 1c).

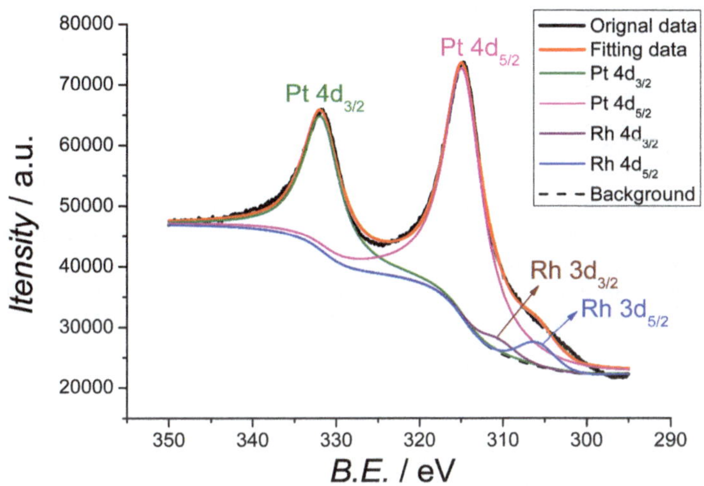

Fig. 4 XPS spectra of THH Pt-Rh NCs in Pt4d and Rh3d regions.

Professor Dr Strasser responded: It is indeed important to know what is on the surface under reactive conditions. We do not have direct experimental evidence, but it appears reasonable to assume that the CO molecules, which are adsorbed at +0.05 V/RHE, are largely adsorbed on Pt surface atoms. The emergence of two separate anodic CO stripping peaks in the voltammetric scan is related to CO oxidation on Pt surface atoms situated in the local environment of a Pt atom ensemble (stripping at larger overpotentials) *versus* CO oxidation on Pt surface atoms located in a local environment interfacing to Ni hydroxide. As mentioned in another reply at these Discussions, I note that the low onset potential of CO stripping for the catalyst pretreated in acidic conditions could be partially due to the formation of undercoordinated Pt adatoms which oxidize CO also at clearly more negative potentials. In the case of the PtNi surfaces pretreated in alkaline conditions, a significant evolution of surface roughness is less likely.

Dr Philip Davies opened the discussion of the paper by Professor Dr Strasser: In recent work[1] we have observed that relatively mild treatments of carbon surfaces can have profound effects on metal deposition on those surfaces. How are your carbon surfaces prepared and have you observed any effect of pretreatment on the nanoparticles?

1 E. Bouleghlimat, P. R. Davies, R. J. Davies, R. Howarth, J. Kulhavy and D. J. Morgan, *Carbon*, 2013, DOI: 10.1016/j.carbon.2013.04.076.

Professor Dr Strasser replied: Our flat carbon surfaces are commercial graphite plates from SGL carbon, which were used as received. Our high surface area carbons were also commercial vulcan XC72R materials. We did not pretreat them before use.

Mr McPherson asked: Would you expect any similarities between your work with PtNi catalysts and the chemistry of the archetypal bimetallic oxidation catalyst PtRu? Indeed, have you examined any of these catalysts using the same techniques? As commercially available "CO tolerant" catalysts sold for the oxidation of small organic molecules, an understanding of any compositional or structural changes during use could have immediate technological relevance.

Professor Dr Strasser replied: We have not examined PtRu so far. Ru-rich alloy particles may not stay single phase as we observed for the cubic PtNi, and if a single phase exists it may not necessarily be cubic any more. As far as the synthesis of shaped particles is concerned a surface segregation based on the different interaction of Ru and Pt with the solvent or capping ligand appears possible, perhaps similar to the way Ni does. Ru is more oxophilic than Pt. Also as far as the changes under reactive conditions are concerned, the surface compositional changes for such a PtRu anode catalyst may be similar to PtNi under reducing, say hydrogen-rich, conditions in the sense that Ru oxide can be reduced to a lower oxidation state, and, once metallic, it may prefer to alloy with Pt into the bulk and leave the surface. Under oxidative environments, again in line with the Pt Ni, I would expect Ru atoms to come preferably to the surface.

Their behavior in their reactive environment in a fuel cell however is certainly very different because one is a cathode, the other an anode electrocatalyst. So the degradation of PtNi catalyst would not occur in PtRu catalysts. The hydrocarbon fuel molecule would provide surface coverage and create a very different situation from the oxygen environment of the PtNi cathode catalyst.

Dr Hughes remarked: XPS is often quoted as having an analysis depth of as much as 10 nm from which 100% of the signal comes. Given that the particles sizes in the paper are 6–9 nm, this suggests that XPS is probably analysing most if not all the volume of the nanoparticles. In light of this, the changes in intensity ratios for Ni : Pt in Fig. 6 seem large (but consistent with the change of signal to noise in the spectra themselves). Is it possible that there might be other explanations? The nanoparticles seem to have some degree of agglomeration in Fig. 2. Could there be changes in the way the agglomerates disperse?

Professor Dr Strasser answered: When we were trying to come up with a reliable value for the probe depth of our XPS methods for our particular particles, we started with the usual IMFP curve, which is admittedly not accurate for nanoparticles, and then looked for earlier studies or corrections valid for Pt alloy particles of comparable size. We found among others a paper by Tao *et al.*[1] which had estimates of the probe depth to 2–3 nm for bimetallic particles of comparable size. So, we feel that we don't probe down to 10 nm from the surface and therefore feel the Ni : Pt ratios are near surface ratios.

1 F. Tao, M. E. Grass, Y. Zhang, D. R. Butcher, J. R. Renzas, Z. Liu, J. Y. Chung, B. S. Mun, M. Salmeron and G. A. Somorjai, *Science*, 2008, **322**, 932–934.

Professor Park enquired: The paper presents the interesting experimental work on surface segregation of PtNi nanoparticles under various gaseous environments. The model nanoparticles used in the study are octahedral and spherical PtNi bimetallic nanoparticles with different sizes. I wonder, which factor, shape or size, is the more critical factor that determines the surface segregation on PtNi nanoparticles?

Professor Dr Strasser replied: We believe it is shape rather than size in this case. However, the detailed synthesis procedure we were using may affect the obtained Pt and Ni distribution as well. We have previously worked with more spherical, cuboctahedral PtNi of similar composition and did not observe any significant surface segregation of Ni towards (111) surface terraces. We note that the particles were not heat treated after synthesis, because they would not survive this treatment. Higher temperature annealing may result in a different Pt and Ni segregation than observed here.

Professor Hutchings said: In your study you use two methods, one that involves surfactants and this leads to spherical particles and another surfactant-free method that leads to octahedral particles. This is an elegant approach. How sure are you that the surfactant is removed by the post-treatment and additionally perhaps using DMF in the surfactant-free methods leads to species that decorate the surface? Perhaps the surface needs to have some surface coverage to aid activity. On way to test this would be to add small amounts of surfactant in a controlled way during the CO oxidation reaction, it would be interesting to observe the effects.

Professor Dr Strasser responded: Yes, I fully agree that would be an interesting experiment. Well, we don't know to what extent we have removed the surfactant or solvent from the spherical or the octahedral particles. The spherical with the long chain capping ligands undergo a mild treatment in air followed by hydrogen. The octahedra undergo washing steps only. It is hard to be sure all surfactants have been removed. I agree adding the capping ligands later, such as DMF, may cause interesting changes to the catalyst activity. Unfortunately, we don't know whether the capping ligand is the potential poison or promoter, or whether it is another carbonaceous molecule that is formed during the annealing (in case of spherical particles). In the case of octahedral particles DMF is not likely to be the poison/promoter, if at all.

Dr Lai remarked: In the electrochemical studies, it was concluded that after the PtNi particles were treated by voltammetric cycling in an acidic solution (which leaves a Pt-rich surface due to leaching of the Ni at the surface), exposure to alkaline media "draws out" the Ni in the bulk of the particle through OH adsorption, leading to $Ni(OH)_2$ surface segregation. If correct, this would be a very interesting finding which impacts on studies of bi-metallic electrocatalysts in general. I was wondering if you had any indication for Ni segregation other than the shoulder (labelled "Ni") in the voltammograms in Fig. 12. Alternatively, this could also be indicative of a Pt surface with a high amount of low-coordination sites (rather than a PtNi surface) on platinum single-crystals, CO stripping in acidic media always occurs in a single peak regardless, and increasing the density of low-coordination sites shifts the peak to less anodic overpotential,[1,2] whereas CO-stripping in alkaline media is much more sensitive to low-coordination sites and shows distinct peaks for different types of sites.[3] Therefore, both the shift of the peak to less positive potentials in $HClO_4$ and the occurrence of the shoulder labelled "Ni" in NaOH in Fig. 12b and c compared to Fig. 12a could also be explained by an increase in low-coordinated Pt sites due to the pre-treatment, without surface segregation of Ni.

1 N. P. Lebedeva, M. T. M. Koper, E. Herrero, J. M. Feliu and R. A. van Santen. *J. Electroanal. Chem.*, 2000, **487**, 37–44.
2 S. C. S. Lai, N. P. Lebedeva, T. H. M. Housmans and M. T. M. Koper, *Top. Catal.*, 2007, **46**, 320–333.
3 G. Garcia and M. T. M. Koper, *Phys. Chem. Chem. Phys.*, 2008, **10**, 3802–3811.

Professor Dr Strasser answered: We don't have direct chemical evidence that the early voltammetric peak labelled "Ni" is related to water-activating Ni hydroxide species/sites in or just below the surface. Low-coordinated sites and defects may contribute to this peak. We point out however, that the alkaline pretreated catalysts do show a much more intense voltammetric peak at a very similar electrode potential. This increase in intensity does not appear to be fully accounted for by defect site formation. Hence our conclusion that the peak of the acid pretreated may also be related to Ni hydroxides.

Dr García had four comments: (a) What is the reason to produce hydrogen from a hydrogen saturated solution? (b) Ni is not a catalytic material for CO oxidation. Therefore, it acts as promoter not as a catalyst. (c) The rotating speed of the rotating disk electrode is not specified. (d) In agreement with Dr Lai, the observed anodic peak at *ca.* 0.5 V (Fig. 12 b and c) was previously observed at the Pt(111) electrode rich in defects and associated to the CO oxidation on low-coordinated sites.[1] Therefore, it is suggested that the small differences between the acid-treated octahedral and the spherical NP catalysts have to be addressed to similar surfaces, *i.e.* the acidic leaching process in conjunction with the 100 cycles seem to homogenize and produce similar atomic surface arrangement that results in a Pt surface with a higher density of defect sites.

1 G. García, P. Rodríguez, V. Rosca and M. T. M. Koper, *Langmuir*, 2009, **25**, 13661–13666.

Professor Dr Strasser replied: (a) This was actually not a typo, we did bubble the solution with hydrogen in order to make sure we have a well established Nernst potential for the working electrode at all times and potentials. The hydrogen gas does not affect the hydrogen evolution catalysis, however, it ensures that the forward and backward scan actually go through the zero current point, so minimizing hysteresis effects (b) Ok, yes, Ni promotes the catalysts on Pt by providing oxygenates. (c) Rotation rate was 1600 rpm. (d) I agree that surface defects formed in the acid-cycled particles may contribute to the smaller peak at around +0.5 V. It does coincide though quite surprisingly with the large peak observed on the alkaline pretreated samples, which is, in our mind, linked to the presence of Ni hydroxides on the surfaces. This led us to believe that it may be related to a surface segregation of sub-surface Ni, given that the experiment was carried out at pH 13.

Professor Dr Suzer opened the discussion of the paper by Dr Yang: Although I find the paper quite interesting and novel, since it is new and probably will lead to a more controlled route for nanoparticle preparation, I have some serious comments about interpretation of XPS data.

Starting on page 9, in the last two paragraphs, and continuing on pages 10 and 11, the authors make claims on the structure of their Ni/Au nanoparticles as having core–shell structure with almost no scientific evidence to back such claims. They state the absence of any observable chemical shift on the Au4f peaks as the sole evidence, which is absolutely unacceptable. If anything, the size-dependent shifts should have been much more significant. They could have however, used some intensity arguments, which might have indicated some clues about the surface enrichment of Ni or Au atoms.

Dr Yang responded: There is a size effect in core level XPS spectra but the Ni core was the same size in all samples and the amount of Au in each nanoparticle was also the same though the shape was different, that is a solid Au nanoparticle or a Au shell around a nanoparticle. Nevertheless no core level shift was observed in the Au 4f. On the other hand in a uniform alloy, a core level shift of ~0.5 eV would be expected, which would be easily observed. We cannot find a value specifically for AuNi but AuCu shows a core level shift of 0.48 eV.[1] So we continue to assert that the lack of any observable core level shift is evidence that the Ni@Au nanoparticles are not a uniform alloy.

We agree that it is not definitive proof – one could argue that there may be a shape-induced core level shift that is equal and opposite to an alloy induced shift but this is unlikely.

We have only taken the lack of a Au 4f core level shift as *evidence* for a core–shell structure and not *proof* as is clear from the cautious wording in the paper: Although we cannot rule out a degree of alloying at the interface, the XPS findings point to an atomic arrangement which is primarily a core–shell structure.

We also note that our subsequent work using high-resolution TEM has proved the assertion of a core–shell structure to be correct.

1 W. Eberhardt, S. C. Wu, R. Garrett, D. Sondericker and F. Jona, *Phys. Rev. B*, 1985, **31**, 8285–8287.

Professor Hermann asked: What is the size distribution of these particles? Are they mono-disperse or is there a large range of sizes?

Dr Yang answered: The reported nanoparticles are not mono-dispersed as described in the paper. This is because we have used helium nanodroplets formed by supersonic jet expansion of pre-cooled helium gas, in which the size of helium nanodroplets follows statistical distributions (depending on the expansion conditions). However, if we expand liquid helium through a very small aperture, the helium nanodroplets will have the same size as the aperture, in which case mono-dispersed nanoparticles will be formed. For example, for the nanoparticles with an average size of 1000 atoms, the width of distribution is only 8% (following the Poisson statistics for the pickup).

Professor Anderson commented: In Fig. 5 of your paper, you show a broad, maybe bimodal size distribution, and state that if particles agglomerated on the surface, this should be evident in the particle shape or grain boundaries in the particles, however, the energy associated with agglomeration of small particles may be enough to anneal such evidence away.

Dr Yang responded: Indeed there is a strong interaction between metal atoms, for example, the Ag–Ag interaction is about 3 eV. However, for two nanoparticles to merge into one, it is necessary to break some metallic bond first and then the energy will be released when the atoms aggregates. This means that although it is almost energetically even, there is a huge barrier for two nanoparticles to merge into one.

Some other evidence is from the synthesis of nanoparticles using wet chemistry methods, in many cases large nanoparticles are actually the aggregation of small particles. People have to use high-temperature treatment to ensure crystalline structure of the particles.

In our experiments, the highest temperature that nanoparticles have experienced is room temperature, at which there is little chance for particles to merge unless it takes a very long time – which is obviously not the case.

Professor Anderson then noted: You invoke a multicenter growth mechanism to produce multiple particles within one droplet, but it seems surprising to me that multiple particles would not merge within the droplet, given the rapid diffusion in superfluid He, and the fact that the droplets are really not very large. Can you comment on the dynamics of particle motion in the droplets?

Dr Yang responded: Multi-centre growth is indicated in the spectroscopy measurement of silver nanoparticles grown in large helium nanodroplets composed of more than 10^6 helium atoms.[1] The dynamics of particle motion in the droplets is unclear but if we assume that particles are spherical before aggregation, then the initial driving force is van der Waals interaction. However, if two nanoparticles are so close that they aggregate within the flying time of helium nanodroplets in the vacuum chamber (a few ms) before hitting the deposition target, then the driving force is likely to be electrostatic interaction because in helium nanodroplets there is no excess energy for two nanoparticles to merge into one spherical particle, and the consequence of this would be the formation of nanoparticle chains. The uneven shape will cause induced dipoles so the interaction is more likely driven by dipole–dipole and/or dipole–neutral interactions.

1 E. Loginov, L. F. Gomez, N. Chiang, A. Halder, N. Guggemos, V. V. Kresin and A. F. Vilesov, *Phys. Rev. Lett.*, 2011, **106**, 233401.

Professor Campbell commented: I suggest that if a He droplet captures two electrons (from any stray electrons that may be in the vacuum chamber), they would localise on two growing metal clusters. This would cause cluster–cluster repulsions and thus explain why you sometimes see two clusters per He droplet.

Dr Yang responded: This is an very interesting point, which it hard to say yes or no. In our experiment there is no direct exposure of filament to the pickup region so the chance is low. In the future, once it is possible, we could apply a DC voltage and then look at the nanoparticles using TEM, which will clearly tell whether it is true or not.

Professor Freund said: My question refers to Fig. 7. Both traces look very much like NiO if one compares with published data.[1] So I would question the assignment to metallic Ni.

1 St. Uhlenbrock *et al.*, *J. Phys.: Condens. Matter*, 1992, **4**, 7973.

Dr Yang replied: Indeed there are indications for the formation of NiO in the XPS spectra. As stated in the paper, we cannot exclude such a possibility as Ni particles are exposed to air when we transferred them for TEM investigation.

Professor Bowker asked: Can you tell us more about this method? Presumably as the He nanodroplet absorb thermally hot metal atoms some He is evaporated – are the clusters naked by the time they hit the surface? What exactly does "soft landing" mean in this case – is the beam velocity very low?

Dr Yang explained: Nanoparticles are formed by the addition of atoms to helium nanodroplets one by one. If the helium nanodroplets sequentially pass through more than one pickup cells, the addition of different types of materials, as in the case of Ni/Au and Ag/Au, will lead to the formation of core–shell nanoparticles.

When atoms aggregate together, the interaction energy will be converted to thermal energy, released by evaporation loss of helium atoms. In this sense, the thermal energy (kinetic energy) of the metal atom is often small when compared with binding energies.

When helium nanodroplets hit a surface, they will be evaporated away instantly, leaving bare particles on the surface. Soft landing in this case is mainly originated from the low velocity of nanoparticles (less than 300 m s^{-1}) but the remaining helium atoms in the droplets can also provide a cushion effect.

Professor Bowker then enquired: Can you tell us anything about adhesion of the particles to the surface – are they only weakly interacting, low interfacial contact area?

Dr Yang responded: We are unsure about the adhesion force of the particle to the surface of TEM substrate. However, as the nanoparticles are relatively small the adhesion force can be quite strong. Taking silver particles as the example, the interacting potential can be as high as 1 eV per atom.

Professor Freund returned to the discussion concerning "soft landing": If one considers 0.1 eV $atom^{-1}$ to be a rather low energy, a particle with 200 atoms will have an energy of 20 eV. May one consider this to be a small energy?

Dr Yang clarified: 20 eV is by no means a small energy but the typical interaction energy between surface and nanoparticles is at the level of 1 eV per atom and the metallic bonding between metal atoms in the nanoparticles is even stronger. In this sense, a landing energy of 20 eV is so low to have a significant impact to either the particles or the surface.

Professor Anderson continued this discussion: That is true, however, a much more important factor is that the metal atoms in the particle will have substantial bond energies with most supports, and this source of energy will far exceed the kinetic energy of landing in most cases. Our experience with landing cluster ions on surfaces is that for clusters in the <25 atom range, we see no differences in cluster morphology or chemical activity for variations in kinetic energy from 0.05 to 1 eV $atom^{-1}$, however, for energies over 5 to 10 eV $atom^{-1}$, there clearly is substrate damage and implantation of cluster atoms into the support (see ref. 1). My feeling is that for depositions below 1 eV $atom^{-1}$, there is undoubtedly transient local heating that may help anneal deposited clusters, however, the final cluster structure should depend mostly on the strength of the metal–support interaction.

1 M. Aizawa *et al.*, *Surf. Sci.*, 2003, **542**, 253–275.

Dr Yang replied: This is a very interesting comment. Let us scale up the problems: assuming we have a big stone rather than a small particle landing on a surface, although the speed could be very low, it might still make a significant damage to the surface, why? In this case it is not any more the energy per atom, and I think the importance lies in the capability of surface and particles to cope with impact energy. So, how quickly can the impact energy be dissipated before damage is made? My feeling is that this is an important aspect too.

Professor Anderson commented: Indeed, exactly this behavior is observed in energy- and size-selected impacts on surfaces. The threshold energy per atom where damage and implantation are seen is lower for large clusters than for atoms or small clusters, presumably because the larger, three-dimensional clusters concentrate the energy and momentum in a small footprint. This effect is shown in ref. 1 in my previous comment.

Professor Behm opened a general discussion of Dr Tian's, Professor Dr Strasser's and Dr Yang's papers: We heard a lot about the use of adsorption energies for mapping out reactivities. In these cases, I am missing the consequences for the entire reaction cycle. For instance, step sites as strongly binding sites are undoubtedly good for breaking C–C bonds, since they stabilize the adsorbed dissociation products. But, as a consequence of this stabilization, these adsorbed reaction intermediates are also harder to remove in a subsequent reaction step. Similar aspects hold true also for bimetallic surfaces/catalysts. For instance, Ni in PtNi surfaces (paper by Professor Dr Strasser) may be helpful to split water by stabilizing adsorbed OH species, but the resulting OH_{ad} is then much harder to react away in a subsequent reaction step. These effects, which are well known as the Sabatier principle, call for an analysis of the full reaction cycle in such cases.

Professor Campbell replied: I agree thoroughly. In this respect, some readers may be interested in the method we introduced for analyzing microkinetic models of multistep mechanisms which allows one to determine quantitatively which elementary steps and which adsorbed intermediates are the most important ones kinetically.[1]

1 C. Stegelmann, A. Andreasen and C. T. Campbell, *J. Am. Chem. Soc.*, 2009, **131**, 8077–8082.

Professor Dr Strasser also responded: I fully agree with Professor Behm's comment that we need to consider the entire reaction sequence, ad- and desorption, to be relevant to the full catalytic process. As for the promoting effect of surface Ni on the activation of water, I agree that the initial formation of $Ni(OH)_x$ results in OH adsorbates or ligands which would be bonded fairly strongly, yet that subsequent OHs would be bonded less strongly and could aid in continuing the full catalysis. In fact, sustained promoted water splitting associated with enhanced hydrogen evolution or CO oxidation was observed in our experiments, which tells us that the water activation process is not transient, and does not come to a complete stop due to OH poisoning on the Ni(OH) material.

Professor Somorjai noted: So far most catalytic studies address chemistry that occurs on a single catalyst and deals with the dynamic nature of surface structure and oxidation states. In order to investigate complex enzyme-like reactions we should construct catalyst architecture instead of studies of single catalysts. Maybe the time has come to start investigating more complex reactions to bridge the gap between heterogeneous and homogeneous catalysis on the one hand and enzyme catalysis on the other.

Professor Freund agreed with Professor Somorjai: It is important to identify new areas of research for the young people in the field. With reference to enzymatic catalysis, indeed we should not leave research in this area to biology-based researches, simply since investigations at the atomic level are necessary, and chemists and physicists could have crucial input.

Dr Vincent added: There is a vast body of work on mechanisms of metalloenzymes which parallels the work in surface science on metal catalysis. This goes beyond mere "snapshots" of enzyme action obtained from X-ray crystallography and includes many rapid-quench or *in situ* spectroscopic studies of enzyme turnover. We and others are probing metalloenzyme mechanisms using *in situ* infrared spectroelectrochemical methods (for example, for hydrogenases in ref. 1). There are certainly interesting parallels between metalloenzyme chemistry and metal surface chemistry and there would be value in exploring common ground. Examples include the use of a bimetallic catalytic site (such as NiFe in hydrogenases for H_2 oxidation) and poisoning of hydrogenase active sites at high potential by hydroxide. However it is important to remember that the protein in metalloenzymes is more than just packaging for the metal catalytic centre, and is often crucial in delivery of electrons and protons and tuning active site function. In most cases it is unlikely to be possible to replicate enzyme function with metals alone.

1 A. J. Healy, P. A. Ash, O. Lenz and K. A. Vincent, *Phys. Chem. Chem. Phys.*, 2013, **15**, 7055–7059.

Professor Hutchings continued this discussion: Nature has developed enzyme catalysts based on readily available metals *e.g.* Fe, Mn, Cu and controls the reactivity by the ligands that surround these active metals. Hence the metal–ligand interaction is critical. We have discussed methods that seek to make ligand-free surfaces by the gas phase generation of clusters and then "soft landing" them onto a surface. However, do we need ligand-free surfaces or should we be looking into purposely doping surface to control the reactivity? This to me is an important point.

Professor Campbell remarked: It would be good for future research in this field if someone would tabulate some literature values for catalytic rates per unit metal surface area for one or two liquid solution-phase test reactions, that give reproducible rates between groups. When one made new NP catalysts, one could then verify the cleanliness of the metal surfaces by comparing rates per area to these benchmark-system rates. The rate of CO oxidation has been widely used as such a benchmark system, but for gas-phase reactivity.

Professor Behm commented: Many papers in the literature provide data that can not be converted into reaction rates, and error bars, including differences between measurements on apparently identical materials, but from different batches, are hardly ever given. This makes a materials-based comparison of activities with other data such as adsorption energies or theory based data problematic and in many cases essentially impossible.

Professor Campbell replied: I agree that this is a problem. Decades ago, an editor-in-chief of the Journal of Catalysis (Michel Boudart, if I remember correctly) started a policy that no paper would be accepted if it did not make an attempt to express the rates in units of moles or molecules of product per unit area (or number of surface atoms) of the active material. Even when we still are arguing about which phase is the active one or if it is at the perimeter sites, one should always give enough information in a paper so that the reader can estimate the rate per unit area of each phase that may prove to be the active one. Thank you and Bravo for that suggestion!

Professor Bowker closed the general discussion: We must remember when comparing biocatalysis and heterogeneous catalysis/electrocatalysis that there is a huge different in the *density* of active sites (per unit volume) and we push rates high by using thermal activation. Humans are greedy and want fast production in the smallest volume units possible considering engineering constraints. Nature is more patient and works under ambient conditions with relatively low rates (per unit volume), with production happening using solar energy, diffused over large areas. Because of those challenging conditions for catalysis, the active site is embedded in a "machinery" of material which directs molecules to the active site, selects molecules for the active site, often has a "trapdoor" mechanism, has supporting scaffolding for the active site, channeling within that scaffolding and an internal electron transport pathway.

Charge competition with oxygen molecules determines the growth of gold particles on doped CaO films

Yi Cui,[a] Kai Huang,[a] Niklas Nilius*[ab] and Hans-Joachim Freund[a]

Received 16th November 2012, Accepted 8th January 2013
DOI: 10.1039/c3fd20130a

The influence of gas-phase oxygen on the growth of Au nanoparticles on Mo-doped CaO films has been investigated by means of low temperature scanning tunnelling microscopy and X-ray photoelectron spectroscopy. Whereas at ideal vacuum conditions, only 2D Au islands develop on the oxide surface, the fraction of 3D deposits increases with increasing O_2 pressure until they become the dominant species in 10^{-6} mbar oxygen. The morphology crossover arises from changes in the interfacial electron flow between Mo donors in the CaO lattice and different ad-species on the oxide surface. In the absence of O_2 molecules, the donor electrons are predominately transferred to the Au ad-atoms, which consequently experience enhanced binding to the oxide surface and agglomerate into 2D islands. In an oxygen atmosphere, on the other hand, a substantial fraction of the excess electrons is trapped by adsorbed O_2 molecules, while the Au atoms remain neutral and assemble into tall 3D particles that are typical for non-doped oxides. Our experiments demonstrate how the competition for charge between different adsorbates governs the physical and chemical properties of doped oxides, so widely used in heterogeneous catalysis.

Introduction

The ability of supported metal particles to promote certain catalytic reactions depends on their elemental composition and the electronic and chemical interplay with the oxide support. Also the particle shape has proven to be of relevance, as certain reactions require the availability of specific active sites on the catalyst surface. A prominent example is the $CO \rightarrow CO_2$ conversion on supported metal catalysts, where the metal deposits are responsible for capturing the CO molecules, while the oxide stabilizes and activates the O_2 species. The actual reaction is then believed to occur at the metal–oxide perimeter, explaining why particles with 2D or raft-like shapes are particularly suited to catalyze this reaction.[1–4]

The geometry of metal particles on oxide supports has been investigated down to the atomic scale by using a variety of microscopic and spectroscopic techniques.[5–8] The metal–oxide interface, *i.e.* the atomic registry and interfacial bond length, has thereby received special attention, as it governs the morphology of the growing aggregates.[9,10] A careful shape characterization is also required in order to rationalize the particles' electronic properties that, to a certain extent, reflect the symmetry of the confining potential and hence the position of the metal ion cores.[11] Whereas particle shapes could be analyzed only before and after a chemical reaction in the past,[6,7,12] *in situ* studies employing synchrotron-based EXAFS and

[a]*Fritz-Haber-Institut der Max-Planck-Gesellschaft, Faradayweg 4–6, 14195 Berlin, Germany. E-mail: nilius@fhi-berlin.mpg.de*
[b]*Carl von Ossietzky Universität Oldenburg, Institut für Physik, D-26111 Oldenburg, Germany*

environmental TEM now provide evidence for changes in coordination numbers and particle shapes directly during a catalytic process.[13–17] In fact, the particles undergo substantial structural modifications at elevated pressure and temperature conditions, being explained with the different thermodynamic stability of bare and adsorbate-covered surfaces and the temporary formation of oxide or hydrate phases at reaction conditions.[18,19]

In this paper, we demonstrate how the presence of a gas phase influences the growth behaviour of supported metal particles and how this effect depends on the state of doping of the oxide beneath. The phenomenon described here is therefore different from a pure environmental effect on existing particles, and sheds light on the interplay between adsorbates and dopants inside the oxide matrix on the growth process of metals. The effect of gas-phase species on the morphology of metal surfaces and nanoparticles has been addressed in several experimental and theoretical studies before.[20–23] The regular Pt(557) surface, for example, was found to develop a high density of kinks and corners in high-pressure CO,[24] but exposes ultra-small Pt nanoclusters in an O_2 environment.[25] Electron micrographs of individual Pt particles, on the other hand, revealed the transformation of (111) and (100) facets into various high-index planes upon O_2 exposure at 670 K.[26,27] Similarly, the (100) facets of Au nanoclusters were shown to reconstruct in a CO environment in order to maximize the Au–CO interaction.[28] The observed restructuring is driven by the enhanced binding of most adsorbates to open, corrugated surfaces that are, however, thermodynamically unfavourable at vacuum conditions due to their high surface free energy.

A second effect on the particle morphology, the presence of dopants in the oxide lattice, has moved into the focus of catalysis research only recently. In a combined scanning tunnelling microscopy (STM)/density functional theory (DFT) study, it was shown that small quantities of Mo embedded in a CaO matrix substantially alter the growth morphology of gold.[29] Whereas mainly 3D deposits developed on the pristine oxide due to the weak metal–oxide adhesion in that case, monolayer Au islands occurred after doping with Mo. The morphology crossover was explained with an electron transfer from the Mo donors into the ad-gold, activating efficient Au–CaO interaction channels.[30] The reverse effect was found upon inserting low-valence dopants, *e.g.* Li, into the CaO lattice. The Li ions generate acceptor-type hole-states in the oxide valence band, which effectively trap excess electrons and prevent their transfer into the ad-gold. The neutral gold follows a 3D growth behaviour, similar to the one on non-doped supports.[31] In this paper, we demonstrate how a variable background of O_2 alters the charge exchange between donors in the CaO lattice and adsorbed Au species. We show that gold grows into tall 3D deposits in the presence of oxygen and discuss possible implications of the underlying charge competition on the catalytic properties of doped metal–oxide systems.

Experimental

The studies were performed in an ultra-high vacuum (UHV) chamber equipped with a low-temperature STM ($T = 9.5$ K) and an X-ray photoelectron spectrometer (ESCALAB 200). The CaO films were prepared by Ca deposition onto a Mo(001) surface in 5×10^{-7} mbar O_2 at room temperature, followed by an annealing step to 1000 K in UHV.[32] The films contained a significant amount of Mo impurities, originating from atom diffusion out of the metal support into Ca substitutional sites of the CaO lattice. Due to a relatively large Mo diffusion barrier, the impurity concentration gradually diminished with film thickness from 25% directly at the interface to almost zero beyond the 20th monolayer (ML).[33] For 20–25 ML thick films used in the experiment, the mean Mo concentration in the near-surface region was estimated to be 2–3%. Practically dopant-free films, being referred to as pristine in the following, could be produced by increasing the CaO thickness to around 50 ML. Both, the self-doped and pristine films displayed a sharp (1×1) square pattern in LEED, indicative for a fully relaxed, (001)-terminated rocksalt structure. XPS measurements confirmed the one-to-one stoichiometry of the CaO films and the presence of Mo dopants at low film thicknesses. The oxide surface consisted of atomically flat domains of up to 100 nm^2 size, separated by predominately [100] oriented line defects, as derived from the STM data.[32] The dislocation lines develop spontaneously in the oxide film in order to remove misfit-induced lattice strain with the metal substrate. The Au particles were prepared by depositing nominally 0.5 ML gold from a thermal evaporator using a deposition rate of 0.5 ML min^{-1}. In order to investigate the impact of oxygen on the Au growth, the samples have been exposed to different O_2 partial pressures at 300 K before, during or after gold exposure.

Results

As discussed in earlier work,[29] Au growth on pristine and Mo-doped CaO films proceeds in opposite manner (Fig. 1a, b). While on pristine films, the Au particles adopt large height-to-diameter ratios indicative for a small metal–oxide adhesion, flat, monolayer islands develop in the presence of Mo-dopants. The drastic change in particle morphology becomes evident in shape histograms, being produced by analysing a few hundred Au deposits on differently prepared oxide films (Fig. 2). Interestingly, the growth behaviour was found to be sensitive not only to the presence of dopants, but also to the gas atmosphere in which metal deposition was carried out. In 5×10^{-8} mbar O_2, the Au growth starts diverging from the ideal 2D regime that is observed on doped films, and 3D deposits become visible in the STM (Fig. 1c and 2c). The alteration of the growth mode initially affects only particles around the oxide dislocation lines. However, with increasing O_2 partial pressure, 3D particles appear even away from the surface defects. In 5×10^{-7} mbar of oxygen, already half of the Au deposits adopt 3D geometries, while the remaining entities still retain a monolayer nature (Fig. 1d). Surprisingly, no intermediate particle shapes are detected, which becomes evident in a bimodal shape distribution (Fig. 2d). At 5×10^{-6} mbar oxygen pressure, 3D shapes entirely dominate the Au particle ensemble,

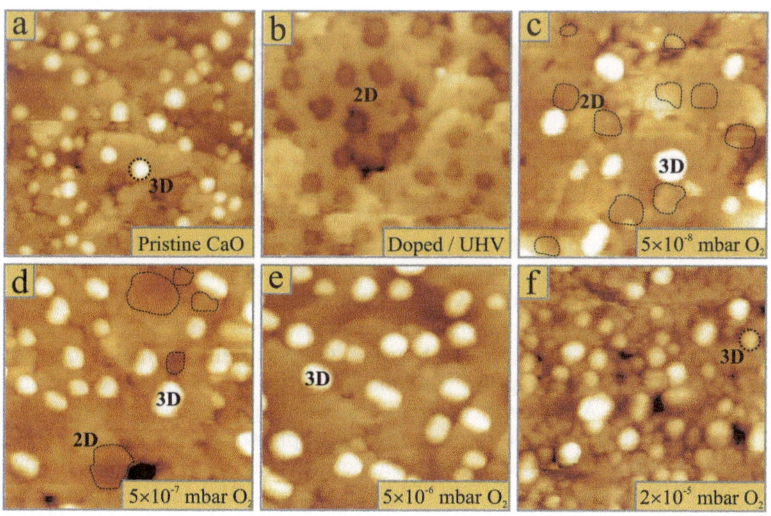

Fig. 1 STM topographic images of 0.5 ML gold deposited on (a) 50 ML CaO without dopants in the surface region and on 25 ML thick, self-doped CaO films (b) in UHV and (c–f) in O_2 pressures increasing from (c) 5×10^{-8} mbar, (d) 5×10^{-7} mbar, (e) 5×10^{-6} mbar to (f) 2×10^{-5} mbar. All images are 50×50 nm^2 in size and have been obtained at 4.0 V sample bias. Note that 2D Au islands are imaged as depressions on the CaO due to their low conductivity.[33] In (c,d), these islands are marked by dashed lines.

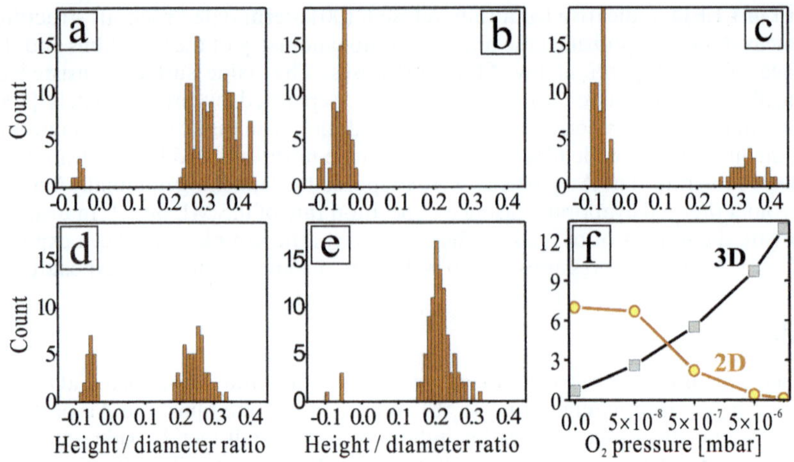

Fig. 2 Histograms of Au-particle shapes for growth conditions identical to the ones shown in Fig. 1: (a) non-doped CaO in UHV, Mo-doped CaO in (b) UHV, (c) 5×10^{-8} mbar, (d) 5×10^{-7} mbar and (e) 5×10^{-6} mbar of oxygen. The occurrence of negative aspect ratios in the histograms relates to the negative contrast of monolayer Au islands in the respective STM images. Note the bimodal shape distributions in panels (c) and (d), reflecting the coexistence of 2D and 3D particles. (f) Evolution of the number density of 2D and 3D deposits as a function of the O_2 pressure during deposition.

while 2D islands have vanished from the histogram. This growth situation does not change anymore at higher O_2 partial pressures, only that the CaO surface starts degrading at such conditions (Fig. 1f). We relate the observed roughening effect to two factors. First, our O_2 gas source might be contaminated with traces of water that interacts with the hydrophilic CaO surface and gradually transforms it into a disordered calcium hydroxide phase. Second, the excess oxygen might promote Mo segregation from the interior of the CaO film to its surface, where it aggregates into a rough Mo oxide overlayer. Such a segregation process is thermodynamically feasible due to the substantial energy gain when changing the oxidation state of Mo ions from +2 in an ideal rocksalt environment to +6 in saturated MoO_3 surface units.[33] As a consequence of the reduced surface quality, also the density of Au particles increases at high O_2 partial pressures, as new nucleation sites becomes available for the incoming Au atoms (Fig. 2f).

In a second set of experiments, we have analysed the influence of pre-dosed oxygen on the growth morphology of gold. For this purpose, the CaO films were exposed to 20 L O_2 prior to Au deposition, which produced large quantities of 0.5 Å-deep depressions in the surface (Fig. 3a). Following earlier DFT calculations,[34,35] we assign those entities to O_2 molecules that lie flat on the surface and therefore appear slightly elongated. This interpretation is supported by manipulation experiments, in which selected O_2 molecules were split into their atomic constituents by applying a 3.0 V pulse with the STM tip[36]. Also pre-dosed oxygen was found to alter the Au growth morphology, although the actual deposition step was performed in UHV. The particle ensembles prepared on a Mo-doped, O_2 precovered CaO have the same bimodal shape distribution as the ones produced in moderate O_2 background pressure (Fig. 3c). Apparently, surface-bound and gas-phase molecules play a similar role, *i.e.* they promote the development of 3D instead of 2D particles even on doped oxide films. Only the efficiency of the geometry crossover is reduced, a trend that is easily explained with the smaller availability of surface oxygen at pre-dosing compared to continuous dosing conditions.

Finally, we have explored the effect of post-oxidizing the Au particles. As discussed before, gold exposure in UHV produces 2D metal sheets on the Mo-doped

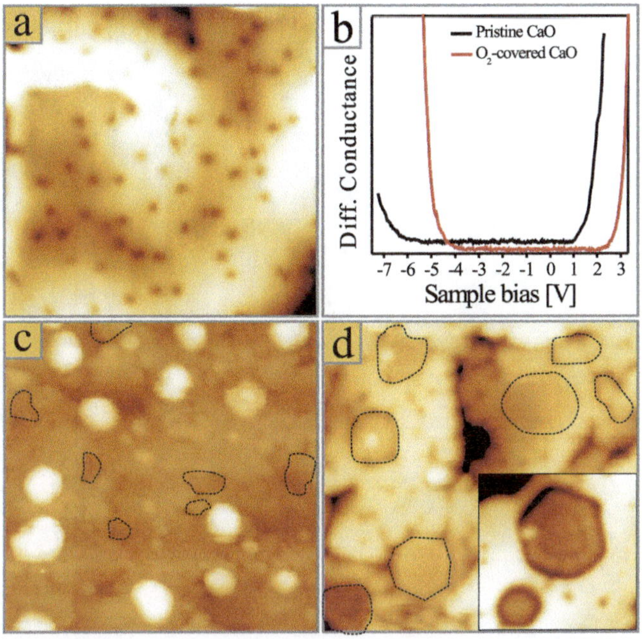

Fig. 3 (a) STM topographic image of a Mo-doped CaO film after exposure to 20 L of oxygen (25×25 nm^2). (b) Conductance spectra showing the onset of valence and conduction band on clean and O_2-covered CaO films. (c) 0.5 ML Au deposited onto a Mo-doped CaO film pre-covered with 400 L O_2 and (d) 0.5 ML Au post-oxidized with 100 L O_2 (40×40 nm^2). The 2D islands are marked with dashed lines. The inset in (d) shows two selected Au monolayer islands with an oxidized rim (10×10 nm^2).

CaO surface.[29] Exposing such an ensemble to 100 L O_2 does not affect the original 2D particle shapes (Fig. 3d). In fact, the main difference between as-grown and post-oxidized Au islands is a characteristic rim in the latter case, being assigned to adsorbed molecules along the low-coordinated island perimeter.[37] On thick, non-doped CaO films, the Au growth morphology was found to be independent of oxygen, as the molecules promote the 3D growth that is already intrinsic to the pristine oxide.

Before discussing the effect of oxygen on the Au growth, we want to highlight another interesting phenomenon. Although 3D particles develop on both, non-doped CaO in vacuum and Mo-doped CaO in an O_2 environment, the resulting particle shapes are rather different. Vacuum-grown particles are mostly hemispherical and do not exhibit well-shaped basal planes or side facets (Fig. 1a). We are nonetheless assured of their crystalline nature and assign them to fcc-semispheres, densely covered with step edges. In contrast, particles grown in an oxygen environment have distinct geometries, as depicted in Fig. 4. The most abundant particle type has a quasi-rectangular basal plane and two roof-like side facets extending along the main axis (Fig. 4b). The particle orientation always follows a [100] or [010] direction of the CaO(001) lattice, indicating an epitaxial relationship between gold and the CaO surface. A second, frequently observed particle type has a truncated pyramidal shape with three large and three small side facets (Fig. 4c). Finally, particles with hexagonal or octahedral base planes and pyramidal appearance are found at oxygen-rich conditions (Fig. 4e). The latter two types have a random orientation with respect to the CaO lattice and are less abundant than the roof-like deposits. In general, the 3D deposits produced in an O_2 environment feature higher crystallinity than the vacuum-grown species, suggesting that the surface oxygen increases the mobility of the Au atoms and facilitates their aggregation.[12]

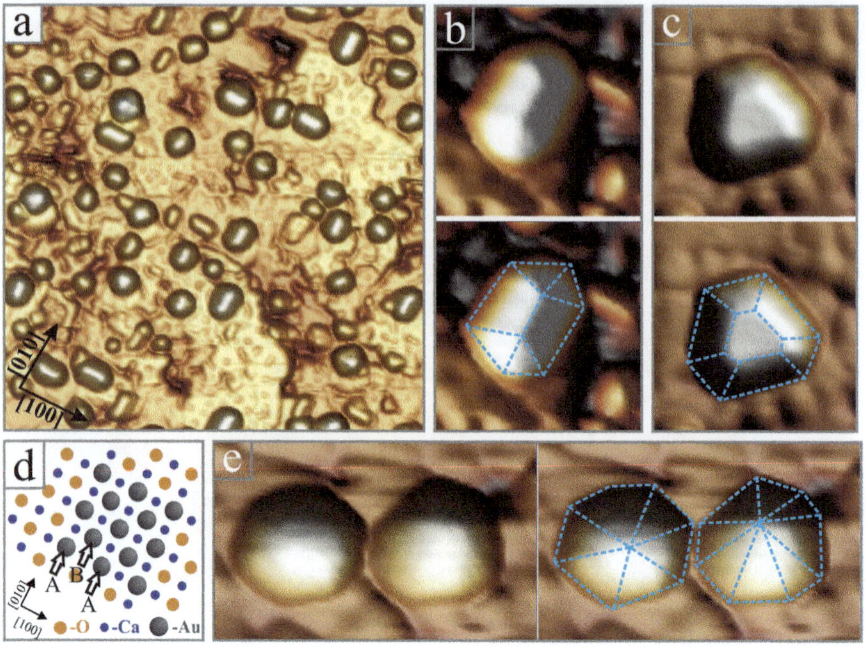

Fig. 4 (a) Pseudo-3D representation of crystalline Au particles grown on a Mo-doped CaO film in 5×10^{-6} mbar O_2 (50×50 nm^2). Enlarged images of four particles with (b) roof-like, (c) truncated pyramidal (8×8 nm^2) and (e) hexagonal and octahedral shapes (8×12 nm^2). The particle edges are retraced with dashed lines. (d) Ball model showing the hcp-Au $(11\bar{2}0)$/CaO(001) interface that gets pseudomorphic upon slight distortion of the Au bulk lattice. The A and B planes of the hcp lattice are indicated with arrows.

Discussion

According to the main experimental observations, we divide our discussion into two parts that are (i) the transition from 2D to 3D deposits upon gold deposition at increasing O_2 pressures and (ii) the improved crystallinity of particles grown in oxygen excess.

The fundamental reason for the development of 2D Au islands on Mo-doped CaO films is an electron transfer from the Mo donors into the Au ad-species.[29,38] The process is energetically favourable as the filled Mo 4d states in the CaO band gap are higher in energy than the Au 6s affinity levels. Thus produced Au^- species experience effective interaction schemes with the CaO support, namely Coulomb attraction to the oxide cations and polaronic stabilization of the Au^- on the ionic surface,[30] both promoting a 2D growth regime on the doped oxide film (see Fig. 1b). The impact of oxygen on the Mo–Au charge exchange now provides a plausible explanation for the modified Au growth behaviour at non-vacuum conditions. Similar to gold, the O_2 molecules are electron acceptors and able to trap Mo excess electrons in their low-lying antibonding orbitals. Taking the electron affinity of the gas-phase species as reference (EA = 0.5 eV),[39] the lowest acceptor level of the molecule lies below the highest occupied Mo 4d gap state, which renders formation of surface O_2^- species feasible. Co-deposition of oxygen and gold leads therefore to a situation in which both adsorbates compete for charges from the Mo impurities. For the experimental Au deposition rate of 0.5 ML min^{-1} and different O_2 partial pressures, the following interplay occurs between the ad-species on the CaO surface. At the lowest O_2 pressure (10^{-8} mbar), the number of both, Au atoms and O_2 molecules increases by roughly 1×10^{13} cm^{-2} per second, assuming a sticking of one. However, as gold is the much stronger electron acceptor (EA = 2.3 eV)[39] most extra charges get trapped by the Au atoms and their growth behaviour changes only moderately (Fig. 1c). Deviations from the 2D growth mode are only observed around the CaO dislocation lines, where empty defect states in the band gap emerge as third opponent in the charge competition.[40] At 10^{-7} mbar O_2, every incoming Au atom already faces 10 O_2 molecules on the surface. Moreover, the charge reservoir established by near-surface Mo donors quickly depletes due to the high number of oxygen acceptors on the surface. However, as not every O_2 molecule immediately sticks and takes an electron, a dynamical situation arises in which the Au atoms are still able to attract a substantial charge fraction from the Mo dopants. This new balance gives rise to a bimodal shape distribution in which 2D and 3D deposits coexist on the CaO surface (Fig. 2c). At 10^{-6} mbar O_2, finally, the ratio between O_2 and Au species approaches 100 : 1 and acquisition of relevant charge quantities becomes difficult for the gold. Consequently, most Au atoms remain neutral and grow into tall 3D deposits right from the beginning. The resulting small interface area with the CaO support hampers the electron transfer also in a later stage, further stabilizing the 3D growth regime of Au in this stage (Fig. 1e).

To support the scenario of charge competition, we give circumstantial evidence for a negative charging of Au and O_2 species on the CaO surface. The charge donation from the Mo impurities can be derived indirectly from changes in the CaO electronic structure. In STM conductance spectra, an up-shift of the oxide valence and conduction band by around 1.0 V is revealed upon oxygen exposure (Fig. 3b), an effect that was also observed after Au deposition onto the doped CaO films.[33] XP-spectra show a similar trend and both, the O1s and Ca2p core levels lower their binding energy upon dosing the electron-accepting species (0.3 and 0.2 eV for Au and O_2, respectively, Fig. 5). We relate this shift to an upward bending of the oxide bands, induced by the enhanced e–e repulsion between adsorbates and oxide electronic states in response to the electron transfer from the Mo donors.[31] The magnitude of the effect is smaller in XPS than in STM due to the extreme surface sensitivity of the latter technique. Note that the presence of neutral or positively-charged adsorbates would lead to a constant or down-shifted state-density, in disagreement with the experimental data. The corresponding shift of the Mo 3d core levels could not be resolved in our XP data because of the low dopant

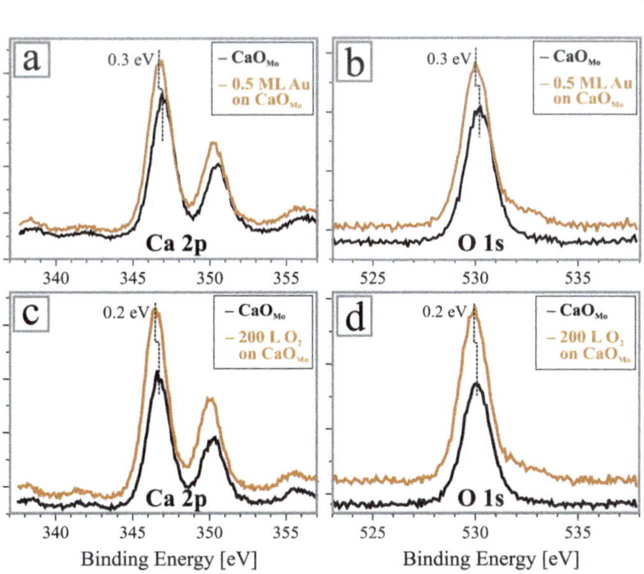

Fig. 5 XP spectra of Ca 2p and O 1s peaks of a Mo-doped CaO film before and after (a,b) Au deposition and (c,d) O_2 exposure. In both cases, the oxide states experience a shift to lower binding energies as a response to the charge transfer into the adsorbates.

concentration and the poor signal-to-noise ratio. This information needs to be obtained with a synchrotron-based experiment in future.

A charge transfer into oxygen and mediated by this, a strengthening of the O_2–CaO interaction, is also in line with the observed thermal stability of O_2 at room temperature and the easiness to image the molecules in the STM even on thick oxide films. On non-doped wide-gap insulators, such as MgO and Al_2O_3, molecular O_2 typically desorbs well below 300 K and is hard to detect in STM even at cryogenic temperature.[41,42] Moreover, recent electron paramagnetic resonance studies indicated the formation of O_2^- species on ultrathin MgO/Ag films, with the only difference that the extra electrons originated from the metal support below and not from donors inside the oxide film.[43] The maximum density of O_2^- species was determined to be 10^{13} cm^{-2}, a value that is likely higher on the more basic CaO surface in our case.[43] We finally note that surface-bound oxygen may not only alter the charge transfer into the gold, but also reduce the Au–CaO adhesion directly, *e.g.* by forming a chemisorbed oxygen layer just below the growing ad-islands.[19,44] The reduced interface interaction in this case would shift the Au equilibrium shape towards 3D, reflecting Wulff's principle that a smaller adhesion leads to larger particle heights.[6,45] We expect, however, that this effect is less relevant than the charge competition between electron-accepting species in our experiment.

In the last paragraph, we discuss the improved crystallinity of Au deposits grown at O_2-rich conditions and try to rationalize the observed particle geometries. The most abundant particle type is a roof-like crystallite oriented along a CaO[100] direction (Fig. 4b). Its elongated shape indicates a strong unidirectional growth and a pseudomorphic relationship with the CaO(001) lattice beneath. As a hexagonal Au(111) interface, being the thermodynamically favoured termination of fcc-gold, is incompatible with an epitaxial growth on the square oxide lattice, we suggest that gold crystallizes in an hcp lattice and develops $(11\bar{2}0)$ interface planes. This growth motive has recently been identified for Ni and Mg deposits grown on the iso-structural MgO(001) surface.[10,46] Due to the two-fold symmetry of the hcp $(11\bar{2}0)$ plane, a dense array of metal–oxygen bonds is formed across the interface that substantially contributes to the metal–oxide adhesion. A similar interface geometry can now be realized for hcp Au $(11\bar{2}0)$ planes on the slightly larger CaO(001) lattice. The required distortion of the Au lattice amounts to 2% and −4% along the two orthogonal CaO[100] directions, respectively, values that are in reach for metal nanoclusters with their high structural flexibility (Fig. 4d). A $(11\bar{2}0)$ interface would also explain the unidirectional growth of gold, as the particles primarily expand along the low-strain direction on the CaO support. Even the side-facets of the roof-shaped Au deposits can be assigned with this model. The most plausible candidates are the hcp (0001) planes, which are tilted by 45° against the surface plane and characterized by particularly low surface free energies due to their closed-packed nature. We finally note that DFT calculations found similar total energies for Au_{40} clusters in either hcp or fcc configuration on the CaO(001), suggesting that both growth modes are energetically accesible.[46]

Apart from the roof-shaped crystallites, none of the other particle types develop an epitaxial relationship with the square CaO(001) lattice (Fig. 4c,e). Evidently, those deposits have other means to interact with the oxide support and are not forced to grow in registry with the oxide lattice. Icosdahedral and pyramidal particle shapes have been observed for various metals (Au, Ag, Pd and Cu) on different oxide supports before, *e.g.* on hexagonal $Al_2O_3(0001)$,[5] square MgO(001)[10] and square $SrTiO_3(001)$,[47,48] especially at high deposition temperature. This universality suggests that such geometries are mainly determined by the low formation energy of (111) facets of fcc metals and follow the classical Wulff construction principle for metal crystallites on weakly interacting supports.[45,49] The ratio between epitaxial-hcp and symmetry-mismatched fcc deposits is again connected to the role of oxygen in the particle growth. At O_2-rich conditions (1×10^{-6} mbar), the ad-gold is unable to acquire charges from the Mo donors and remains neutral. Interfacial Au–O_{CaO}

bonds are thus a crucial means to anchor the particles on the CaO surface, a fact that promotes the formation of epitaxial hcp(11$\bar{2}$0) interface planes. With an enhanced electron flow to the gold at lower O_2 pressures, charge-mediated interaction schemes become available. As a result, the particles gain structural flexibility and adopt growth shapes that are energetically more favorable than the hcp configuration. In the limit of zero oxygen, the Mo excess charges entirely govern the growth on the CaO surface and the gold spreads into large, monolayer islands.[29]

We note that other oxygen-induced effects may contribute to the particle shapes observed in our experiment. Oxygen is a known surfactant[50] and increases the mobility of Au adatoms by forming intermediate metal–oxygen units. A higher diffusivity of the incoming species naturally improves the crystallinity of the ad-particles. In addition, O_2 adsorption may selectively stabilize certain Au facets,[44] modifying the particle equilibrium shape as compared to UHV conditions.[25,26] Given the weakness of the oxygen–Au bond,[39] we believe however that direct adsorption of O_2 species is only of minor importance for the growth morphology of the Au particles.

Summary

Using a combined STM and XPS approach, we have demonstrated that the equilibrium shape of nanoparticles on doped CaO films results from a competition for electrons between Au atoms and oxygen species. At vacuum conditions, excess electrons provided by Mo ions in the oxide lattice are transferred to incoming Au atoms, which subsequently agglomerate into flat 2D islands in order to maximize the metal–oxide adhesion. Simultaneous oxygen supply from the gas phase interrupts this electron flow, as the surface O_2 species are able to trap extra electrons as well. Once the majority of electrons are stabilized by O_2 molecules, the Au growth changes from a 2D to a 3D mode, the latter being characteristic for non-doped oxides. The oxygen-induced morphology crossover is irreversible and the Au particles retain a 3D shape even after the oxygen has been removed from the system.

Our paper highlights the complexity of charge-transfer processes between dielectric supports, ad-metals and molecules, a combination that typically occurs in heterogeneous catalysis. The direction of the charge flow is given by the actual potential situation at the catalyst surface, the relative energies of competing donor and acceptor levels and finally the electro-negativity of the involved molecules. The efficiency of the charge transfer therefore depends not only on intrinsic catalyst properties but also on the composition of the surrounding gas phase. The highly dynamic nature of charge-transfer processes calls for an exploration of catalytically-relevant systems directly at operation conditions. Our findings also disclose a possibility to alter the growth mode of metals simply by changing the gas atmosphere during deposition, which opens a new pathway to optimize structure and morphology of supported metal catalysts.

Acknowledgements

The authors are grateful for financial support from the DFG within the Cluster of Excellence 'Unicat'. Y.C. thanks the Humboldt Foundation for a fellowship.

References

1 L. M. Molina and B. Hammer, *Appl. Catal., A*, 2005, **291**, 21–31.
2 M. Chen and D. W. Goodman, *Acc. Chem. Res.*, 2006, **39**, 739–746.
3 I. X. Green, W. Tang, M. Neurock and J. T. Yates, *Science*, 2011, **333**, 736–739.
4 A. A. Herzing, C. J. Kiely, A. F. Carley, P. Landon and G. J. Hutchings, *Science*, 2008, **321**, 1331–1335.
5 K. H. Hansen, T. Worren, S. Stempel, E. Lægsgaard, M. Bäumer, H. J. Freund, F. Besenbacher and I. Stensgaard, *Phys. Rev. Lett.*, 1999, **83**, 4120–4123.

6 M. Bäumer and H.-J. Freund, *Prog. Surf. Sci.*, 1999, **61**, 127–198.
7 C. R. Henry, *Prog. Surf. Sci.*, 2005, **80**, 92–116.
8 G. J. Hutchings, *Chem. Commun.*, 2008, 1148–1164.
9 G. Pacchioni and N. Rosch, *J. Chem. Phys.*, 1996, **104**, 7329–7337.
10 S. Benedetti, P. Myrach, A. di Bona, S. Valeri, N. Nilius and H.-J. Freund, *Phys. Rev. B: Condens. Matter Mater. Phys.*, 2011, **83**, 125423.
11 X. Lin, N. Nilius, H. J. Freund, M. Walter, P. Frondelius, K. Honkala and H. Häkkinen, *Phys. Rev. Lett.*, 2009, **102**, 206801–206804.
12 D. Starr, S. Shaikhutdinov and H.-J. Freund, *Top. Catal.*, 2005, **36**, 33–41.
13 J.-D. Grunwaldt and B. Clausen, *Top. Catal.*, 2002, **18**, 37–43.
14 P. L. Hansen, J. B. Wagner, S. Helveg, J. R. Rostrup-Nielsen, B. S. Clausen and H. Topsøe, *Science*, 2002, **295**, 2053–2055.
15 G. Jacobs, J. A. Chaney, P. M. Patterson, T. K. Das and B. H. Davis, *Appl. Catal., A*, 2004, **264**, 203–212.
16 S. Mostafa, F. Behafarid, J. R. Croy, L. K. Ono, L. Li, J. C. Yang, A. I. Frenkel and B. R. Cuenya, *J. Am. Chem. Soc.*, 2010, **132**, 15714–15719.
17 M. W. Small, S. I. Sanchez, N. S. Marinkovic, A. I. Frenkel and R. G. Nuzzo, *ACS Nano*, 2012, **6**, 5583–5595.
18 M. D. Ackermann, T. M. Pedersen, B. L. M. Hendriksen, O. Robach, S. C. Bobaru, I. Popa, C. Quiros, H. Kim, B. Hammer, S. Ferrer and J. W. M. Frenken, *Phys. Rev. Lett.*, 2005, **95**, 255505–255508.
19 T. Schalow, M. Laurin, B. Brandt, S. Schauermann, S. Guimond, H. Kuhlenbeck, D. E. Starr, S. K. Shaikhutdinov, J. Libuda and H. J. Freund, *Angew. Chem., Int. Ed.*, 2005, **44**, 7601–7605.
20 S. Giorgio, M. Cabié and C. R. Henry, *Gold Bull.*, 2008, **41**, 167–173.
21 B. J. McIntyre, M. Salmeron and G. A. Somorjai, *J. Vac. Sci. Technol., A*, 1993, **11**, 1964–1968.
22 G. A. Somorjai and G. Rupprechter, *J. Phys. Chem. B*, 1999, **103**, 1623–1638.
23 K. P. McKenna and A. L. Shluger, *J. Phys. Chem. C*, 2007, **111**, 18848–18852.
24 F. Tao, S. Dag, L.-W. Wang, Z. Liu, D. R. Butcher, H. Bluhm, M. Salmeron and G. A. Somorjai, *Science*, 2010, **327**, 850–853.
25 Z. Zhu, F. Tao, F. Zheng, R. Chang, Y. Li, L. Heinke, Z. Liu, M. Salmeron and G. A. Somorjai, *Nano Lett.*, 2012, **12**, 1491–1497.
26 M. Cabié, S. Giorgio, C. R. Henry, M. R. Axet, K. Philippot and B. Chaudret, *J. Phys. Chem. C*, 2010, **114**, 2160–2163.
27 G. Rupprechter and H.-J. Freund, *Top. Catal.*, 2000, **14**, 3–14.
28 H. Yoshida, Y. Kuwauchi, J. R. Jinschek, K. Sun, S. Tanaka, M. Kohyama, S. Shimada, M. Haruta and S. Takeda, *Science*, 2012, **335**, 317–319.
29 X. Shao, S. Prada, L. Giordano, G. Pacchioni, N. Nilius and H. J. Freund, *Angew. Chem., Int. Ed.*, 2011, **50**, 11525–11527.
30 G. Pacchioni, L. Giordano and M. Baistrocchi, *Phys. Rev. Lett.*, 2005, **94**, 226104–226107.
31 X. Shao, N. Nilius and H. J. Freund, *J. Am. Chem. Soc.*, 2012, **134**, 2532–2534.
32 X. Shao, P. Myrach, N. Nilius and H. J. Freund, *J. Phys. Chem. C*, 2011, **115**, 8784–8789.
33 X. Shao, N. Nilius, P. Myrach, H. J. Freund, U. Martinez, S. Prada, L. Giordano and G. Pacchioni, *Phys. Rev. B: Condens. Matter Mater. Phys.*, 2011, **83**, 245407–245411; X. Shao, N. Nilius and H. J. Freund, *Phys. Rev. B: Condens. Matter Mater. Phys.*, 2012, **85**, 115444–115451.
34 N. U. Zhanpeisov, V. Staemmler and M. Baerns, *J. Mol. Catal. A: Chem.*, 1995, **101**, 51–60.
35 L. Wang, *Appl. Surf. Sci.*, 2011, **257**, 5499–5502.
36 B. C. Stipe, M. A. Rezaei, W. Ho, S. Gao, M. Persson and B. I. Lundqvist, *Phys. Rev. Lett.*, 1997, **78**, 4410–4413.
37 P. Frondelius, H. Häkkinen and K. Honkala, *Angew. Chem., Int. Ed.*, 2010, **49**, 7913–7916.
38 F. Stavale, X. Shao, N. Nilius, H. J. Freund, S. Prada, L. Giordano and G. Pacchioni, *J. Am. Chem. Soc.*, 2012, **134**, 11380–11383.
39 D. R. Lide (Ed), *CRC Handbook of Chemistry and Physics*, (Boca Raton, CRC Press, 2010).
40 H.-M. Benia, P. Myrach, A. Gonchar, T. Risse, N. Nilius and H.-J. Freund, *Phys. Rev. B: Condens. Matter Mater. Phys.*, 2010, **81**, 241415–241418.
41 R. M. Jaeger, J. Libuda, M. Bäumer, K. Homann, H. Kuhlenbeck and H. J. Freund, *J. Electron Spectrosc. Relat. Phenom.*, 1993, **64–65**, 217–225.
42 H.-J. Freund, Th. Klüner, R. Wichtendahl, S. Thiel, M. Adelt, W. Drachsel, M. Bäumer, H. Kuhlenbeck, T. Risse, K. Al-Shamery, M. Kampling, H. Hamann, 'Metal–Ligand Interactions in Chemistry, Physics and Biology', *NATO ASI Ser.*, Vol. E, p. 91–128, (Kluwer, Academic Publ, 2000).
43 A. Gonchar, T. Risse, H. J. Freund, L. Giordano, C. Di Valentin and G. Pacchioni, *Angew. Chem., Int. Ed.*, 2011, **50**, 2635–2638.

44 J. M. Gottfried, K. J. Schmidt, S. L. M. Schroeder and K. Christmann, *Surf. Sci.*, 2002, **511**, 65–82.
45 G. Wulff, *Z. Kryst.*, 1901, **34**, 449–530.
46 R. Ferrando, G. Rossi, F. Nita, G. Barcaro and A. Fortunelli, *ACS Nano*, 2008, **2**, 1849–1856.
47 F. Silly and M. R. Castell, *Phys. Rev. Lett.*, 2006, **96**, 086104–086107.
48 F. Silly and M. R. Castell, *Appl. Phys. Lett.*, 2005, **87**, 213107–213103; F. Silly, A. C. Powell, M. G. Martin and M. R. Castell, *Phys. Rev. B: Condens. Matter Mater. Phys.*, 2005, **72**, 165403–165408.
49 N. Cabrera, *Surf. Sci.*, 1964, **2**, 320–345.
50 H. Wolter, K. Meinel, C. Ammer, K. Wandelt and H. Neddermeyer, *Phys. Rev. B: Condens. Matter*, 1997, **56**, 15459–15470.

Preparation and structure of a single Au atom on the $TiO_2(110)$ surface: control of the Au–metal oxide surface interaction

Kiyotaka Asakura,*[a] Satoru Takakusagi,[a] Hiroko Ariga,[a] Wang-Jae Chun,[b] Shushi Suzuki,[c] Yuichiro Koike,[a] Hiromitsu Uehara,[a] Kotaro Miyazaki[a] and Yasuhiro Iwasawa[d]

Received 23rd November 2012, Accepted 12th December 2012
DOI: 10.1039/c2fd20131c

Three-dimensional Au structures on bare and organic-compound-modified $TiO_2(110)$ surfaces were interrogated by Au L_3-edge polarization dependent total reflection fluorescence X-ray absorption fine structure (PTRF-XAFS) spectroscopy. On the bare $TiO_2(110)$ surface, icosahedral Au_{55} nanoclusters were the main product found. When the surfaces were modified with *ortho* or *meso* mercaptobenzoic acid (*o*-MBA or *m*-MBA), Au was atomically dispersed. Sulfur atoms in the *o*- and *m*- MBA formed strong covalent bonds with Au to produce stable Au-MBA (*o*- and *m*- forms) surface complexes. On the other hand, only oxygen atoms on the surface did not make a strong enough interaction to stabilize the Au species. We discuss how the Au species formed on the modified $TiO_2(110)$ surface and the possibility to control the Au structure by the surface modification method.

Introduction

Control of the structure of transition metal species on oxide surfaces is important for preparation of well-defined catalysts. For this purpose, we have to understand the origins of the metal–metal oxide interactions so that we can regulate the interaction. Recently, we have discussed this point using polarization dependent total reflection fluorescence X-ray absorption fine structure (PTRF-XAFS) studies of single crystal oxide surfaces.[1] XAFS is an oscillatory feature that appears near and above the X-ray absorption edge. XAFS provides information about the local environment around the X-ray absorbing atoms. Usually XAFS is applied to a powder sample to obtain the local structure of supported metal catalysts directly.[2] However, because the structural information derived from XAFS is averaged over all spatial directions, it is dicult to deduce the nature of the metal–support interactions directly. The XAFS signal ($\chi(k)$) is polarization dependent, as shown in eqn(1) and (2) where

[a]*Catalysis Research Center, Hokkaido University Kita 21-10, Kita Sapporo 001-0021, Japan. E-mail: askr@cat.hokudai.ac.jp; Fax: +81-11-706-9113; Tel: +81-11-706-9113*
[b]*Graduate School of Arts and Sciences, International Christian University, Mitaka, Tokyo 181-8585, Japan. E-mail: wchun@icu.ac.jp; Fax: +81-422-33-1449; Tel: +81-422-33-3293*
[c]*Department of Crystalline Materials Science, Graduate School of Engineering, Nagoya University, Nagoya 464-8603, Japan. E-mail: shushi@apchem.nagoya-u.ac.jp; Fax: +81- 52-789-5299; Tel: +052-789-5299*
[d]*Innovation Research Center for Fuel Cells, Department of Engineering Science, Graduate School of Information Engineering Science, The University of Electro-Communications, 1-5-1 Chofugaoka, Chofu, Tokyo 182-8585, Japan. E-mail: iwasawa@pc.uec.ac.jp; Fax: +81-(0)42-443-5483; Tel: +81-(0)42-443-5921*

θ_j and $\chi_j(k)$ is the angle between the bond and polarization and XAFS oscillation for each bond, respectively. This polarization dependency can be leveraged in XAFS on single crystal oxide surfaces to obtain three-dimensional structural information and the metal support interaction can be determined.[1]

$$\chi(k) = \sum_j 3\cos^2\theta_j \cdot \chi_j(k) \text{ for K} - \text{edge} \tag{1}$$

$$\chi(k) = \sum_j (0.7 + 0.9\cos^2\theta_j) \cdot \chi_j(k) \text{ for L}_{2,3} - \text{edge} \tag{2}$$

We have studied Ni, Mo, Cu, and Pt dispersed on Al_2O_3 (0001) and TiO_2 (110) single crystals. Overall, these PTRF-XAFS studies have demonstrated that the metal species binds to the crystal surface through the metal–anion bond or metal–O bond.[9]

The TiO_2(110) surface is a typical oxide surface that has several dierent adsorption sites, as shown in Fig. 1. These are referred to as three-fold hollow sites, bridge and atop sites of surface oxygen atoms, step edges and kinks, 5-fold Ti sites, and oxygen defect sites.[3] In our previous XAFS studies we did not observe evidence for metal–cation bonds within the covalent bond range. We concluded that even if there are interactions between the adsorbed metal species and surface cations (Ti), they are not strong enough to fix the metal species. Early transition metals form bonds with surface oxygen atoms to give monomers, dimers, or chains, depending on the surface structure and preparation conditions.[4,5,6] The late transition metals also interact with oxygen dangling bonds. Atomically dispersed Ni was found at the kink site of a step where two oxygen dangling bonds were present.[7,8] The Ni–O bond distances for the upper and lower terrace oxygen atoms were 0.199 and 0.204 nm, respectively. The Ni adsorbed at a Ti vacancy site, and the variation in Ni–O bond lengths for the upper and lower terrace oxygen atoms, reflected the dierent Ti–O bond lengths in the TiO_2 rutile crystal. Single-layer Ni clusters are stabilized by multiple bonds to surface oxygen atoms.[9] Single-layer Ni was seen to deform to an fcc (110)-like structure to increase contact with the surface oxygen atoms. About 90% of the Ni atoms had averaged Ni–O bond distances of 2.3 ± 0.1 Å.

On the other hand, it was difficult to obtain a monoatomically dispersed Cu. When we deposited Cu on the bare TiO_2(110), we could not obtain Cu monomer species. Instead, three-dimensional Cu clusters formed, even at 0.02 ML where we obtained Ni monomer species as mentioned above.[8] This was because the Cu–O binding energy was much smaller than that of Ni–O. The formation enthalpies of Cu_2O and CuO

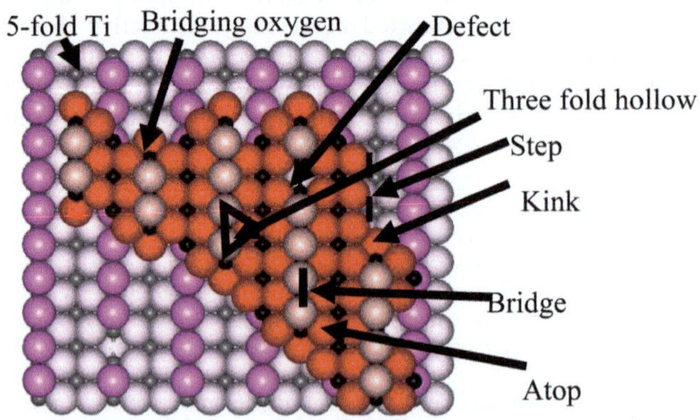

Fig. 1 The TiO_2(110) model structure. Large and small spheres represent O and Ti, respectively. There are several adsorption sites on TiO_2(110) as labeled in the figure.

were −157.3 and −168.6 kJ mol^{-1}, respectively, while that of NiO was −239.7 kJ mol^{-1}.[10] We were able to obtain Cu monomers in three dierent ways. The first approach was to deposit the Cu(dpm)$_2$ (dpm = dipivaloylmethane) complex. The Cu(dpm) species attached to the bridge site of the bridging oxygens, and was stabilized by the dpm ligand to form a tetrahedral structure.[11,12] The monomer Cu species could be transformed into Cu trimers and hexamers with a low temperature reduction treatment.[13] The second way was achieved through surface modification by an anchored S-atom-containing organic molecule. Since Cu is a soft Lewis acid, it should have a larger anity toward sulfur (soft base) than oxygen (hard acid) and atomically dispersed Cu can be obtained. We used thiophene carboxylic acid (TCA) to modify the surface.[13] Because the carboxylic acid stably adsorbed on the TiO_2 (110) surface, the TCA was fixed firmly on it through the carboxylic acid moiety and the S atom in the thiophene ring was directed upwards. Consequently, a S-modified TiO_2(110) surface was obtained. Cu deposited on this surface was atomically dispersed and had a sandwiched structure between S and O with a linear arrangement.[14] Recently we successfully fixed Cu using mercaptobenzoic acids (*o*-, *m*-, *p*-MBA)-modified TiO_2(110).[14] Our third approach was to control diusion of the Cu atom on the TiO_2(110) surface. As mentioned above, the Cu should have stronger interactions with bridging oxygen sites than with five-fold Ti sites. Cu may then diuse along on the bridging oxygen rows. Through this facile diusion, Cu atoms can easily migrate to find a Cu island and aggregate to give three-dimensional clusters. If the Cu diusion was suppressed, Cu nucleation would occur locally before reaching a larger island, leading to smaller cluster formation. At the extreme case, atomically dispersed metal species would be created. We attempted to block Cu diusion by depositing acetic anhydride on the TiO_2 surface; the acetic anhydride dissociatively absorbs on TiO_2 to produce two types of acetates.[15] Acetate species A was located on the 5-fold Ti sites in a bridge-bidentate structure. On the other hand, acetate species B incorporated one of the TiO_2 bridging oxygen atoms, with its original carbonyl oxygen adsorbed on a five-fold Ti site. Acetate B species were therefore oriented perpendicular to the bridging oxygen row. Acetate B can be converted to acetate A by leaving the defect. On this surface, Cu diusion was considerably restricted by the presence of acetate B and/or the oxygen defect. Finally, the Cu atoms were trapped by the nearby acetate and became fixed on the surface through bonds with two oxygen atoms: the acetate B carbonyl and the bridging oxygen atoms.[16] In the latter two methods, the atomically dispersed Cu atoms were always stabilized by two bonding motifs: sandwiched by S and the bridging oxygen or by the acetate and the bridging oxygen.

Metallic Au is chemically stable and less reactive. Recently, Au nanoclusters have drawn much attention because of their low temperature activity toward CO oxidation.[17] It was quite dicult to obtain atomically dispersed Au species because of the low anity of surface oxygen. When Au was deposited on the TiO_2(110) (1 × 1) and (1 × 2) surfaces, the main product formed was an icosahedral Au_{55}.[18,19] Even after surface modification with TCA, atomically dispersed Au could not be obtained, although the size of the Au clusters became smaller, as shown later. We further enhanced the Au–surface interaction by modification of the surface with *o*- and *m*-MBA compounds to obtain atomically dispersed Au species. In this paper, we report the local structure of Au as determined by PTRF-XAFS, and compared the local structures with other surface Au species to discuss the principles of Au atomic dispersion.

Experimental

Optically polished Nb-doped (0.05 wt%) TiO_2(110) samples (20 × 20 × 1 mm^3, Furuuchi Co., Japan) were cleaned by immersion in 10% HF solution for 10 min, followed by annealing in air at 700 °C for 1 h according to previous work.[20,21] The cleaned surface was immersed in 1 mM ethanol solutions of either *o*-MBA or *m*-MBA (Toronto Research Chemicals Inc, Canada) at room temperature for 24 h to modify the TiO_2(110) surface with the corresponding MBA monolayers.

The prepared sample was then transferred to the UHV PTRF-XAFS chamber. Au was evaporated onto the MBA-modified $TiO_2(110)$ surfaces by resistive heating of a tungsten filament wrapped with Au wire (99.99+% purity, Nilaco Co., Japan). Au coverage was estimated from X-ray photoelectron spectroscopy (XPS) peak area ratios of Au $4f_{7/2}$ to Ti $2p_{3/2}$. 1 ML was defined as 5.2×10^{14} cm^{-2} corresponding to the $TiO_2(1 \times 1)$ unit cell. TCA was deposited on the Au surface as reported previously.[14] PTRF-XAFS experiments were carried out at BL9A of the Photon Factory at the Institute of Material Structure Science (KEK-IMSS-PF). X-Rays were monochromatized with a Si(111) double-crystal and focused using a pair of bent conical mirrors.[22] The beam size on the sample was regulated with a pinhole ($\phi = 0.4$ mm) to reduce undesirable X-ray irradiation other than on the sample. The total reflection conditions were adjusted using a UHV-compatible high-precision six-axis goniometer as reported elsewhere.[1,23] Considering the rectangular unit cell of the $TiO_2(110)$ surface,[3] XAFS measurements were carried out in three dierent orientations relative to the electric vector($\vec{E}$) of the incident X-rays. PTRF-XAFS spectra for the orientation parallel to the surface were obtained from the $\vec{E}$d// [001] and $\vec{E}$// [1$\bar{1}$0] directions. We acquired the structure in the perpendicular direction mainly from the // [110] XAFS signal. The Au L_α fluorescence was detected by a 19-element Ge solid-state detector (GL0110S, Canberra, Meriden, USA). Self-absorption was negligible because the coverage was smaller than 1 ML.

XAFS analysis was carried out by a REX 2000 program (Rigaku Co. Tokyo, Japan).[24] The XAFS oscillations were extracted using a spline smoothing method and normalized by an edge height. The preliminary analysis was carried out by a least-squares curve fitting method using FEFF derived parameters:

$$\chi(k) = \frac{\sum_j (0.7 + 0.9\cos^2\theta_j) S_j F_j(k_j) e^{-2k_j^2\sigma_j^2} \sin(2k_j r_j + \varphi_j(k_j))}{k_j r_j^2}$$
$$k_j = \sqrt{k^2 - 2m\Delta E_j/\hbar^2} \tag{3}$$

directions, S_j, σ_j, r_j, ΔE_j are the amplitude reduction factor, Debye–Waller factor, bond distance, and the energy shift in the origin of the photoelectron kinetic energy of the j-th bond, respectively. $F_j(k)$ and $\phi_j(k)$ are the backscattering amplitude and the phase shift functions, respectively. We used the same fitting parameters within the same coordination shell. The final structure was determined by an iterative method using the FEFF8.04 code[25,26] based on a real-space model structure.[27,28] The FEFF calculation was carried out with a Hedin–Lundqvist potential. The Debye–Waller factor for Au–Au was fixed to one value for each polarization direction. The polarization dependence expressed in eqn (2) was also considered in the simulation. To quantify the goodness of the fits we carried out a chi squared test using the following equation:

$$X^2 = \frac{\nu}{(\nu - 1)N_{\text{data}}} \sum \frac{[\chi_{\text{obs}}(k) - \chi_{\text{cal}}(k)]^2}{[\sigma_{\text{obs}}(k)]^2} \tag{4}$$

where $\chi_{obs}(k)$ and $\chi_{cal}(k)$ are the observed and calculated XAFS oscillations, respectively. N_{data} is the number of data points, and $\sigma_{obs}(k)$ is the measurement error, estimated from the standard deviations of the observed data. To avoid confusion between the XAFS oscillation and the chi square test values, we here used symbol X^2 instead of chi squared (χ^2).

Results

First, we briefly reviewed the Au structure on the bare $TiO_2(110)$ surface reported elsewhere.[19,20] There was little polarization dependence in the Au XAFS oscillations, indicating that three-dimensional Au nanoclusters were formed. The Au–Au distance of 2.70 Å was shorter than that of bulk Au (2.88 Å). The shorter bond distance resulted from the formation of small ~1 nm diameter clusters.[29] A real space simulation showed that Au_{55} icosahedral structures were the main Au product formed on the bare $TiO_2(110)$ surface.

It was difficult to create an atomically dispersed Au. When Au was deposited on a $TiO_2(110)$ (1 × 2) surface, which contains more Ti^{3+} and has a rougher surface, we found that the same Au_{55} icosahedral species formed on this $TiO_2(110)$ surface.[20] When the surface was modified with TCA, which helped to stabilize the Cu atoms on the $TiO_2(110)$ surface, we observed a reduction in the size of Au clusters formed on the surface.

Fig. 2 shows the XAFS oscillations after the deposition of Au on the TCA-modified $TiO_2(110)$ surface. The amplitude of XAFS oscillations were smaller than those of Au deposited on the bare $TiO_2(110)$ surface, even though Au–Au bonds were formed and agglomeration occurred, as mentioned in the Introduction. The Au–Au distance was 2.68 Å, with smaller Au–Au coordination numbers relative to the unmodified (or bare) surface, as shown in Table 1, indicating that smaller nanoparticles were formed on the TCA-covered surface than on the bare surface. We also observed a polarization dependence. The polarization dependence of the

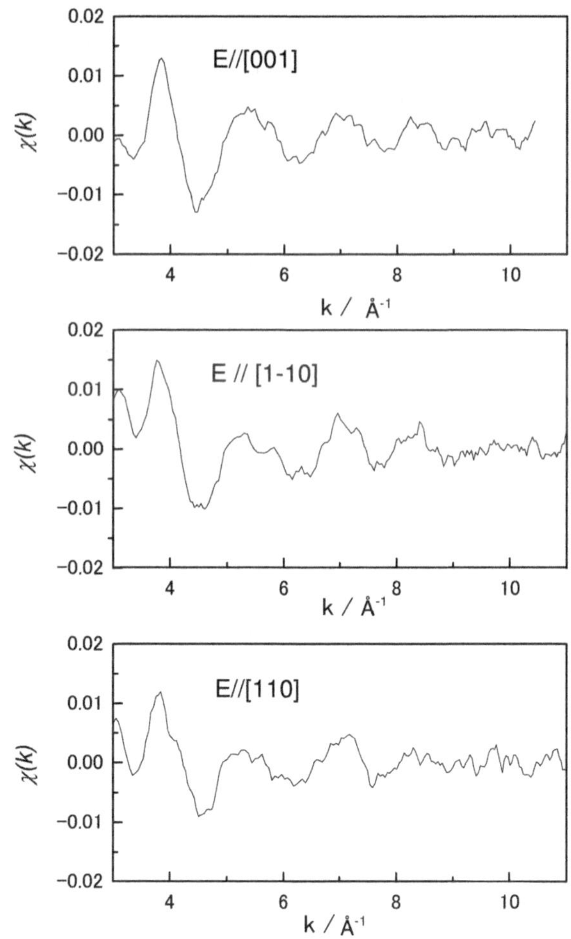

Fig. 2 XAFS oscillations $\chi(k)$ of Au on a TCA-modified $TiO_2(110)$ surface with dierent polarization dependencies, as indicated in the figure.

coordination numbers in Table 1 indicated that two-layer fcc-like small Au_{10} (~6 Å) nanoclusters were likely to be stabilized on the TCA surface. The TCA modified surface experienced stronger interactions with Au than the bare $TiO_2(110)$ surface through the S–Au bond to stabilize the two-dimensional structure. However, the Au–TCA interaction was not strong enough to stabilize the Au monomer, as was seen with the Cu–TCA interaction.

Then we modified the $TiO_2(110)$ surface with *o*-MBA. Fig. 3 shows the polarization-dependent XAFS oscillations of the Au species deposited on the $TiO_2(110)$ surface modified with *o*-MBA. For this surface, we observed totally dierent Au spectra when compared with those prepared on the TCA-modified $TiO_2(110)$ surfaces. The preliminary curve fitting analysis indicated the presence of Au–S bond distances of 2.32 Å. We did not observe Au–Au interactions. We carried out the real space analysis based on the FEFF8 program. Fig. 4 shows the model structure that could reproduce the XAFS oscillations in three directions, as shown in Fig. 5. In this structure, Au was sandwiched between sulfur and a bridging oxygen atom. The Au–S and Au–O bond lengths were 2.32 and 2.10 Å, respectively. The S–Au–O and C–S–Au bond angles were 155 and 143°, respectively. The angular deviation between the S–Au bond direction and the [001] direction was 59°. The atomically dispersed Au structure was thermally stable up to 473 K and no Au–Au bond was observed in the PTRF-XAFS. The Au was probably able to be eciently captured by the S moiety in the *o*-MBA modified surface because of the S-atom positioning relative to the TiO_2 surface; S was seen to locate at about 4 Å above the surface in the proposed model structure which would be appropriate for the Au capture. On the other hand, the TCA modified surface had a S atom at 6.5 Å above the native surface. The monodentate form of the carboxylic structure was required to trap Cu atoms. An alternative reason for the superior Au trapping ability of the *o*-MBA surface could be that the S–Au bond formed on the MBA surface was stronger than that in TCA. To check this point, we modified the TiO_2 surface with *m*-MBA, which eectively moves the S atom to a position about 7 Å away from the $TiO_2(110)$ surface when bridge–bidentate carboxylate adsorption was assumed. Fig. 6 shows the XAFS spectra of Au dispersed on the *m*-MBA-covered $TiO_2(110)$. The observed spectra are similar to those of Au on the *o*-MBA-covered surfaces. The Au was atomically dispersed on the *m*-MBA-covered $TiO_2(110)$ even though the S position was farther away from the surface. We carried out the real space analysis of the Au structure using the FEFF8 code and the fixed MBA molecular structure except the rotation around the S–C(benzene ring) bond. We found that the Au was again sandwiched by the sulfur and bridging oxygen, similar to the arrangement of Au on the *o*-MBA-modified $TiO_2(110)$ as shown in Fig. 7. Au–S and Au–O bond lengths of 2.32 and 2.10 Å, respectively, were deduced. The S–Au–O and C–S–Au bond angles were 166 and 147°, respectively. The angle between the S–Au direction and the [110] direction was 61°. In this case, we postulated a monodentate *m*-MBA structure on the $TiO_2(110)$ surface. Otherwise the structure shown in Fig. 7 would be impossible and a strained structure would have to be considered. We concluded that the Au stabilization by *o*-MBA was not due to the suitable geometrical position to trap the Au, but was instead due to the formation of a strong S–Au bond.

Table 1 Curve fitting results of Au on the TCA-modified $TiO_2(110)$ surface with three dierent polarization directions

	Bond	*N*	*R*/Å	*N*(expected)[a]
E//[1$\bar{1}$0]	Au–Au	5.4	2.67	5.2
E//[001]	Au–Au	5.0	2.67	5.2
E//[110]	Au–Au	3.9	2.68	3.9

[a] Coordination number calculated from a Au10 nanocluster.

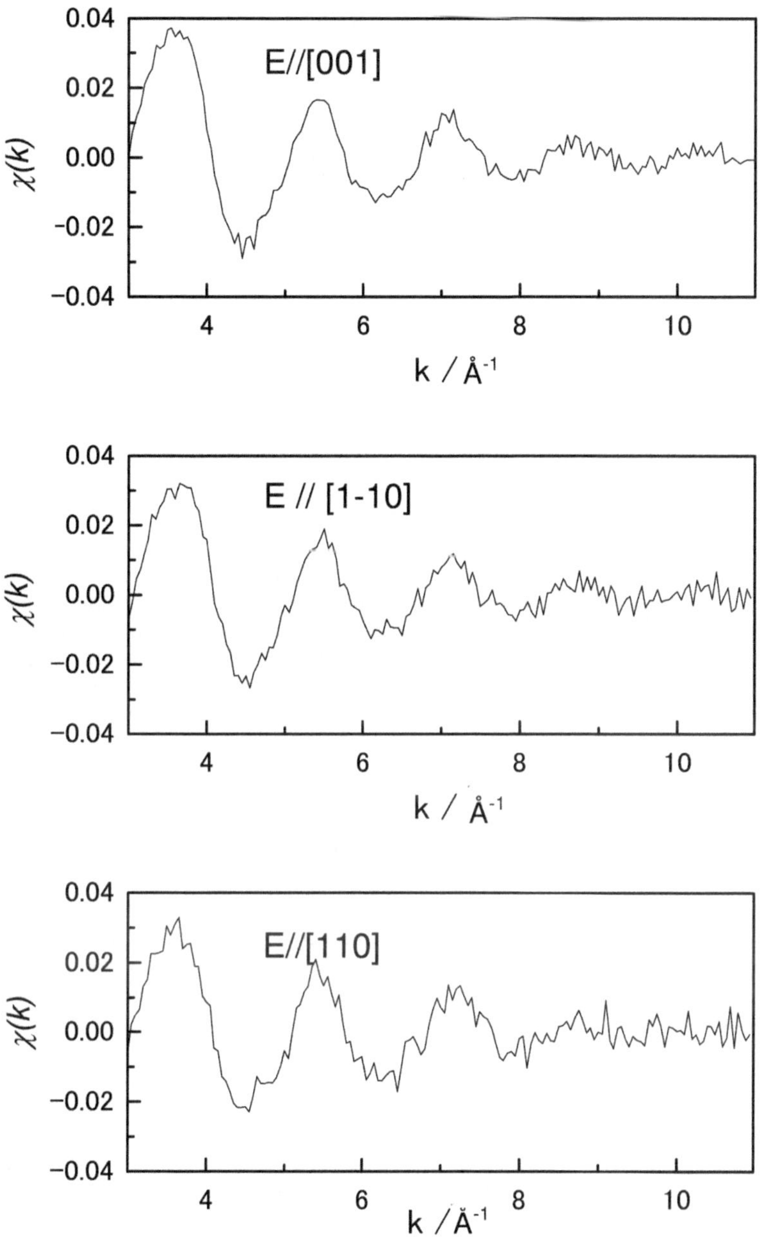

Fig. 3 XAFS oscillations $\chi(k)$ of Au on *o*-MBA-modified $TiO_2(110)$ surface with dierent polarization dependencies, as indicated in the figure.

Discussion

Structure of surface Au species

Au deposited on the bare $TiO_2(110)$ surface easily aggregated to yield a three-dimensional structure. Although rather small Au nanoparticles were formed on the TCA-modified surface, atomically dispersed Au was dicult to obtain. The *o*- and *m*-MBA modified surfaces could provide atomically dispersed Au species. The Au–S bond

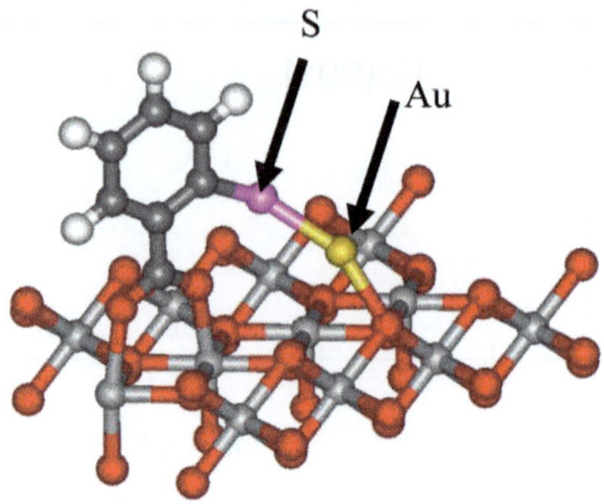

Fig. 4 A model structure for Au on the *o*-MBA modified $TiO_2(110)$ surface. Yellow, purple and red spheres represent S, Au, and O, respectively.

distances in both *o*- and *m*-MBA were 2.32 Å. Table 2 shows the Au–S and Au–O bond distances in some selected compounds. The Au–S bond distances were found to be in the range of 2.29–2.34 Å, corresponding well to the ones found in the Au species on the *o*- and *m*-MBA modified surfaces. Thus the Au was fixed on the surface through Au–S and Au–O covalent bonds. In these compounds the X–Au–S angle was nearly linear and the O–Au–S angle on the $TiO_2(110)$ surface was $160 \pm 10°$.

The reason why thiophene could not fix the atomic Au was based on the different affinity of Au to S atoms in TCA and in *o*- or *m*-MBA. To our knowledge, no Au–thiophene compounds are reported in the literature; however, there are many examples of Au–mercaptobenzene compounds and Au–hydrogenated thiophene compounds called THT (tetrahydrothiophene) which should have dierent bonding feature from that of TCA, as shown in Table 2. Further evidence for the stronger anity of MBA to that of TCA can be found by examining the surface adsorption properties of mercaptobenzene and thiophene. Alkyl mercaptan and mercaptobenzene can be adsorbed on the Au surface up to full coverage to form a self-assembled thin film.[30,31] On the other hand, only a small amount of thiophene was observed to adsorb on the defect sites of the Au (111) surface by vapor deposition at room temperature.[32,33] Dishner *et al.* reported that the immersion method from ethanol solution could provide the thiophene film adsorbed on Au(111).[34] However, Sako *et al.* revealed that the thiophene was converted to butyl mercaptan in the immersion method.[35] Thus, molecular thiophene does not display stable adsorption characteristics on the Au surface. The Au atom can react with the S atom in mercaptobenzene, but not in thiophene. The mercapto moiety can make such a strong bond with Au that it can cleave an O (carboxylate of *m*-MBA)–Ti bond, transforming the MBA bridge–bidentate form to the monodentate form. Such monodentate species can be found in the Cu attached on the thiophene modified surface.[19]

The structure and morphology of Au species on the surface was determined by the balance of anion–Au and Au–Au interactions. As Hu *et al.* reported that Au–O formation enthalpy was more positive than the other metal oxide formation enthalpies, Au on the bare $TiO_2(110)$ grew three-dimensionally,[36] which corresponds well with our observations. The TCA-modified TiO_2 (110) failed to disperse the Au for the same reason. The Au–S interaction was stronger than that of Au–O because of the soft base–acid interactions. MBA could atomically disperse Au species. The Au–TCA interaction was moderate, resulting in smaller Au cluster formation, but not atomically dispersed.

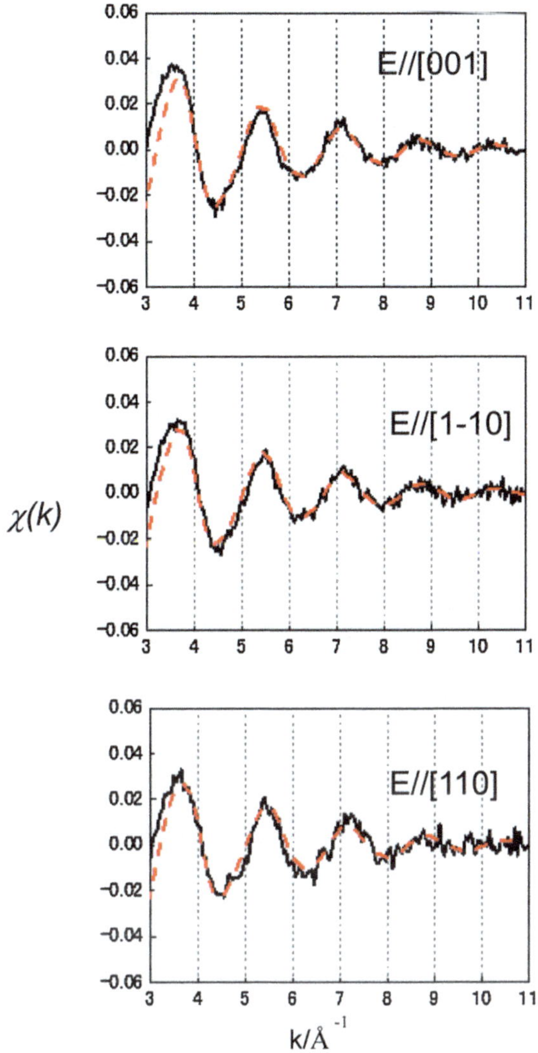

Fig. 5 Comparison between the observed XAFS oscillations $\chi(k)$ (solid line) of Au on the *o*-MBA-modified $TiO_2(110)$ surface and those calculated (broken line) based on the model structure illustrated in Fig. 4.

The formation mechanisms of atomically dispersed Au and small Au cluster on the modified $TiO_2(110)$

On the bare $TiO_2(110)$ surface, deposited Au migrates to seek a stable absorption site. The Au–O bond is too weak to stabilize the Au monomer and the Au continues to diuse until finding an Au island to aggregate and form three-dimensional clusters. In the case of *o*-MBA, the S dangling bond is directed toward the Au trapping site. The deposited Au is easily trapped by the S–Au–O bonding motif and stabilized. In *m*-MBA, the S atom has no bond directed toward the trapping site. The *m*-MBA exists in the following equilibrium between monodentate and bridge–bidentate adsorption structures.

$$\text{m-MBA(bridge–bidentate)} \leftrightarrows \text{m-MBA(monodentate)} \quad (5)$$

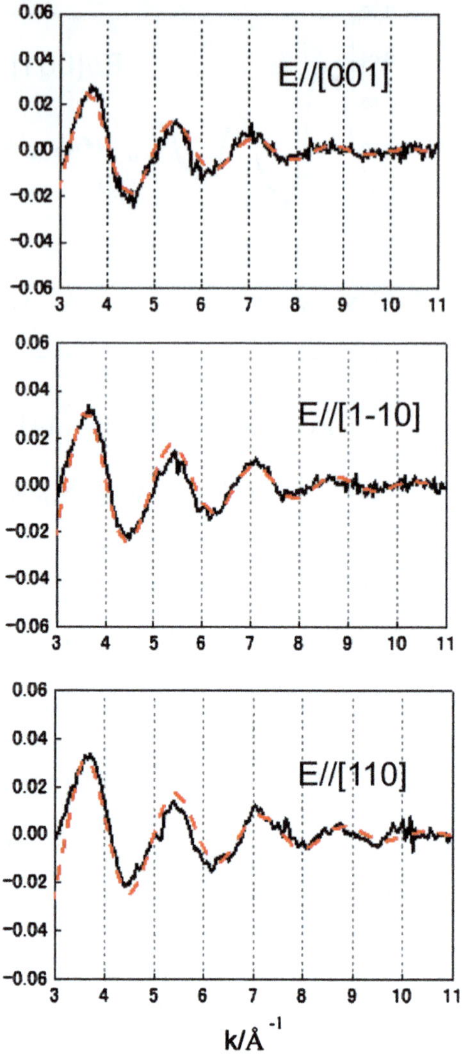

Fig. 6 Comparison between the observed XAFS oscillations $\chi(k)$ (solid line) of Au on the *m*-MBA-modified $TiO_2(110)$ surface and those calculated (broken line) based on the model structure illustrated in Fig. 7.

The equilibrium between monodentate and bridge–bidentate carboxylate structures has been demonstrated by the STM observation of formic acid on $TiO_2(110)$.[48] The monodentate *m*-MBA can change the orientation that can trap Au by bonding with the thiol species.

$$\text{m-MBA(bridge–bidentate)} + \text{Au} + 2\text{O(bridging)} \leftrightarrows \text{m-MBA(monodentate)–Au–O(bridging)} + \text{HO(bridging)} \quad (6)$$

where O(bridging) means a bridging oxygen atom.

In equilibrium (6) the product side is more favorable, and the forward reaction in equilibrium (5) will proceed. Thus, the Au is trapped efficiently on the *m*-MBA-modified $TiO_2(110)$ surface. In the case of TCA, similar equilibria might be present, but because of the weak interactions between thiophene and Au, the product side would not be favorable in equilibrium (7) and trapped Au would not be stable.

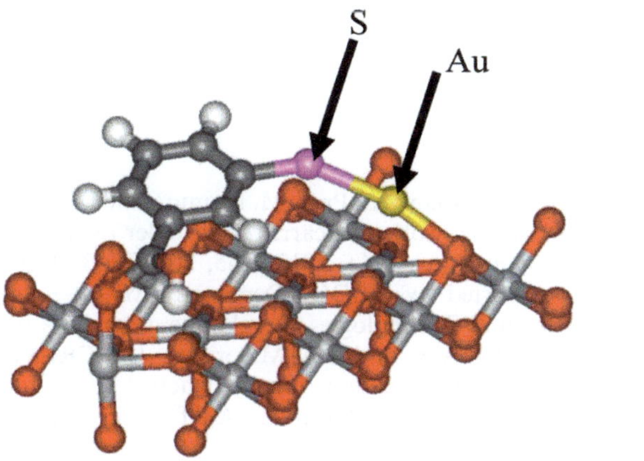

Fig. 7 A model structure of Au on the *m*-MBA modified $TiO_2(110)$ surface. Yellow, purple and red spheres represent S, Au, and O, respectively.

Table 2 Au–S and Au–O distances in selected Au species

Compounds	Au–S/Å	Au–O/Å	Note	Ref
$Au_{25}SR_{15}$	2.33		R = Phenylethyl	37
$R_3PAu\{SC(OMe){=}NC_6H_4NO_2\text{-}4\}$	2.31		R = Et, Cy, Ph[a]	38
$[(t\text{-}C_4H_9O)_3SiSAu]_4$	2.29		Au_4S_4	39
$Au_2Cl_2(\mu\text{-dppf})$	2.30		Linear S–Au–P	40
$(C_6H_4(COOH)S)$–Au-P $PhMe_2)$	2.31		Linear S–Au–P	41
$[Au(C_2H_3O_2)(C_{18}H_{15}P)]$		2.063	Linear O–Au–P	42
$[Au(NO_3)(CNEt)]$		2.033	Linear O–Au–C	43
$Au(C_7H_5O_2)\{PPh_3\}]\cdot C_6H_6$		2.034	Linear O–Au–P	44
$[Au(ppy)(tsc)]\cdot 1.5H_2O$	2.276	2.057	Square	45
$[Au(C_6F_5)_3(THT)]$	2.362		Square	46
$(AuTHT^{b})_{\infty}$	2.306,2.335		S–Au–S	47

[a] Et = Ethyl, Cy = Cyclohexyl, Ph = Phenyl. [b] THT = tetrahydrothiophene.

$$\text{TCA(bridge–bidentate)} + \text{Au} + \text{O(bridging)} \leftrightarrows \text{TCA(monodentate)–Au–O(bridging)} \quad (7)$$

However, the Au diffusion is limited by the TCA interaction, and Au aggregation occurs locally before reaching the larger Au islands. Consequently, smaller, two-layer Au clusters are formed on the TCA surface.

Conclusions

To stabilize atomic Au and small Au clusters on oxide surfaces, the Au–anion interaction is the most important. The Au–O bond is too weak to stabilize Au monomers, and Au atoms easily aggregate into a three-dimensional metal nanocluster. The Au is strongly stabilized by interaction with the mercapto moiety (–SH) of an MBA molecule, which is fixed to the surface through its carboxylic acid functionality. The Au–S bonds in Au- *o*-MBA and Au- *m*- MBA are both covalent with a bond length of 2.32 Å. The Au–S bond formed in TCA is not strong enough to fully stabilize atomic Au, but does serve to hinder Au diusion on the oxide surface, which results in smaller Au clusters than on the bare oxide. The surface modification is useful to control the structures of Au species which have small interaction with the surface oxygen atoms.

Acknowledgements

The authors express their thanks to Prof. M. Nomura and the PF staff for their helpful technical support. The work was carried out under the approval of the Photon Factory Advisory Committee, (No. 2012G165, 2010G128, 2010G601). This work was financially supported by NEDO (Demonstrative Research on Solid Oxide Fuel Cells/Fundamental Technologies Development/Analysis of Structure, Reactions, and Material Transfer of MEA Materials), JST-CREST (Single molecule and atomic level reactions (Creation of nanostructure catalysts and materials for environmental conservation)), and Grants-in-Aid for Scientific Research for Priority Fields (20047001, 2215002).

Notes and references

1 K. Asakura, in *Catalysis Book Series*, ed. J. G. Catalano, 2012, vol. 24, pp. 281.
2 Y. Iwasawa, ed. *X-ray absorption spectroscopy for catalysts and surfaces*, World Science, Singapore, 1996.
3 U. Diebold, *Surf. Sci. Rep.*, 2003, **48**, 53.
4 W. J. Chun, K. Asakura and Y. Iwasawa, *Chem. Phys. Lett.*, 1998, **288**, 868.
5 W. J. Chun, K. Asakura and Y. Iwasawa, *J. Phys. Chem. B*, 1998, **102**, 9006.
6 W. J. Chun, K. Asakura and Y. Iwasawa, *Catal. Today*, 2001, **66**, 97.
7 Y. Koike, K. Ijima, W. J. Chun, H. Ashima, T. Yamamoto, K. Fujikawa, S. Suzuki, Y. Iwasawa, M. Nomura and K. Asakura, *Chem. Phys. Lett.*, 2006, **421**, 27.
8 Y. Koike, W. J. Chun, K. Ijima, S. Suzuki and K. Asakura, *Mater. Trans.*, 2009, **50**, 509.
9 Y. Koike, K. Fujikawa, S. Suzuki, W. J. Chun, K. Ijima, M. Nomura, Y. Iwasawa and K. Asakura, *J. Phys. Chem. C*, 2008, **112**, 4667.
10 D. D. Wagman, W. H. Evans, V. B. Parker, R. H. Schumm, I. Halow, S. M. Bailey, K. L. Churney and R. L. Nuttall, *J. Phys. Chem., Ref. Data*, 1982, 11 Supplement No. 2.
11 Y. Tanizawa, T. Shido, Y. Iwasawa, M. Nomura, W.-J. Chun and K. Asakura, *J. Phys. Chem. B*, 2003, **107**, 12917.
12 Y. Tanizawa, W. J. Chun, T. Shido, K. Asakura and Y. Iwasawa, *J. Synchrotron Radiat.*, 2001, **8**, 508.
13 W. J. Chun, Y. Koike, K. Ijima, K. Fujikawa, H. Ashima, M. Nomura, Y. Iwasawa and K. Asakura, *Chem. Phys. Lett.*, 2007, **433**, 345.
14 S. Takakusagi, H. Nojima, H. Ariga, K. Miyazaki, T. Wada, W.-J. Chun, M. Nomura, Y. Iwasawa and K. Asakura, submitted to *Phys. Chem. Chem. Phys.*
15 H. Ashima, W.-J. Chun and K. Asakura, *Surf. Sci.*, 2007, **601**, 1822.
16 W. J. Chun, Y. Koike, H. Ashima, K. Kinoshita, K. Ijima, K. Fujikawa, S. Suzuki, M. Nomura, Y. Iwasawa and K. Asakura, *Chem. Phys. Lett.*, 2009, **470**, 99.
17 G. J. Hutchings and J. K. Edwards, in *Frontiers of Nanoscience*, ed. L. J. Roy and J. P. Wilcoxon, Elsevier, 2012, vol. 3, pp. 249–293.
18 W. J. Chun, K. Miyazaki, N. Watanabe, Y. Koike, S. Takakusagi, K. Fujikawa, M. Nomura and K. Asakura, *J. Ceram. Soc. Jpn.*, 2011, **119**, 890.
19 W. J. Chun, K. Miyazaki, N. Watanabe, Y. Koike, S. Takakusagi, K. Fujikawa, M. Nomura, Y. Iwasawa and K. Asakura, *J. Phys. Chem.*, 2013, **117**, 252.
20 Y. Yamamoto, K. Nakiajima, T. Ohsawa, Y. Matsumoto and H. Koinuma, *J. Appl. Phys.*, 2005, **44**, L511.
21 R. Nakamura, N. Ohashi, A. Imanishi, T. Osawa, Y. Matsumoto, H. Koinuma and Y. Nakato, *J. Phys. Chem. B*, 2005, **109**, 1648.
22 M. Nomura and A. Koyama, KEK Report 98-194, 1998.
23 W. J. Chun, Y. Tanizawa, T. Shido, Y. Iwasawa, M. Nomura and K. Asakura, *J. Synchrotron Radiat.*, 2001, **8**, 168.
24 K. Asakura, in *X-ray Absorption Fine Structure for Catalysts and Surfaces*, ed. Y. Iwasawa, World Scientific, Singapore, 1 edn, 1996, vol. 2, pp. 33.
25 A. L. Ankudinov, B. Ravel, J. J. Rehr and S. D. Conradson, *Phys. Rev. B: Condens. Matter Mater. Phys.*, 1998, **58**, 7565.
26 A. L. Ankudinov, C. E. Bouldin, J. J. Rehr, J. Sims and H. Hung, *Phys. Rev. B: Condens. Matter*, 2002, **65**, 104107.

27 W. J. Chun, K. Asakura and Y. Iwasawa, *J. Phys. Chem. B*, 1998, **102**, 9006.
28 W. J. Chun, K. Asakura and Y. Iwasawa, *Catal. Today*, 1998, **44**, 309.
29 J. T. Miller, A. J. Kropf, Y. Zha, J. R. Regalbuto, L. Delannoy, C. Louis, E. Bus and J. A. van Bokhoven, *J. Catal.*, 2006, **240**, 222–234.
30 A. Ulman, *An introduction to Ultrathin organic films from BL to SAM*. Editor, Academic Press, San Diego, 1991.
31 H. A. Biebuyck, C. D. Bain and G. M. Whitesides, *Langmuir*, 1994, **10**, 1825.
32 F. Elfeninat, C. Fredriksson, E. Sacher and A. Selmani, *J. Chem. Phys.*, 1995, **102**, 6153.
33 G. Liu, J. A. Rodriguez, J. Dvorak, J. Hrbek and T. Jirsak, *Surf. Sci.*, 2002, **505**, 295.
34 M. H. Dishner, J. C. Hemminger and F. J. Feher, *Langmuir*, 1996, **12**, 6176.
35 E. O. Sako, H. Kondoh, I. Nakai, A. Nambu, T. Nakamura and T. Ohta, *Chem. Phys. Lett.*, 2005, **413**, 267.
36 M. Hu, S. Noda and H. Komiyama, *Surf. Sci.*, 2002, **513**, 530.
37 M. Zhu, C. M. Aikens, F. J. Hollander, G. C. Schatz and R. Jin, *J. Am. Chem. Soc.*, 2008, **130**, 5883.
38 S. Y. Ho, E. C.-C. Cheng, E. R. T. Tiekink and V. W.-W. Yam, *Inorg. Chem.*, 2006, **45**, 8165.
39 W. Wojnowski, B. Becker, J. Saßmannshausen, E. M. Peters, K. Peters and H. G. Von Schnering, *Z. Anorg. Allg. Chem.*, 1994, **620**, 1417.
40 F. Canales, M. C. Gimeno, A. Laguna and P. G. Jones, *J. Am. Chem. Soc.*, 1996, **118**, 4839.
41 U. Helmstedt, S. Lebedkin, T. Hocher, S. Blaurock and E. Hey-Hawkins, *Inorg. Chem.*, 2008, **47**, 5815.
42 P. G. Jones, *Acta Crystallogr., Sect. C: Cryst. Struct. Commun.*, 1984, **40**, 1320.
43 T. J. Mathieson, A. G. Langdon, N. B. Milestone and B. K. Nicholson, *J. Chem. Soc., Dalton Trans.*, 1999, 201.
44 P. G. Jones, *Acta Crystallogr., Sect. C: Cryst. Struct. Commun.*, 1985, **41**, 905.
45 D. Fan, C.-T. Yang, J. D. Ranford, J. J. Vittal and P. Foo Lee, *Dalton Trans.*, 2003, 3376.
46 J. Coetzee, W. F. Gabrielli, K. Coetzee, O. Schuster, S. D. Nogai, S. Cronje and H. G. Raubenheimer, *Angew. Chem., Int. Ed.*, 2007, **46**, 2497.
47 S. Ahrland, B. Noren and A. Oskarsson, *Inorg. Chem.*, 1985, **24**, 1330.
48 M. Aizawa, Y. Morikawa, Y. Namai, H. Morikawa and Y. Iwasawa, *J. Phys. Chem. B*, 2005, **109**, 18831.

27. W. J. Chun, K. Asakura and Y. Iwasawa, J. Phys. Chem. B, 1998, 102, 9006.
28. W. J. Chun, K. Asakura and Y. Iwasawa, Catal. Today, 1998, 44, 309.
29. J. T. Miller, [illegible] J. R. Regalbuto, [illegible] Delannoy, C. Louis, E. Bus and J. A. van Bokhoven, J. Catal., 2006, 240, 222.
[illegible]
[illegible]
[illegible]

Operando atomic structure and active sites of TiO_2(110)-supported gold nanoparticles during carbon monoxide oxidation

Marie-Claire Saint-Lager,* Issam Laoufi and Aude Bailly

Received 21st December 2012, Accepted 4th January 2013
DOI: 10.1039/c2fd20157g

It is well known that gold nanoparticles supported on TiO_2 act as a catalyst for CO oxidation, even below room temperature. Despite extensive studies, the origin of this catalytic activity remains under debate. Indeed, when the particle size decreases, many changes may occur; thus modifying the nanoparticles' electronic properties and consequently their catalytic performances. Thanks to a state-of-the-art home-developed setup, model catalysts can be prepared in ultra-high vacuum and their morphology then studied in *operando* conditions by Grazing Incidence Small Angle X-ray Scattering, as well as their atomic structure by Grazing Incidence X-ray Diffraction as a function of their catalytic activity. We previously reported on the existence of a catalytic activity maximum observed for three-dimensional gold nanoparticles with a diameter of 2–3 nm and a height of 6–7 atomic planes. In the present work we correlate this size dependence of the catalytic activity to the nanoparticles' atomic structure. We show that even when their size decreases below the optimum diameter, the gold nanoparticles keep the face-centered cubic structure characteristic of bulk gold. Nevertheless, for these smallest nanoparticles, the lattice parameter presents anisotropic strains with a larger contraction in the direction perpendicular to the surface. Moreover a careful analysis of the atomic-scale morphology around the catalytic activity maximum tends to evidence the role of sites with a specific geometry at the interface between the nanoparticles and the substrate. This argues for models where atoms at the interface periphery act as catalytically active sites for carbon monoxide oxidation.

Introduction

It is now well established that gold nanoparticles (GNPs) supported on rutile TiO_2 act as catalysts for the oxidation of carbon monoxide (CO), even below room temperature.[1] Despite extensive studies, the origin and the mechanisms of the catalytic activity remain unclear. Most of the questions were focused on the activation of the oxygen molecule since there is experimental evidence that the limiting step of the reaction is the oxygen dissociation.[2] Depending on the active sites, as well as the substrate type, several mechanisms were proposed. Some models assume that the reaction takes place fully on the GNPs' surface, where oxygen adsorption and dissociation can occur. In that case the active sites are mainly identified as low-coordinated gold atoms.[3] The GNPs' catalytic activity can also be due to a quantum size effect linked to a metal–non metal transition appearing for clusters with a thickness smaller than 1 nm.[4] Haruta *et al.* first suggested that the reaction

Institut Néel, CNRS et Université Joseph Fourier, BP 166, F-38042 Grenoble cedex 9, France. E-mail: marie-claire.saint-lager@grenoble.cnrs.fr

could take place at the interfacial perimeter between the GNPs and the substrate.[5] This assumption was recently supported by studies that stress the crucial role of this perimeter in the CO oxidation, but the reaction mechanism is still strongly debated.

Kotobuki *et al.* demonstrated with temporal analysis of reactive products that active oxygen species can be stored on Au/TiO_2 catalysts at 350 K by exposure to thermal O_2 pulses.[6] They showed that the oxygen storage capacity and the CO oxidation activity of these catalysts scale with the perimeter length of the interface between the TiO_2 support and the GNPs. In the continuity of this work, Widmann *et al.* proposed a mechanism where CO adsorbed on the GNPs reacts with activated surface lattice oxygen species at the perimeter of the Au–TiO_2 interface.[7] At room temperature only the perimeter sites are involved, whereas at higher temperatures (>350 K) these oxygen species can be provided by migration on TiO_2 *via* oxygen vacancies. The authors concluded that those highly stable active oxygen species allow the facile dissociation of the oxygen molecule without assistance of any intermediate complex.

By means of a combination of infrared kinetic spectroscopic measurements and Density Functional Theory (DFT) calculations, Green *et al.* evidences that dual Ti–Au catalytic sites located at the perimeter interface of 3-nm GNPs supported on TiO_2 play a key role in the CO oxidation.[8] They indicate that O–O bond scission is activated by the formation of a CO–O_2 complex at these dual sites. They also stress the possibility for CO to adsorb either on Au or on TiO_2 sites thus giving two different sources of CO for the oxidation reaction, whereas the oxygen activation takes place around the perimeter area. These experiments, carried out at low temperature (about 110 K), make it possible to distinguish which CO adsorption sites will first deliver the CO for the reaction indicating that the TiO_2 sites are energetically more favorable at such a low temperature. The CO adsorbed on top of the GNPs will only be able to diffuse to the active perimeter area at higher temperatures.

Fujitani *et al.* also stressed the importance of the nanoparticle-support periphery.[9] But according to the fact that CO oxidation on supported GNPs is greatly influenced by moisture, they suggest that reaction intermediates are rather OH^- species.

A lot of attention was also dedicated to visualize GNPs at the atomic scale in real space and under reactive conditions. Yoshida *et al.* showed that the adsorption of carbon monoxide at room temperature induces the reconstruction of {100} facets of GNPs supported on CeO_2 during the CO oxidation.[10] The aberration-corrected environmental Transmission Electron Microscopy (TEM) images evidence that the carbon monoxide is preferentially adsorbed on the on-top sites of the gold atoms; the outermost gold layer forming an undulating hexagonal lattice. This reconstruction is observed for total pressures as high as 2×10^3 Pa. Although CO was thought to preferentially adsorb on the low-coordination sites, such as steps, edges and corners, this study evidences that atomically clean reconstructed {100} facets of roughly 4 nm-diameter GNPs are also able to adsorb it.

This illustrates that gaseous environments can strongly modify the GNPs' atomic structure at the nanoscale. But deep structural changes can occur even in Ultra-High Vacuum (UHV) thus modifying the nanoparticles' electronic properties and consequently their catalytic properties. By means of aberration-corrected TEM, Walsh *et al.* studied carbon-supported gold nanoparticles in the size range for which they are usually known to be catalytically active, *i.e.* 2 to 10 nm. In UHV the multiply twinned icosahedral and decahedral structures turn out to be the most stable ones.[11] These highly strained GNPs (especially at the surface) are expected to be much more active than the single crystalline ones. Indeed, based on powder X-ray diffraction, Cunningham *et al.* found that, for $Au/Mg(OH)_2$, only catalysts with size distributions below 1 nm were active for CO oxidation. They identified two structural forms (cubo-octahedral and icosahedral) and observed that only the icosahedral one is catalytically active.[12] However such a strained structure with a great number of low-coordinated atoms at the surface does not seem to be stable under annealing, or even under microscope-electron-beam-induced sintering with the appearance of atomically clean surface facets in order to minimize the surface energy.[13] In any case, a lattice contraction is generally observed in very small particles. A decrease of the Au–Au distance with decreasing GNP size below a few nanometers was measured by Extended X-ray Absorption Fine Structure (EXAFS). According to the authors this effect could be responsible for the catalytic properties of the gold nanoparticles.[14]

This brief overview attests that all the recent developments allowing the observation of the structure at the atomic scale in reactive conditions pave the way towards a thorough understanding of the catalytic properties of supported GNPs. The present work comes within this scope. Thanks to a state-of-the-art home-developed setup, model catalysts can be prepared in UHV and their morphology then studied in *operando* conditions by Grazing Incidence Small Angle X-ray Scattering (GISAXS), as well as their atomic structure by Grazing Incidence X-ray Diffraction (GIXRD) as a function of their catalytic activity.[15,16] It was used to characterize gold nanoparticles supported on TiO_2(110) during CO oxidation, by recording at the same time the catalytic activity, GISAXS patterns and GIXRD data. We previously reported that the reaction rate dependence on the GNPs diameter presents a maximum for diameters of about 2 nm.[17,18] Above the maximum, the reaction rate fits well a power law of the diameter,[18] which is consistent with low-coordinated active sites located at the GNP interface periphery. However, this cannot explain the activity fall observed for GNPs smaller than 2 nm. The analysis of the evolution of the GNPs' atomic-scale structure and morphology around this maximum should yield a better understanding of the key parameters involved in their catalytic properties. Moreover, in contrast to atomic structures determined by X-ray measurements performed on powders, the use of nanoparticles epitaxially grown on a TiO_2(110) single crystal reveals information on their shape, their atomic structure and their lattice parameter parallel and perpendicular to the substrate surface. In this frame, GISAXS and GIXRD lead to the definition of what can be called an 'average' particle analyzed in the reciprocal space and representing a wide collection of nanoparticles. In that sense it is a good complement to local probes such as TEM or Scanning Tunneling Microscopy (STM). Moreover these X-ray techniques can easily be operated in *operando* conditions, the simultaneous determination of the reaction rate allowing a direct correlation with the geometrical parameters of the average GNP.

Experimental method

The experiments were performed in a home-developed setup consisting of an UHV preparation chamber connected to a batch reactor running from UHV up to reactive conditions at ambient pressures and allowing both X-ray scattering and catalytic activity measurements in static conditions.[16,17] The setup is operated on the "2 + 2" diffractometer of the GMT station at the beamline BM32 of the European Synchrotron Radiation Facility (ESRF).[15] The experimental procedure was previously detailed;[17,18] we simply summarize the main features hereafter. The GNPs were synthesized in the UHV preparation chamber (base pressure of 5.10^{-8} Pa). The TiO_2(110) single crystal, used as a substrate, was first bombarded with 750 eV argon ions and then annealed at 1000 K under an oxygen partial pressure of 10^{-3} Pa in order to restore the surface stoichiometry.[19] Gold nanoparticles were grown by UHV vapor deposition at 300 K from a Knudsen cell. As calibrated by a quartz microbalance, the average thickness of the gold deposit ranged between 0.05 and 3 atomic monolayers (1 ML corresponding to 0.235 nm, equivalent to one Au(111) atomic layer). Once prepared, the Au/TiO_2 (110) samples were transferred under UHV into the batch reactor in order to perform X-ray and catalytic activity measurements during the carbon monoxide oxidation. The measurement procedure involved three steps; each lasting approximately two to three hours: (1) Au/TiO_2(110) was first studied in UHV (2.10^{-7} Pa) at room temperature (RT); (2) then during annealing at 473 K in 2.10^3 Pa of oxygen; (3) and finally during CO conversion into CO_2 that was started by adding 20 Pa of CO to the oxygen, while keeping the sample at 473 K. The CO and CO_2 partial pressures in the reactor were deduced respectively from the 28 and 44 ionization currents measured by a mass spectrometer. The gas composition was followed as soon as the sample was introduced in the batch reactor and all along the experimental procedure. During the step (3), the reaction rate was defined as the number of CO molecules converted into CO_2 per gold atom per second on the basis of the total number of gold atoms.[17] The GISAXS patterns were generally acquired at the beginning and at the end of each step. In between, diffraction spectra were collected along several directions in the reciprocal space.

To enhance the GNPs' signal, X-ray measurements were performed with a photon energy of 18 keV at a grazing incidence close to the critical angle (0.13° for TiO_2 at 18 keV). A shutter, opened only during X-ray data collection, was mounted in order to avoid long time exposures to the X-ray beam and consecutive sample surface damage. Finally, no significant difference could be observed in either the scattering or the catalytic activity with and without photons on the sample. For GIXRD measurements, a standard NaI scintillator detector was used and the GISAXS patterns were collected on a 2D camera (Princeton Instruments, pixel size of 56.25 μm).[16]

The structure of bulk gold is face centered cubic (fcc) and two kinds of epitaxy relationships are usually observed on TiO_2(110) : $(112)_{Au}//(110)_{TiO_2}$ and $(111)_{Au}//(110)_{TiO_2}$, depending on the substrate temperature during the GNPs growth.[20] However, by means of GIXRD in UHV, Lazzari *et al.* found that these two epitaxial relationships are nearly equiproportional whatever the temperature.[21] For the two epitaxy orientations, gold grows with the $<110>_{Au}$ direction of the dense rows aligned with the $<001>_{TiO_2}$ of the oxygen ones, the interatomic distances being 0.296 nm for TiO_2 and 0.2886 nm for gold. As usual in surface crystallography, the z direction is perpendicular to the surface, that means along the $<110>_{TiO_2}$ direction, and the x direction has the smallest lattice parameter that is along $<001>_{TiO_2}$. The unit cell parameters are thus equal to $a = 0.296$ nm, $b = c = 0.6492$ nm. The present diffraction measurements show that the $(111)_{Au}//(110)_{TiO_2}$ epitaxy is predominant, especially under reactive conditions. For convenience the gold lattice is described in its hexagonal cell. The schematic representation of the fcc gold reciprocal space (with Miller indices: H_{Au}, K_{Au}, L_{Au}) over the TiO_2(110) one (with Miller indices: H_{TiO_2}, K_{TiO_2}, L_{TiO_2}) is given in Fig. 1a for $L = 0$. Fig. 1b is a 3D schematic representation of the gold reciprocal space for the two kinds of stacking usually observed along the <111> trigonal axis.

For each gold deposit, the mean geometrical parameters of the GNPs, such as the diameter D, the height H and the interparticle distance L, were deduced from the quantitative analysis of the GISAXS patterns performed using the IsGISAXS software.[22] These measurements are detailed elsewhere.[16,17,18] Considering the whole set of samples measured *in operando*, the best fit of the GISAXS patterns was obtained assuming a truncated sphere to model the GNPs shape. The geometrical parameters are an average value of those obtained for the two main TiO_2(110) surface directions. In contrast to the interparticle distance L, the diameter D was found identical in these two directions with decreasing value when the deposited equivalent gold thickness diminishes. Fig. 2 sums up the main results concerning the dependence of the CO conversion rate as a function of the mean geometrical parameters deduced from the quantitative analysis of the GISAXS patterns recorded in real time.[17]

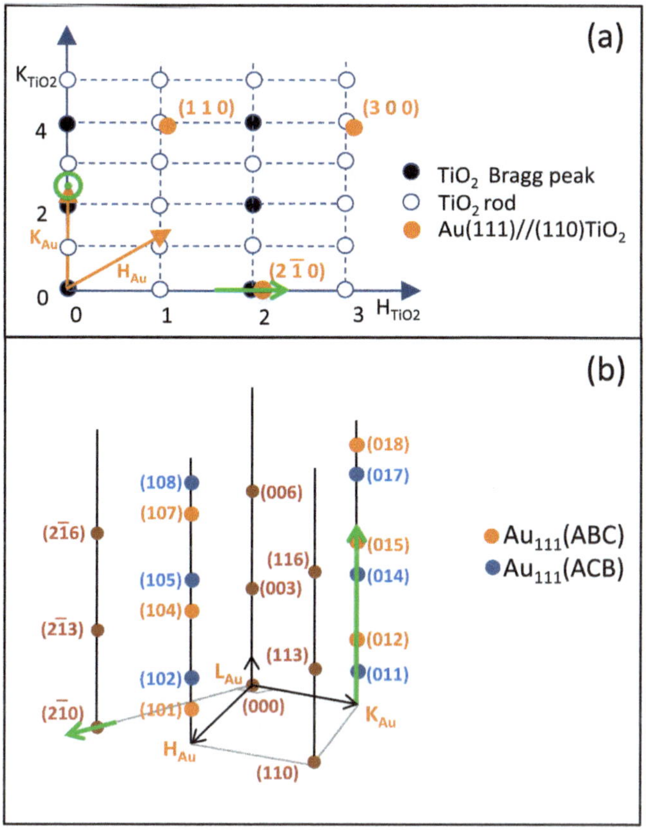

Fig. 1 Schematic representations of the reciprocal space of $(111)_{Au}//(110)_{TiO_2}$ (a) at $L = 0$ with in black the TiO_2 peaks (filled circles: Bragg peaks and empty circles: cut of the surface rods) and in orange the gold Bragg peaks. H_{TiO_2} and K_{TiO_2} are the Miller indices in the frame of the $TiO_2(110)$ surface. The hexagonal cell of the fcc bulk gold is represented in orange. (b) 3D representation of the gold reciprocal space. The orange spots represent the Bragg peaks of the ABC stacking and the blue ones those of the ACB one. The dark red spots are peaks common to the two stackings. In the two schemes the green arrows indicate the directions along which the diffracted intensity was collected for Fig. 3.

Results

Evolution of the diffraction spectra under reactive gases

The evolution of the diffraction spectra obtained during the three steps of the experimental procedure is illustrated in Fig. 3 for a sample with an equivalent gold thickness of 0.1 ML (Fig. 3a and 3c) and another one with 0.5 ML (Fig. 3b and 3d). Two directions of the reciprocal space are represented: parallel to the surface along H and perpendicularly to it along the rod $(0\ 1)_{Au}$.

For the H-scan, the Au $(2\ {-1}\ 0)_{Au}$ Bragg peak is expected at $H_{TiO_2} = 2.05$ (Fig. 3a and 3b). In UHV, its position is found at a higher value, and the shift is larger when the gold thickness decreases (corresponding to smaller particles). For 0.1 ML, the quantitative GISAXS analysis gave 1 nm for the diameter and 0.6 nm for the height and the Au $(2\ {-1}\ 0)_{Au}$ peak was found at $H_{TiO_2} = 2.11 \pm 0.02$. It corresponds to a lattice contraction of 3.4%. In the case of the 0.5 ML deposit, the nanoparticles are larger; the diameter being 2.2 nm and the height 1.5 nm. The gold maximum is found at $H_{TiO_2} = 2.07 \pm 0.01$ indicating that there is still a small lattice contraction of 0.8%. When annealing at 473 K and also after introducing CO for starting the oxidation

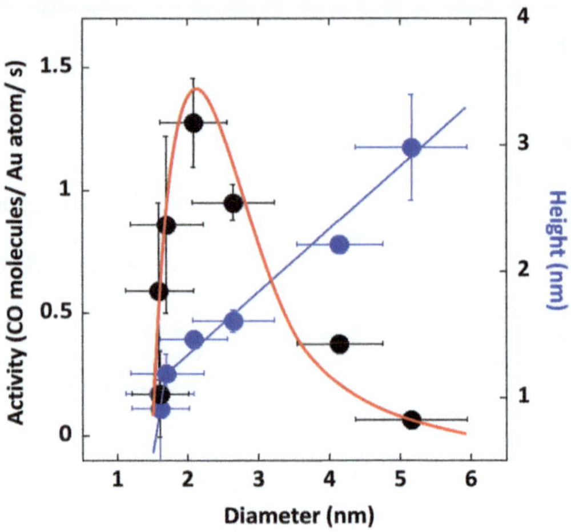

Fig. 2 Variation of the CO conversion reaction rate into CO_2 per Au atom per second (black) and of the mean height H (blue) as a function of the mean diameter D of the GNPs measured *in operando*. The full lines are guides for the eye.

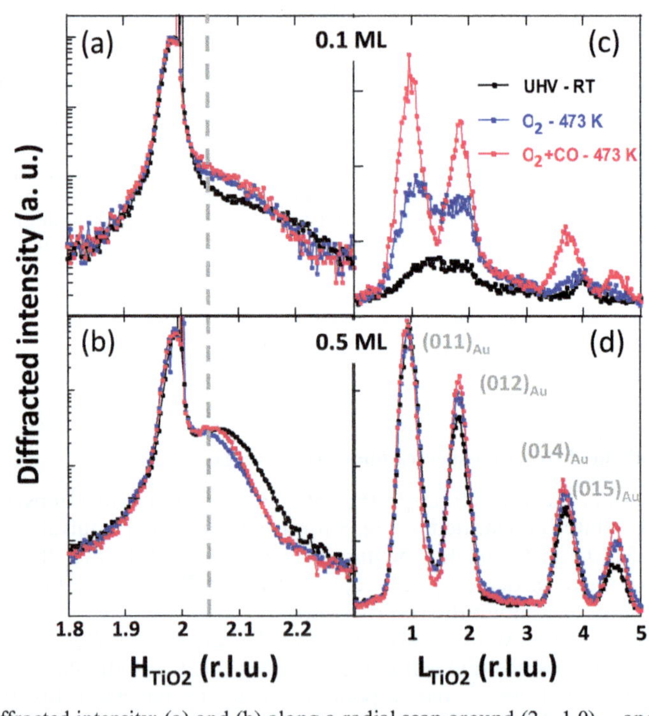

Fig. 3 Diffracted intensity: (a) and (b) along a radial scan around $(2\ -1\ 0)_{Au}$ and (c) and (d) along the $(01)_{Au}$ rod for two samples with gold equivalent thicknesses of: (a) and (c) 0.1 ML and (b) and (d) 0.5ML during the three steps of the experimental procedure. (1) in UHV at RT (black), (2) under 2.10^3 Pa of oxygen at 473 K (blue) and (3) after adding 20 Pa of CO while keeping $T = 473$ K (red). In (a) and (b) the thin peak centered at $(200)_{TiO_2}$ with a broad and intense foot is the TiO_2 Bragg peak, the shoulder on the right is the $(2\ -1\ 0)_{Au}$ Bragg peak that is expected at $(2.05\ 0\ 0)_{TiO_2}$ for bulk Au, as indicated by the grey dashed line.

reaction, strong changes are observed for the thinnest deposit, whereas at 0.5 ML there is only a gentle evolution. Above 1 ML, the spectra remained independent of the environment. This behavior can be correlated to what was deduced from GISAXS patterns measured at the same time.

Under reaction conditions (step 3), for the 0.1 ML sample, the diameter grows up to 1.7 nm and the height to 1.2 nm, as deduced from GISAXS, and the gold peak shifts to $H_{TiO_2} = 2.06 \pm 0.02$, a value closer to the bulk gold one. For the 0.5 ML deposit, the peak reaches the position of the bulk value. In that case, we can observe that there is a subtle difference between step (2) and step (3); the relaxation seeming slightly larger under oxygen than under reactive conditions.

Along the $(0\ 1)_{Au}$ rod and during the step (1), in UHV and at room temperature, the intensity collected for 0.1 ML, with a height of about 3 atomic planes, presents only one main broad peak (Fig. 3c). For the 0.5 ML, the height corresponds to at least 6 atomic planes and in that case, the diffraction spectra exhibit well defined peaks at $L_{Au} = 1, 2, 4, 5$ (Fig. 3d) characteristic of the two sequences ABC and ACB of the fcc stacking. In the case of the 0.1 ML deposit, the fcc ordering is progressively induced by gas adsorption at 473 K. Indeed, under oxygen at 473 K (step 2), the main broad peak starts to split into two peaks. Then, under reaction conditions at 473 K, the nanoparticles height increases up to 6 atomic planes. In the same conditions, the peaks of the 0.5 ML deposit only slightly increase and become a little bit thinner.

Au–Au distances and catalytic activity

The interatomic distances d_{Au-Au} in the gold nanoparticles during reaction were deduced from the diffraction spectra. Perpendicularly to the surface, it was obtained by averaging the position of the peak at $L = 2$ and 4 along the $(0\ 1)_{Au}$ rod with the one for $L = 3$ of the $(2\ -1)_{Au}$ rod. This choice allowed to cancel the shift of the peaks due to defects in the fcc stacking. In the surface plane, it was calculated by deconvoluting the gold peaks from the TiO_2 one in the radial scan around $(200)_{TiO_2}$. Besides, we verified that the width of the gold peaks is consistent with the size deduced by GISAXS. The variation of the Au–Au distances is reported in Fig. 4 as a function of the particles diameter. They were normalized to the bulk gold values that are

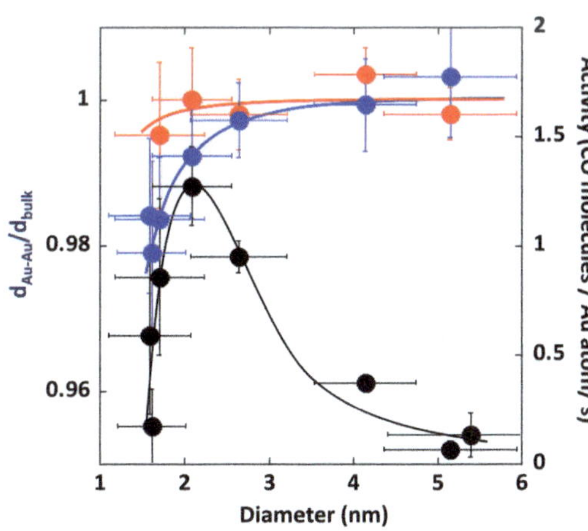

Fig. 4 Variation of the CO conversion reaction rate into CO_2 per Au atom per second (black) and of the interatomic distance d_{Au-Au}, normalized to the bulk gold value, (blue and red: perpendicular and parallel to the surface, respectively) as a function of the GNPs' mean diameter D. All the parameters were measured *in operando* during CO oxidation (step (3)): the catalytic activity by mass spectrometry, the diameter D was deduced from GISAXS and d_{Au-Au} from the diffraction measurements. The full lines are guides for the eye.

$d_\perp = a_0/\sqrt{3}$ and $d_{//} = a_0/\sqrt{2}$; the bulk Au cubic lattice parameter a_0 being 0.408 nm. A contraction is observed for sizes below ≈ 2–3 nm, as already reported by using EXAFS.[14] But in the present case, the Fig. 4 shows that this effect is clearly anisotropic: it is almost negligible in the surface plane whereas it is quite significant perpendicularly to it. This can be correlated to the shape of the nanoparticles, since the aspect ratio *H*/*D* is less than 1 and decreases in this region from 0.7 to 0.5. Interestingly, the comparison with the variation of the catalytic activity for CO oxidation in Fig. 4 indicates that this contraction coincides with the fall of the catalytic activity below the optimum diameter.

Geometry of the active sites

Looking at the geometrical parameters of the nanoparticles corresponding to the drop of the activity below the optimum diameter, we observed that the slope of the variation of the mean height H as a function of the diameter becomes more abrupt (Fig. 2). This means that the nanoparticles become flatter in this particular region. Actually, in the region around the maximum of activity, the value of the mean diameter varies slowly while the height changes more significantly. It is thus interesting to plot the variation of the GNP geometrical parameters as a function of the particle height and to correlate it to that of the activity for CO oxidation such as in Fig. 5. It shows that the diameter is more or less constant in the region of the maximum activity indicating that it is not the essential parameter. At first sight, the models, where the reaction occurs at the interface perimeter between the GNPs and the TiO_2 surface, would expect that the variation of the CO conversion rate per Au atom would follow the same variation as the fraction of Au atoms on the perimeter.[9] Fig. 5 shows that it is not the case here, since this fraction continuously increases when the particle size decreases, including while crossing the maximum of activity. In contrast, the aspect ratio *H*/*D* presents a behavior similar to the activity with a maximum for the same value of the height indicating that the geometry of the particles at the interface is a crucial parameter. The schemes inserted in Fig. 5 evidence that the maximum of activity corresponds to a sharp angle θ between the GNP surface tangent and the substrate. The fall on both sides of the

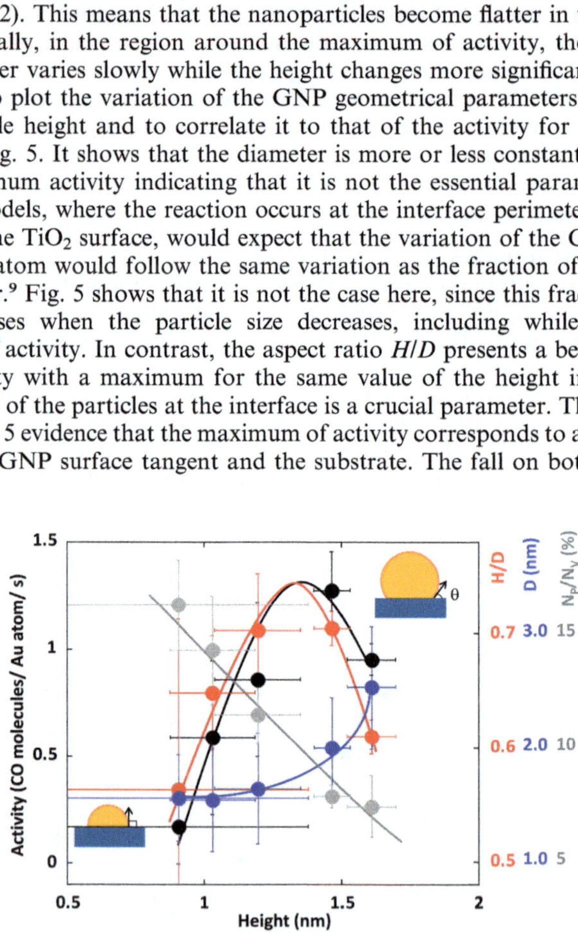

Fig. 5 Comparison of the variation of the GNPs geometrical parameters around the catalytic activity maximum as a function of the height. (black) Reaction rate of the CO conversion into CO_2 per Au atom per second, (blue) diameter D of the GNPs, (grey) ratio between N_P and N_V, the number of atoms at the interface perimeter between the nanoparticles and the TiO_2 substrate and the total number of atoms within the GNP, respectively, and (red) the aspect ratio *H*/*D*. A schematic representation of the supported nanoparticles is given with, on the left, $H/D = 0.5$ and on the right, $H/D = 0.7$ as observed close to the catalytic activity maximum. θ is the angle between the GNP surface tangent and the substrate.

maximum corresponds to θ becoming closer to a right angle. This suggests that the active sites could result from the combined action of several atoms near the perimeter. Thereby, a kind of cavity is created where low-coordinated Au atoms and the titania substrate can interact simultaneously with the reaction entities.

To get more information we need to use an atomic model to describe the GNPs' structure. Most of the images obtained by TEM with atomic resolution evidenced a shape close to cubo-octahedrons for particle sizes in the range of the present work and moreover with a (111) face in contact with the substrate surface.[23,24,25] Consistently, as shown hereabove, the diffraction data collected for samples under reaction conditions reveal a fcc structure with the <111> direction perpendicular to the surface that is compatible with such cubo-octahedrons.[26] However, we did not find any trace of signal, indicating well-defined facets on the recorded GISAXS patterns, which would present oriented rods corresponding to the (111) and (100) planes as observed for well-ordered faceted supported clusters.[27] This may be due to a misorientation of the gold nanoparticles or to particles too small to present facets large enough to be seen by GISAXS. Therefore the truncated sphere used to model the nanoparticles for quantitative analysis of the GISAXS patterns is a quite good approximation. The two parameters needed to define the corresponding truncated cubo-octaedrons are m and l, the number of atoms on the edges and the number of atomic planes parallel to the surface, respectively. Their relationships with the diameter D and the height H measured by GISAXS were established in ref. 18 in line with reference:[28]

$$3m - 2 \cong N_{\mathrm{D}} = (D - 2r_{\mathrm{Au}})/(d_{//} + 1)$$

and

$$l = (H - 2r_{\mathrm{Au}})/(d_{\perp} + 1),$$

N_{D} being the number of atoms along the diameter D and $r_{\mathrm{Au}} = 0.026$ nm, the gold atomic radius as seen by X-rays. The variation of the interatomic distance $d_{\mathrm{Au-Au}}$ mentioned above is neglected here and the bulk values for $d_{\perp}$ and $d_{//}$ are used.

As shown in Fig. 6, the experimental data do not exactly match with integer values of m and l. They represent the center of the size distribution of the GNPs on the TiO_2 substrate. The width of this distribution in the two directions is represented by error bars in the insert of Fig. 6. For the height, it can be interpreted as a distribution of particles with different number of atomic layers; it varies noticeably as a function of the particle size. For the flattest nanoparticles, the error bar is large and due to a wide distribution of the size on the sample. For the biggest ones, the truncated sphere becomes a less-good approximation of the GNPs' shape that begins to present well defined facets, enlarging the value of the error. In the middle range, that means close to the optimum size for the activity, the particle height has a thin distribution ($l = 7 \pm 0.5$ at the maximum). As previously reported this is a consequence of the sintering induced by the temperature and then by the reaction itself.[17,18] Concerning the dispersion width of the diameter, it is roughly constant and corresponds to ±2 atoms. Thus, it cannot be attributed to a distribution of octahedrons with several m integers. Indeed, a distribution with m and $m + 1$ atoms on the edge would correspond to diameters with N_{D} and $N_{\mathrm{D}} + 3$ respectively, and would thus result in a bimodal distribution for the diameter, that is definitively not observed. This indicates that there are defects and that the GNPs shape is not ideal. For instance, a twin structure was evidenced for nanoparticles exposed to 1 vol. % of CO in air (100 Pa) by High Resolution TEM (HRTEM).[25] However, we can observe that all the experimental data are localized close to an integer value of m (less than ±1 atom), that means with a shape not far from the ideal cubo-octahedron model.

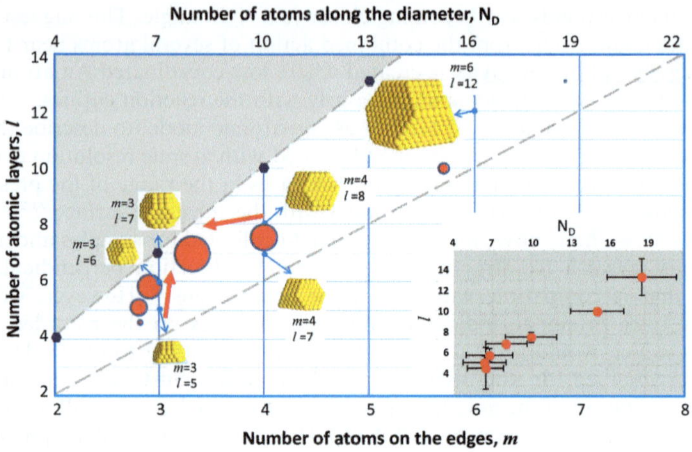

Fig. 6 Reaction rate of the GNPs as a function of the size of the corresponding truncated cubo-octahedron with, on the vertical axis the number of atomic layers l, and on the bottom horizontal axis, m, the number of atoms on the edges and on the upper one, N_D, the number of atoms along the diameter; the scales for l, N_D and m are proportional to the value of $d_\perp$, $d_{//}$ and $3d_{//}$, respectively. The red spots are proportional to the reaction rate for CO oxidation, as deduced from the experimental data. They are located at the m and l values deduced from the mean D and H, as explained in the text. Schematic representations of cubo-octahedrons are given for several couples of (m, l) pointed by the blue arrows (see text for the red arrows). The couples (m, l) corresponding to complete cubo-octahedrons are marked by dark blue diamonds. On the lower grey dashed line, the aspect ratio H/D is 0.5 and it is 0.8 on the upper one, that crosses the dark blue diamonds. The part of the diagram above it has been shaded since the corresponding (m, l) couples cannot be assigned to the cubo-octahedral geometry. In the insert, the experimental data were also plotted as a function of l (vertical axis) and N_D (upper horizontal axis) with their error bars calculated from the width of the Gaussian distribution of the GNPs size given by the GISAXS analysis.

Fig. 6 shows that the data below the optimum have m close to 3, when the number of layers l grows, the reaction rate increases (vertical red arrow in Fig. 6). Above the optimum size for the (almost) same number of layers, the activity increases when m goes down to 3 (horizontal red arrow in Fig. 6). For higher values of m, the main factor prevailing for the activity fall is the rapid decrease of the fraction of potentially active sites at the surface relative to the total number of atoms in the particles. Finally, the maximum of activity occurs for particles close to a cubo-octahedron with $m = 3$. The main feature of this GNPs morphology is the presence of edges parallel and close to the substrate surface. They connect a (111) hexagonal face of the upper part of the particles and a (100) square face that makes a sharp angle of 55° with the TiO_2 surface. The combination of such an angle, together with low-coordinated gold atoms on the edges in the vicinity of the substrate surface constitute a particular geometry that could be very active sites for the CO oxidation.

Conclusion

In this work we studied the structural properties and the morphology of the GNPs around the maximum observed for the CO oxidation reaction rate as a function of the particles diameter, thanks to *in operando* GISAXS, GIXRD and activity measurements. For GNPs bigger than 2–3 nm, that is above the optimum size for the catalytic activity, the fcc structure is already established during the gold growth on $TiO_2(110)$. For the smallest ones, the fcc ordering appears only after annealing at 473 K and under reactive gas pressure. Before adding 20 Pa of CO, starting the oxidation reaction, the GNPs were exposed to 2×10^3 Pa of oxygen at 473 K.

The diffraction data present a subtle difference between the two steps with the relaxation of the Au–Au distance that seems slightly larger under oxygen than under reactive conditions. This may be due to oxygen interaction with outermost atoms of the GNPs which could provoke a dilatation of the Au–Au distances (the measured value is an average over the particle). As already reported, under oxygen pressure, gold is probably pre-oxidized, thus producing $Au^{\delta+}$ sites on the GNP surface.[29] Such positively charged Au sites were also evidenced under high pressure oxygen at room temperature by X-ray Photoelectron Spectroscopy (XPS),[30,31] as well as by X-ray Absorption Near-Edge Structure (XANES).[32] This gold activation was attributed to X-rays.[31] Such a pre-oxidation step was shown to modify the reaction pathway at low temperature (120 K), since it deactivates the most reactive sites.[29] However, in most of the cases, CO introduction quickly reduced gold to its metallic form,[31,32] all the more since the temperature is equal to 473 K in the present study. The interaction of oxygen with GNPs was also observed by HRTEM leading to rounded GNPs.[33] In a more recent work, Uchiyama *et al.* drew the diagram of the GNPs morphology grown on CeO_2 as a function of the O_2 and CO partial pressures.[24] According to it, the pressure conditions of step (2) of the present study correspond to the domain where the GNPs are rounded. In reaction conditions (*i.e.* after CO introduction) the diagram indicates that the GNPs should be well faceted.

Under reaction conditions, the GNPs have the fcc bulk structure with the $<111>_{Au}$ direction perpendicular to the $TiO_2(110)$ surface. However, for diameters below 2–3 nm, they are anisotropically strained with a contraction perpendicularly to the surface that can reach 2% whereas it is negligible in the parallel direction. This contraction of d_{Au-Au} coincides with the fall in the catalytic activity below the optimum size. A deeper structural analysis is in progress to determine if more significant structural changes occur in this size range to explain the fall of the GNPs' catalytic activity.

In the present work, we show that the maximum of activity corresponds to a sharp angle between the GNP surface tangent and the substrate. This work confirms the dominating role of the sites at the interface but with a specific geometry involving substrate atoms and several atoms of the GNPs. The temperature being 473 K, the diffusion on the GNPs surface as well as on the TiO_2 surface cannot be the limiting factor, as it is for studies done at much lower temperatures (around 100 K).[7,29] This would require models involving the existence of intermediate complexes on mixed adsorption sites to activate the oxygen molecules.

These results on $TiO_2(110)$-supported GNPs show that the combination of GISAXS, GIXRD and catalytic activity measurements is a very promising way to examine at the atomic scale nanoparticles with sizes around 2 nm and below while they are catalyzing a given reaction.

Acknowledgements

The technical staffs (SERAS and MCMF) of Institut Neel-CNRS and of the BM32 CRG beamline at ESRF are acknowledged for their assistance. We also thank Rémi Lazzari and Jacques Jupille of the INSP for their help during the synchrotron experiments.

References

1 M. Haruta, N. Yamada, T. Kobayashi and S. Iijima, *J. Catal.*, 1989, **115**, 301.
2 V. A. Bondzie, S. C. Parker and C. T. Campbell, *Catal. Lett.*, 1999, **63**, 143.
3 M. Mavrikakis, P. Stoltze and J. K. Nørskov, *Catal. Lett.*, 2000, **64**, 101.
4 M. Valden, X. Lai and D. W. Goodman, *Science*, 1998, **281**, 1647.
5 M. Haruta, S. Tsubota, T. Kobayashi, H. Kageyama, M. J. Genet and B. Delmon, *J. Catal.*, 1993, **144**, 175; M. Haruta, *Catal. Today*, 1997, **36**, 153; M. Haruta, *Faraday Discuss.*, 2011, **152**, 11.

6 M. Kotobuki, R. Leppelt, D. A. Hansgen, D. Widmann and R. J. Behm, *J. Catal.*, 2009, **264**, 67.
7 D. Widmann and R. J. Behm, *Angew. Chem., Int. Ed.*, 2011, **50**, 10241.
8 I. X. Green, W. Tang, M. McEntee, M. Neurock and J. T. Yates, Jr, *Science*, 2011, **333**, 736.
9 T. Fujitani and I. Nakamura, *Angew. Chem., Int. Ed.*, 2011, **50**, 10144.
10 H. Yoshida, Y. Kuwauchi, J. R. Jinschek, K. Sun, S. Tanaka, M. Kohyama, S. Shimada, M. Haruta and S. Takeda, *Science*, 2012, **335**, 317.
11 M. J. Walsh, K. Yoshida, A. Kuwabara, M. L. Pay, P. L. Gai and E. D. Boyes, *Nano Lett.*, 2012, **12**, 2027.
12 D. A. H. Cunningham, W. Vogel, H. Kageyama, S. Tsubota and M. Haruta, *J. Catal.*, 1998, **177**, 1.
13 M. J. Walsh, K. Yoshida, M. L. Pay, P. L. Gai and E. D. Boyes, *ChemCatChem*, 2012, **4**, 1638.
14 J. T. Miller, A. J. Kropf, Y. Zha, J. R. Regalbuto, L. Delannoy, C. Louis, E. Bus and J. A. van Bokhoven, *J. Catal.*, 2006, **240**, 222.
15 M.-C. Saint-Lager, A. Bailly, P. Dolle, R. Baudoing-Savois, P. Taunier, S. Garaudée, S. Cuccaro, S. Douillet, O. Geaymond, G. Perroux, O. Tissot, J.-S. Micha, O. Ulrich and F. Rieutord, *Rev. Sci. Instrum.*, 2007, **78**, 083902.
16 M.-C. Saint-Lager, A. Bailly, M. Mantilla, S. Garaudée, R. Lazzari, P. Dolle, O. Robach, J. Jupille, I. Laoufi and P. Taunier, *Gold Bull.*, 2008, **41**, 159.
17 I. Laoufi, M.-C. Saint-Lager, R. Lazzari, J. Jupille, O. Robach, S. Garaudée, G. Cabailh, P. Dolle, H. Cruguel and A. Bailly, *J. Phys. Chem. C*, 2011, **115**, 4673.
18 M.-C. Saint-Lager, I. Laoufi, A. Bailly, O. Robach, S. Garaudée and P. Dolle, *Faraday Discuss.*, 2011, **152**, 253.
19 J.-M. Pan, B. L. Maschhoff, U. Diebold and T. E. Madey, *J. Vac. Sci. Technol. A*, 1992, **10**, 2470.
20 F. Cosandey, L. Zhang and T. E. Madey, *Surf. Sci.*, 2001, **474**, 1.
21 R. Lazzari, G. Renaud, J. Jupille and F. Leroy, *Phys. Rev. B*, 2007, **76**, 125412.
22 R. Lazzari, *J. Appl. Crystallogr.*, 2002, **35**, 406.
23 S. Giorgio, M. Cabié and C. R. Henry, *Gold Bull.*, 2008, **41**, 167.
24 T. Uchiyama, H. Yoshida, Y. Kuwauchi, S. Ichikawa, S. Shimada, M. Haruta and S. Takeda, *Angew. Chem., Int. Ed.*, 2011, **50**, 10157.
25 Y. Kuwauchi, H. Yoshida, T. Akita, M. Haruta and S. Takeda, *Angew. Chem., Int. Ed.*, 2012, **51**, 7729.
26 R. Van Hardeveld and F. Hartog, *Surf. Sci.*, 1969, **15**, 189.
27 M. Rauscher, R. Paniago, T. H. Metzger, Z. Kovats, J. Domke, H. D. Pfannes, J. Schulze and I. J. Eisele, *J. Appl. Phys.*, 1999, **86**, 6763.
28 A. Carlsson, A. Puig-Molina and T. V. W. Janssens, *J. Phys. Chem. B*, 2006, **110**, 5286.
29 I. X. Green, W. Tang, M. McEntee, M. Neurock and J. T. Yates, Jr, *J. Am. Chem. Soc.*, 2012, **134**, 12717.
30 P. Jiang, S. Porsgaard, F. Borondics, M. Köber, A. Caballero, H. Bluhm, F. Besenbacher and M. Salmeron, *J. Am. Chem. Soc.*, 2010, **132**, 2858.
31 K. Dumbuya, G. Cabailh, R. Lazzari, J. Jupille, L. Ringela, M. Pistor, O. Lytkena, H.-P. Steinrück and J. M. Gottfried, *Catal. Today*, 2012, **181**, 20.
32 J. A. van Bokhoven, C. Louis, J. T. Miller, M. Tromp, O. V. Safonova and P. Glatzel, *Angew. Chem., Int. Ed.*, 2006, **45**, 4651.
33 S. Giorgio, S. Sao Joao, S. Nitsche, D. Chaudanson, G. Sitja and C. R. Henry, *Ultramicroscopy*, 2006, **106**, 503.

CO and O overlayers on Pd nanocrystals supported on TiO_2(110)

Chi Ming Yim,[a] Chi Lun Pang,[a] David S. Humphrey,[a] Christopher A. Muryn,[b] Karina Schulte,[c] Rubén Pérez[de] and Geoff Thornton*[a]

Received 30th November 2012, Accepted 17th December 2012
DOI: 10.1039/c2fd20137b

We have prepared a model catalytic system by depositing Pd onto a TiO_2(110) surface held at ~720 K. Scanning tunneling microscopy (STM) reveals well-defined Pd nanocrystals consisting of (111) top facets with {111} and {100} side facets. The Pd nanocrystals go down to about 10 nm in width and 1.3 nm in height. Top facets can be imaged with atomic resolution, indicating the absence of TiO_x encapsulation. The model catalyst was probed by exposure to CO and O_2. By varying the CO exposure, different CO overlayers were formed on the (111) top facets, with coverages ranging from 0.33 to 0.75 of a monolayer. Near edge X-ray absorption fine structure (NEXAFS) measurements at 300 K reveal that at around 0.5 ML coverage, CO is oriented with the molecular axis more or less normal to TiO_2(110). Dosing small amounts of O_2 separately on a Pd/TiO_2(110) surface led to an overlayer of $p(2 \times 2)$-O formed on the (111) top facet of the Pd nanocrystals at 190 K.

1 Introduction

The study of metal nanoparticles on oxide supports has received considerable attention, for reasons both practical and fundamental. On the practical side, the heterogeneous catalysts used in industrial applications usually consist of nanometer-sized metal particles distributed on an oxide support. In order to improve their performance, a better understanding of their properties is necessary. One method of achieving this is to investigate appropriate model catalyst systems. On the fundamental side, nanometer-sized metal particles have been shown to exhibit properties deviating from their corresponding single crystal surfaces. These differences have been shown to relate to the unique phenomena occurring in those systems such as kinetic interplay between facets,[1] spillover and reverse spillover of adsorbates between metal particles and their supports,[2] and the strong metal–support interactions (SMSI).[3]

Scanning probe methods such as STM and non-contact atomic force microscopy (NC-AFM) have been employed extensively over the past decade to investigate metal nanoparticles on metal oxide surfaces.[1,4–8] Issues such as initial metal

[a]*Department of Chemistry and London Centre for Nanotechnology, University College London, 17–19 Gordon Street, London, WC1H 0AH, UK. E-mail: g.thornton@ucl.ac.uk*
[b]*School of Chemistry & Photon Science Institute, The University of Manchester, Oxford Road, Manchester, M13 9PL, UK*
[c]*The MAX-IV Laboratory, Lund University, P.O. Box 118, 221 00 Lund, Sweden*
[d]*Lawrence Berkeley National Laboratory, 1 Cyclotron Road, Mail Stop 66-200, Berkeley, California, CA 94720, USA*
[e]*Universidad Autonoma de Madrid, Departamento de Fisica Teorica de la Materia Condensada, Facultad de Ciencias, Modulo C-5, E-28049 Madrid, Spain*

adsorption site, spillover, and thermal stability have been studied. Despite this interest, only one of these studies has addressed the adsorption of gas molecules on supported nanoparticles. Højrup Hansen *et al.*[8] studied NO and O_2 adsorption on Pd nanoparticles on an ultrathin alumina film, itself supported on NiAl(110). However, because their images were taken under a partial pressure of NO and O_2, their observations were limited to saturation coverages of these adsorbates. Under-saturated overlayers are important as further reaction is much more likely.

Here, we address the issue of gas adsorption on supported metal nanoparticles by studying the coverage-dependent adsorption of CO on Pd nanocrystals supported on TiO_2(110). Pd/TiO_2 systems attract much attention in surface science due to their use in a wide range of processes including nitrite reduction in drinking water,[9] acetylene hydrogenation[10] and automobile catalysis where interaction with CO is an important consideration.[11] Preparing chemically active, well-defined Pd nanoparticles on rutile TiO_2(110) is a difficult task that is hindered by the propensity for the Pd particles to become encapsulated with a TiO_x layer ($x < 2$) after annealing. The onset temperature for encapsulation is about 800 K.[4] In addition to degrading the chemical reactivity (such as CO uptake),[4,5] encapsulation also drastically alters the structure of the Pd islands.[7] Similar encapsulation-induced structural changes were also discovered for Pt/TiO_2(110).[6]

STM experiments reveal the presence of well-ordered CO overlayers on top of the Pd nanoparticles. These overlayers have different structures depending on the CO coverage. Similar overlayers formed on the Pd(111) surface have also been observed using low energy electron diffraction (LEED),[12] X-ray photoelectron spectroscopy (XPS)/electron energy loss spectroscopy (EELS)[13] and STM[14] in the past, with their stability studied by Loffreda *et al.*[15] using DFT theory. Here, we analyse the stability of the CO overlayers formed on the Pd nanoparticles using an exchange-correlation functional that includes van der Waals (vdW) non-local correlations. The vdW functional proposed by Dion *et al.*[16,17] has been applied to the chemisorption of CO on the (111) surfaces of late 4d and 5d transition metals.[18] This functional reduces or completely solves the well-known discrepancy between theory and experiment regarding CO site preference and improves the value of the adsorption energies. In our study, we have used the optimised vdW functional optB86b, recently introduced by Klimeš *et al.*,[19,20] that improves significantly the accuracy of the original vdw-DF and reaches better than chemical accuracy for the S22 data set.[20]

2 Experimental and theoretical methods

Synchrotron radiation spectroscopies at MAX-Lab

To investigate the bonding geometry of CO on the TiO_2(110)-supported Pd nanoparticles, we performed near-edge X-ray absorption fine structure (NEXAFS) measurements at Beamline I311 at MAX-lab. The beamline is described elsewhere.[21] The analysis chamber is equipped with a SCIENTA-SES200 hemispherical analyzer for photoemission (XPS) and NEXAFS in Auger yield mode. The angle between the analyzer and the incident photon beam is 55°.

The TiO_2(110) sample was cleaned by cycles of Ar^+ sputtering and annealing to ~1000 K. The sample temperature was checked using a K-type thermocouple spot-welded near the edge of the sample, which was mounted on a tantalum (Ta) sample holder. The sample cleanliness and long-range order were verified with XPS and LEED, respectively. Pd nanoparticles were formed by depositing Pd onto TiO_2(110) with the sample held at 800 K. Pd was evaporated from an *Omicron EFM3* electron-beam evaporator. Pd 3d and Ti 2p photoemission spectra ($h\nu = 620$ eV) were recorded to confirm the presence and chemical nature of Pd nanoparticles and to verify that no encapsulation had taken place on the Pd nanoparticles. All XPS were taken at normal emission. Binding energies were measured with respect to the Fermi edge (E_F) determined from the Ta sample plate in direct electrical contact with the sample. Research-grade CO was dosed by backfilling the preparation chamber, with its purity verified with a quadrupole mass spectrometer.

Carbon K-edge NEXAFS ($h\nu = 280$–320 eV) measurements were carried out to examine the bonding geometry of CO on the $Pd/TiO_2(110)$ surface. The degree of polarisation in the C K-edge region was estimated to be 0.94. NEXAFS data were recorded with the angle of incidence ranging between 40 and 90 degrees by monitoring the carbon KLL Auger yield. All NEXAFS and XPS measurements were carried out at 300 K.

STM experiments

The STM experiments were carried out with an *Omicron GmbH* low temperature STM housed in a bath cryostat within an ultrahigh vacuum (UHV) chamber (base pressure $= 2 \times 10^{-11}$ mbar). The adjoining preparation chamber was equipped with facilities for electron beam heating, ion sputtering, Pd deposition, XPS, LEED, and quadrupole mass spectroscopy (QMS).

The $TiO_2(110)$-(1×1) samples (*Pi-Kem*) were prepared through cycles of argon-ion sputtering (1 kV, drain current ~7 μA) and annealing to 950 K. Sample cleanliness and long-range order were verified with XPS and LEED. Pd islands were formed by physical vapor deposition (PVD) onto the as-prepared $TiO_2(110)$ surface with the sample held at 700–800 K. The Pd source consisted of Pd wire (*Advent*, 99.95%) wrapped around a tungsten filament that was resistively heated. CO (*Laborgase*, 99.998%) was dosed mainly with a directional doser placed ~50 mm away from the STM stage or *via* backfilling with a leak valve in the preparation chamber. In one case, CO was first dosed from the residual vacuum. The directional doser consisted of a 0.25 inch stainless steel tube welded to a leak-valve. O_2 (*SIP Analytical Ltd.*, 99.999%) was dosed using the directional doser. The purity of both CO and O_2 were checked using QMS.

STM measurements were carried out with electrochemically etched tungsten tips (*Advent*, purity = 99.95 %, diameter = 0.25 mm). The tips were conditioned by vacuum annealing at 200 °C and voltage pulses during scanning. STM images were acquired at 124–190 K.

Computational details

The DFT calculations for the CO overlayers on a Pd(111) surface have been performed using the Vienna *ab initio* simulation package (VASP, version 5.2.12) with the projector-augmented wave (PAW) method and a cutoff of 500 eV for the plane-wave expansion of the wavefunctions.[22,23] For the exchange-correlation functional, we have used the optB86b vdW functional implemented in VASP by Klimes *et al.*[20] using the algorithm of Román-Pérez and Soler.[24] A spin-restricted approach has been used since spin polarisation effects have been found to be negligible.

Geometry optimisations were performed on a supercell structure using periodic boundary conditions. The (111) surface is modeled by a four layer slab, with a CO adlayer on one side of the slab and a vacuum space equivalent to six ideal bulk metallic layers. We use the lattice constant of 3.946 Å calculated theoretically for Pd, that is close to the experimental value (3.82 Å). We have considered three different surface unit cells to simulate the different coverages: $(\sqrt{3} \times \sqrt{3})R30°$-1CO (0.33 ML), c($4 \times 2$)-2CO (0.5 ML), and ($2 \times 2$)-3CO (0.75 ML). The two uppermost substrate layers and the CO molecules are allowed to relax, while the bottom two layers of Pd atoms were held fixed in their bulk positions. All atomic coordinates of the adsorbed species and the metal atoms in the relaxed metal layers were optimised to a force of less than 0.01 eV $Å^{-1}$ on each atom. Brillouin zone integration was performed using Monkhorst–Pack grids of $12 \times 12 \times 1$, $8 \times 8 \times 1$ and $4 \times 4 \times 1$, for the $(\sqrt{3} \times \sqrt{3})$, ($2 \times 2$), and c($4 \times 2$) unit cells, respectively, and a Methfessel–Paxton smearing of 0.2 eV.

3 Results and discussion

NEXAFS measurements of CO on Pd/TiO_2(110)

A Pd/TiO_2(110) surface was prepared by depositing Pd onto the TiO_2(110) surface with the sample held at 800 K. This resulted in well-ordered Pd nanoparticles distributed on the TiO_2(110) surface, with their (111) top facets oriented parallel to the TiO_2(110) surface, as evidenced by the emergence of Pd(111)-(1 × 1) spots in LEED.[25] Ti 2p XPS peaks taken before and after the Pd dose showed no significant increase in the amount of Ti^{3+} and Ti^{2+} species, indicating that no encapsulation takes place on the Pd nanoparticles. Using the change in integrated intensity of Ti $2p_{3/2}$ peaks, the effective thickness of Pd present on the TiO_2(110) surface was calculated to be 10.4 ± 3.7 Å. Note that this value will depend on the morphology as well as the size distribution of the nanoparticles and is therefore only a nominal value.

After growing Pd nanoparticles on the TiO_2(110) surface, NEXAFS spectra of TiO_2(110)/Pd/CO were recorded in the photon energy range 280–320 eV at different angles of incidence from the Pd/TiO_2(110) surface before and after exposure of 2000 L CO (1 L = 1 langmuir = 1.33×10^{-6} mbar s) at 300 K. The kinetic energy window of the analyzer was centred at 265 eV with a width of 4.4 eV. To remove unwanted signals from the substrate, we normalised the NEXAFS spectra by subtracting the spectra before CO exposure from those recorded on the CO-covered surface. The resulting difference spectra are displayed in Fig. 1a.

Each spectrum in Fig. 1a consists of a peak at ~287.2 eV convoluted with a step-like feature. The peak corresponds to the π^* resonance, in which the 1s electrons in carbon atoms are photo-excited to the unoccupied π^* orbitals of CO. STM data (not shown) indicate that nearly 90% of the TiO_2(110) substrate is covered with Pd islands. Moreover, no CO adsorption on the TiO_2(110) surface was observed at 300 K. On this basis, we conclude that only CO bonded to the Pd islands contributes to the X-ray absorption signal. To determine the orientation of CO molecules on the Pd islands, taking threefold symmetry of the Pd islands into account, we numerically fitted the area of the π^* resonance peak as a function of incident angle using the appropriate equations in ref. 26. The best fit to the data in Fig. 1b corresponds to

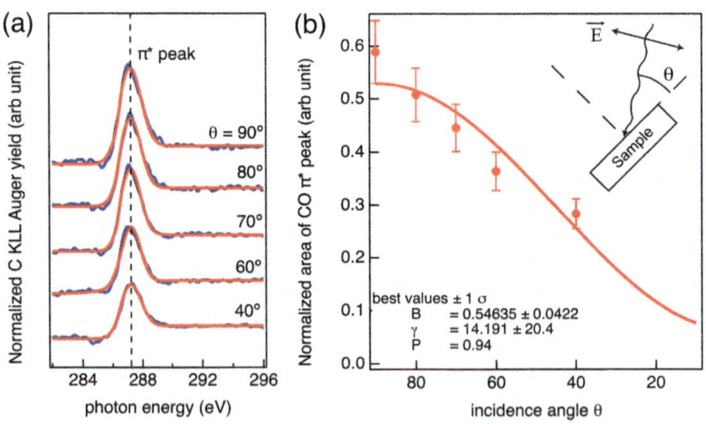

Fig. 1 (a) Normalised carbon KLL NEXAFS as a function of photon energy measured at different incident angles on CO-covered Pd nanoparticles grown on TiO_2(110), recorded at room temperature. Only the region of the π^* resonant peak is shown. Blue dots are experimental data and red solid lines are numerical fits. (b) Normalised area of the CO π^* resonant peak as a function of incident angle. Dots are experimental data, and the solid line is a numerical fit to the experimental data using the appropriate equations in Stöhr and Outka's paper ref. 26. The best-fit values shown are the arbitrary constant B, degree of polarisation P (= 0.94), and the angle between the C–O axis and the surface normal γ.

a bond angle, γ, of $14.2 \pm 22.1°$. Hence, CO bonds more or less vertically on the top facet of the Pd islands.

CO overlayers at different coverages in STM

Pd was deposited on the rutile $TiO_2(110)$ substrate with the sample held at a temperature of ~720 K. This led to the formation of Pd nanoparticles distributed on the $TiO_2(110)$ surface, as shown in Fig. 2a. The shape of the particles is indicative of well-ordered nanocrystals comprised of a (111) top facet with {111} and {100} side facets. Their average heights and diameters are 2.38 ± 0.44 nm and 22.0 ± 5.2 nm, respectively. Occasionally, the top facet can be imaged with atomic resolution, as shown in Fig. 2b. The nearest-neighbour separation of bright protrusions is about 270 ± 10 pm, consistent with the Pd lattice constant, 275 pm, thus validating the suggestion that the top facet is a Pd(111)-(1 × 1) face. It also demonstrates that the Pd nanocrystals formed in this way are not encapsulated.[7] Note that due to the spatial limit of STM, we could not detect any lattice contraction of the (111) top facet as suggested in the micro-calorimetric measurements of CO on $Fe_3O_4(111)$ film-supported Pd nanoparticles by Fischer–Wolfarth *et al.*[27]

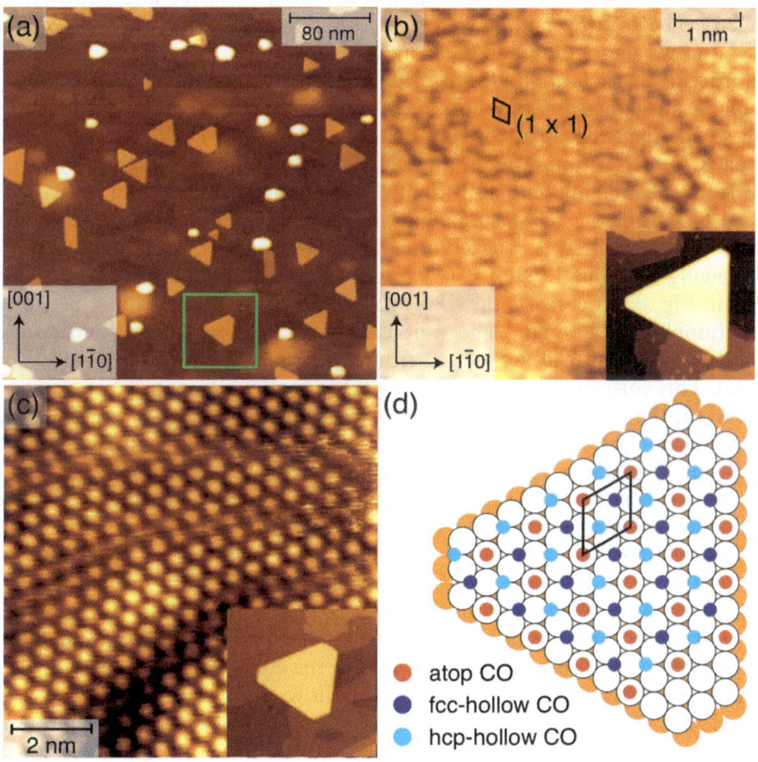

Fig. 2 (a) STM image (390 × 390 nm², 1.6 V, 10 pA) of Pd islands distributed on the $TiO_2(110)$ surface. The islands are pseudo-hexagonal and have mean measured heights and diameters of 2.38 ± 0.44 nm and 22.0 ± 5.2 nm, respectively. (b) Atomically-resolved STM image acquired on a Pd nanocrystal (5.5 × 5.5 nm², −0.3 V, 0.2 nA) with a coverage of < 0.3 ML of CO. Due to an adventitious tip change, in contrast to most images, the underlying Pd(111)-(1 × 1) top facet was resolved. (c) Molecularly resolved STM image (9 × 9 nm², 1.1 V, 10 pA) recorded on top of one of the Pd islands following a saturation dose of CO at 124 K. This leads to the formation of a (2 × 2)-3CO ordered phase giving a CO coverage of 0.75 ML. This image was taken from the island indicated with the green square in (a) and inset in (c). (d) The proposed model for (2 × 2)-3CO, with CO occupying atop, as well as fcc- and hcp- hollow sites. All STM data were recorded at 124 K. The $TiO_2(110)$ azimuths are indicated.

To dose CO onto the surface *via* the directional doser, the STM is paused and the tip retracted by 20 nm. Fig. 2c shows a STM image of the surface following CO dosing with nominal pressure and duration of 1×10^{-7} mbar and 3 s, respectively, at a temperature of 124 K. This dose is equivalent to a nominal 0.23 L exposure. This procedure led to saturation of CO on the {111} top facet of the Pd islands. As the doser was directed towards the surface only 50 mm away, the local pressure around the sample was estimated to be 100 times higher than that measured by the ion gauge. The estimation was made based on the exposure required to form a saturated CO overlayer without directional dosing.

Zooming into the island indicated with a green square in Fig. 2a reveals the presence of an ordered CO layer as shown in Fig. 2c. The ordered structure consists of hexagonally packed bright dots with a nearest neighbour distance of 550 ± 16 pm. The principal axes of this overlayer run parallel to those of the Pd(111) top facet so that the structure has a (2×2) arrangement. According to Rose *et al.*'s STM work on the native Pd(111) single crystal surface,[28] the (2×2) configuration is found when the CO coverage is 0.75 ML, where a monolayer (ML) corresponds to 1.527×10^{15} molecules cm^{-2}. Each primitive unit cell contains three CO molecules, so the overlayer is described as (2×2)-3CO. In their model, one CO molecule occupies the atop site while the other two CO molecules occupy the fcc and hcp hollow sites (Fig. 2d). The overlayer in Fig. 2c is strikingly similar to that reported by Rose *et al.*[28] and we therefore made the same assignment.

In order to characterise under-saturated CO overlayers, we investigated the evolution of CO structures as a function of coverage by dosing small quantities of CO in a stepwise manner. Because the Pd nanocrystals can be easily contaminated by residual CO, every time after we saturated the sample with CO, we warmed the STM from the imaging temperature of 124 K up to 200 K in order to desorb physisorbed CO from the STM stage and the cryostat. After this, a freshly-prepared Pd/TiO_2(110) sample was inserted into the microscope and held at 124 K.

Following a nominal 0.01 L CO exposure, an ordered structure was formed. As shown in Fig. 3a, it consists of hexagonally-packed unit cells with a measured unit cell parameter of 470 pm. The unit cell is rotated by 30° with respect to the Pd(111)-(1×1) unit cell, as judged from the step-edges of the Pd island (inset of Fig. 3a). This corresponds to a $(\sqrt{3} \times \sqrt{3})$R30° configuration, a structure previously reported for the single crystal surface that has a coverage of 1/3 ML.[28] The model for this overlayer has CO occupying fcc-hollow sites and is shown in Fig. 3b.

With our experimental setup it was difficult to prepare a Pd/TiO_2(110) surface covered with exactly 0.5 ML. When deviating from this coverage, the CO becomes too mobile to image in STM. To circumvent this, we adopted the following approach: first we saturated the Pd/TiO_2(110) surface with CO at 150 K, then warmed it to >200 K before bringing it back to 150 K. This led to a surface with the (111) top facet of every Pd nanocrystal covered with a c(4×2)-2CO overlayer. On one of the Pd nanocrystals (Fig. 3c), CO molecules arrange themselves to form a zigzag pattern. Therefore, we propose that in the c(4×2)-2CO structure in Fig. 3c, CO molecules occupy both fcc- and hcp- hollow sites, leading to a hollow-hollow (*hh*-) c(4×2)-2CO (Fig. 3d). This structure is reminiscent of similar images observed by Rose *et al.*[28] for CO adsorbed on the native Pd(111) surface.

Fig. 3c also shows a facet edge on the right-hand side. Note that the structure deviates from that of the (111) top facet: the protrusions along the facet edge appear with varying brightness in STM. In the *hh*-c(4×2)-2CO structure proposed for the (111) top facet, CO at both fcc- and hcp- hollow sites appear with equal brightness. As such, we believe that those protrusions arise from the CO adsorbing at the sites on the {111} side facet, or less likely, along the edge of the top facet (which comprises only atop sites and bridge sites). Judging from their positions and relative brightness, we attribute those protrusions to CO that occupy different adsorption sites on the {111} side facet. The brighter protrusions correspond to CO that occupy hcp-hollow sites, and the dimmer protrusions to CO that occupy fcc-hollow sites

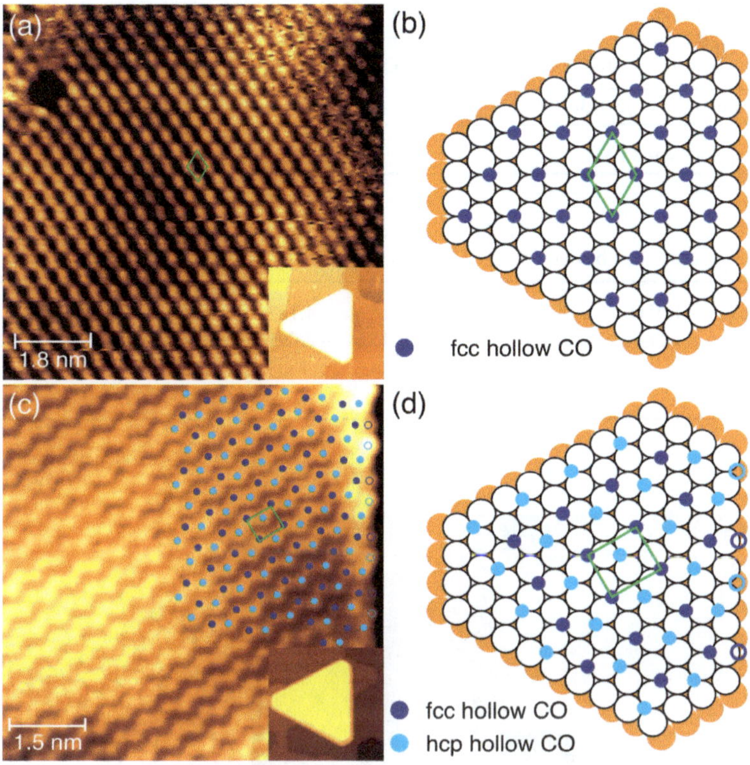

Fig. 3 (a) Molecularly resolved STM image (9 × 9 nm^2, 0.9 V, 0.05 nA) recorded on a Pd nanocrystal (width = 22.3 nm, height = 3.0 nm) at 124 K. The (111) top-facet of the Pd nanocrystal was covered with 0.33 ML of CO. (b) A ball model showing how CO bonds to the (111) top facet of a nanocrystal in the proposed ($\sqrt{3} \times \sqrt{3}$)R30°-1CO structure. (c) Molecularly resolved STM image (7.6 × 7.6 nm^2, 0.5 V, 0.1 nA) recorded on a Pd nanocrystal (width = 29.1 nm, height = 2.7 nm) at 150 K. Its (111) top facet was covered with 0.5 ML of CO, prepared with non-directional dosing method (see text). (d) A ball model of *hh*-c(4 × 2)-2CO ordered structure. CO occupying the fcc- and hcp- hollow sites on the {111} side facet are indicated with open circles. The TiO_2(110) azimuths are shown in Fig. 2.

(marked with open circles in Fig. 3). This assignment is based on the higher vertical positions of the hcp-hollow sites on the {111} side facets compared with the fcc-hollow sites. This bonding of CO on the {111} side facet is also illustrated in Fig. 3d.

We can compare the overlayers observed in STM with their adsorption energies computed using the VASP code. The latter are shown in Table 1 where it can be seen that the lowest energy adsorption sites are in line with those calculated in the earlier work.[15] In other words, at 0.33 ML, fcc hollow sites are occupied, while at 0.5 ML both fcc and hcp hollow sites are occupied. At 0.75 ML atop sites along with fcc and hcp hollow sites are filled. The inclusion of vdW contributions increases the adsorption energies (up to 0.3 eV molecule^{-1} for the 0.75 ML coverage) but does not modify the relative stability of the different sites for a given coverage.

O_2 adsorption on Pd nanocrystals

We also investigated O_2 adsorption on the Pd nanocrystals with STM. Following a nominal O_2 exposure of 0.005 L on the Pd/TiO_2(110) surface by directional dosing at 190 K, an oxygen-induced ordered structure was formed on the (111) top facet of the Pd nanocrystals, as shown in Fig. 4a. The ordered structure consists of hexagonally

Table 1 PBE + vdW calculated adsorption energies E_{ads} (eV) and adsorption energies per molecule E_{ads} mol^{-1} (eV) for the chemisorption of CO on Pd(111), at coverages 0.33 ML [($\sqrt{3} \times \sqrt{3}$)R30°-1CO], 0.5 ML [c(4 × 2)-2CO] and 0.75 ML [(2 × 2)-3CO]

Structure	($\sqrt{3} \times \sqrt{3}$)R30°-1CO				c(4 × 2)-2CO		(2 × 2)-3CO	
Coverage	$\theta = 0.33$ ML				$\theta = 0.50$ ML		$\theta = 0.75$ ML	
Sites	atop	bridge	hollow fcc	hollow hcp	bridge-bridge	fcc + hcp	atop + bridge-bridge	atop + fcc + hcp
E_{ads} (eV)	−1.5860	−2.0023	−2.1958	−2.1594	−4.1834	−4.2923	−3.7242	−5.3351
E_{ads} mol^{-1} (eV)	−1.5860	−2.0023	−2.1958	−2.1594	−2.0917	−2.1462	−1.2414	−1.7784

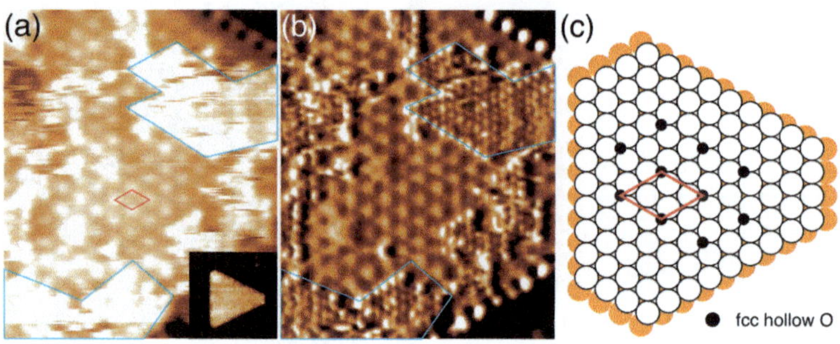

Fig. 4 (a) Atomically resolved STM image (8.0 × 9.3 nm^2, 0.45 V, 0.15 nA) recorded on a Pd nanocrystal (diameter = 15.0 nm, height = 1.9 nm) at 190 K. Its (111) top facet was partially covered with < 0.25 ML of atomic oxygen, leading to the formation of a p(2 × 2)-O structure. Edge structures are also observed on the top- and bottom-right of the image. (b) The corresponding current map of (a) illustrates that the unresolved areas (highlighted in blue) in (a) are the (111) top-facet regions not covered with atomic O, as the (111)-(1 × 1) unit cells are resolved. (c) Ball model of the p(2 × 2)-O ordered phase, in which atomic O is proposed to adsorb at the fcc-hollow sites.

packed bright dots with a nearest neighbour distance of 550 ± 20 pm. The principal axes of this overlayer run parallel to those of the Pd(111) top facet so that the structure has a (2 × 2) arrangement. As O_2 thermally dissociates to form atomic-O pairs on Pd(111) at 160 K,[29] we believe that the (2 × 2) ordered structure on the {111} top facet consists of atomic oxygen, leading to p(2 × 2)-O. In the ball model in Fig. 4c, we propose that in the p(2 × 2)-O structure, oxygen atoms adsorb at the fcc-hollow sites. These sites have been shown to be most favoured on the (111) surface of Pd single crystals.[29,30] Note that some areas (highlighted in blue) in the STM topography image are not well resolved (Fig. 4a). However, those areas are resolved in the current map recorded simultaneously (Fig. 4b). In the current map, the areas which are unresolved in Fig. 4a can be seen to consist of hexagonally packed bright dots with a nearest neighbour distance of 271 ± 20 pm, consistent with the Pd lattice constant of 2.75 Å. From this, we conclude that those areas are parts of the top facet not covered with atomic oxygen. Apart from the top-facet region, we also observed some edge structures on the oxygen-covered Pd nanocrystals, as shown in the top- and bottom-right of the images in Fig. 4a and 4b. Those edge structures look very similar to one of those reported in Højrup-Hansen *et al.*'s work.[8] As yet we do not have a model for this observed edge structure.

4 Summary

We have prepared a model catalytic system by forming Pd nanocrystals on rutile $TiO_2(110)$. Those nanocrystals, with average diameter of ~22 nm and height of ~2 nm, consist of (111) top facets with {111} and {100} side facets. The top facets can be imaged with atomic resolution, confirming the Pd(111)-(1 × 1) long range order expected and also demonstrating that the nanocrystals are not encapsulated. By varying the CO exposure, a rich variety of CO overlayers were formed on the (111) top facets with coverages ranging from 1/3 ML to the saturated 3/4 ML (2 × 2)-3CO overlayer. Importantly, some of these overlayers were under-saturated and are therefore likely to be more reactive than the saturated (2 × 2)-3CO layer. DFT calculations of adsorption energies indicate that for all the structures observed, fcc-hollow sites are occupied. Above 0.33 ML, first hcp hollow sites are occupied and then additionally the atop site at 0.75 ML. By dosing a small amount of O_2 on $Pd/TiO_2(110)$ at 190 K, we observed the formation of the p(2 × 2)-O ordered structure as well as oxygen-induced edge structures on the island edges. Island edges and island–oxide interface regions have long been posited as active sites for catalysis and our model system offers the opportunity to study any such effect directly with STM. This work shows that adsorbate overlayers can be the same on nanoparticles as those observed on metal single crystal surfaces, even below saturation coverage. However, in a future publication we will present data that evidences significant differences in behaviour.

Acknowledgements

This work was supported by EPSRC (UK), COST1104 and European Research Council Advanced Grant ENERGYSURF. RP acknowledges the support of the Spanish Projects MAT2011-023627 (MINECO), PR2011-0402 (MEC) and Fundación CajaMadrid.

References

1 V. P. Zhdanov and B. Kasemo, *Surf. Sci. Rep.*, 2000, **39**, 25–104.
2 W. C. Conner and J. L. Falconer, *Chem. Rev.*, 1995, **95**, 759–788.
3 S. J. Tauster, *Acc. Chem. Res.*, 1987, **20**, 389–394.
4 Z. Chang and G. Thornton, *Surf. Sci.*, 2000, **459**, 303–309.
5 M. Bowker, P. Stone, R. A. Bennett and N. Perkins, *Surf. Sci.*, 2002, **497**, 155–165.
6 O. Dulub, W. Hebenstreit and U. Diebold, *Phys. Rev. Lett.*, 2000, **84**, 3646–3649.
7 R. A. Bennett, C. L. Pang, N. Perkins, R. D. Smith, P. Morrall, R. I. Kvon and M. Bowker, *J. Phys. Chem. B*, 2002, **106**, 4688–4696.
8 K. Højrup Hansen, Ž. Šljivančanin, E. Laegsgaard, F. Besenbacher and I. Stensgaard, *Surf. Sci.*, 2002, **505**, 25–38.
9 W. Gao, J. Chen, X. Guan, R. Jin, F. Zhang and N. Guan, *Catal. Today*, 2004, **93–95**, 333–339.
10 W. Kim, *J. Catal.*, 2004, **226**, 226–229.
11 H. Imagawa, T. Tanaka, N. Takahashi, S. Matsunaga, A. Suda and H. Shinjoh, *J. Catal.*, 2007, **251**, 315–320.
12 H. Ohtani, M. A. V. Hove and G. A. Somorjai, *Surf. Sci.*, 1987, **187**, 372–386.
13 S. Surnev, M. Sock, M. G. Ramsey, F. P. Netzer, M. Wiklund, M. Borg and J. N. Andersen, *Surf. Sci.*, 2000, **470**, 171–185.
14 M. K. Rose, T. Mitsui, J. Dunphy, A. Borg, D. F. Ogletree, M. B. Salmeron and P. Sautet, *Surf. Sci.*, 2002, **512**, 48–60.
15 D. Loffreda, D. Simon and P. Sautet, *Surf. Sci.*, 1999, **425**, 68–80.
16 M. Dion, H. Rydberg, E. Schröder, D. C. Langreth and B. I. Lundqvist, *Phys. Rev. Lett.*, 2004, **92**, 246401; M. Dion, H. Rydberg, E. Schröder, D. C. Langreth and B. I. Lundqvist, *Phys. Rev. Lett.*, 2005, **95**, 109902.
17 D. C. Langreth, M. Dion, H. Rydberg, E. Schröder, P. Hyldgaard and B. I. Lundqvist, *Int. J. Quantum Chem.*, 2005, **101**, 599–610.
18 P. Lazić, M. Alaei, N. Atodiresei, V. Caciuc, R. Brako and S. Blügel, *Phys. Rev. B*, 2010, **81**, 045401.

19 J. Klimeš, D. R. Bowler and A. Michaelides, *J. Phys.: Condens. Matter*, 2010, **22**, 022201.
20 J. Klimeš, D. Bowler and A. Michaelides, *Phys. Rev. B*, 2011, **83**, 195131.
21 R. Nyholm, J. N. Andersen, U. Johansson, B. N. Jensen and I. Lindau, *Nucl. Instrum. Methods Phys. Res., Sect. A*, 2001, **467–468**, 520–524.
22 G. Kresse and J. Furthmüller, *Comput. Mater. Sci.*, 1996, **6**, 15–50.
23 G. Kresse, *Phys. Rev. B*, 1996, **54**, 11169–11186.
24 G. Román-Pérez and J. Soler, *Phys. Rev. Lett.*, 2009, **103**, 096102.
25 M. Bowker, P. Stone, P. Morrall, R. Smith, R. A. Bennett, N. Perkins, R. Kvon, C. L. Pang, E. Fourre and M. Hall, *J. Catal.*, 2005, **234**, 172–181.
26 J. Stohr and D. A. Outka, *Phys. Rev. B*, 1987, **36**, 7891–7905.
27 J.-H. Fischer-Wolfarth, J. Farmer, J. Flores-Camacho, A. Genest, I. Yudanov, N. Rösch, C. Campbell, S. Schauermann, and H.-J. Freund, *Phys. Rev. B*, 2010, **81**, 241416.
28 M. K. Rose, T. Mitsui, J. Dunphy, A. Borg, D. F. Ogletree, M. B. Salmeron and P. Sautet, *Surf. Sci.*, 2002, **512**, 259–266.
29 M. K. Rose, A. Borg, J. C. Dunphy, T. Mitsui, D. F. Ogletree and M. B. Salmeron, *Surf. Sci.*, 2004, **561**, 69–78.
30 J. Blanco, C. González, P. Jelinek, J. Ortega, F. Flores, R. Pérez, M. K. Rose, M. B. Salmeron, J. Méndez, J. Wintterlin and G. Ertl, *Phys. Rev. B*, 2005, **71**, 113402.

Fabrication of complex model oxide catalysts: Mo oxide supported on $Fe_3O_4(111)$

Chanut Bamroongwongdee, Michael Bowker,†* Albert F. Carley, Philip R. Davies,* Robert J. Davies and Dyfan Edwards

Received 30th November 2012, Accepted 17th December 2012
DOI: 10.1039/c2fd20134h

Industrial catalysts for the oxidation of methanol to formaldehyde consist of iron molybdate [$Fe_2(MoO_4)_3$]. Using a variety of techniques we have previously shown that the surface of these catalysts is segregated in MoO_3, and in order to understand the relationship between surface structure and reactivity for these systems we have begun a surface science study of this system using model, single crystal oxides. Model catalysts of molybdenum oxide nanoparticles and films on an $Fe_3O_4(111)$ single crystal were fabricated by the hot-filament metal oxide deposition technique (HFMOD), where molybdenum oxides were produced using a molybdenum filament heated in an oxygen atmosphere. Low energy electron diffraction (LEED), X-ray photoelectron spectroscopy (XPS), and scanning tunnelling microscopy (STM) have been used to investigate molybdenum oxide nanoparticles and films deposited on $Fe_3O_4(111)$. The molybdenum oxide film forms in the highest oxidation state, +6, and is remarkably stable to thermal treatment, remaining on the surface to at least 973 K. However, above ~573 K cation mixing begins to occur, forming an iron molybdate structure, but the process is strongly Mo coverage dependent.

Introduction

Formaldehyde, one of the most important industrial chemicals, is produced from methanol using either silver or iron molybdate catalysts. Over iron molybdate, methanol undergoes a partial oxidation reaction, forming formaldehyde and water.

$$CH_3OH + \tfrac{1}{2}O_2 \rightarrow HCHO + H_2O \tag{1}$$

The industrial iron molybdate catalysts are Mo–Fe–O mixed oxides with an excess of Mo compared to $Fe_2(MoO_4)_3$. Whereas stoichiometric ferric molybdate has a Mo/Fe ratio of 1.5, in the catalyst the ratio is typically 2 or higher. This excess of Mo is needed to maintain the active phase due to Mo loss occurring at hot spots during the industrial process which can result in a reduction of the lifetime of the catalyst to around twelve months. In contrast, an excess of iron in the molybdate catalyst results in deeper oxidation.[1–4]

The nature of the active site in the catalyst is a subject of debate and there are two schools of thought regarding the origin of catalytic activity. The first is that it is due only to the stoichiometric $Fe_2(MoO_4)_3$ phase, the other is that the mixed oxides phase with an excess of Mo is responsible. Clearly then, deducing the role of

Wolfson Nanoscience Laboratory, Cardiff Catalysis Institute, School of Chemistry, Cardiff University, Park Place, Cardiff, CF10 3AT, UK

† and at Centre for Catalytic Science, Research Complex at Harwell (RCaH), Rutherford Appleton Laboratory, Harwell, Oxon, OX11 0QX, UK

molybdenum at the surface of the catalyst and its chemical state is important for gaining a deeper understanding of the system.

Temperature programmed desorption (TPD) studies have been carried out by several authors,[5,6] comparing the reactivity of methanol over Fe_2O_3, MoO_3 and the industrial iron molybdate catalyst. Single crystal MoO_3 gave more methanol desorption than formaldehyde,[5] whereas polycrystalline MoO_3 gave more formaldehyde. Reaction with Fe_2O_3 results in complete combustion to CO_2 and H_2O, whereas over MoO_3, formaldehyde was found to be the only carbon-containing product.[6] The industrial catalyst sample also gave rise to formaldehyde uniquely, but was found to be much more active, with conversion beginning at 150 °C compared with 270 °C for MoO_3. The combustion observed for reaction over Fe_2O_3 points towards formate as an intermediate species, whilst for the Mo-containing compounds, the involvement of a methoxy surface intermediate is suggested. The inference here, therefore, is that the surface is dominated by the molybdenum oxide species, which we,[7] and subsequently Routray *et al.*[8] have confirmed using low-energy ion scattering (LEIS) and transmission electron microscopy (TEM).

Although a great deal of research has been conducted into iron molybdate catalysts,[1] little has been undertaken on the surface science of these compounds. We have reported[9] some preliminary work, carried out in collaboration with Freund's group at the Fritz Haber Institute in Berlin, in which Mo layers were deposited on thin films of single crystalline $Fe_3O_4(111)$ grown on a Pt(111) single crystal. The main conclusion was that ordered mixed structures were formed after annealing above 900 K, in which the Mo and Fe cations are mixed. We present here results detailing the deposition of molybdenum oxide films on a bulk iron oxide single crystal surface for use as a model catalyst in the investigation of iron molybdate catalysis, with the aim of identifying the nature of, and formation of the active phase and, ultimately, of the active site for selective methanol oxidation.

Experimental

All experiments were performed in a UHV system built by Omicron Vacuum Physik capable of performing scanning tunnelling microscopy (STM), X-ray photoelectron spectroscopy (XPS), low energy ion scattering (LEIS/ISS) and low energy electron diffraction (LEED). The system comprises three separate chambers pumped by a combination of four turbomolecular pumps, three titanium sublimation pumps and three ion pumps, resulting in a base pressure of $<1 \times 10^{-9}$ mbar.

The Fe_3O_4 sample was cleaned by Ar^+ bombardment at 1 keV, followed by annealing in 1×10^{-7} mbar oxygen, typically at 873 K, although temperatures in the range 673–973 K were used for certain experiments. Surface cleanliness was monitored by XPS and ISS, and gas purity analysed using a quadrupole mass spectrometer. All gas exposures are quoted in Langmuirs ($1\ L = 10^{-6}$ torr s). XP spectra were recorded using an Al Kα photon source and an analyser pass energy of 50 eV unless stated otherwise. Binding energies were calibrated to the O(1s) peak at 530.2 eV. Ar^+ bombardment of the Ag(111) sample was carried out at ambient temperature at 600 eV and the sample subsequently annealed at 723 K for 30 minutes. XP spectra were calibrated to the Ag($3d_{5/2}$) peak at 368.1 eV.[10]

The iron oxide sample used was an $Fe_3O_4(111)$ single crystal (Pi-Kem Ltd). The sample was mounted on a standard Omicron molybdenum plate *via* spot-welded Ta strips. A thermocouple was attached to the sample plate holder for temperature measurement. All STM images were analysed using WSxM software.[11]

Molybdenum oxide films were grown by a hot filament metal oxide deposition (HFMOD) technique.[12,13] A current, typically 3.2 A, was passed through a coil of 0.25 mm Mo wire (Goodfellow, 99.95%) in an oxygen environment, with a typical pressure of 4×10^{-6} mbar. The molybdenum oxide films were then annealed at 873 K in oxygen (1×10^{-7} mbar), see discussion below, and the surface composition analysed by XPS and/or ISS. For small MoO_x coverages corresponding to less than a monolayer, surface concentrations were calculated from the area of the Mo(3d) photoelectron peaks using the methods described in detail elsewhere.[14,15] For higher coverages the thickness of the MoO_x films were calculated from the attenuation of the Fe(2p) peaks. A plot of film thickness against deposition time indicates uniform film growth, Fig. 1.

Results

1. Clean Fe_3O_4 surfaces

STM images of the clean $Fe_3O_4(111)$ surface at relatively low magnification (Fig. 2a) show terraces tens of nm in width separated by steps of ~0.5 nm height (Fig. 2e), consistent with the expected height of the repeat unit for Fe_3O_4 in the [111] direction.[16,17] However, even at this low resolution it is clear that there is further structure within the terraces and a closer examination reveals a mixture of surfaces. Most of the surface (Fig. 2b and c) is covered in a hexagonal structure with an inter-atomic distance of ~0.62 nm consistent with the p(2 × 2) LED patterns observed previously for the (111) surfaces of Fe_3O_4 thin films.[18,19] More recent work[20] has demonstrated that the topography of the magnetite surface can be sensitive to the preparation conditions, particularly the final annealing step. Our conditions and STM images are similar to those reported by Lennie *et al.*,[21] who refer to the hexagonal surface structure as "Termination A" and assign it to a layer of oxygen capping an octahedral iron layer. The same surface was observed by Flynn *et al.*,[22] who, on the basis of scanning tunnelling spectroscopy data and subsequent chemical reactions, assigned it to an $Fe_3O_4(111)$ surface terminated with $Fe^{2+/3+}$ ions in tetrahedral sites ("Fe_{tet1}"). In our case the "Termination A" surface is accompanied by areas that are not atomically resolved, with a step height between the two layers of 0.12 ± 0.03 nm (Fig. 2f, profile (iii)), the wide range in values being due to unresolved structure within the second surface termination. This differs significantly from the 0.38 and 0.05 nm step heights reported by Lennie *et al.* for what they term "Termination B" which they assigned to an "Fe_{tet2}–Fe_{oct}–Fe_{tet1}" layer. However, our results are more consistent with this model than with that of Flynn's; the removal of the capping oxygen from the Fe_{oct1} layer in Lennie's model would give a step down to the iron layer of ~0.11 nm with the additional unresolved structure in the lower layer being accounted for by the presence of defects necessary to stabilize the excess positive charge. In Flynn's model however, the expected step heights from the tetrahedrally coordinated iron layer would be 0.06 or 0.18 nm. LEED patterns of the clean surface (Fig. 3), show only a p(2 × 2) structure consistent with the dominant Termination A.

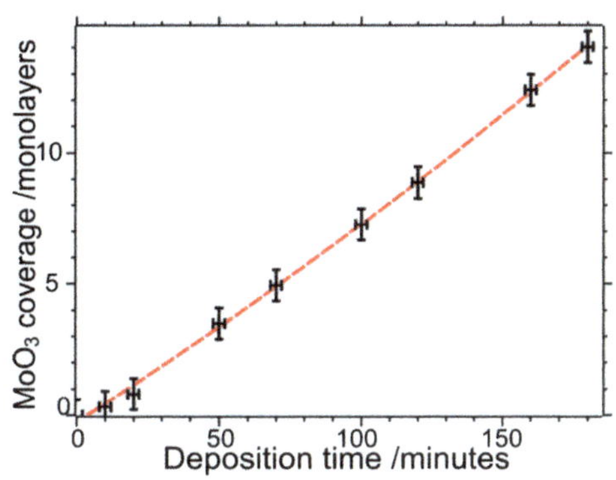

Fig. 1 MoO_x film thickness as a function of the deposition time, determined from the Mo(3d) XPS signal. Filament current: 3.8 A; oxygen pressure: 8.0 × 10^{-6} mbar.

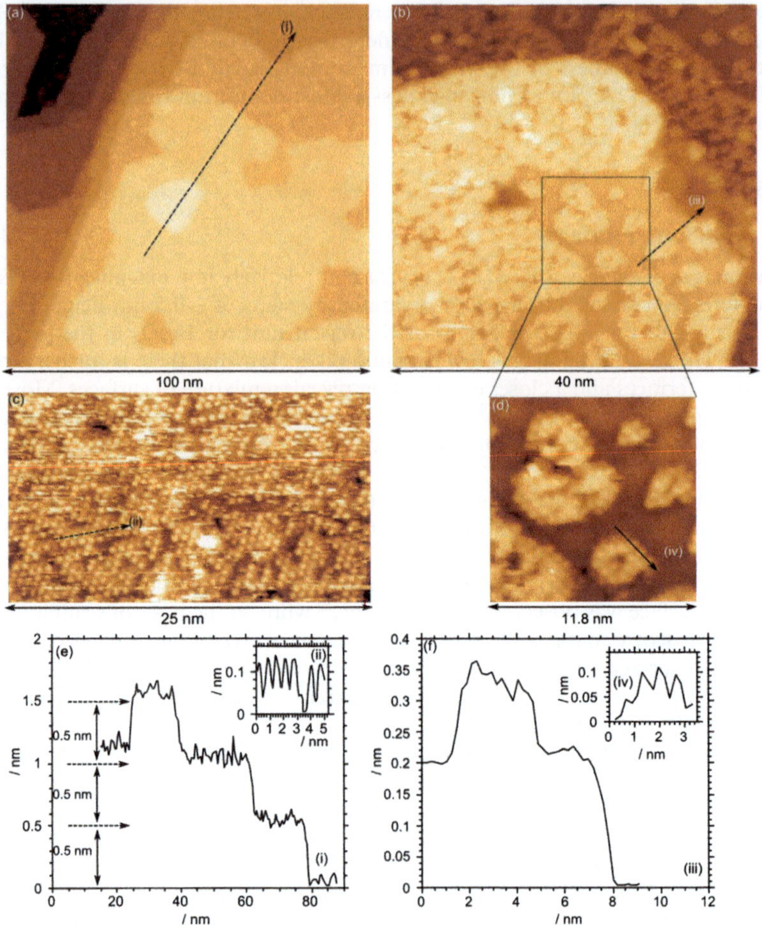

Fig. 2 STM images from the clean $Fe_3O_4(111)$ surface after sputtering and annealing in 1×10^{-7} mbar of oxygen at 873 K. (a) Large-scale image in which one can clearly observe individual terraces, separated by single height steps (~0.5 nm). (b)–(d) Higher magnification views showing the complex nature of the surface. Line profiles are identified with Roman numerals and drawn in (e) and (f). ($V_b = -1.0$ V, $I_t = 0.465$ nA).

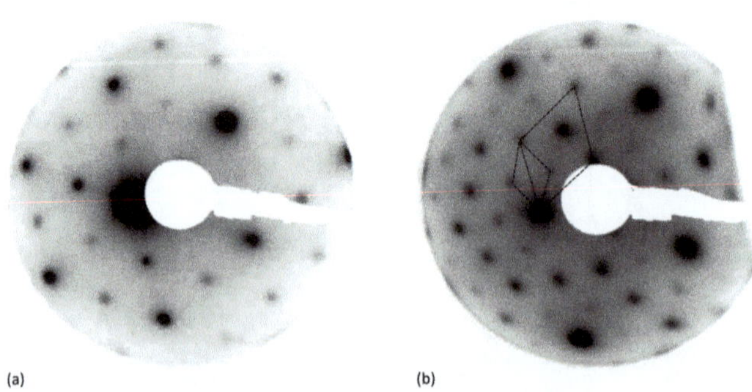

Fig. 3 LEED pattern recorded from the $Fe_3O_4(111)$ single crystal surface. (a) Clean surface at 70 eV after annealing in oxygen pressure of 10^{-7} mbar for 30 min at 873 K. (b) After deposition of MoO_x for 50 min and annealing to 973 K for 30 min in oxygen pressure of 10^{-7} mbar.

2. Deposition of molybdenum oxide on Ag(111) and Fe_3O_4 surfaces

Mass spectrometric studies of MoO_3 vaporization have shown a mixture of gaseous ions of general formula Mo_nO_{3n}(g) (n = 3, 4, 5) with the most abundant being $Mo_3O_9^+$, $Mo_4O_{12}^+$ and $Mo_5O_{15}^+$.[23,24] Previous investigations of HFMOD show the deposition of a mixture of MoO_3 and MoO_2, with exact ratios dependent on conditions.[13] To confirm the nature of the deposited films under the present conditions, molybdenum oxide was first deposited on a Ag(111) surface which offers a non-reducible support to the incoming MoO_x, and the possibility of oxidation *via* a surface reaction is unlikely, since the dissociative sticking probability of oxygen at this surface is extremely low. XP spectra of the resulting films show a single peak in the O(1s) region at 530.5 eV which was invariant with coverage (not shown) and the expected doublet in the Mo(3d) region with a Mo($3d_{5/2}$) binding energy of 231.3 eV, Fig. 4a. The latter was also unchanged with coverage and is at a binding energy that has been assigned previously to Mo^{5+} but probably consists in the present case of a mixture of oxidation states from Mo^{6+} to Mo^{4+}.[25] In contrast, the initial deposition of molybdenum oxide on the magnetite surface, Fig. 4b, results in an Mo($3d_{5/2}$) binding energy of 232.4 eV characteristic of Mo^{6+}. This remained constant for coverages up to ~3 monolayers but then decreased with further exposure reaching a binding energy of 231.3 eV after the deposition of 8 or more monolayers of molybdenum oxide (data not shown). The O(1s) peak was unchanged from the clean surface binding energy of 530.2 eV, though the peak broadened slightly at the highest coverages of molybdenum. Annealing of the molybdenum oxide covered surfaces in the presence of oxygen (10^{-7} mbar, 30 min) at temperatures of up to 973 K did not lead to a decrease in the overall intensity of the molybdenum peaks but even at 473 K resulted in a shift in binding energy to 232.2 eV. The clear implication

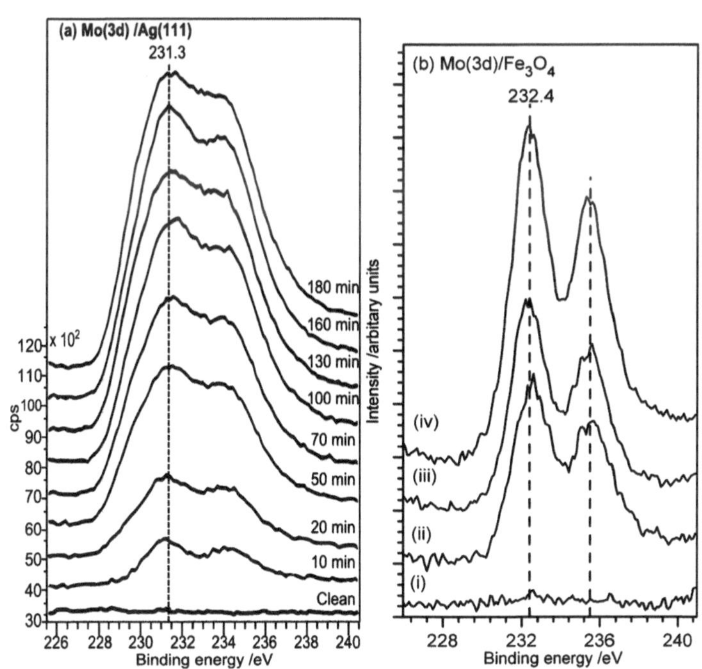

Fig. 4 XP spectra showing the Mo($3d_{5/2}$) and Mo($3d_{3/2}$) peaks after increasing exposures to MoO_x. The Mo($3d_{5/2}$) peak binding energy of 232.4 eV is characteristic of a Mo^{6+} species. The surface concentration of Mo is calculated from the XP peak area using methods described previously.[14,15] (a) MoO_x deposition on Ag(111). (b) MoO_x deposition on clean Fe_3O_4(111), (i) clean surface; (ii) $\sigma_{Mo} = 3.7 \times 10^{14}$ cm^{-2}, (iii) $\sigma_{Mo} = 4.5 \times 10^{14}$ cm^{-2}, (iv) $\sigma_{Mo} = 6.2 \times 10^{14}$ cm^{-2}.

of these data is that the initial deposition of molybdenum oxide involves a mixture of oxidation states which are converted upon adsorption at the iron oxide surface to become fully oxidized. However, there is no corresponding shift in the Fe(2p) spectra to indicate any reduction of the $Fe^{3+/2+}$, either because the changes are too small to be detected or due to the ability of the reduced iron oxide surface to dissociate the oxygen that is available in the gas phase. Once the molybdenum oxide layer thickness exceeds ~3 monolayers this oxidation pathway for the molybdenum oxide is closed off, and reduced Mo states begin to appear at the surface. To ensure that the surfaces are fully oxidized the MoO_x/Fe_3O_4 surfaces were annealed at 873 K for 30 min in 1×10^{-6} mbar O_2. Three concentrations of molybdenum oxide were studied in detail and are reported below: 3.7×10^{14}, 4.5×10^{14} and 6.2×10^{14} cm^{-2}, (expressed as the surface concentration of molybdenum), Fig. 4b. The unreconstructed oxygen-terminated (111) surface of magnetite has an oxygen surface density of 1.31×10^{15} cm^{-2}.

LEED patterns recorded from the magnetite surface after deposition of the molybdenum oxide showed no extra features, only an increase in the background intensity. Annealing in oxygen at 10^{-7} mbar for 30 min at temperatures up to ~873 K led to much sharper LEED patterns but no additional features were present. However, a new LEED structure was observed for higher coverages of molybdenum ($>1 \times 10^{15}$ cm^{-2}) annealed to 973 K, Fig. 3b. A $\begin{bmatrix} 2 & 2 \\ -2 & 4 \end{bmatrix}$ pattern (with reference to the hexagonal close packed oxygen) was observed. This corresponds to a $(2\sqrt{3} \times 2\sqrt{3})R30°$ structure with a unit cell dimension of 1.02 nm.

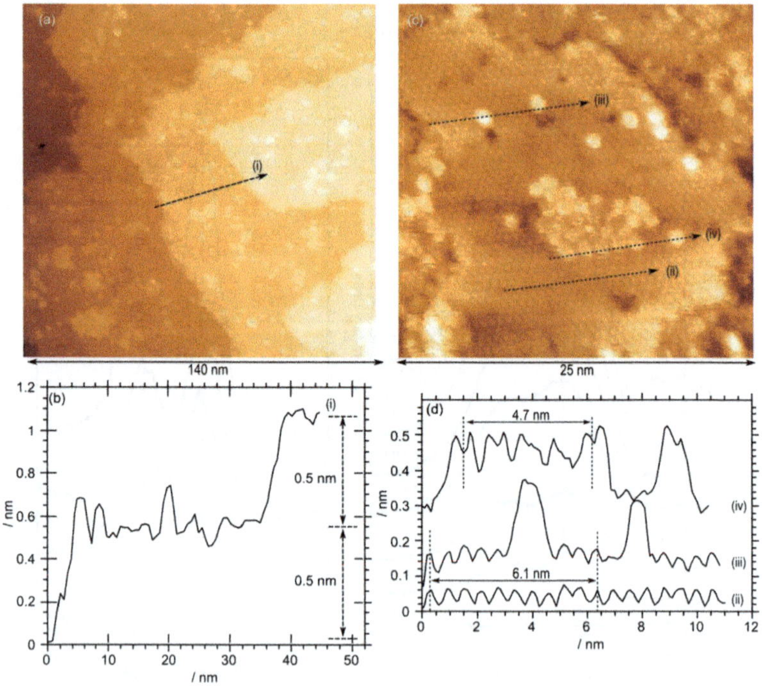

Fig. 5 STM image after exposure of a clean $Fe_3O_4(111)$ single crystal surface to molybdenum oxide followed by annealing in oxygen at 10^{-7} mbar. Total Mo concentration calculated from Mo(3d) XP spectra = 3.7×10^{14} cm^{-2}. Profile (i) shows the 0.6 nm periodicity of the underlying surface; profiles (ii) and (iii) show the 0.15 nm height of the adsorbed features and the 1.2 nm periodicity of the islands. ($V_b = -1.0$ V, $I_t = 0.465$ nA).

After annealing in oxygen at 873 K, the STM images of the lowest concentration of MoO_x studied (3.7×10^{14} cm^{-2}) show a surface dominated by Termination A, Fig. 5. The adsorbed molybdenum oxide appears as bright, approximately circular features scattered across the surface; there are some individual such units, but the majority of them are in the form of islands. The features have a very regular diameter of approximately 1 nm and an apparent height of approximately 0.25 nm. For many of the features a "hollow" is present in the centre (see for example Fig. 5d, profile (iv)). There is clear evidence of order within the molybdenum oxide islands, with regular spacing between features of ~1.2 nm, double that of the underlying substrate. By extrapolating the substrate lattice over the island and the individual features, it is apparent that the molybdenum oxide sits directly on top of one of the bright features from the substrate. The ordered areas are comprised of a p(2 × 2) structure with regards to the underlying Fe layer, which would therefore correspond to a p(4 × 4) structure with regards to the oxygen layer. We were unable to obtain confirmation of this structure from LEED, possibly because of the small domain size.

At the higher molybdenum surface concentration of 4.5×10^{14} cm^{-2}, extended rounded islands are evident covering approximately 1/3 of the surface (Fig. 6a). There are steps of ~0.5 nm between terraces, together with islands within the terraces with heights of approximately 0.3 nm. Higher magnification images, Fig. 6b, show that other areas of the image retain the (2 × 2) structure seen at the

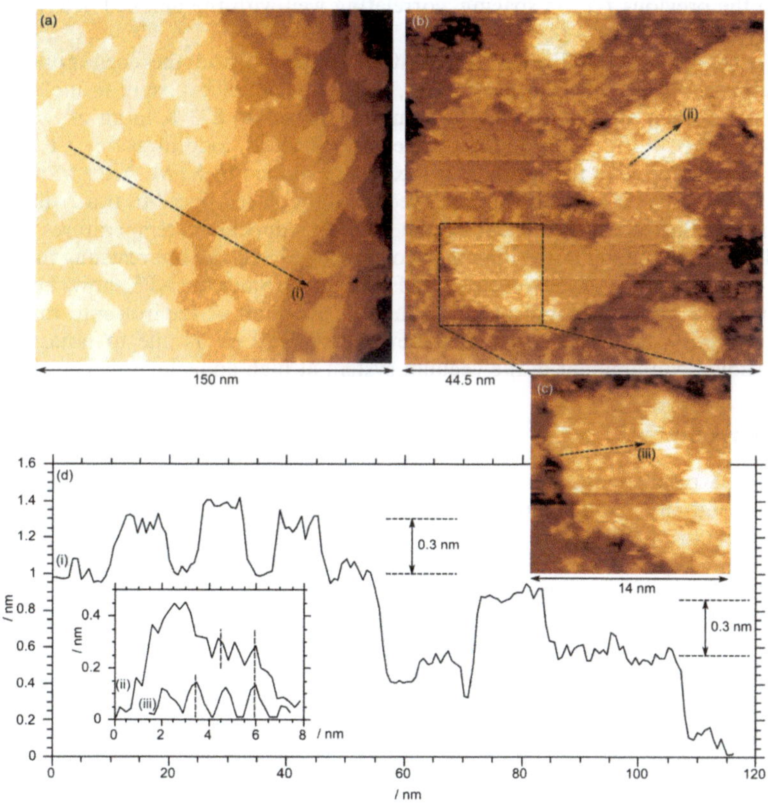

Fig. 6 STM image after exposure of a clean $Fe_3O_4(111)$ single crystal surface to molybdenum oxide followed by annealing in oxygen at 10^{-7} mbar. Total Mo concentration calculated from Mo(3d) XP spectra = 4.5×10^{14} cm^{-2}. Profile (i) shows the typical 0.3 nm step height; profiles (ii) and (iii) show the 0.6 nm and 0.12 nm periodicity of the underlying structures. (V_b = +2.5 V, I_t = 0.525 nA).

clean surface (*e.g.* profile (ii)), albeit with a large number of scattered bright features that correspond well to molybdenum oxide groups. In addition, the raised areas exhibit a large hexagonal lattice unit cell with a periodicity of 1.2 nm (Fig. 6, inset c) corresponding to the (4 × 4) structure characteristic of the molybdenum oxide islands seen at lower coverage. A small number of brighter areas are also present at this concentration. These appear to be approximately 0.15 nm above the (4 × 4) structures and constitute a second layer. Interestingly, these do not exhibit the (4 × 4) structure but instead show features with a 0.6 nm periodicity, suggesting a return to the (2 × 2) structure. The second layer areas also show individual bright features that are poorly resolved but similar to the individual molybdenum oxide groups observed at lower terraces.

Further exposure of the surface to molybdenum oxide increases the molybdenum surface concentration to 6.2×10^{14} cm^{-2}. STM images show that the large islands have propagated across the surface, and now cover the majority of the surface, Fig. 7a. Step heights between terraces are approximately 0.4 nm, with individual features on top of the terraces with heights of approximately 0.2 nm. High resolution images of the molybdenum area show an underlying structure with a 0.6 nm periodicity. Fig. 7a shows the surface after annealing to 873 K, Fig. 7b and c show the surface after further annealing to 973 K in an oxygen pressure of 10^{-7} mbar. The STM images reveal a smoother surface with terraces separated by 0.5 nm steps but still with scattered features on some terraces. A close up of the terraces, Fig. 7e, reveals a well ordered structure with a periodicity of 1.03 ± 0.02 nm rather than the previous 1.2 nm spacing, correlating well with the unit cell determined by LEED.

Fig. 8 shows Low Energy Ion Scattering spectra and the corresponding XP spectra of the clean iron oxide surface, the surface after exposure to molybdenum oxide and then after annealing the resulting surface to 873 K in an oxygen atmosphere of 1×10^{-6} mbar. The surface molybdenum concentration is 6.2×10^{14} cm^{-2}, comparable to that of the third exposure studied by STM. After MoO_x deposition, the Fe peak is no longer observed, being replaced by a Mo peak; the oxygen signal also increases. Upon heating, the Fe peak reappears at temperatures of 593 K and above, becoming larger than the Mo peak by 673 K, but not reaching the level of the clean surface, and showing a lower Fe : O ratio. Since the XPS shows no desorption of the MoO_x this indicates that there is diffusion of the iron into the molybdenum oxide layer. The peak that develops at 680 eV is due to potassium, an occasional contaminant in the vacuum system at high annealing temperatures.

Discussion

The XPS results from the silver surface indicate that the deposition of the molybdenum initially involves a mixture of molybdenum oxidation states, but on magnetite these are oxidized on adsorption to Mo^{6+}. Although higher coverages on the magnetite begin to show reduced molybdenum states, these are fully oxidized by annealing in oxygen. In all our experiments therefore we believe we have studied a MoO_3-like state. At low coverages these are manifested in the STM as 0.15 nm high features located on the sites previously assigned by Lennie *et al.* to capping oxygen on Termination A. The molybdenum features have diameters of ~1.0 nm and there is evidence for a depression in the centre (Fig. 5c and d, profile (iv)) but we were unable to obtain high resolution images of them. To account for this large size we propose that each feature is an Mo_3O_9 trimer – these are known to be a constituent of the gas phase evaporation of molybdenum oxides;[23] analogous W_3O_9 trimers have previously been reported[26] on TiO_2 surfaces and imaged by STM. The adsorption site for the trimers was shown by STM to be on top of the bright features assigned by Lennie *et al.* to "capping" oxygen adatoms. The model proposed in Fig. 9 details how the size of Mo_3O_9[27] would steer it towards the formation of a (4 × 4) structured adlayer. A more densely packed structure would appear

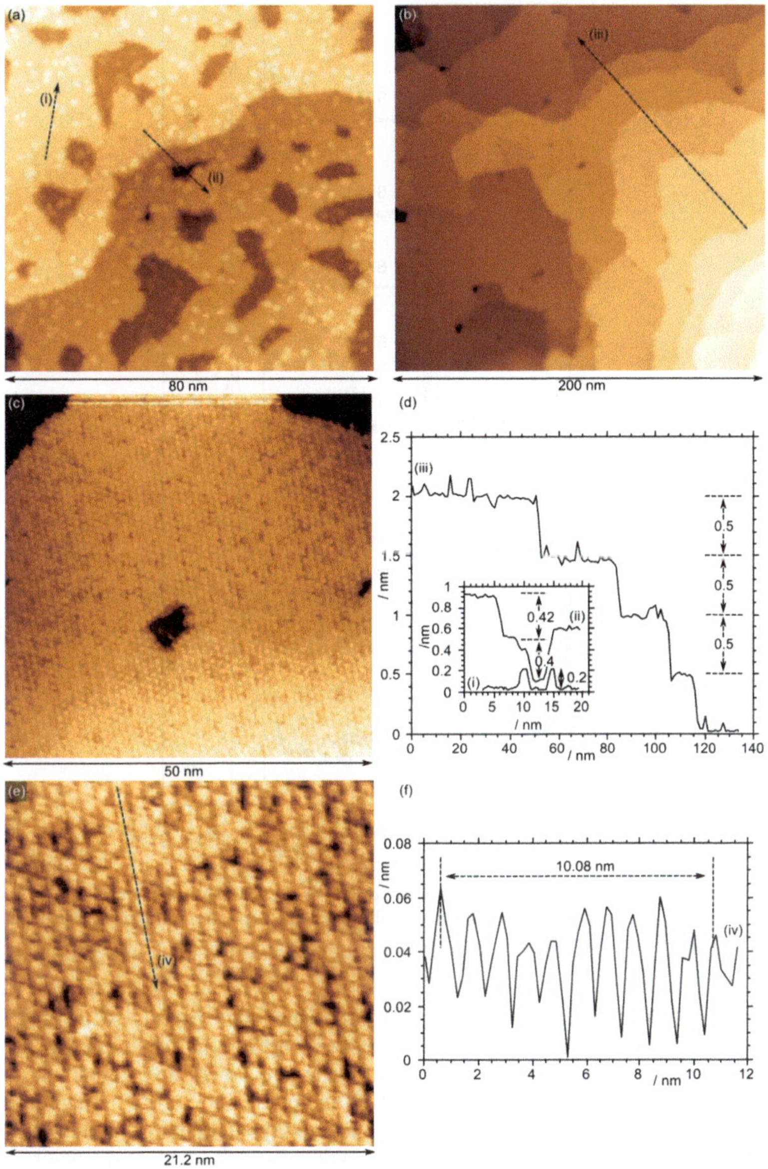

Fig. 7 STM images of $Fe_3O_4(111)$ with a total Mo concentration calculated from Mo(3d) XP spectra of 6.2×10^{14} cm^{-2}. (a) and (b) STM of MoO_x overlayer after annealing in oxygen at 873 K. (c) and (d) After further annealing at 973 K. Profile (i) shows the typical 0.3 nm step height; profiles (ii) and (iii) show the 0.6 nm and 0.12 nm periodicity of the underlying structures. (V_b = +2.5 V, I_t = 0.525 nA).

to be unlikely. At higher molybdenum oxide coverages, the majority of the surface is covered in a second layer, which appears to have returned to the (2 × 2) structure of the clean surface. The LEIS scattering reveals that following annealing above 473 K, Fe is present in the topmost surface layer. This corresponds to the temperature range in which the conversion and formaldehyde selectivity are at their highest when methanol is reacted over the mixed iron molybdate catalyst and in future work we will explore the reactivity of methanol with these surfaces. We know that from earlier work on powdered catalysts[28] that as the bulk loading of Mo increases, so the

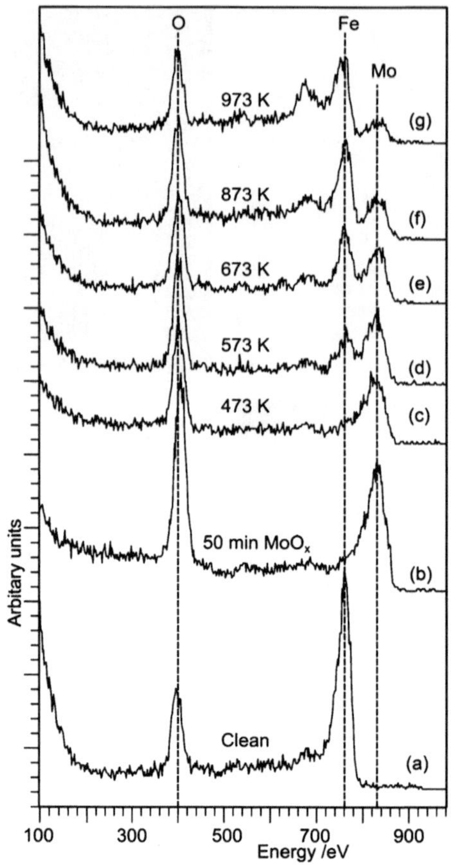

Fig. 8 LEIS spectra of an iron oxide surface exposed to the MoO_x flux from a hot Mo filament and after subsequent annealing: (a) clean, (b) following Mo exposure, and after annealing at (c) 473 K, (d) 573 K, (e) 673 K, (f) 873 K, (g) 973 K.

selectivity to CO_2 (typical of iron oxide) diminishes very rapidly to be largely replaced by CO production, with formaldehyde beginning to dominate at bulk Mo : Fe ratios as low as 0.2. It will be important to assess how the single crystal materials behave in relation to bulk catalyst behavior, and to relate this to *surface* structure. These results show that molybdenum stays in the surface region of the crystal. This agrees with earlier work on powdered catalysts systems, where it was shown by TPD measurements[2] and by LEIS and TEM[3] that the surface is enriched in Mo. We have also recently shown,[29] in a closer analogy to the present paper, that when Mo is chemically dosed onto either an Fe_2O_3 or Fe_3O_4 powder, it stays in the surface region. For all these materials, the presence of molybdenum in the surface layer is crucial to produce high selectivity to formaldehyde, since iron oxide alone is a combustor.

Acknowledgements

Dr David Willock is thanked for helpful discussions regarding the magnetite crystal structure. RJD is grateful to the University of Wales for a studentship partly funded by the Sir Charles Wright Endowment fund and CB was funded by the Thai government. DE was supported by grant number EP/E03974X/1 from the EPSRC in the UK, for which we are grateful.

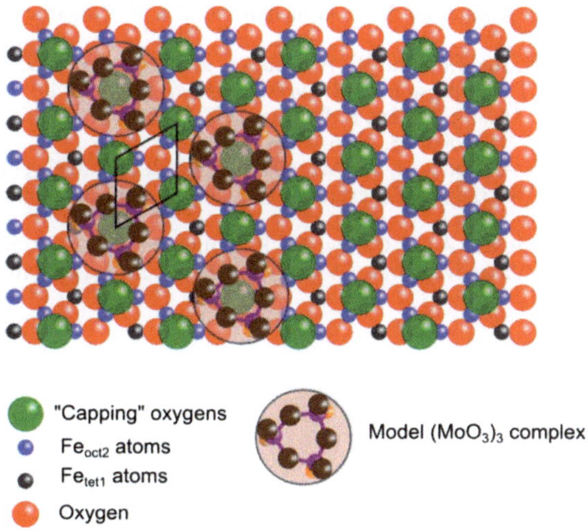

Fig. 9 Schematic model of the cyclic $(MoO_3)_3$ trimers adsorbed on the A Termination[21] of an $Fe_3O_4(111)$ surface. The $(MoO_3)_3$ clusters are shown having the plane of the rings located on top of the capping oxygens. The hexagonal surface unit cell of $Fe_3O_4(111)$ is indicated. The van der Waals contour of the $(MoO_3)_3$ clusters is based on dimensions calculated by Jang and Goddard.[27]

References

1 A. P. V. Soares, M. F. Portela and A. Kiennemann, *Catal. Rev. Sci. Eng.*, 2005, **47**, 125–174.
2 A. Andersson, M. Hernelind and O. Augustsson, *Catal. Today*, 2006, **112**, 40–44.
3 A. P. V. Soares, M. F. Portela, A. Kiennemann and L. Hilaire, *Chem. Eng. Sci.*, 2003, **58**, 1315–1322.
4 E. Söderhjelm, M. P. House, N. Cruise, J. Holmberg, M. Bowker, J.-O. Bovin and A. Andersson, *Top. Catal.*, 2008, **50**, 145–155.
5 W. E. Farneth, F. Ohuchi, R. H. Staley, U. Chowdhry and A. W. Sleight, *J. Phys. Chem.*, 1985, **89**, 2493–2497.
6 M. Bowker, R. Holroyd, A. Elliott, P. Morrall, A. Alouche, C. Entwistle and A. Toerncrona, *Catal. Lett.*, 2002, **83**, 165–176.
7 M. House, M. Shannon and M. Bowker, *Catal. Lett.*, 2008, **122**, 210–213.
8 K. Routray, W. Zhou, C. J. Kiely, W. Grünert and I. E. Wachs, *J. Catal.*, 2010, **275**, 84–98.
9 J. J. Uhlrich, J. Sainio, Y. Lei, D. Edwards, R. Davies, M. Bowker, S. Shaikhutdinov and H.-J. Freund, *Surf. Sci.*, 2011, **605**, 1550–1555.
10 V. Kaushik, *J. Electron Spectrosc. Relat. Phenom.*, 1991, **56**, 273–277.
11 I. Horcas, R. Fernandez, J. Gomez-Rodriguez, J. Colchero, J. Gomez-Herrero and A. Baro, *Rev. Sci. Instrum.*, 2007, **78**, 013705.
12 M. A. Bica de Moraes, B. C. Trasferetti, F. P. Rouxinol, R. Landers, S. F. Durrant, J. Scarmínio and A. Urbano, *Chem. Mater.*, 2004, **16**, 513–520.
13 S. F. Durrant, B. C. Trasferetti, J. Scarmínio, C. U. Davanzo, F. P. M. Rouxinol, R. V. Gelamo and M. A. Bica de Moraes, *Thin Solid Films*, 2008, **516**, 789–793.
14 A. F. Carley and M. W. Roberts, *Proc. R. Soc. London, Ser. A*, 1978, **363**, 403–424.
15 A. F. Carley, P. R. Davies, R. V. Jones, K. R. Harikumar, G. U. Kulkarni and M. W. Roberts, *Surf. Sci.*, 2000, **447**, 39–50.
16 M. E. Fleet, *Acta. Cryst. B.*, 1981, **37**, 917–920.
17 B. Wechsler, D. Lindsley and C. Prewitt, *Am. Miner.*, 1984, **69**, 754–770.
18 A. Barbieri, W. Weiss, M. A. Van Hove and G. A. Somorjai, *Surf. Sci.*, 1994, **302**, 259–279.
19 M. Ritter and W. Weiss, *Surf. Sci.*, 1999, **432**, 81–94.
20 A. Sala, H. Marchetto, Z.-H. Qin, S. Shaikhutdinov, T. Schmidt and H.-J. Freund, *Phys. Rev. B*, 2012, **86**, 155430.
21 A. R. Lennie, N. G. Condon, F. M. Leibsle, P. W. Murray, G. Thornton and D. J. Vaughan, *Phys. Rev. B*, 1996, **53**, 10244–10253.
22 K. T. Rim, J. P. Fitts, T. Müller, K. Adib, N. Camillone III, R. M. Osgood, S. Joyce and G. W. Flynn, *Surf. Sci.*, 2003, **541**, 59–75.

23 E. F. Fialko, A. V. Kikhtenko, V. B. Goncharov and K. I. Zamaraev, *J. Phys. Chem. A*, 1997, **101**, 8607–8613.
24 J. Berkowitz, M. G. Inghram and W. A. Chupka, *J. Chem. Phys.*, 1957, **26**, 842–846.
25 J.-G. Choi and L. T. Thompson, *Appl. Surf. Sci.*, 1996, **93**, 143–149.
26 O. Bondarchuk, X. Huang, J. Kim, B. D. Kay, L.-S. Wang, J. M. White and Z. Dohnálek, *Angew. Chem., Int. Ed.*, 2006, **45**, 4786–4789.
27 Y. H. Jang and W. A. Goddard, *J. Phys. Chem. B*, 2002, **106**, 5997–6013.
28 M. P. House, A. F. Carley, R. Echeverria-Valda and M. Bowker, *J. Phys. Chem. C*, 2008, **112**, 4333–4341.
29 M. Bowker, C. Brookes, A. F. Carley, M. P. House, M. Kosif, G. Sankar, I. Wawata, P. P. Wells and P. Yaseneva, *Phys. Chem. Chem. Phys.*, 2013, DOI: 10.1039/c3cp50399b.

General discussion

Dr Yim opened the discussion of the paper by Dr Nilius: You showed in Fig. 1 that the CaO film gets roughened when you deposited gold at an oxygen pressure of 2×10^{-5} mbar. Does this roughening process also take place when you placed the CaO film in the same oxygen-rich environment without depositing any gold?

Dr Nilius replied: Yes, the roughening effect is only a function of the pressure. For this reason, we believe that contaminants in the O_2 gas (*e.g.* water) are responsible for this. Our chamber is not optimized for high pressure applications.

Dr Yim then asked: Can the roughened area of the CaO film be resolved in STM?

Dr Nilius answered: No, the rough areas are apparently amorphous and there is nothing to see in STM.

Dr Yang enquired: How does the CaO–Au interaction compare with the metallic bonding between Au and Au?

Dr Nilius responded: The binding of Au to doped CaO is unusually strong for a metal–oxide interaction (up to 3.5 eV). It thus enters the region of the cohesion energy inside a metal.

Professor Thornton queried: Does Mo diffuse through the selvedge when exposed to oxygen to form a MoO_x surface, and does this give rise to the roughness seen in the STM image?

Dr Nilius answered: The Mo-doped CaO films are very sensitive to annealing in oxygen, which leads to surface roughening as seen in STM and disappearance of the LEED pattern. In addition, Mo^{6+} species appear in XPS. Our interpretation is that oxygen in the gas phase forms a strong stimulus for Mo segregation to the surface, where it becomes oxidized to MoO_3-type aggregates.

Professor Thornton then asked: Does gold sit on a MoO_x or CaO surface?

Dr Nilius replied: We are unable to answer this question, as we have tried to suppress Mo segregation in our experiments. All data shown in the paper are produced on CaO terminated surfaces, with very little or no Mo in the top most layer.

Professor Bowker asked for clarification: Is it that under vacuum Mo is not at the surface, but it is at the surface under oxygen?

Dr Nilius responded: As we can identify surface and subsurface Mo species in the STM, we are pretty confident that almost no Mo resides at the CaO surface upon vacuum annealing to 1000–1100 K. After annealing in oxygen (5×10^{-7} mbar), however, the surface became so rough that the LEED pattern disappeared and no STM experiments could be performed anymore. We ascribe this roughening to the segregation of Mo to the surface, triggered by the oxygen adsorbates.

Dr Whiston noted: It should be possible to test your hypothesis about the role of Mo in controlling the morphology change of Au nanoparticles on CaO by comparing other basic supports under the same conditions, such as MgO. Have you tried this and if you do not observe the same effect can you explain why?

Dr Nilius answered: We have repeated our experiment with MgO films grown on the very same Mo(001) single crystal. Although these films were of high structural quality as well, all Au deposits had pronounced 3D shapes and monolayer islands were essentially absent. We explain this difference with the smaller lattice parameter of MgO (0.42 nm) as compared to CaO (0.48 nm) which renders Mo diffusion into the film, and hence self doping, difficult. Interestingly, also intentional doping of MgO with another transition metal (Cr) did not induce the 3D to 2D growth transition of adsorbed gold. We analyzed this surprising result in detail with the help of DFT calculations and found that Cr is a rather bad charge donor (just a single electron can be released), while Mo ions may donate up to three electrons towards surface Au species. The relevant data has been published in refs. 1 and 2.

1 F. Stavale, X. Shao, N. Nilius, H.-J. Freund, S. Prada, L. Giordano and G. Pacchioni, *J. Am. Chem. Soc.*, 2012, **134**, 11380–11383.
2 X. Shao, P. Myrach, N. Nilius, H.-J. Freund, U. Martinez, S. Prada, L. Giordano and G. Pacchioni, *Phys. Rev. B*, 2011, **83**, 245407.

Professor Hutchings remarked: This is a fascinating study. By controlling the oxygen pressure the nanoparticles can either be in two dimensions, *i.e.* a monolayer raft, or as a three-dimensional nanoparticle. Is there any way in which you could make an intermediate bilayer structure? Bilayer Au nanoparticles have been considered to be important in CO oxidation and so this methodology might provide a way to specifically make such structures and test the reaction mechanism.

Dr Nilius responded: It looks like that the charge competition between Au adatoms and oxygen molecules on the surface of a Mo-doped CaO film knows only one winner. That means, either the Au atoms take the electrons and aggregate into ideal monolayer islands, or the oxygen gets the charge and the Au grows in a 3D fashion. The bimodal shape distribution of Au particles grown in the presence of oxygen clearly reveals that no intermediate particle shapes are formed (see Fig. 2 of our paper). We therefore conclude that the mechanism of charge competition is unsuitable to tune particle shapes in an arbitrary manner.

Professor Campbell asked: What do you know about the nature of the chemical bond of this O_2 species to the surface? Is its high electron affinity that you observed predicted by DFT or other calculations or measurements of its electronic structure?

Dr Nilius replied: The binding of O_2 to doped CaO is predicted to be governed by a charge transfer from the Mo levels into antibonding oxygen orbitals (DFT). As a consequence, the molecule develops attractive Coulomb forces to Ca ions in the surface. Detailed calculations are presently being carried out in the Sauer group.

Professor Campbell enquired: Can you estimate the heat of adsorption of the molecularly adsorbed O_2 that you so nicely showed captures electrons on the CaO(100) surface?

Dr Nilius replied: The O_2 molecules are stable up to 350–400 K, and I therefore expect an adsorption energy of the order of 1.0 eV. TDS measurements are on the way.

Professor Behm asked: What is the critical length of the interaction of a Au island and a Mo dopant, which affect the O_2 adsorption on the Au island. How many dopant atoms are expected to interact with the Au island to yield these effects, and is there direct (experimental) evidence for that?

Dr Nilius answered: The question on how many active dopants are actually needed to switch the growth of Au from 3D to 2D behavior is interesting, but not easily answered at this stage. Based on DFT calculations and tunneling spectroscopy data, we have estimated the maximum number of transfer electrons within a 2D ad-island to be comparable to the number of perimeter atoms, *i.e.* every edge gold is able to take one extra charge. Whether this charge maximum is really needed to stabilize the 2D growth and how many dopants are required in a certain volume below the island is not known. To answer this question, we have to measure more reliable dopant profiles in our films and to determine the critical distance over which electrons can be transferred from the dopants into the gold, for instance *via* angle-resolved XPS. Earlier DFT studies have correlated the interaction length between dopants and ad-gold with the tunneling length of electrons in a dielectric matrix, which amounts to 1–2 nm approximately.

Professor Bowker noted: You say that the Au nanoparticles are negatively charged, but although you show some XPS spectra, you don't show any for the gold. Presumably this charging should be see in the Au4f spectra?

Dr Nilius responded: So far, we were unable to see negatively charged Au species in XPS, as the spectra were dominated by the neutral atoms in the inner parts of the islands and by strong final-state effects. We will try to redo these experiments for single Au adatoms bound to Mo-doped CaO, making use of a synchrotron facility.

Professor Bowker commented: You say that the perimeter atoms are the most strongly bound, which is unusual. I guess that means that the diffusing metal atoms should be those from the centre of the island.

Dr Nilius apologised for this misunderstanding. We know that the perimeter atoms are the ones that are charged, while inner atoms in the Au islands are neutral. This effect reflects the tendency of the excess charges to maximize their mutual distance, which implies that they localize preferentially in the outer-most Au atoms. We could show this effect *via* tunnelling spectroscopy, see ref. 1.

1 X. Lin, N. Nilius, M. Sterrer, P. Koskinen, H. Häkkinen and H.-J. Freund, *Phys. Rev. B*, 2010, **81**, 153406.

Professor Anderson enquired: This result that the particle structure is determined by competition between gold and O_2 for electrons is quite interesting. I assume that you have also tried the experiment of preadsorbing oxygen on the doped oxide surface, and then depositing Au. What particle growth mode does this result in? In a similar vein, if you have 2D islands stabilized by electron transfer from the dopant atoms, what happens if you expose this sample to O_2 after particle growth? Is the O_2 able to stick, and is it able to displace gold as the electron acceptor?

Dr Nilius answered: We have deposited Au also on oxygen pre-covered CaO surfaces (400 L) and found that more than 50% of the Au clusters were unable to take their 2D shapes but adopted 3D geometries. Apparently, the electrons stored in the oxygen over-layer could not be acquired by the incoming Au atoms, which remained charge neutral and aggregated into 3D deposits. In addition, an ensemble of 2D islands formed at vacuum conditions could not be transformed into a set of 3D deposits simply by post-depositing high amounts of oxygen. After such a treatment, we still found monolayer Au islands on the surface, which displayed however a characteristic brim structure that might be indicative for Au-oxidation by the incoming oxygen. Such a phenomenon has indeed been predicted by theory.[1] Conversion of all 2D islands into 3D particles could be achieved, however, *via* moderate heating of the sample.

1 H. Häkkinen *et al.*, *Angew. Chem.*, 2010, **49**, 7913.

Dr Yang returned to a previous comment that CaO–Au interaction is weaker than the Au-Au interaction: This indicates that thermodynamically, when over-layers of Au clusters are formed, they are unlikely to be converted to 2D geometry. On the other hand, it suggests that (a) 2D gold layer is formed by the migration of Au atoms on the surface of CaO; (b) 3D gold islands can be formed due to the lack of Mo atom underneath.

Dr Nilius responded: I completely agree that the thermodynamically favored state is still 3D clusters even on doped CaO. However, this can only be reached up heating up the system to 600–800 K, while at room temperature 2D islands prevail. Apparently, there is a kinetic hindrance to reach the 3D equilibrium, and also the high interface energy due to the dopants stabilizes the 2D islands a lot.

Professor Attard opened the discussion of the paper by Professor Asakura: Although you succeed in tethering the small gold atoms using this method, does not the sulfur ligand deactivate the gold or certainly change its catalytic properties?

Professor Asakura replied: I have not yet investigate the catalytic activity. It will be interesting to see how the sulfur compound changes the Au activity for CO oxidation.

Professor Behm remarked: The Au–anion bond was said to be most important for the formation of individual Au atoms. But it seems that other effects are important as well, since for CeO_2 a significant fraction of Au is atomically dispersed in the CeO_2 surface regime, at least this is the state of the literature, while this is not the case for other oxides such as TiO_2 in the stable state. So it seems that there are other important effects, in addition to the Au–O bond.

Professor Asakura responded: We have checked the effect of defects and (1×2) reconstruction for Au dispersion at as low loading as 0.03 ML. As far as we know, we always have the Au cluster with a few nm. Most of them are highly symmetric like icosahedral. The CeO_2 may be our next target to be revealed in the metal–support interaction. As you suggested, as CeO_2 can disperse the Au atomically we can get structural information around Au. Thank you for your suggestion.

Professor Thornton commented: Gold atoms are stabilised at oxygen vacancies on $TiO_2(110)$ and may be another way forward to obtaining isolated gold species.

Professor Asakura responded: Up to now we have not obtained evidence for the direct interaction between metal (Ni, Cu, Pt, Au) and Ti cation by PTRF-EXAFS.[1] For example, in the low coverage of Ni, we found the step site oxygen atoms made a bond with Ni.[2] When we prepared the defective surface and deposited Ni on it, we found Ni cluster bound with oxygen atoms.[3] In case of Au we tried to disperse Au on several surfaces at low coverage. At the lowest possible limit of coverage for measuring Au, we have not successfully obtained Au–O nor Au–Ti. except Au–Au interaction.[4,5] As mentioned in the paper, Au aggregation easily occurred to form three-dimensional clusters due to the weak interaction with oxygen. We may have Au monomers on the $TiO_2(110)$ surface with very low coverage but it is out of range for the detection of present experimental ability.

1 K. Asakura, *Polarization-dependent Total Reflection Fluorescence Extended X-ray Absorption Fine Structure and its Application to Supported Catalysis*, RSC Publishing, Cambridge, 2012, pp. 281–322.

2 Y. Koike, K. Ijima, W. J. Chun, H. Ashima, T. Yamamoto, K. Fujikawa, S. Suzuki, Y. Iwasawa, M. Nomura and K. Asakura, *Chem. Phys. Lett.*, 2006, **421**, 27–30.

3 H. Uehara, M. H. bin Hanaffi, Y. Koike, K. Fujikawa, S. Suzuki, H. Ariga, S. Takakusagi, W.-J. Chun, Y. Iwasawa and K. Asakura, *Chem. Phys. Lett.*, 2013, **570**, 64–69.

4 W.-J. Chun, K. Miyazaki, N. Watanabe, Y. Koike, S. Takakusagi, K. Fujikawa, M. Nomura, Y. Iwasawa and K. Asakura, *J. Phys. Chem. C*, 2013, **117**, 252–257.

5 W.-J. Chun, K. Miyazaki, N. Watanabe, Y. Koike, S. Takakusagi, K. Fujikawa, M. Nomura and K. Asakura, *J. Ceram. Soc. Jpn.*, 2011, **119**, 890–893.

Dr Nilius asked: Does the benzoic acid used for stabilization of Au adatoms on TiO_2 form a regular adsorption pattern on the surface (as seen in LEED for example) and what conclusion can you draw about the spatial distribution of the gold atoms from this? Do you think that your approach of stabilizing monomeric gold *via* MBA has universal character and would work for any other oxide material?

Professor Asakura answered: Unfortunately we did not have a clear LEED pattern for MBA adsorbed surface and the surface with Au deposited on it. I believe that by the selection of an appropriate anchoring moiety (COOH in case of TiO_2) we can apply the method to other oxide surfaces.

Dr Philip Davies queried: Since sulfur has the potential to significantly reduce the activity of the gold, have you considered using other heteroatoms in the hydrocarbon "tethers", amines or alcohols for example?

Professor Asakura replied: I do not know the catalytic effect of sulfur on Au and now we are investigation catalysis of the species. Another possible tethering ligand is –CN (weak base).

Professor Hutchings remarked: You are making isolated supported Au atoms with your approach. However, the test reaction of CO oxidation, suggested by you in discussion, may not be the most appropriate reaction since a single Au atom may not be able to activate CO and O_2 effectively. Hardacre and co-workers[1] have shown that Au dimers are effective for methanol carbonylation. Is there any way that your approach could make Au dimers as they could be very interesting as potential catalysts? Of course small clusters would be even more interesting.

1 A. Goguet, C. Hardacre, I. Harvey, K. Narasimharao, Y. Saih and J. Sa, *J. Am. Chem. Soc.*, 2009, **131**, 6973.

Professor Asakura answered: The Au monomer prepared here is stable and it is not so easy to construct a dimer from this Au monomer. The other way we are now trying is to prepare polynuclear Au (Au_{13}, Au_{10} and Au_6) stabilized with ligands and after the deposition on the modified TiO_2 surface, the Au species is activated by removing the ligand. In the same way we can have dimers by the preparation and deposition of dimer molecules.

Professor Freund commented in response: When one considers the question: What is the smallest cluster size of a given metal that could be responsible for a given reaction, it may be useful to resort to data from the gas phase, where reactions, in particular, CO oxidation has been studied and TOFs have been revealed as a function of the number of atoms in the cluster.[1]

1 H. Schwarz, *Angew. Chem. Int. Ed.*, 2005, **44**, 2336.

Professor Bowker enquired: Your sample has a high level of Nb doping – why is this and are you sure it is not in the surface or sub-surface region?

Professor Asakura replied: We use Nb-doped $TiO_2(110)$ in order to obtain electric conductivity without defects. We cannot detect Ti^{3+} on the $TiO_2(110)$ doped surface. In addition we can not see any Nb in the XPS analysis so that we do not think there is the segregation of Nb. We observed $TiO_2(110)$ 1×1 LEED pattern and the Nb seems not to alter the adsorption properties of the carboxylic acid.

Professor Freund remarked on the preparation of model catalysts on $TiO_2(110)$: One should recognize that TiO_2 should be referred to as a self-doped material due to the presence of Ti-interstitials and of oxygen vacancies.

Professor Asakura responded: As far as I know the Nb is located at Ti and less Ti^{3+} is present in the XPS when doped with Nb.

Professor Behm remarked: Considering that supported Au catalysts are generally pre-treated by calcination at 400 °C, I wonder how relevant mechanisms involving (existent) oxygen vacancies are for real catalysis.

Professor Asakura replied: I think pretreatment of Au under oxygen is to prevent the Au migration and sintering compared to the hydrogen reduction of Au which created larger particles. I guess that the interaction between the metal and support may be larger in the oxidative conditions. This is not the case with the theoretical calculation, see for example ref. 1.

1 K. Okazaki, Y. Morikawa, S. Tanaka, K. Tanaka and M. Kohyama, *Phys. Rev. B*, 2004, **69**, 235404.

Dr Yang asked: In the paper it claims that on the bare $TiO_2(110)$ surface, icosahedral Au_{55} nanoclusters were the main product found. Indeed, Au_{55} is more stable than Au particles of similar size; so under certain conditions, for example, electron impact, Au_{55} clusters are favoured. However, using the preparation method in this article, there is no mechanism to destroy Au particles once they are formed. So I wonder if Au clusters have a distribution, and clusters such Au_{54} and Au_{56} *etc.* would have a similar probability as Au_{55}?

Professor Asakura replied: EXAFS is in principle an average technique so that Au particles should have a size distribution. Moreover we could not distinguish Au_{54} and Au_{56} from Au_{55} because of the limit of accuracy of coordination numbers. We should have the error bar 10–20 pm. We have checked Au_{13} and Au_{169} (2-shell and-4 shell clusters) but Au_{55} is the most appropriate.

Professor Anderson noted: Magic numbers like 55 atoms are important for isolated gas phase clusters, where they may represent geometric or electronic shell closures of highly symmetric structures. For clusters on surfaces, I think that these gas phase magic numbers are largely irrelevant because the clusters are bound with substantial energies to the surface, which distorts the clusters relative to the symmetric gas-phase structures. Furthermore there is usually some electron density transfer accompanying the bonding, which, of course, can result in very different orbital filling patterns compared to an isolated cluster. I assume that the fact that Professor Asakura is fitting his results using an icosaheadral cluster model really just means that such an Au_{55} cluster has about the same coordination and bond length parameters as the average clusters present on the support.

Professor Asakura responded: I agree that oxide affects the structure of metal clusters on top of the oxides. However, in the case of Au where interaction is not so strong, the structure is less affected by the oxide surface and in the first approximation we can describe it in the isolated clusters. At the moment more fine structure both geometrical and electronic can not be discussed in the present EXAFS data.

Professor Bowker opened a general discussion of Dr Nilius' and Professor Asakura's papers, addressing Professor Asakura: Where do metal atoms bind on oxide surfaces? In your paper we hear that oxygen sites are needed, whereas in the paper by Dr Nilius the interaction appears to be with cationic sites. In reality we would imagine that the particle wants to interact with as many sites as possible and does so by an oscillatory polarisation interaction with *both* cationic and anionic sites at the surface.

Professor Asakura answered: I claimed our paper that the important interaction is that between the anion and metal with a covalent bond. Another point I claimed in the paper was that the metal–cation bond with less than 2.8 Å was invisible in EXAFS, which means the metal–cation covalent bond was not present. It is quite interesting the electrostatic modification of cluster electronic state that must induce the inhomogeneity of clusters and can be one of the origins of unique catalytic activity of Au.

Professor Freund commented: According to calculations by Häkkinen *et al.*,[1] neutral Au on undoped CaO resides on the surface oxygen of CaO, while it changes to Ca sites when more negatively charged in the presence of Mo as a dopant. With reference to determining the charge on Au with XPS, my suggestion is not rely on the XPS shift alone, but to consider the Auger parameter based on core–core–core Auger transitions in order to analyze initial and final state effects.[2]

1 H. Häkkinen *et al.*, *Angew. Chem. Int. Ed.*, 2013, **52**, 1424.
2 G. Hohlneicher *et al.*, *J. Electron. Spectrosc. Relat. Phenom.*, 1985, **37**, 209.

Dr Nilius remarked: The binding mechanism between negatively-charged gold and the CaO support is not limited to Coulomb attraction and cationic lattice sites are therefore not the only relevant binding sites on the surface. In fact, the potential landscape for adsorption is rather flat and almost the same binding energies are found for cationic, anionic and hollow sites on the CaO(001). The finding has been explained with the high structural flexibility of the CaO surface that allows for huge polaronic lattice distortions in presence of the Au(−) species. The relevant data is published in ref. 1.

1 X. Shao, S. Prada, L. Giordano, G. Pacchioni, N. Nilius and H.-J. Freund, *Angew. Chem., Int. Ed.*, 2011, **50**, 11525.

Professor Bowker commented: In consideration of charge transfer, I wonder if that is an inaccurate description of the interaction between a metal nanoparticle and the surface. Isn't the interaction more simply described as polarisation, and does not the interface reflect the laterally oscillatory pattern of charge? Obviously this would particularly affect *thin* island-type particles 1 or 2 monolayers thick.

Professor Hutchings continued the discussion: It is possible to make Au_{55} and to support this material. Lambert and co-workers[1] have shown this but when Au_{55} is supported it fragments into a broad range of other structures between 1–3 nm in diameter. It would be nice to think we can make these clusters in the gas phase and then support them, but when they interact with the support they may not retain their structural identity. How can we build such supported structures and fine tune such processes?

1 M. Turner, V. B. Golovko, O. P. Vaughan, P. Abdulkin, A. Berenguer-Murcia, M. S. Tikhov, B. F. Johnson and R. M. Lambert, *Nature*, 2008, **454**, 981.

Professor Bowker closed the general discussion: We need to develop the rules which dictate the size and morphology of nanoparticles on surfaces. These must include the inter-related concepts of hydrophobicity/hydrophilicity, acid–base strength, oxophobicity/oxophilicity, electron affinity of nanoparticles, and hence how these can be manipulated in catalyst preparation to give the desired product.

Dr Robert Davies opened the discussion of the paper by Dr Saint-Lager†: The paper describes a correlation between particle geometry and activity, specifically that "maximum activity corresponds to a sharp angle θ between the GNP surface tangent and the substrate." What experimental factor dictates the θ angle? And how accurately can it be controlled?

Dr Bailly responded: The GNPs morphological parameters under reactive conditions ($CO + O_2$ mixture at 473 K) were deduced from GISAXS by modeling the nanoparticles by a truncated sphere (see ref. 21 of the paper). We checked two kinds of shape – cylinder and truncated sphere – both leading to similar values. The truncated sphere model was finally chosen because it led to slightly better fits. The parameters deduced from the fit are the mean diameter D and the mean height H, from which we deduced the contact angle θ. The latter is governed by the initial amount of gold deposited in UHV and the reaction-induced sintering.

Indeed, we can mention that the study of gold growth on $TiO_2(110)$ in UHV evidenced that it is 2D below a critical amount of metal (about 0.1 ML gold equivalent thickness). It stays quasi-2D up to about 0.25 ML (this is the gold amount corresponding to the maximum of the catalytic activity). For higher coverages, it turns to a 3D growth.[1] Our GISAXS measurements are consistent with these results.[2] Moreover we evidenced a reaction-induced sintering with a maximum corresponding to the one of the catalytic activity for CO oxidation. The reaction acts as a thermal annealing leading to an irreversible change of the clusters towards the equilibrium shape which is quasi-spherical in the 2 nm size range.[3] These two effects (growth kinetics and reaction-induced sintering) could explain the occurrence of the H/D maximum corresponding to the maximum of catalytic activity.

1 U. Diebold, *Surf. Sci. Rep.*, 2003, **48**, 53.
2 Laoufi *et al.*, *J. Phys. Chem. C*, 2011, **115**, 4673.
3 Saint-Lager *et al.*, *Faraday Discuss.*, 2011, **152**, 253.

Professor Asakura said: Your SAXS data show not so large contraction of Au–Au distance in the small nanocluster, though the EXAFS shows a larger contraction of Au–Au. Is there any possibility that the SAXS only detects a minor larger Au nanoparticle coexisting on the TiO_2 surface?

Dr Bailly replied to Professor Asakura: It should be mentioned that we were able to detect the signal of particles as small as 1 nm by X-ray diffraction (black curve in Fig. 3(a)), and even smaller by GISAXS (see Fig. 2 and 3 and Table 1 of Laoufi *et al.*[1]) The Au–Au distances reported in Fig. 4 are directly obtained by diffraction measurements during CO oxidation, that is to say in *operando* conditions. In these conditions, the size of the nanoparticles is significantly bigger than the one obtained just after the GNPs growth in UHV. For instance the diameter grew from 1 to 2 nm in the case of the 0.1 ML deposit represented in Fig. 3(a) for which the shift of the gold peak is much larger in UHV than under reactive conditions, corresponding to an in-plane parameter with a larger contraction in UHV. In that sense, there is no obvious contradiction between the EXAFS and *operando* GISAXS results. Additionally, as explained in your paper, atomic gold and very small clusters are poorly stabilized on a bare $TiO_2(110)$ surface, which is consistent with our own results. It is thus very interesting to find TiO_2 treatments to stabilize these small entities.

1 Laoufi *et al.*, *J. Phys. Chem. C*, 2011, **115**, 4673.

† Dr Saint-Lager's paper was presented by Dr Bailly, Institut Néel, CNRS et Université Joseph Fourier, Grenoble cedex 9, France.

Professor Behm enquired: I wonder about the state of the oxide samples after the treatment (sputtering plus subsequent annealing at 1000 K in 10^{-5} mbar O_2). Is the O_2 partial pressure to avoid thermal reduction of the sample, what is the color of the samples after this treatment?

Dr Bailly answered: The annealing at 1000 K in presence of oxygen, just after cleaning by an Ar^+ sputtering, led to a (1 × 1)-TiO_2(110) surface and allowed to keep the oxygen stoichiometry of the TiO_2(110) surface. The sample color did not change and stayed pale yellow and transparent all along the experiments. It is well known that in these conditions, the sample surface is stable and not reactive in oxidizing conditions.[1] This is a crucial point for the present *operando* measurements in order to avoid a strong disturbance induced by a possible substrate restructuring under gas exposure.

1 U. Diebold, *Surf. Sci. Rep.*, 2003, **48**, 53.

Dr Nilius remarked: In your contribution, you make a very strong statement on the precise size and shape of the Au nanoparticles that show maximum reactivity in the CO conversion. Because there is typically a rather broad distribution of particle geometries on oxide surfaces (due to defects, inhomogeneities, different terraces sizes), there must be a big error bar on the data, in contrast to the actual plot shown in the inset of Fig. 6. In addition, it would be helpful to have more detailed information on how the contact angle θ was determined. Is this value based on some assumptions on the particle shape or a direct result of the GIXD measurements?

Dr Bailly responded: As explained in the text, the error bars in the inset of Fig. 6 are the width of the Gaussian distributions of the geometrical parameters, as deduced directly from the GISAXS experimental data fitting. Determination of the contact angle has been communicated in the first question raised by Dr Robert Davies.

Professor Anderson remarked: Your results are quite beautiful, however, your methods are essentially sensitive to the volume-weighed average environment of the gold in the samples. If catalysis happened equally on all gold sites, then it would be sensitive to the area-weighed structure of the distribution of particles, *i.e.*, more sensitive to smaller particles. For this particular system, however, there is evidence that much smaller clusters are quite important in the catalysis, and that large three-dimensional particles may simply be spectators. My question is whether there is some way, perhaps using a selective poisoning technique, that would allow you to determine whether you are studying the real active species, or if you are simply looking at large inactive gold particles.

Dr Bailly responded: The question of knowing whether we measure the same objects with the different techniques or not (mass spectrometry, GISAXS and GIXS) is a crucial point. As previously reported,[1] a rapid evolution of the GISAXS patterns was observed as a response to the beginning of the CO conversion (evidenced by mass spectrometry). It is attributed to a reaction-induced sintering. The GISAXS analysis showed that the amplitude of this sintering is directly correlated to the catalytic activity with a maximum at the same time for both phenomena occurring exactly in the same conditions. Although the mechanism at the origin of this sintering is not yet understood, it indicates that GISAXS is directly sensitive to the active species. We also verified that the size of the nanoparticles measured by diffraction and GISAXS is the same in order to make sure that both X-ray techniques "see" the same nanoparticles.

1 Saint-Lager *et al.*, *Faraday Discuss.*, 2011, **152**, 253.

Dr Torrente asked: Fig. 5 of this paper shows that the catalytic activity of GNP in the CO oxidation reaction is a function of the *H/D* ratio, showing a maximum at *H/D* values ~0.75. The discussion indicates that this is directly related to the contact angle θ between the GNP and the support. If the contact angle is a function of the GNP wettability capability of the support, how can this angle change by only varying the metal loading? Previous studies (*e.g.* ref. 1) show that the contact angle can be tuned by varying the preparation method, however, in this paper, all catalysts are prepared in the same way.

Is it possible that the metal–support interaction changes due to the lattice strain of the smaller particles and it becomes negligible for bigger ones?

Additionally, and making reference to the previously mentioned paper by Haruta,[1] a higher wettability (high contact angle) is desired in GNP for enhancement of its CO oxidation activity. This is contradictory to what is claimed in your paper (top of page 9). Do you have any further information which could justify this disagreement?

1 M. Haruta, *CATTECH*, 2002, **6**, 102.

Dr Bailly replied: Concerning the catalytic activity for CO oxidation, the present results seem to be contradictory to Haruta's paper. But he mentioned that the nanoparticles with the lowest wettability (which means with a quasi-spherical shape) are less active for CO oxidation than the hemispherical ones, but also much larger than those studied here (a few tens of nm compared to a few nm in our case). This point is clear if we compare the scale of the two TEM images of the nanoparticles in the two cases with diameters close to a few nm for the most active ones while the diameter of the spherical ones is in the range of a few tens of nm. In the present work, the two parameters (shape and size) can be decoupled since, as shown in Fig. 5, the diameter stays in the range of 2 nm and only slightly varies around this optimum value while the *H/D* ratio significantly changes similarly to the rate of CO conversion into CO_2. Regarding the evolution of the *H/D* ratio, this has been communicated in the first question raised by Dr Robert Davies.

Professor Hutchings returned to the discussion point raised by Professor Anderson: I feel the real conclusion is that the activity you observe scales with the gold loading up to what corresponds to particles with sizes of 2 nm. It could be that, of course, indeed the 2 nm particles are active and hence your hypothesis concerning the angle at the interface is consistent with the activity you observe. However, it could be that there is another species that scales with the gold concentration, for example small clusters. These species could agglomerate at high gold concentrations and so one would observe the same correlation. It is possible to use other techniques to determine if small clusters are indeed present or absent? Aberration-corrected microscopy of Au/TiO_2 catalysts shows that single atoms, clusters comprising both monolayer and multilayers can be present.

Dr Bailly replied: The existence of a maximum of catalytic activity associated to particles with a size around 2 nm strongly questions the role of very small clusters, since following the curve of Fig. 2, the activity drops when the size decreases below this optimum diameter. To explain such a variation with very small clusters, we would imagine that their concentration varies similarly with a maximum corresponding to a distribution of "large" particles with a diameter of 2 nm. Nevertheless we previously showed that during the procedure of the exposure to reactive gases, the mean volume per island increases with an evolution depending on the gas nature. Under oxygen at 473 K, it continuously increases for decreasing sizes and exhibits a maximum under the CO + O_2 mixture at 473 K.[1] As a result, the smallest 3D clusters disappear, leaving place to an assembly of bigger 3D nanoparticles with a narrower size distribution.[2] It should also be noted that most of the previous microscopy studies were done in UHV. This *operando* study clearly shows that in this case, the active entities differ from the ones obtained in UHV just after gold deposition. One could thus imagine performing *in situ* STM or environmental TEM studies.

1 Saint-Lager *et al.*, *Faraday Discuss.*, 2011, **152**, 253.
2 Laoufi *et al.*, *J. Phys. Chem. C*, 2011, **115**, 4673.

Professor Bowker noted: The method of metal vapour deposition actually produces fairly narrow particle size distributions, often much narrower than from wet chemical methods.

Professor Thornton asked Dr Bailly: The conditions used for preparation probably give rise to a rough surface. Was this picked up in the crystal truncation rod data?

Dr Bailly replied: We actually did not record the crystal truncation rods during these *operando* studies. In contrast to STM, X-ray diffraction does not need conducting samples. This is why it is not necessary to reduce the TiO_2 surface by annealing in UHV. An annealing in oxygen is more advantageous because the sample is more stable in reactive conditions. As a drawback, the surface of this unreduced $TiO_2(110)$ sample is less well-known than the reduced one (extensively studied by STM) and probably rougher.[1]

1 G. Cabailh *et al.*, *Phys. Rev. B*, 2007, **75**, 241403.

Professor Bowker commented: I congratulate you on such careful measurements, and it's interesting to note that such well defined materials give very similar curves for the activity dependence on particle size to catalysts prepared in very different ways. Presumably this means that they have similar active sites. Concerning the activity dependence – the activity is practically 0 at ~5 nm – why is this?

Dr Bailly replied: The activity dependence practically falls to zero for 5 nm-sized GNPs since it is normalized to the total number of gold atoms. The surface-to-volume ratio thus decreases while the size increases and consequently, the activity falls. This fall is quicker than what is expected for active sites corresponding to the whole surface,[1,2] indicating that only some kinds of sites are active at the GNPs surface.

1 N. López *et al.*, *J. Catal.*, 2004, **223**, 232.
2 Saint-Lager *et al.*, *Faraday Discuss.*, 2011, **152**, 253.

Professor Hutchings added: Another important factor that should be considered is the nature of surface defects as these will be important in the observed catalysis. These could well play an important role in the deposition of the gold species. Can you comment on this?

Dr Bailly replied: The density of the different kind of defects strongly depends on the surface preparation. We did not perform any study on the clean $TiO_2(110)$ surface. We found that the nanoparticles' surface density is more or less isotropic just after gold deposition in UHV, as seen from the blue curves of Fig. 4 in ref. 1, indicating that the growth started at random punctual defects such as oxygen vacancies. Thus it would be interesting to compare the particles density on the TiO_2 substrate after Ar^+ sputtering followed by an annealing at 1000 K either in UHV or under 10^{-5} Pa of oxygen. Unfortunately the allocated synchrotron beamtime was too short to perform such a study. Anyway we can remark that the density of particles becomes strongly anisotropic after thermal annealing at 473 K and the reaction-induced sintering (green and red curves of Fig. 4) indicating that the nanoparticles are pinned on step edges.

1 Saint-Lager *et al.*, *Faraday Discuss.*, 2011, **152**, 253.

Professor Anderson opened the discussion of the paper by Professor Thornton: In Fig. 2a, you show two very distinct types of particles – the triangular flat particles you are focusing on, but also roughly an equal number what appear to be tall, thin, irregular pillars. Can you explain how you get such a distinctly bimodal set of structures for an apparently quite homogeneous support?

Professor Thornton replied: As shown in Fig. 1 here, the smaller features in Fig. 2(a) in the paper are also triangular in shape. They have a width of <10 nm and a typical height of 6 nm. This bimodal growth mode probably arises from different nucleation sites, such as single and multiple oxygen vacancies. We note that extremely high aspect ratio Pd wires can also be grown on $TiO_2(110)$ when the deposition temperature is increased to 900 K.[1] These nanowires have an average length and cross section of 1000 nm and 5×5 nm^2.

1 D. S. Humphrey, G. Cabailh, C. L. Pang, C. A. Muryn, S. A. Cavill, H. Marchetto, A. Potenza, S. S. Dhesi and G. Thornton, *Nano Lett.*, 2009 **9**, 155.

Professor Bowker commented: The particle shapes produced here (especially the flat, triangular shapes) are likely to be formed because deposition was carried out at high temperature. Ambient temperature deposition tends to produce smaller, more rounded particles.

Professor Hermann noted: You point out as a result from your studies that CO adsorption at (111) facets of 20 nm Pd nanoparticles is structurally identical to what was found for extended Pd(111) single crystal surfaces. This is not surprising in view of the fact that 20 nm Pd nanoparticles contain of order 300 000 atoms and, therefore, can be considered already bulk-type.

Professor Thornton responded: Well, that is what you might expect. However, we have observed that the CO adsorption behaviour on some Pd nanocrystals are quite different from those found on the extended Pd(111) surface. We believe that this is induced by the morphology of the support upon which the nanoparticles are supported. We will discuss this in detail in a later publication.

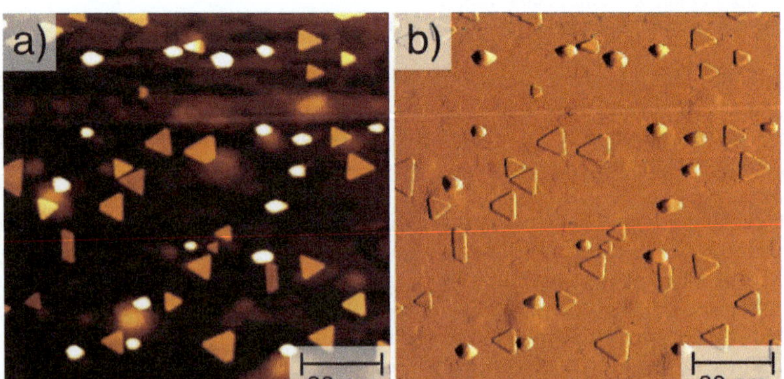

Fig. 1 (a) Same as Fig. 2a in the paper. (b) The corresponding current map of (a) showing that the tall, thin features that cannot be resolved in (a) are also triangular in shape and have a width of <10 nm and a typical height of 6 nm.

Professor Campbell remarked: You show beautiful homogeneous surfaces of TiO_2 by STM with however two very different Pd island types: large flat ones that have their 111 face parallel to the oxide surface, and the other much higher Pd islands that are smaller parallel to the surface. These could be explained as arising from small regions on the TiO_2 surface that have slightly different surface structure, and thus have nucleated a different epitaxial relationship for Pd/TiO_2 growth, where the Pd face that is parallel to the TiO_2 surface has a much higher surface energy than Pd(111). This would allow the sides of these particles, which can be lower-energy faces, to grow at the expense of the top facet, resulting in these smaller but thicker islands of Pd.

Professor Thornton answered: The main heterogeneity on a TiO_2(110) surface arises from steps and O vacancies. It seems more likely that the latter play a role in the nucleation of the Pd nanoparticles, with two different motifs giving rise to the two types of nanoparticle *via* the mechanism you describe.

Professor Freund said: I wanted to remind you that we saw very similar atomically-resolved STM images of Pd on a thin alumina film grown on NiAl(110).[1] We used IR to determine adsorption sites on the terminating (111) and (100) facets. Can I suggest to explore IR on your systems, now that C. Wöll's group has successfully demonstrated that one may perform such studies on bulk oxide surfaces.

1 K. H. Hansen *et al.*, *Phys. Rev. Lett.*, 1999, **83**, 4120.

Professor Thornton responded: Your images of Pd nanoparticles on an alumina film do indeed look very similar to those in our Discussion paper. Our purpose of including such an image (Fig. 2b) was simply to confirm the lack of encapsulation of the nanoparticles. We agree that an infrared study on the TiO_2(110)/Pd/CO system would be a valuable complement to our STM and NEXAFS study.

Mr Zhao asked: In experimental study of Pd catalysts, CO chemisorption is frequently used as a method to determine the ratio of surface atoms per total Pd atoms (Pd dispersion). There is argument on the stoichiometric ratio of CO : Pd. One important reason is that CO has varied bonding modes on the Pd surface (atop, bridge, hollow, *etc.*). According to your work, could you please comment on this argument? Do you think CO chemisorption is still a proper way to determine Pd dispersion?

Professor Thornton replied: Under the conditions we employ, the stoichiometry varies with coverage from 0.33 : 1 to 0.75 : 1, at least on the top facets of the nanoparticles. However, for catalytic studies much higher pressures and temperatures are employed. This will lead to a stoichiometric ratio.

Professor Bowker remarked: In Fig. 2b there appear to be islands of CO and clean surface. Calorimetric measurements imply a continuous decrease in heat of adsorption on metals for CO adsorption with increasing coverage, so we would think that CO molecules would cover the (111) surface on the particle in an homogeneous way.

Professor Thornton responded: In an STM study of CO on the native Pd(111) surface, Rose *et al.*[1] observed that the $(\sqrt{3} \times \sqrt{3})R30°$-1CO structure nucleates at defect sites. The ad-CO structure then expands with increasing CO coverage, and forms a complete $(\sqrt{3} \times \sqrt{3})R30°$-1CO layer at 0.33 ML CO. As the nanocrystal in Fig. 2b was covered with <0.3 ML CO, formation of small $(\sqrt{3} \times \sqrt{3})R30°$-1CO islands on the (111) top facet is therefore not surprising. This behaviour is, of course, also consistent with the calorimetry data.

1 M. K. Rose, T. Mitsui, J. Dunphy, A. Borg, D. F. Ogletree, M. B. Salmeron and P. Sautet, *Surf. Sci.*, 2002, **512**, 48.

Professor Neurock commented: Many catalytic systems operate under conditions where the surfaces are fully covered with adsorbates. The higher partial pressures of CO in CO oxidation or in Fischer–Tropsch synthesis will result in surface coverages which exceed a full monolayer where the coordinatively unsaturated edge and corner metal atoms can uptake more than a single CO.[1] As such the influence of the cluster size, morphology as well as pressure effects all become important.

1 B. T. Loveless, C. Buda, M. Neurock and E. Iglesia, *J. Am. Chem. Soc.*, 2013, **135**, 6107–6121.

Dr Beaumont said: Angular-dependant NEXAFS is a useful tool for determining adsorbate geometry on surfaces of known symmetry and has been employed to great effect for the study of adsorbates on single crystals. Employing this technique for nanoparticles would be really interesting – but presents difficulties in analysis since there is (as you raised during the presentation) a substantial (>25% of the geometric surface) contribution from sites not aligned flat/parallel to the surface as assumed in the method of analysis used from Stöhr and Outka. Nevertheless, with only very limited deviation outside the error bars, it must be noted the data fit very well the expected curve for near perpendicular adsorption geometry of CO. In the presentation and discussion of this it was mentioned that you have since tried to model the data as a sum of a number of contributions from the top facets and the side facets, each in different orientations, to obtain better answers. This would be interesting, however I wondered if you could comment on the likely size of the error bars in such a calculation – it would be very interesting if other (more complex) molecules' adsorption geometries could be studied in this way too but I'm concerned the errors would make the analysis of the data too insensitive to discriminate different geometries. Sadly the ideal experiments would also be on smaller rather than larger nanoparticles where significant changes may be expected and this would worsen not improve this issue. Can you offer any thoughts on this from your experience of trying to handle angular-dependant NEXAFS data when dealing with nanoparticles?

Professor Thornton answered: There is extremely limited utility of the angular dependence of NEXAFS for the general case of an adsorbate on a "three-dimensional" nanoparticle. The adsorbate would be expected to have a random orientation relative to the electric vector of the incoming X-ray beam. In other words, the adsorbate would appear to behave like a gas-phase molecule. For the case of an oriented nanoparticle such as the moiety we have on $TiO_2(110)$, the angular dependence is more useful, but nevertheless limited in value. The data are more suitable for simulation than for the type of fitting procedure described by Stöhr and Outka for single crystal surfaces. We can go beyond the simple approximation for the NEXAFS analysis used in our Discussion paper and simulate the NEXAFS from side and top facets.

In our simulation, we consider a nanoparticle that has a (111) top facet as well as equally sized {111} and {100} side facets. We assume that all facets are covered with the same amount of CO, which bonds with its molecular axis parallel to the facet-normal. On this basis, we simulated the measurement of the CO π^* resonance peak as a function of the polar angle θ of the incident photon beam and investigated how factors such as the particle's aspect ratio and size, and the polarization of the photon beam influence the overall absorption signal. We found that a particle with larger total area of the side facets will lead to a higher offset at $\theta = 0°$ in the NEXAFS curve. In addition, when the total area of the side facets is comparable or larger than that of the top facet, both changing the area ratio between the {111} and {100} side facets and rotating the particle with respect to its top facet normal can cause noticeable change to the NEXAFS curve. We also investigated the particle size dependence of the contribution of CO at defect sites (edge and corner sites) and found that they start to give an observable contribution when the nanoparticle diameter is below 5 nm.

Note that our simulation can only provide a general idea of how factors such as the particle's size and aspect ratio can influence the overall shape of the NEXAFS curve. In reality, a NEXAFS curve recorded from an experiment is an ensemble average of the X-ray absorption signals coming from adsorbates on many nanoparticles, which can have different sizes, shapes and orientations from each other. In order to investigate the bonding geometry of adsorbates on supported nanoparticles using angular dependence NEXAFS, one first needs to prepare a system consisting of well-defined nanoparticles with a narrow size distribution.

Professor Bowker queried: Are you sure there is no encapsulation? In Fig. 2a the titania surface around the nanoparticles appears to be roughened. Are these just steps on surface?

Professor Thornton answered: The structure seen below the Pd nanoparticles is due to steps on the substrate. The absence of encapsulation is evidenced first by the observed area of the clean nanoparticles in Fig. 2b, the lack of additional Ti^{3+} contribution to the XPS spectra after depositing Pd, and finally the adsorption of CO on the Pd nanoparticles. The latter would not be expected if the nanoparticles were encapsulated.

Dr Nilius opened the discussion of the paper by Dr Philip Davies: W_3O_9 clusters are known to form ring-type configurations only on ideal, weakly interacting oxide surfaces but prefer linear structures upon charge transfer from a defect for example. This phenomenon has been suggested as an interesting means to monitor charge exchange on oxide surfaces. Did you observe a ring-to-linear transition also for Mo_3O_9 clusters on the Fe_3O_4 support?

Dr Philip Davies replied: We didn't observe any recognizable linear molybdenum oxide species in the STM but we were only able to get stable STM images after annealing the deposited films in oxygen. After this treatment the molybdenum oxides appear to interact very strongly with the substrate. It would be interesting to study this further at lower temperatures.

Dr Yim enquired: You showed in the LEIS spectra (Fig. 8) that the $(2\sqrt{3} \times 2\sqrt{3})$ R30° structure shown in the STM picture in Fig. 7c should consist of iron, molybdenum and oxygen. Do you have any idea of its atomic arrangement?

Dr Philip Davies answered: We do not have atomically-resolved STM images of this surface structure. However, the LEED pattern we observe is identical to that obtained for molybdenum oxide films grown on an Fe_3O_4 film on Pt(111).[1] The model proposed for that structure was based on an iron molybdate, $Fe_2(MoO_4)_3$ with the surface enriched in molybdenum.

1 J. J. Uhlrich, J. Sainio, Y. Lei, D. Edwards, R. Davies, M. Bowker, S. Shaikhutdinov and H.-J. Freund, *Surf. Sci.*, 2011, **605**, 1550.

Professor Asakura remarked: Mo oxide cluster is formed during deposition. If the filament temperature is changed, does the cluster size change accordingly?

Dr Philip Davies responded: We have not been able to obtain good quality images of the surface before annealing in oxygen and so we are not able to comment on the nature of the molybdenum clusters formed immediately on adsorption. However, there is some information on the nature of the molybdenum oxides in the gas phase as a function of filament temperature in refs. 11 and 12 in the paper.

Professor Bowker noted in response: The effective volatility of Mo is higher in oxygen. From other work in the literature a variety of species desorb from Mo in an oxygen environment and Mo_3O_9 is the most common (depending on oxygen pressure and filament temperature). It may be that it is such a species that dominates upon landing on the surface and it is this that gives the trimer ring structures (a six-membered ring of three Mo and three O).

Professor Freund referred to the work originating from PNNL on W_3O_9 on $TiO_2(110)$:[1] As far as I know, when WO_3 is evaporated, the trimer W_3O_9 is the most abundant gas phase species. The PNNL group was able to resolve the binding size on the $TiO_2(110)$ surface. May the information be used to deduce information in the present case of Mo_3O_9 on $Fe_3O_4(110)$?

1 O. Bondarchuk *et al.*, *Angew. Chem., Int. Ed.*, 2006, **45**, 4786.

Dr Philip Davies answered: There are several similarities between the observations of Bondarchuk *et al.* and the present paper; for example Bondarchuk *et al.* were also unable to image the deposited oxide without a pre-annealing step. The features they assign to W_3O_9 on TiO_2, have exactly the same *footprint* in the STM images as the individual moieties we see here, providing support for an assignment to Mo_3O_9.

Professor Bowker added in response: The W trimers in the work referred to at PNNL were tilted because the TiO_2 (110) is very corrugated and the molecules tilt into corrugation. In the current paper they are symmetric because the iron oxide surface is less corrugated and is six-fold symmetric.

Dr Nilius asked: Did the Mo_3O_9 clusters in your experiments behave as isolated entities (could you move them around with the STM tip for example) or did they interact with the Fe_3O_4 surface rather strongly, *e.g.* by sharing oxygen species?

Dr Philip Davies answered: We did not observe any movement of the Mo_3O_9 clusters under our STM tip. As far as we can tell they are strongly bound to the sites where the "capping" oxygens of the "Termination A" layer would be situated.

Professor Behm opened a general discussion of Dr Saint-Lager's, Professor Thornton's and Dr Philip Davies' papers: I would like to point out an aspect which is often lost in discussions of experiments, in particular if they are not their own experiments (also with reference to the paper by Dr Saint-Lager). This is the role of the reaction conditions and the catalyst pre-treatment. In this paper, the different mechanistic proposals for CO oxidation were nicely put together, and it looks like that there is a big discrepancy. But when looking at the experimental conditions it becomes clear that these are very different. The experiments by the Yates group were done at 120 K, where even weakly adsorbing species are stably adsorbed on the surface, while our reaction experiments were done at temperatures above room temperature, where the steady-state concentration of the weakly bound species is extremely low and where surface lattice oxygen can be reactively removed (inhibited at low temperatures). When considering these differences, different mechanistic proposals become more consistent; they mainly indicate that under different reaction conditions different reaction pathways are dominant (rate determining). Similar aspects hold true also for the role of water impurities, which will affect the reaction in extremely low concentrations in the feed gas, well below those in most gas supplies.

Dr Yang addressed Professor Thornton: The general topics in this Discussion are about nanocatalysts and molecular systems; however, another important aspect of catalysis is the state of the reactants. In recent years there have been a lot of effort on mode-selective chemistry on surfaces and there are quite a few articles published in *Science* which have combined laser and STM on the single crystal surfaces. I wonder if it is possible to couple laser light into your system and consider investigating mode-selective chemistry?

Professor Thornton replied: This would indeed be an interesting experiment, although our system is not currently set up for experiments of that kind.

Professor Bowker made a general comment: Concerning the special reactivity of gold nanoparticles – there has now been an enormous amount of work in this area, from theory to catalyst synthesis, from single crystals to powders. Is there any agreement between fundamental and catalytic studies? Why are small particles important, but large ones are no good (per surface atom). Is there any consensus?

Professor Hutchings responded: In my opinion there are likely to be features concerning the size of the nanoparticles that are reaction specific. For example small nanoparticles and possibly clusters may be the active species for ambient temperature CO oxidation, but in this case the presence of water is essential and the sites at the interface between the gold and the support are considered to be key in the activation of oxygen. Theoreticians have been studying this. However, for the oxidation of alcohols it is possible that much larger nanoparticles are active and the small nanoparticles can be associated with non-selective chemistry.

Dr Bailly added in response: This is a hard task to determine the agreements between fundamental and catalytic studies. However we can mention that for the size dependence of the catalytic activity for CO oxidation of gold nanoparticles supported on $TiO_2(110)$, the experimental data follow a general behavior whatever the catalyst synthesis: by chemical means on powders or by physical means on single crystals. It passes through a maximum for a diameter around 2–3 nm.[1] It is not clear if it corresponds to intrinsic properties of the active sites. It more probably results from a specific equilibrium state of the morphology obtained for the Au/$TiO_2(110)$ system in reaction conditions.

1 Laoufi *et al.*, *J. Phys. Chem. C*, 2011, **115**, 4673.

Professor Hutchings made a general comment: When we are discussing the mechanism of the CO oxidation reaction on Au nanoparticles it is worth noting that there maybe three distinct mechanisms depending on the reaction temperature. There is a very low temperature mechanism operating at *ca.* 120 K, then there is a reaction mechanism operating at sub-ambient to just above ambient temperature, and this is the mechanism that we have been discussing. It is possible that there is a further mechanism at higher reaction temperatures. In the mechanism operating at ambient temperature water is a crucial component and hence we need to consider the role of water in this reaction.

Professor Neurock continued the discussion on CO oxidation: As Professor Behm nicely pointed out that while we often discuss the subtle differences in how CO oxidation proceeds, there is growing consensus on the mechanism as well as the sites that carry out the oxidation of CO and other reactants over Au/TiO_2. These reactions can be carried out over nm Au particles of Au and even bulk Au provided that the reaction is carried out in the presence of base.

I would add that we can learn a great deal by comparing the results reported in the literature for gas phase catalysis, aqueous phase catalysis and electrocatalysis over supported Au as all three are active in the oxidation of CO as well as a series of alcohols. Behm,[1] for example, previously showed that the rate of CO oxidation for Au/TiO_2 can be correlated with the number of perimeter sites. Hayden[2] showed a very similar correlation for electrocatalytic oxidation of CO over Au/TiO_2. Both results suggest that the reactions are carried out at the Au/TiO_2 interface which can activate O_2 in gas phase catalysis and water in electrocatalysis to generate the active oxidant. The adsorbed O^-, OH^- or OOH^- intermediates that form on Au in these reactions are weakly bound and tend to hold on to their negative charge. As such they are rather basic and can readily undergo nucleophilic addition to the adsorbed CO as well as aldehyde molecules. At low temperatures, O_2 can react directly with CO or dissociate to form O adatoms that react with CO.

The presence of ppm levels of water can assist CO oxidation by reacting with oxygen to form surface hydroxyl or hydroperoxy intermediates that readily react with CO.[3] CO oxidation can also proceed in aqueous media over bulk Au particles, if the reaction is carried out in the presence of base or hydrogen peroxide.[4,5] CO oxidation in electrocatalysis even occurs in the absence of O_2 in both acidic as well as basic media. In acidic media, higher potentials are required to electrocatalytically activate water to generate the active oxidant on Au. In alkaline media, the OH^- intermediates readily oxidize CO over Au electrodes.

Alcohol oxidation over Au is very similar to that for CO oxidation. It is readily carried out catalytically over supported bulk Au particles[6] as well as electrocatalytically over Au electrodes in the presence of alkaline media.[7] Both the catalytic as well as the electrocatalytic oxidations proceed *via* a base-promoted dehydrogenation of the alcohol on Au to form the corresponding aldehyde. The aldehyde subsequently undergoes a nucleophilic attack by surface or solution phase hydroxyl intermediates ultimately resulting in the formation acid products. We recently published two review papers which describe the similarities and differences in the oxidation of CO as well as alcohols over supported Au in catalysis and electrocatalysis.[8,9]

1 M. Kotobuki, R. Leppelt, D. A. Hansgen, D. Widmann and R. J. Behm, *J. Catal.*, 2009, **264**, 67–76.
2 B. E. Hayden, D. Pletcher, M. E. Rendall and J. P. Suchsland, *J. Phys. Chem. C*, 2007, **111**, 17044–17051.
3 M. Date and M. Haruta, *J. Catal.*, 2001, **201**, 221–224.
4 M. A. Sanchez-Castillo, C. Couto, W. B. Kim and J. A. Dumesic, *Angew. Chem.*, 2004, **43**, 1140–1142.
5 W. C. Ketchie, Y. L. Fang, M. S. Wong, M. Murayama and R. J. Davis, *J. Catal.*, 2007, **250**, 94–101.
6 B. N. Zope, D. D. Hibbitts, M. Neurock and R. J. Davis, *Science*, 2010, **330**, 74–78.
7 Y. Kwon, J. Schouten and M. T. Koper, *ChemCatChem*, 2011, **3**, 1176–1185.
8 A. Wieckowski and M. Neurock, *Adv. Phys. Chem.*, 2011, 1–16.
9 B. Braunchweig, D. Hibbitts, M. Neurock and A. Wieckowski, *Catal. Today*, 2013, **202**, 197–209.

Professor Bowker responded to Professor Neurock: I guess that if O atoms on the oxide peripheral to the Au nanoparticles are active, then it's likely that if they are hydrogenated they will be more reactive, since they are more weakly bound in the lattice.

Professor Neurock replied: The OH groups that form on Au are a little more weakly held to the Au than the atomic oxygen intermediates. The O^- and OH^- intermediates, however, are both weak enough and show barriers that are rather low thus indicating that they are both reactive as is shown in Fig. 9. The reactivity is then dictated by the higher coverages of oxygen that result from the gas-phase oxygen *versus* the lower coverage of hydroxyl intermediates that form during reaction.

Professor Campbell commented: Regarding the role of water promoter in oxidation catalysis by Au nanoparticles, I have proposed a mechanism for this in a previous Faraday Discussion,[1] and comments therein: (1) Water dissociates on the oxide to produce surface hydroxyls, (2) H migrates from the oxide onto the Au particles or Au/oxide interface (at very low concentration), (3) O_2 inserts into H_{ad} bound to Au to make Au–OOH_{ad} (*i.e.*, $O_{2,g} + H_{ad} \rightarrow OOH_{ad}$), which is exothermic by ~80 kJ mol^{-1}. This adsorbed hydroperoxy species should readily dissociate to make adsorbed OH plus an O adatom ($OOH_{ad} \rightarrow OH_{ad} + O_{ad}$), which we estimated to be exothermic by >175 kJ mol^{-1} on Au. Thus, water (or H_2, or any H source) should act like a catalyst for activating O_2 to produce O_{ad} on Au, which is well known to be a very effective oxidant.

1 C. T. Campbell, J. C. Sharp, Y. X. Yao, E. M. Karp and T. L. Silbaugh, *Faraday Discuss.* 2011, **152**, 227–239.

Professor Behm said: I would like to comment on the active oxygen species in the CO oxidation reaction on supported Au catalysts, and specifically on Au/TiO_2 catalysts. Our TAP reactor measurements provide clear proof that whatever the active species for reaction at room temperature and above may be, it must be stable up to 400 °C on a time scale of hours.

Professor Attard remarked: I think that in discussing this sort of chemistry involving adsorbed OH and the possible role of water, the gaseous dioxygen sets the electrochemical potential and that depending on the magnitude of this parameter, the catalytic rate is set. Moreover, the nature of the bonding of OH to the surface may change from an adsorbed anion (hydroxide) to a covalently bonded OH depending on the electrochemical potential and hence, the strength of the M–OH bond should be thought of as a tuneable parameter depending on the O_2/H_2O partial pressures, not necessarily a "fixed" value.

Professor Bowker commented: Au doesn't self poison with CO like other metals, maybe if we selectively poison other metals they could make good CO oxidation catalysts.

Professor Bowker then addressed Professor Neurock: Although you suggest that OH groups are important for CO oxidation catalysis, and there is some evidence to support that view, it seems to me that these careful experiments show that, although the results are very similar to those on catalysts, there is no water dosed here and OH groups are probably not present.

Professor Neurock answered: It is fairly well established[1–4] that ppm levels of water accelerate the oxidation of CO. While it is difficult to fully exclude the possibility of trace amounts of water, our previous studies for CO oxidation[5] were carefully carried out in order to eliminate water and there were no observable OH peaks present in the IR data. As such, we would indeed attribute the reactivity in our previous study to the oxidation of CO by O_2 at the Au–Ti^{4+} interface. In the present study, water and OH groups are formed as part of the reaction and can as such aid in enhancing oxidation activity.

1 M. Date and M. Haruta, *J. Catal.*, 2001, **201**, 221–224.
2 J .T. Calla and R. J. Davis, *Ind. Eng. Chem. Res.*, 2005, **44**, 5403–5410.
3 H. H. Kung, M. C. Kung and C. K. Costello, *J. Catal.*, 2003, **216**, 425–432.
4 M. Ojeda, B. Z. Zhan and E. Iglesia, *J. Catal.*, **285**, 92–102.
5 I. Green, W. Tang, M. Neurock and J. T. Yates, *Science*, 2011, **333**, 736–739.

Professor Hutchings observed: We have spent a long time discussing CO oxidation but this is not the only reaction catalysed by gold. For example alcohol oxidation is also catalysed by gold nanoparticles and can involve larger particles of the type being synthesized by any methods we have discussed today. Initially it was found that base had to be present but more recently base-free catalysis has been observed with gold nanoparticles. Clearly there is a rich field of catalytic chemistry for further study in this case.

Professor Neurock closed the general discussion with a response to Professor Hutchings: Much of what we learn for the oxidation of CO can be carried over to the oxidation of alcohols. Alcohol oxidation proceeds by the same mechanisms as that for CO. The alcohol can react with the weakly bound $O_2^{\cdot}$, $O^{\cdot}$, $OH^{\cdot}$ or $OOH^{\cdot}$ intermediates at the Au–TiO_2 interface to form an aldehyde. As in CO oxidation, the unsaturated C=O bond of the aldehyde subsequently undergoes a nucleophilic attack by a basic surface oxygen species resulting in the formation of an organic acid. The base-free oxidations that Professor Hutchings and others are carrying out are really very interesting.

Stability and migration barriers of small vanadium oxide clusters on the $CeO_2(111)$ surface studied by density functional theory

Joachim Paier,* Thomas Kropp, Christopher Penschke and Joachim Sauer

Received 12th February 2013, Accepted 26th February 2013
DOI: 10.1039/c3fd00012e

By virtue of periodic density functional theory, we investigate structure and thermodynamic stability of $(VO)_k$ and $(VO_2)_k$ ($k = 1, 2, 3$) clusters deposited on the $CeO_2(111)$ surface, which serve as models for the very active sub-monolayer vanadia catalyst on a ceria support. We find V always completely oxidized (oxidation state +5) and coordinated to four O atoms. As a consequence, Ce^{4+} is (partially) reduced to Ce^{3+}. Thus, localized Ce-4*f* states are populated, which requires an onsite U-term (PBE+U) to avoid over-delocalization of *f*-electrons. Importantly, trimers of VO_2 were found to be extraordinarily stable (agglomeration energy: −1.68 eV), whereas aggregation of VO species on $CeO_2(111)$ is thermodynamically clearly unfavourable (agglomeration energy: 3.45 eV). As a consequence a large area of the V_nO_m phase diagram (for relevant temperatures) is dominated by the VO_2 trimer. The latter is less active towards reduction/oxidation than the active monomer and dimer of VO_2, which are not present in the phase diagram at all, although directly observed by recent STM measurements. This suggests that kinetic effects hinder VO_2 to grow into larger oligomers. The lowest migration energy barrier we found is as high as 1.95 eV, which indicates that adsorbed monomeric VO_2 is "kinetically locked" at low temperatures and explains why monomers are stabilized on the ceria surface.

Introduction

Supported vanadium oxides represent very active and selective catalysts for many reactions of industrial relevance, as *e.g.* the oxidative dehydrogenation (ODH) of alcohols.[1–5] However, activity and selectivity of the vanadium oxide catalyst is strongly influenced by the rather complex interplay between several factors, most importantly (i) the nature of support material,[3,4,6,7] particularly its reducibility, as well as (ii) loading of the support.[8] On the one hand, reducible supporting oxides, such as ceria, titania, and zirconia exhibit much higher turn-over frequencies for ODH reactions than inert, irreducible oxides, as *e.g.* silica and alumina.[4,7] On the other hand, catalysts with loadings of vanadia corresponding to the so-called monolayer or sub-monolayer surface coverage are pronouncedly more active than crystalline V_2O_5 nanoparticles.[8]

For a structural characterization of coexisting monomeric and polymerized vanadia of the aforementioned monolayer species, which wet the support, predominantly Raman and UV/Vis spectroscopy,[3,5,9–12] as well as X-ray absorption spectroscopy have been used.[13,14] Various model systems of vanadia catalysts, as *e.g.* single

Humboldt-Universität zu Berlin, Institut für Chemie, Unter den Linden 6, 10099 Berlin, Germany. E-mail: joachim.paier@chemie.hu-berlin.de

crystals,[15,16] thin films,[17–19] and clusters supported on planar metal oxide substrates[20–25] have been extensively studied both experimentally and theoretically.[26–31]

Ceria is an easily reducible support material ideally suited for oxidation reactions following a Mars-van Krevelen mechanism.[32] From a surface science viewpoint, it is also perfectly suited for thin film experiments, since well defined surfaces can be grown on metal substrates.[33,34] The structure and reactivity of vanadia catalysts supported on $CeO_2(111)$ has been studied using photoelectron spectroscopy and temperature-programmed desorption (TPD).[23–25] However, these studies could not resolve its atomic structure.

Recently, important structural as well as electronic structure details with respect to monolayer vanadia catalysts supported on ceria were published.[35–38] Employing density functional theory (DFT) combined with experimental results using scanning tunneling microscopy (STM), infrared reflection absorption spectroscopy (IRAS), and X-ray photo emission spectroscopy (XPS), Baron *et al.*[35] showed for the first time that $O{=}V^{5+}O_n$, *i.e.* vanadyl terminated species, are formed on the $CeO_2(111)$ surface upon V deposition at various loadings. The nuclearity of vanadia species, which sintered into larger polymers upon annealing to 700 K, could be directly observed by STM and their vibrational properties could be monitored by IRAS.

Upon aggregation, the vanadyl stretching frequency was found to be blue-shifted by approximately 25 cm^{-1} relative to the V=O stretching mode of the monomer. Furthermore, XPS of the vanadia/ceria samples revealed a Ce-4*f* peak indicating vanadium in its highest oxidation state (+5). These findings have been entirely supported by DFT, predicting V or any VO_n species deposited onto $CeO_2(111)$ to be stabilized in the +5 oxidation state.[37]

Reactivity towards ODH of methanol of these model catalysts was studied by TPD experiments.[36] At very low loadings of vanadium, at which predominantly monomeric vanadium oxide species were stabilized by $CeO_2(111)$, an extraordinary increase of the methanol oxidation activity ("α-peak" at low desorption temperature) could be detected.[36,39] However, this low temperature TPD peak vanishes upon increasing the V loading on the ceria support inducing formation of trimers and larger oligomers. Thus, as mentioned in Ref. 35, reactivity of larger vanadia clusters shows strong similarity to the reactivity of vanadia clusters supported on alumina and silica.[20,22,39] DFT calculations generated clear evidence for a true cooperative effect between catalyst and support: Ceria is accommodating the electrons originating from V in its 4*f* states and vanadia is apparently releasing the oxygen atom, which is energetically more costly in pure ceria. More specifically, it could be shown that structure relaxation effects are largely responsible for the high reactivity, as demonstrated by reactivity descriptors such as oxygen defect formation energy[40] or hydrogenation energy.[41]

A comprehensive theoretical investigation of reactivity trends as a function of size of VO and VO_2 oligomers was performed by Penschke *et al.*[38] They confirmed the experimental observations, that larger species, specifically VO_2 oligomers, show significantly larger O defect formation energies and gain less energy upon hydrogenation than monomeric VO_2 at the $CeO_2(111)$ surface. These findings indicate lower reactivity toward ODH reactions. Overall, they could establish (i) a direct relation between thermodynamic stability and nuclearity of the vanadia clusters, and (ii) an inversely proportional relation between stability and activity completely in line with experiment. The authors conjectured that stabilization of VO_2 monomers at the ceria surface is due to kinetic effects, *i.e.* large migration barriers, which fits the observation of a kinetically limited growth.[35]

Computation of these barriers, which has not been accomplished up to now, is central aspect of the present work. We will discuss DFT+U results of two migration pathways for monomeric VO_2 at the $CeO_2(111)$ surface. We found that barriers are very large due to deformation of the very stable $O{=}VO_3$ tetrahedron as well as its reduction upon migration. The conjecture about kinetic effects stabilizing vanadia monomers on ceria has therefore been confirmed by transition state theory (TST) based on DFT+U. Therein neglected temperature effects will certainly modify barrier heights quantitatively, but still we believe that TST provides a qualitatively correct picture.

Computational details

Computational settings applied in this study follow closely those given in Ref. 38. Calculations were performed using the projector augmented-wave (PAW)[42,43] method as implemented in the Vienna *ab initio* simulation package (VASP).[44,45] We take care of corrections for the onsite Coulomb correlation of occupied f orbitals *via* the DFT+U approach based on the Perdew–Burke–Ernzerhof PBE exchange–correlation functional, *i.e.* PBE+U. The spin-polarized calculations employed a Hubbard U parameter of 4.5 eV for the Ce *f*-electrons. A kinetic energy cutoff of 600 eV for the plane-wave basis set was employed. Structural optimizations (including transition state structures) were performed until maximal forces acting on the relaxed atoms were smaller than 0.02 eV $Å^{-1}$. Vibrational frequencies use central differences for the force derivative with atomic displacements of ±0.015 Å.

We use a large $p(4 \times 4)$ surface unit cell employing a cell vector of 15.518 Å (lattice constant of 3.879 Å). The cell was cut in [111] orientation from the bulk phase of CeO_2 optimized using the PBE+U(4.5) method (see Da Silva *et al.*[46]). The present study uses two different slab models, one consisting of 9 atomic layers ($Ce_{48}O_{96}$) for calculations on thermodynamic stability, and a second one consisting of 6 atomic layers ($Ce_{32}O_{64}$) for the computationally more involved nudged elastic band (NEB) calculations as implemented in VASP following the algorithms discussed in Ref. 47 and 48. Refinement of transition state structures was accomplished by the so-called climbing image method as implemented by Henkelman and coworkers.[49] Sampling of the Brillouin zone was restricted to the Γ point. We use asymmetric slab models, *i.e.* atomic positions of the lowest lying trilayer were fixed in bulk positions. The vacuum layer was set to 10 Å. Careful tests showed that using six instead of nine layers results in an increased absolute value for the VO_2 adsorption energy by 0.12 eV only, *i.e.* 2.35%. Structural parameters are virtually identical.

All optimized structures were proved to be minima by the absence of imaginary vibrational frequencies. Similarly, transition state structures were identified by a single imaginary mode as well as by following along the normal mode eigenvector down-hill the internal reaction coordinate. Frequency calculations employ partial Hessians. For consistency reasons, the supported vanadium oxides are labeled V_nO_m, with n and m corresponding to the content of the surface cell on V and O atoms *in addition* to the $CeO_2(111)$ surface. PBE+U corrected for dispersion effects use the method suggested by Grimme[50,51] (for the van der Waals parameters, *cf.* Ref. 38).

Results

Structural details of oligomeric VO and VO_2 at $CeO_2(111)$

Most important minimum energy structures of species considered in this work are shown in Fig. 1. As stated in previous work,[35–37] the essential structural motif is a slightly distorted vanadate tetrahedron having the vanadium atom in its center, coordinated to four oxygen ligands, *i.e.* three V–O–M bonds, with M being either V or Ce and a single vanadyl (V=O) bond. Importantly, per VO unit three surface Ce^{4+} are reduced to Ce^{3+}, whereas per VO_2 only one Ce^{3+} is formed (depicted as dark blue balls in Fig. 1). The influence on the total energy upon varying Ce^{3+} sites at the surface as well as varying spin alignments, *i.e.* ferromagnetic *versus* antiferromagnetic states, was found to be small. For further discussions on this issue we refer

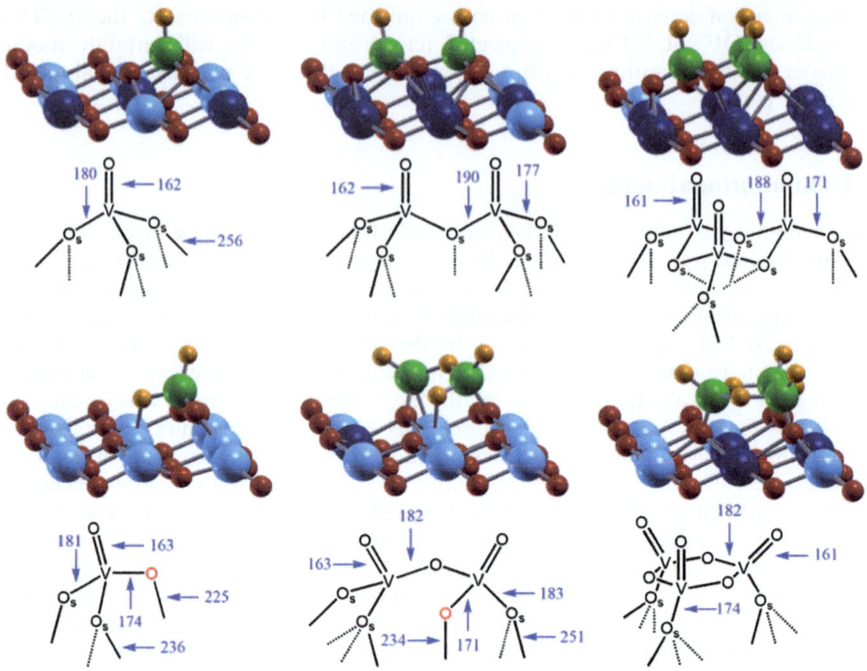

Fig. 1 Minimum energy structures of VO (top) and VO_2 (bottom) species represented as ball and stick models. For clarity reasons, only a (3x3) cut-out of the first O–Ce–O trilayer is shown. Color code: Ce^{4+}: light blue; Ce^{3+}: dark blue; V^{5+}: green; surface O^{2-}: red; O^{2-} originating from VO and VO_2: orange. Corresponding schematic representations are given below. Bond distances are given in pm. The "special" O atom in VO_2 and V_2O_4 is highlighted red and "O_S" refers to a surface oxygen atom. Ball-stick models were created using XCrysDen.[52]

to Ref. 38, but emphasize that changes in the energy due to the above mentioned effects were estimated to be smaller than 320 meV (= 30 kJ mol^{-1}).

Most important difference between VO- and VO_2-based systems adsorbed on $CeO_2(111)$ is the way how oligomers are built. The VO dimer and trimer exclusively use surface oxygen atoms for this purpose, whereas vanadium atoms in V_2O_4 and V_3O_6 are linked by the oxygen atoms of the VO_2 moieties. We emphasize the existence of a special, "interphase" oxygen atom in VO_2 and V_2O_4 distinct to other anchoring oxygen atoms. It connects vanadium to a surface cerium atom with a bond length of 225 pm. This special oxygen atom is vertically displaced from the surface and the associated V–O bond distance of 174 pm is shorter compared to the distance of V to a regular surface O atom (see Fig. 1).

Another structural characteristic of VO_2 monomer and dimer is a "pseudo-oxygen-vacancy" formed upon adsorption. The importance of this structural feature was revealed by mechanistic studies on the oxidation of methanol at the $VO_2 \cdot CeO_2(111)$ model catalyst.[53] The pseudo-vacancy is well seen in top-view representations of the surface cell (see M1 in Fig. 6 or Fig. 1 and 5 in Ref. 38). It is not related to the aforementioned special O atom, but results from a regular surface O atom coordinated to V. Upon formation of the $O{=}VO_3$ tetrahedron, the surface O needs to adopt a pronouncedly upraised position relative to the surface, which opens a "hole" in the hexagonal O–Ce–O surface network, which was also found for V_2O_4.

Concerning VO oligomers, the $O{=}VO_3$ tetrahedron is located such that V coordinates in hollow position relative to three surface oxygen atoms, which results in a trigonal local symmetry. As a consequence, V is approximately located atop a subsurface oxygen atom. In V_3O_6, the gas-phase VO_2 ad-species create a hexagonal ring of three O–V (gas-phase) moieties (orange O atoms in Fig. 1), and V atoms directly bind to surface oxygen atoms. In this case no pseudo-vacancy is formed.

Simulations of STM images based on the Tersoff-Hamann[54,55] approximation are presented in Fig. 2. The images probe the occupied valence bands from −2.0 to −0.5 eV with respect to the Fermi level. As depicted in the densities of states (Fig. 3 and 4 in Ref. 38), these states predominantly consist of O-2*p* orbitals.

The images show VO and VO_2 units as bright blobs in agreement with experiment (*cf. e.g.* Ref. 35). VO clusters appear slightly smaller than VO_2 clusters. This effect is slightly more pronounced for respective dimers. In Fig. 2, surface oxygen atoms emerge as grey blobs, but these details are hardly resolved in experimental images (compare with Fig. 1a in Ref. 35). Note that the dark areas close to VO_2 and V_2O_4 indicate the above mentioned *missing* surface oxygen atom, which strongly reconstructs in order to coordinate to V, that is, it opens a "pseudo vacancy". Moreover, the experimentally observed distance of spots within trimers approximately equal to the lattice constant of CeO_2 is confirmed by the simulation. However, according to Fig. 2, unambiguous discrimination between VO and VO_2 species merely based on STM appears hardly possible.

Thermodynamic stability

Relative stabilities of the various vanadium oxide clusters are obtained using PBE+U as well as dispersion-corrected PBE+U+D total energies. Agglomeration energies are obtained by virtue of eqn (1),

$$a\{VO\cdot(CeO_2)_{48}\} + b\{VO_2\cdot(CeO_2)_{48}\} \rightarrow \{V_{a+b}O_{a+2b}\cdot(CeO_2)_{48}\} + (a + b - 1)\,\{(CeO_2)_{48}\}, \quad (1)$$

and compiled in Table 1. The agglomeration of VO_2 units corresponds to an exothermic process and is therefore thermodynamically favourable (dimerization: −0.30 eV; trimerization: −1.68 eV). Correcting for dispersion effects lowers these energies by 90 and 310 meV, respectively. In contrast, agglomeration of VO units

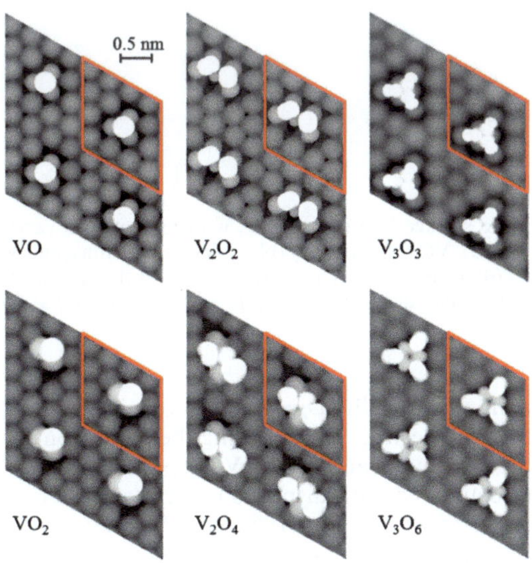

Fig. 2 Simulated STM images of various adsorbed vanadium oxide clusters. The occupied states between −2 and −0.5 eV with respect to the Fermi level have been probed (mainly O-2*p*). The unit cell is represented by the red frame. The vanadia clusters are seen as bright blobs, whereas surface oxygen atoms emerge as grey blobs.

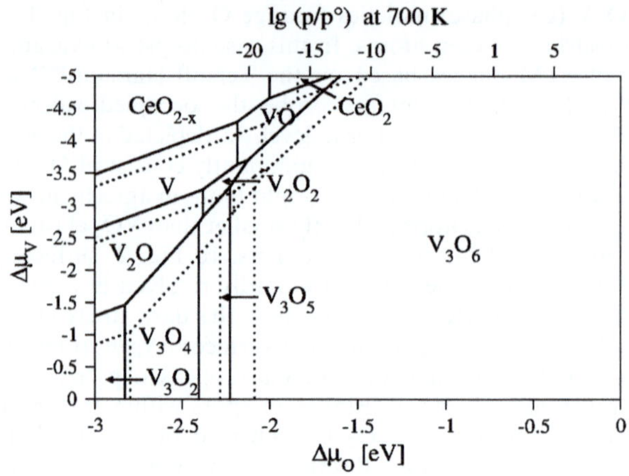

Fig. 3 Phase diagram of various adsorbed vanadium oxide clusters, calculated with (full lines) and without (dotted lines) dispersion correction. The figure shows the most stable species as a function of $\Delta\mu_V$ and $\Delta\mu_O$, with the latter translated into a pressure scale for $T = 700$ K.

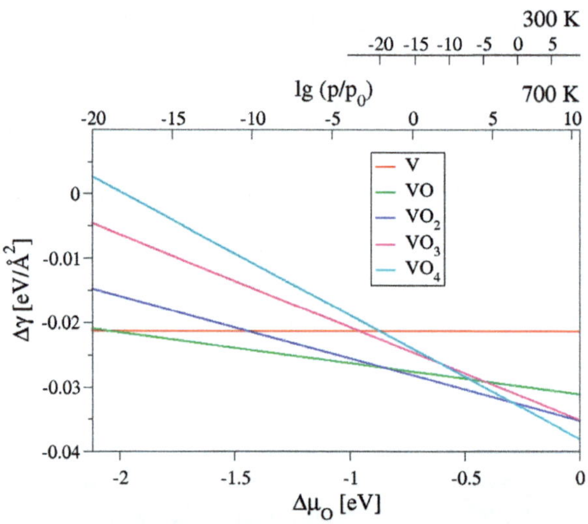

Fig. 4 Variation in the surface free energy, $\Delta\gamma$, with respect to a varying oxygen chemical potential, $\Delta\mu_O$, for V (red), VO (green), VO_2 (blue), VO_3 (magenta), and VO_4 (cyan) deposited on $CeO_2(111)$ using a (4x4) super cell. The scale opposite to the $\Delta\mu_O$-axis represents its translation into the corresponding partial pressure (p/p_0 with $p_0 = 1$ atm) at 700 as well as 300 K.

proceeds strongly endothermically. Dimerization requires 0.91 eV, whereas for the trimerization 3.45 eV are needed. Upon adding the correction for dispersion, these energies slightly decrease by 70 and 320 meV, respectively, hence energies decrease to a similar extent compared with dispersion-corrected results for VO_2. Sign and magnitude of agglomeration energies of "mixed" VO and VO_2 clusters depend on the mixing ratio of respective species. A single VO unit agglomerates with a single VO_2 with 0.0 eV as corresponding reaction energy, which decreases by 110 meV, *i.e.* becomes slightly exothermic, upon inclusion of dispersion effects. Depending on whether VO_2 or VO prevails, aggregation runs exothermic ($2VO_2$ + VO: −0.44 eV) or endothermic (VO_2 + 2VO: 1.00 eV; see Table. 1).

Table 1 Agglomeration energies without (ΔE_{Agg}) and with dispersion effects (ΔE_{Agg}+D) given in eV

Reactants	Product	ΔE_{Agg}	ΔE_{Agg}+D
2 VO	V_2O_2	0.91	0.84
3 VO	V_3O_3	3.45	3.13
2 VO_2	V_2O_4	−0.30	−0.39
3 VO_2	V_3O_6	−1.68	−1.99
VO + VO_2	V_2O_3	0.00	−0.11
2 VO + VO_2	V_3O_4	1.00	0.84
VO + 2 VO_2	V_3O_5	−0.44	−0.66

The phase diagram of V_nO_m clusters deposited on $CeO_2(111)$ is depicted in Fig. 3. V_nO_m is supposed to be in equilibrium with the pristine surface, V bulk and O_2 (in gas phase). Within the framework of "*ab initio* thermodynamics"[56] stability of different adsorbates on a surface is based on the surface free energy, as a function of the Gibbs free energy. The variation of the surface free energy $\Delta\gamma$ is calculated according to

$$\Delta\gamma = \frac{1}{A}\cdot\left[\Delta E_f - n\cdot\Delta\mu_{\mathrm{V}}(a, T) - m\cdot\Delta\mu_{\mathrm{O}}(p, T)\right]. \tag{2}$$

ΔE_f is calculated according to

$$\Delta E_f = E(\mathrm{V}_n\mathrm{O}_m\cdot\mathrm{CeO}_2) - E(\mathrm{CeO}_2) - n\cdot E(V) - \frac{m}{2}\cdot E(\mathrm{O}_2). \tag{3}$$

Eqn (2) and (3) use A as the area of the surface unit cell, *n* and *m* represent the number of vanadium and oxygen atoms added to the surface, $\Delta\mu_{\mathrm{V}}$ and $\Delta\mu_{\mathrm{O}}$ are the respective chemical potentials of vanadium and oxygen relative to the chemical potential at $T = 0$ K. The equations refer to solid V and gas-phase O_2. For solid V, the PBE total energy of metallic vanadium (bcc unit cell) was used (see also Ref. 38).

As shown in Fig. 3, V_3O_6 is the most stable species over a wide range of oxygen partial pressure and vanadium activity. At 700 K, the annealing temperature used in Ref. 35, V_3O_6 is the most stable phase for $p(O_2) > 10^{-10}$ atm. At strongly negative chemical potentials of oxygen, reduced species of V_3O_6 (*e.g.* V_3O_5 and V_3O_4) become more stable. As expected, under UHV (approximately 10^{-11} to 10^{-14} atm) and very low chemical potentials for V, the clean ceria surface is preferred thermodynamically. Most importantly, the active VO_2 species (monomer and dimer) are not visible on the diagram at all. This implies that these species have a significantly lower stability compared to the others. As stated in Ref. 38, correcting for dispersion effects does not incur qualitative changes in terms of relative stability of the V_nO_m phases considered in this work, which is indicated in Fig. 3 by the full lines. However, upon inclusion of dispersion effects, it becomes slightly more difficult to create oxygen defective systems.

We re-visited the stability of the monomeric VO_n ($n = 0 - 4$) species studied by Popa *et al.*[37] using a larger $p(4 \times 4)$ unit cell to study coverage induced effects, *i.e.* possible interactions between the adsorbates (see Fig. 4). We confirm the findings presented in Ref. 37. Interestingly, at low temperatures (see pressure scale for 300 K in Fig. 4) VO_2 is the most stable one, whereas for high temperatures its reduced form, VO, gains stability. Formally, at 700 K VO_2 is stable within a pressure range of 10^{-1} up to 10^6 atm, and hence indicating that VO_2 is present under mildly reducing conditions. However, the upper boundary of this range represents very high pressures challenging the ideal gas assumption applied in *ab initio* thermodynamics.

Migration pathways of $VO_2 \cdot CeO_2(111)$

Path 1. NEB calculations for the first migration pathway (M1 – TS2 – M3 – TS4, *cf.* Fig. 5 and 6) investigated in this work start out from the minimum energy structure found for VO_2 at the $CeO_2(111)$ surface (see Fig. 1). We will refer to this structure as M1 throughout this work. Relative energies and barriers reported in this section represent energy differences relative to M1.

NEB requires interpolation between two sets of coordinates corresponding to respective minimum energy structures. Therefore, to generate the structure of migrated VO_2, M1 was translated along one quarter of a cell vector. By virtue of subsequent NEB calculations, transition state structure TS2 was found. TS2 can be reached upon twisting the VO_4 tetrahedron of M1. Importantly, along this mode, it is the special, interphase oxygen, which is detached from the Ce atom. The rotational direction in M1 is indicated by an orange arrow at the special oxygen atom (*cf.* M1 in Fig. 6). The barrier of this step amounts to 0.77 eV (= 74 kJ mol^{-1}).

Relaxation of TS2 such that the special oxygen atom could fully re-establish its coordination at the "target" Ce atom yielded the local minimum M3, 0.22 eV (= 21 kJ mol^{-1}) higher in energy than M1. This small energy difference is plausible, since M3 is structurally closely related to M1. Upon formation of M3, the pseudo-vacancy as described in previous sections, is closed and re-created at a different position, as easily seen by comparing M1 with M3.

Running NEB on a number of images interpolating between M3 and the translated M1 structure, we found a transition state structure TS4 associated with a barrier of 1.95 eV (= 188 kJ mol^{-1}). Its imaginary mode is represented by inversion of the planar $O{=}VO_2$ moiety while the approaching/detaching surface oxygen atoms oscillate in the surface. Here, vanadium accommodates the oxidation state +4, as monitored by the projected spin-density (V^{4+} coloured dark green in Fig. 6). Starting out from TS4, M1 can be reached by the neighbouring surface oxygen atom coordinating to the VO_3 moiety (indicated by orange arrows). This step recreates the pseudo-vacancy. In addition, an analogous local minimum M4 was found, which is similar to gas-phase VO_3 or VO_3^+, *i.e.* structurally related to an oxygen defective VO_4 tetrahedron forming a trigonal pyramid.[57] In TS4 and M4 the bond distance between V and the anchoring surface O is equal to 179 pm and the vanadyl bond is equal to 163 pm. In M4 the bond between V and the special O amounts to 171 pm and is therefore 1 pm shorter compared with the corresponding bond in TS4.

Path 2. For the second migration path, M1 was translated along a quarter of the second cell vector. We found a very symmetric transition state structure TS5, corresponding to a barrier of 2.73 eV (= 263 kJ mol^{-1}). The mode transforming M1 into TS5 (indicated by a purple arrow in M1) relates to a rocking mode of the VO_2 moiety

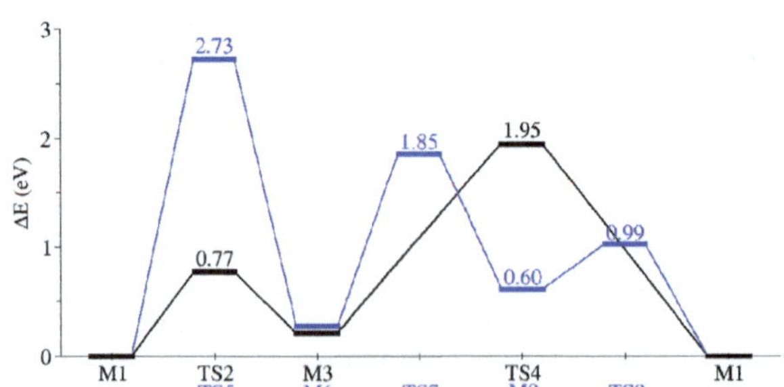

Fig. 5 Proposed energy profile for two different migration pathways of $VO_2 \cdot CeO_2(111)$. Barrier heights are given in eV. Local minima on the PES are indicated with "M", whereas transition state structures are indicated with "TS".

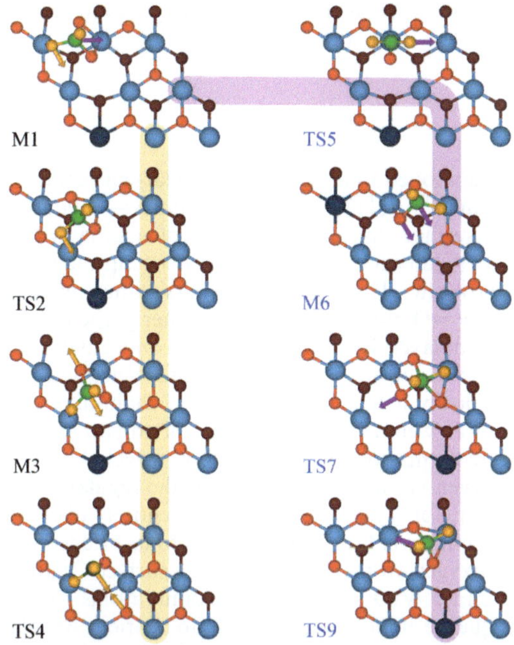

Fig. 6 Top views of structures corresponding to the proposed migration pathways depicted in Fig. 5. Arrows (orange: pathway 1; purple pathway 2) indicate the direction of atomic motion creating the ensuing stationary point. Colour code similar to Fig. 1, except for V^{4+} (dark green). Subsurface oxygen is shown in dark red and surface oxygen is shown in red. Ball-stick models were created using VESTA.[58]

(originally from gas phase). TS5 is reached by detaching the special anchoring O from Ce and a simultaneous flip of the vanadyl O atom towards the surface. In TS5, both aforementioned O atoms adopt a bond distance to V of 163 pm and can therefore be identified as V=O bonds. The O=V=O bond angle amounts to 110°. In fact, the species can be viewed as a (closed-shell) VO_3^- (two vanadyl O, V^{5+}, O^{2-} as anchoring oxygen). The corresponding imaginary mode consists of a rocking motion of the O=V=O group and simultaneous "asymmetric" movements of two surface O atoms underneath. When O=V=O tilts towards the surface, the nearby O approaches the V atom, while the other one moves back into the surface.

Upon relaxation of TS5, one of the vanadyl oxygen atoms coordinates to cerium, which re-establishes the VO_4 tetrahedron adopting the local minimum structure M6, 0.28 eV (= 27 kJ mol^{-1}) higher in energy than M1. Note that the site of Ce^{3+} is altered compared to previous structures, but extensive test calculations showed that therefrom incurred variations in the total energy are small.[38] The tetrahedron moves along the surface as indicated by purple arrows. In addition, the VO_4 tetrahedron deforms, opening an additional coordination site at the V nucleus. A surface O atom binds to V creating the O=VO_4 motif of TS7 (fivefold coordinated V), resulting in a barrier of 1.58 eV (= 152 kJ mol^{-1}). It is noteworthy that the coordination pattern of V in TS7 is akin to the situation in solid V_2O_5, resulting in very similar V–O bond distances. The associated imaginary mode is represented by an asymmetric stretching motion of O–V–O, whereas the first O heals (see purple arrow in TS7) and the second one re-opens a pseudo-vacancy.

Relaxation of TS7 drives the system into M8, 0.60 eV (= 58 kJ mol^{-1}) higher in energy than M1. The rocking motion transforming M8 into M1 essentially preserves the structure of the VO_4 tetrahedron and is, therefore, only associated with a small barrier of 0.33 eV (= 32 kJ mol^{-1}). For this step the special oxygen detaches from the

Ce atom restoring V=O, whereas the former vanadyl oxygen coordinates to a neighbouring Ce atom, thereby re-creating the special, "interphase" oxygen atom (purple arrow in M8).

Discussion

Based on the aforementioned findings, a rather clear-cut picture can be drawn. VO and VO_2 adsorb on the $CeO_2(111)$ surface such that V=O bonds are formed pointing away from the surface. Speaking in terms of a growth mode, VO_2 behaves distinctly from VO. Additional VO species just occupy a neighbouring trigonal surface oxygen hollow site. Thus, growing occurs rather regularly by placing one VO next to another. V=O bonds are oriented rather perpendicularly to the surface forming a flat sub-monolayer. In larger VO_2 units, connectivity changes qualitatively upon growth. The dimer shows a structurally similar characteristic compared with the monomer. Specifically it forms a rather open, chain-like structure, featuring a special oxygen atom as well as a pronouncedly lifted surface oxygen atom bound to V. The strong reorientation of the O atom, which binds to the V nucleus, opens a "pseudo-oxygen vacancy", and perturbs the regular Ce–O surface network significantly. However, for V_3O_6, the situation is entirely different. It is built from three gas-phase VO_2 units creating a ring with V atoms bridged by gas-phase oxygen atoms. None of the oxygen atoms coordinates to a surface Ce atom, as found for the monomer and the dimer. V_3O_6 binds *via* vanadium atoms to three surface oxygen atoms. As a consequence none of them needs to reorient or relax, *i.e.* the regular Ce–O network is largely preserved, and no pseudo-vacancies are formed. Although V=O bonds in V_3O_6 are rather tilted towards the surface, still it adopts a flat ring structure at the $CeO_2(111)$ surface. This flat growth mode of vanadia oligomers on the ceria surface was observed experimentally according to the apparent heights detected by STM for higher V coverage at $CeO_2(111)$.[35] Thus, our PBE+U results completely support this observation.

The experimentally observed blue-shift of the IR active vanadyl stretching frequency upon formation of larger VO_n aggregations was studied by DFT for VO oligomers in Ref. 35 and for VO_2 oligomers in Ref. 38. Both calculated shifts, comparing VO and VO_2 monomers with respective trimers, fit very well to the observed value of 27 cm^{-1}, and hence based on the IR frequencies only, discrimination between VO and VO_2 is not possible.

However, larger VO and VO_2 species are very distinct in terms of stability. While aggregation of VO_2 moieties runs strongly exothermic, the aggregation of VO is thermodynamically not beneficial. Thus, as presented in Table. 1, DFT predicts deposited VO_2 to aggregate into very stable trimers, dominating the phase diagram for not too negative vanadium chemical potentials and for oxygen partial pressures greater than 10^{-11} atm (UHV) at elevated temperatures (700 K, the temperature applied to sinter vanadium oxide on $CeO_2(111)$, *cf.* Ref. 35).

However, as discussed by Penschke *et al.* in Ref. 38, the penalty for such a high stability of V_3O_6 is a loss in activity, which was also observed by TPD.[36] Our previous studies on activity clearly indicate that the VO_2 monomer and dimer are particularly active towards O defect formation and hydrogenation, a finding exactly matching the aforementioned TPD experiment. By virtue of STM applied to the catalyst at low coverage, monomers and dimers were observed as the predominant species, whereas for higher coverages larger oligomers, *e.g.* trimers and heptamers could be detected. A distinction between monomers and trimers can be easily made, since monomers appear as bright blobs, whereas trimers appear as trefoils, *i.e.* bright blobs in three-fold symmetry. The simulated STM images presented in this work perfectly agree with the observation (see Fig. 2).

The active VO_2 species (monomer and dimer), although observed experimentally, do not appear in the phase diagram at all. This is explained by kinetic effects, *i.e.* large migration barriers of VO_2. Two distinct migration pathways of a VO_2

monomer at the $CeO_2(111)$ surface were studied by means of NEB. Indeed, we were successful in locating required transition state structures and local minima, and we found extraordinarily high barriers preventing VO_2 to migrate and hence to dimerize, trimerize *etc.*

The first pathway investigated follows a two-step mechanism: (i) the VO_4 tetrahedron moves rather rigidly preserving its coordination sphere and the "special" O binds to a different surface Ce ion. Note that along this pathway the (previous) special oxygen atom remains always coordinated to V. This step is associated with a lower barrier of 0.77 eV. (ii) Subsequent stepwise substitution (O detachment/attachment) of anchoring surface oxygen atoms requires a very high activation energy of 1.95 eV (= 188 kJ mol^{-1}). In the second step V^{5+} was reduced to V^{4+}. Note that initializing electron spin density at a nearby Ce atom in order to create the "missing" Ce^{3+} in TS4/M4 always yields $V^{4+}(3d^1)$ and $Ce^{4+}(4f^0)$. The same was found upon slight distortion of the structures. Considering the findings of previous work on mixed metal V/Ce gas-phase clusters[59] and the DFT+U studies on bulk $CeVO_4$,[60] it is plausible that M4 or TS4 have to be high in energy, since the coordination number of V in M4/TS4 is low, but still V adopts oxidation state +4. Low coordinated V, as found in gas-phase clusters, solid $CeVO_4$ or vanadia clusters deposited on ceria, prefers the oxidation state +5. As shown by the aforementioned test calculations, TS4 and M4 do not represent artefacts, but appear to be robust even under structural distortion.

For the second pathway, the following sequence of steps was found: (i) First, the special O is detached from surface Ce creating a VO_3 species (TS5), but V remains in oxidation state +5. We found again that detachment of O from V is associated with a very high barrier, independent of the reduction of V as observed in the first pathway. (ii) Second, a re-coordination occurs, forming $O{=}VO_4$ and gaining 2.45 eV in energy with subsequent formation of a five-fold coordinated V (1.58 eV barrier). (iii) O coordinated to V detaches and heals the pseudo-vacancy. This re-coordination process has a high barrier, similar to the second step of the first mechanism. Subsequently, the VO_4 tetrahedron relaxes and re-opens the pseudo-vacancy and (iv) finally, a VO_4 rocking motion (presumably a step with a very low barrier, comparable to (i) of the first migration pathway) transforms the system back into the starting structure (M1).

Importantly, we emphasize that temperature effects, entirely neglected in our study, will certainly affect barriers. At elevated temperatures the oxygen ion mobility of the CeO_2 support is most likely sufficiently high to influence migration processes of vanadium oxide clusters deposited on it. Experimentally, monomeric vanadate species sinter upon heating to 700 K and ultimately form vanadia trimers, as observed by Baron *et al.* (see Fig. 1 in Ref. 35). We believe that under realistic conditions, the support "helps" the $O{=}VO_3$ tetrahedron to migrate, and therefore barriers should be substantially lower than our predictions, *i.e.* 1.86 eV. Nevertheless, the barriers obtained using PBE+U generate clear evidence for the previously conjectured kinetic hindrance of migrating vanadate tetrahedrons on the ceria surface. It is known that barriers obtained using GGA functionals are usually underestimated.[61,62] Thus, the energy barriers presented in this work most likely represent even lower boundaries to barriers obtained using *e.g.* hybrid functionals, which are expected to be more accurate.[63] However, it would be interesting to compare energy barriers with enthalpy barriers including temperature effects by virtue of harmonic frequencies, but we expect that the aforementioned effects do not alter our findings qualitatively.

We mention that current studies on the oxidation mechanism of methanol at the $VO_2{\cdot}CeO_2(111)$ interface,[53] provide strong evidence that the pseudo-vacancy plays an important role. We firmly believe that it is this very structural feature of the ceria surface, created upon adsorption/formation of vanadia, which drives the system towards the theoretically determined high activities in Mars-van Krevelen processes.

The large relaxation effects upon oxygen defect formation in the active, monomeric VO_2 species in fact heal the pseudo-vacancy.

Summary

Based on TST using PBE+U, we found that small VO_2 species, although thermodynamically unstable, are stabilized at the $CeO_2(111)$ surface, because of huge migration energy barriers. This is mainly due to (i) deformation of the vanadate tetrahedron as well as (ii) required re-coordination of O to V. Re-coordination was found to go along with activation energies as large as 2.73 eV (= 263 kJ mol^{-1}). However, the lowest energy barrier we found amounts to 1.95 eV (= 188 kJ mol^{-1}). Specifically, we studied two possible pathways: (i) V features threefold coordination with oxygen atoms, with or without simultaneous reduction of V^{5+} to V^{4+}, or (ii) V adopts a fivefold coordination, with V in its highest oxidation state (+5). Intermediate steps, associated with rather rigid rotation or only slight deformation of the vanadate tetrahedron, correspond to moderate barriers of 0.77 eV (= 74 kJ mol^{-1}) only. Thus, it appears that any reaction steps involving detachment of O atoms coordinated to V are extremely unlikely.

Acknowledgements

The authors are grateful for financial support from the Deutsche Forschungsgemeinschaft (DFG, SFB 546). Grants for compute time at the high-performance compute clusters HLRN (Berlin, Hannover) operated by the North-German Supercomputing Alliance and JUROPA operated by the Supercomputer Centre at the Forschungszentrum Jülich are gratefully acknowledged.

References

1 I. E. Wachs and B. M. Weckhuysen, *Appl. Catal., A*, 1997, **157**, 67–90.
2 A. Khodakov, B. Olthof, A. T. Bell and E. Iglesia, *J. Catal.*, 1999, **181**, 205–216.
3 B. M. Weckhuysen and D. E. Keller, *Catal. Today*, 2003, **78**, 25–46.
4 I. E. Wachs, *Catal. Today*, 2005, **100**, 79–94.
5 B. Kilos, A. T. Bell and E. Iglesia, *J. Phys. Chem. C*, 2009, **113**, 2830–2836.
6 B. Olthof, A. Khodakov, A. T. Bell and E. Iglesia, *J. Phys. Chem. B*, 2000, **104**, 1516–1528.
7 M. A. Bañares, M. V. Martinez-Huerta, X. T. Gao, I. E. Wachs and J. L. G. Fierro, in *Stud. Surf. Sci. Catal.*, ed. A. Corma, F. V. Melo, S. Mendioroz and J. L. G. Fierro, Elsevier, Amsterdam, 2000, vol. 130D, p. 3125.
8 T. Kim and I. E. Wachs, *J. Catal.*, 2008, **255**, 197–205.
9 E. L. Lee and I. E. Wachs, *J. Phys. Chem. C*, 2008, **112**, 6487–6498.
10 H.-S. Kim, S. A. Zygmunt, P. C. Stair, P. Zapol and L. A. Curtiss, *J. Phys. Chem. C*, 2009, **113**, 8836–8843.
11 D. E. Keller, T. Visser, F. Soulimani, D. C. Koningsberger and B. M. Weckhuysen, *Vib. Spectrosc.*, 2007, **43**, 140–151.
12 M. A. Banares and G. Mestl, in *Advances in Catalysis, Vol 52*, ed. B. C. Gates and H. Knozinger, 2009, vol. 52, pp. 43–128.
13 D. E. Keller, S. M. K. Airaksinen, A. O. Krause, B. M. Weckhuysen and D. C. Koningsberger, *J. Am. Chem. Soc.*, 2007, **129**, 3189–3197.
14 M. Ruitenbeek, A. J. van Dillen, F. M. F. de Groot, I. E. Wachs, J. W. Geus and D. C. Koningsberger, *Top. Catal.*, 2000, **10**, 241–254.
15 K. Devriendt, H. Poelman and L. Fiermans, *Surf. Sci.*, 1999, **433**, 734–739.
16 R. P. Blum, H. Niehus, C. Hucho, R. Fortrie, M. V. Ganduglia-Pirovano, J. Sauer, S. Shaikhutdinov and H. J. Freund, *Phys. Rev. Lett.*, 2007, 99.
17 S. Surnev, M. G. Ramsey and F. P. Netzer, *Prog. Surf. Sci.*, 2003, **73**, 117–165.
18 H. Niehus, R. P. Blum and D. Ahlbehrendt, *Surf. Rev. Lett.*, 2003, **10**, 353–359.
19 A. C. Dupuis, M. Abu Haija, B. Richter, H. Kuhlenbeck and H. J. Freund, *Surf. Sci.*, 2003, **539**, 99–112.

20 N. Magg, B. Immaraporn, J. B. Giorgi, T. Schroeder, M. Baumer, J. Dobler, Z. L. Wu, E. Kondratenko, M. Cherian, M. Baerns, P. C. Stair, J. Sauer and H. J. Freund, *J. Catal.*, 2004, **226**, 88–100.
21 S. Kaya, Y. N. Sun, J. Weissenrieder, D. Stacchiola, S. Shaikhutdinov and H. J. Freund, *J. Phys. Chem. C*, 2007, **111**, 5337–5344.
22 Y. Romanyshyn, S. Guimond, H. Kuhlenbeck, S. Kaya, R. P. Blum, H. Niehus, S. Shaikhutdinov, V. Simic-Milosevic, N. Nilius, H. J. Freund, M. V. Ganduglia-Pirovano, R. Fortrie, J. Doebler and J. Sauer, *Top. Catal.*, 2008, **50**, 106–115.
23 G. S. Wong, M. R. Concepcion and J. M. Vohs, *J. Phys. Chem. B*, 2002, **106**, 6451–6455.
24 T. Feng and J. M. Vohs, *J. Catal.*, 2004, **221**, 619–629.
25 J. M. Vohs, T. Feng and G. S. Wong, *Catal. Today*, 2003, **85**, 303–309.
26 M. Calatayud, B. Mguig and C. Minot, *Surf. Sci.*, 2003, **526**, 297–308.
27 A. Vittadini, M. Casarin, M. Sambi and A. Selloni, *J. Phys. Chem. B*, 2005, **109**, 21766–21771.
28 T. K. Todorova, M. V. Ganduglia-Pirovano and J. Sauer, *J. Phys. Chem. C*, 2007, **111**, 5141–5153.
29 M. M. Islam, D. Costa, M. Calatayud and F. Tielens, *J. Phys. Chem. C*, 2009, **113**, 10740–10746.
30 J. Döbler, M. Pritzsche and J. Sauer, *J. Phys. Chem. C*, 2009, **113**, 12454–12464.
31 V. Shapovalov and H. Metiu, *J. Phys. Chem. C*, 2007, **111**, 14179–14188.
32 P. Mars and D. W. van Krevelen, *Chem. Eng. Sci.*, 1954, **3, Supplement 1**, 41–59.
33 D. R. Mullins, P. V. Radulovic and S. H. Overbury, *Surf. Sci.*, 1999, **429**, 186–198.
34 J. L. Lu, H. J. Gao, S. Shaikhutdinov and H. J. Freund, *Surf. Sci.*, 2006, **600**, 5004–5010.
35 M. Baron, H. Abbott, O. Bondarchuk, D. Stacchiola, A. Uhl, S. Shaikhutdinov, H.-J. Freund, C. Popa, M. V. Ganduglia-Pirovano and J. Sauer, *Angew. Chem., Int. Ed.*, 2009, **48**, 8006–8009.
36 M. V. Ganduglia-Pirovano, C. Popa, J. Sauer, H. L. Abbott, A. Uhl, M. Baron, D. Stacchiola, O. Bondarchuk, S. Shaikhutdinov and H.-J. Freund, *J. Am. Chem. Soc.*, 2010, **132**, 2345–2349.
37 C. Popa, M. V. Ganduglia-Pirovano and J. Sauer, *J. Phys. Chem. C*, 2011, **115**, 7399–7410; 2012, **116**, 18572–18573(Erratum).
38 C. Penschke, J. Paier and J. Sauer, *J. Phys. Chem. C*, 2013, **117**, 5274–5285.
39 B. Beck, M. Harth, N. G. Hamilton, C. Carrero, J. J. Uhlrich, A. Trunschke, S. Shaikhutdinov, H. Schubert, H.-J. Freund, R. Schlögl, J. Sauer and R. Schomäcker, *J. Catal.*, 2012, **296**, 120–131.
40 J. Sauer and J. Döbler, *Dalton Trans.*, 2004, 3116–3121.
41 X. Rozanska, R. Fortrie and J. Sauer, *J. Phys. Chem. C*, 2007, **111**, 6041–6050.
42 P. E. Blöchl, *Phys. Rev. B: Condens. Matter*, 1994, **50**, 17953–17979.
43 G. Kresse and D. Joubert, *Phys. Rev. B: Condens. Matter Mater. Phys.*, 1999, **59**, 1758–1775.
44 G. Kresse and J. Furthmuller, *Phys. Rev. B: Condens. Matter*, 1996, **54**, 11169–11186.
45 G. Kresse and J. Furthmuller, *Comput. Mater. Sci.*, 1996, **6**, 15–50.
46 J. L. F. Da Silva, M. V. Ganduglia-Pirovano, J. Sauer, V. Bayer and G. Kresse, *Phys. Rev. B: Condens. Matter Mater. Phys.*, 2007, **75**, 045121.
47 G. Mills, H. Jonsson and G. K. Schenter, *Surf. Sci.*, 1995, **324**, 305–337.
48 H. Jonsson, G. Mills and K. W. Jacobsen, in *Classical and Quantum Dynamics in Condensed Phase Simulations*, ed. J. Berne, G. Ciccotti and D. F. Coker, World Scientific, 1998.
49 G. Henkelman, B. P. Uberuaga and H. Jonsson, *J. Chem. Phys.*, 2000, **113**, 9901–9904.
50 S. Grimme, *J. Comput. Chem.*, 2006, **27**, 1787–1799.
51 T. Kerber, M. Sierka and J. Sauer, *J. Comput. Chem.*, 2008, **29**, 2088–2097.
52 A. Kokalj, *Comput. Mater. Sci.*, 2003, **28**, 155–168.
53 T. Kropp, J. Paier and J. Sauer, *in preparation*, 2013.
54 J. Tersoff and D. R. Hamann, *Phys. Rev. Lett.*, 1983, **50**, 1998–2001.
55 J. Tersoff and D. R. Hamann, *Phys. Rev. B*, 1985, **31**, 805–813.
56 K. Reuter and M. Scheffler, *Phys. Rev. B: Condens. Matter*, 2001, **65**, 035406.
57 A. Bande and A. Lüchow, *Phys. Chem. Chem. Phys.*, 2008, **10**, 3371–3376.
58 K. Momma and F. Izumi, *J. Appl. Crystallogr.*, 2011, **44**, 1272–1276.
59 L. Jiang, T. Wende, P. Claes, S. Bhattacharyya, M. Sierka, G. Meijer, P. Lievens, J. Sauer and K. R. Asmis, *J. Phys. Chem. A*, 2011, **115**, 11187–11192.
60 J. L. F. Da Silva, M. V. Ganduglia-Pirovano and J. Sauer, *Phys. Rev. B: Condens. Matter Mater. Phys.*, 2007, **76**, 125117.
61 Y. Zhao and D. G. Truhlar, *J. Phys. Chem. A*, 2005, **109**, 5656–5667.
62 Y. Zhao, N. Gonzalez-Garcia and D. G. Truhlar, *J. Phys. Chem. A*, 2005, **109**, 2012–2018.
63 K. Yang, J. J. Zheng, Y. Zhao and D. G. Truhlar, *J. Chem. Phys.*, 2010, 132.

Mechanistic insights into the partial oxidation of acetic acid by O_2 at the dual perimeter sites of a Au/TiO_2 catalyst

Isabel Xiaoye Green,[a] Wenjie Tang,[b] Matthew Neurock*[ab] and John T. Yates, Jr.*[ab]

Received 2nd January 2013, Accepted 25th January 2013
DOI: 10.1039/c3fd00002h

The unique interfacial sites of Au nanoparticles supported on TiO_2 are known to catalyze the activation of oxygen and it's addition to small molecules including H_2, CO, NO and propylene. Herein we extend these ideas and show that the unique Au–Ti dual perimeter sites that form at the Au/TiO_2 interface can also catalyze more demanding C–H and C–O bond activation reactions involved in the deoxygenation organic acids such as acetic acid. We have shown previously that acetic acid can be partially oxidized on a Au/TiO_2 catalyst to form a novel gold ketenylidene ($Au_2{=}C{=}C{=}O$) intermediate. In the present work we use *in situ* infrared spectroscopy and first-principle density functional theory (DFT) to examine the mechanism and the kinetics by which this reaction proceeds. The reaction was found to be localized at the dual perimeter sites of the Au/TiO_2 catalyst, where O_2 was activated. In contrast to Au/TiO_2, no ketenylidene formation was observed on a similar Au/SiO_2 catalyst or a TiO_2 blank sample. The reaction involves the activation of multiple C–H bonds as well as the C–O bond in the adsorbed CH_3COO species. C–O bond scission is postulated to occur at the TiO_2 sites, while C–H bond scission occurs on Au sites, both near the active $Au–Ti^{4+}$ dual perimeter sites. $^{18}O_2$ isotopic labeling indicated that the O moiety of the ketenylidene species originates from the acetic acid during the oxidation process involving molecular O_2. The rate-limiting step was found to be the C–O bond scission resulting in an apparent overall activation energy of 1.72 eV as determined from DFT calculations. This is in very good agreement with the experimentally measured apparent activation energy of 1.7 ± 0.2 eV. A deuterium kinetic isotope effect of $\sim$4 indicates that C–H bond activation is kinetically involved in the overall acetate oxidation reaction.

I Introduction

The interfacial bonding between the metal and its support in heterogeneous catalysis often controls the dispersion, morphology and stability of the metal particles that form on the support. These structural and size effects can subsequently influence the catalytic behavior of these materials. In addition, the metal/support interface can also offer unique sites or alter the electronic properties of the metal that can significantly influence its catalytic performance. The special properties of the interface between metal nanoparticles and the support have been proven to be quite important in controlling the catalytic behavior of supported Au nanoparticles for

[a]*Department of Chemistry, University of Virginia, Charlottesville, VA 22904, USA. E-mail: mn4@virginia.edu; mn4@virginia.edu*
[b]*Department of Chemical Engineering, University of Virginia, Charlottesville, VA 22904, USA*

a range of different catalytic systems, including CO oxidation,[1,2] olefin epoxidation,[3,4] water gas shift catalysis,[5] and NO oxidation[6] just to name a few. There has been an on-going debate as to the factors that control the unique catalytic activity of supported Au catalysts which include the electronic properties of nanometer Au clusters,[7] the charge state of Au,[8–11] the presence of corner and edge sites[12–14] and interfacial metal-support sites.[1,2,5,15–18] Much of the recent spectroscopic and theoretical results suggest that the enhanced reactivity is the result of sites that reside at the metal/support interface.[1,2,15–19]

Most of the prior work has focused on the oxidation of small molecules such as CO[2,20–22] and H_2,[15] and the epoxidation of propene[4,23] over gold nanoparticles supported on TiO_2 where the initial adsorption and activation of O–O bond is thought to occur at dual Au–Ti^{4+} sites that reside at the Au and TiO_2 interface. Herein we extend these ideas and show that these same interfacial sites can also carry out the activation of C–H and C–O bonds, leading to the selective de-oxygenation of organic acids. In addition to the interesting science, the results may be of great importance in understanding and controlling the selective catalytic de-oxygenation pathways of biomass-derived organic acids and esters. Despite the importance of this chemistry, very little is known about the elementary surface transformations that control the catalytic de-oxygenation of organic acids.

In the present work we use *in situ* reaction studies by Fourier Transform Infrared spectroscopy (FTIR) together with first-principle density functional theory (DFT) calculations to examine the partial oxidation and de-oxygenation of acetic acid over supported Au particles. In a short initial communication, we reported spectroscopic evidence for the partial oxidation of acetic acid at the dual perimeter sites of Au/TiO_2 to form a gold ketenylidene (Au_2CCO) species.[24] Acetic acid dissociatively adsorbs on the TiO_2 support to form acetate (CH_3COO/TiO_2), which is subsequently oxidized to form the ketenylidene species (C=C=O) that is bound to the Au nanoparticle rather than the TiO_2 support. The change in adsorption sites during reaction indicates that the active sites for this partial oxidation correspond to the Au–Ti^{4+} dual perimeter sites which surround the ~3 nm Au nanoparticles on TiO_2. Herein, we combine detailed *in situ* infrared studies, isotopic labeling and kinetic isotope effect studies, and first-principle DFT simulations to investigate the reaction kinetics and to provide insights into the acetic acid oxidation mechanism, which includes C–O bond dissociation along with three C–H bond activation steps at the Au/TiO_2 dual perimeter sites. The C–H bond activation steps appear to be enhanced by the presence of atomic O species that form at the perimeter sites.

II Experimental and theoretical methods

A Experimental methods

The high vacuum transmission FTIR cell used in the present studies was described in detail earlier.[2,15] A stainless steel vacuum cell with two differentially-pumped KBr windows for transmission IR measurements is employed. The cell is pumped with a 55 L s^{-1} turbomolecular pump to maintain a base pressure of $<10^{-8}$ Torr. A reentrant Dewar with vertical motion (*z*-direction), and with a pair of Ni clamps attached, was used to hold a tungsten mesh in the center of the cell, where the IR beam was focused. The tungsten mesh by itself allows 70–80% of the IR beam to be transmitted. Catalysts were pressed into the tungsten mesh as 7 mm diameter circular spots. The mesh can fit two samples vertically (*z*-direction) at one time; thus a Au/TiO_2 catalyst and a TiO_2 blank sample were monitored together within the same experiment by moving the manipulator in the *z*-direction. A reference IR spectrum of the empty tungsten mesh and the gas phase was taken in each experiment and was subtracted from the IR spectra taken through the catalysts to eliminate the influence from the gas phase and the mesh. A type K thermocouple was spot-welded on the top center of the tungsten mesh, providing 0.1 K temperature resolution. Heating and cooling of the samples were carried out by a combination of liquid N_2 cooling and resistive heating through the tungsten mesh.

The TiO_2 blank sample, Aeroxide TiO_2 P25 with ~70% anatase and ~30% rutile, was donated by Evonik Degussa Corporation. The Au/TiO_2 catalyst was synthesized *via* the deposition precipitation (DP) method with TiO_2 P25, $HAuCl_4 \cdot 3H_2O$, and urea, following the protocol of Zanella *et al.*[25] The DP process was carried out in the dark, at 353 K, under vigorous stirring, for 8 hours. After separation from the liquid phase by centrifuging, the precipitate was washed 5 times, each time with 100 mL of double-deionized water at 323 K, to remove Cl^- ion residue. Final drying and reduction of the catalyst was done in a 100 mL h^{-1} air flow at 373 K for 24 h. This process produced a Au/TiO_2 catalyst with a Au particle average diameter of ~3 nm and a 8 wt% maximum loading.[2,25] No residual chloride was detected by Auger spectroscopy.

To test the necessity of the Au–Ti^{4+} dual perimeter sites for the catalytic reaction, we also examined the reaction over a Au/SiO_2 sample with ~3 nm Au particle diameter. The Au/SiO_2 sample was generously provided by Professor R. Zanella from Universidad Nacional Autonoma de Mexico (UNAM), following the synthesis protocol described elsewhere.[26] Experiments related to the Au/SiO_2 catalyst were done separately from the Au/TiO_2 and TiO_2 catalysts, with similar pre-treatment and experimental parameters.

Before each use, the catalysts were treated at 473 K in vacuum for 30 min, in 18 Torr of O_2 for 210 min, followed by evacuation for 10 min. This treatment removes most of the IR-visible hydrocarbon impurities accumulated on the catalysts' surface during synthesis and assembly. Complete removal of the hydroxyl species requires more aggressive annealing at elevated temperatures (above 673 K for TiO_2[27,28] and well above 1000 K for SiO_2[29]). Such high temperature procedures were not performed to avoid Au particle aggregation. This treatment can also remove acetate from the Au/TiO_2 catalyst surface, but not for TiO_2 or Au/SiO_2, as will be discussed in Section IIIC.

Acetic acid (CH_3COOH, 99.9%) was purchased from Fisher Scientific. Isotope labeled acetic acid species, acetic acid-1-^{13}C ($CH_3{}^{13}COOH$, 99 atom% ^{13}C) and acetic acid-D_4 (CD_3COOD, 99.9 atom% D), were purchased from Sigma-Aldrich. Freeze–pump–thaw cycles were performed before use on all three kinds of acetic acid. Urea (99.5%) and $HAuCl_4 \cdot 3H_2O$ were purchased from Acros. Research purity $^{16}O_2$ gas (99.998%) was purchased from Matheson TRI-GAS. $^{18}O_2$ gas (99% isotopic purity) was purchased from Isotec, Inc.

The IR spectra were taken with a Bruker Tensor 27 FTIR spectrometer. An external liquid N_2 cooled MCT detector was employed. The entire IR optical system pathway was purged with H_2O- and CO_2-free air. Each spectrum is taken with 2 cm^{-1} resolution and an average of 128 interferograms were acquired in 52 s.

B Computational methods

In order to provide insights into the elementary steps and catalytic sites involved in the activation of acetic acid and its partial oxidation, we carried out first-principle periodic DFT calculations on a model Au/TiO_2 interface. A 3-atom-wide and 3 atomic layer high Au nanorod structure was employed with the edge Au atoms on top of the bridge oxygen of a rutile TiO_2(110) slab in order to simulate the ~3 nm diameter Au particles supported by TiO_2 used experimentally. This model comprises a closed packed 3D Au structure with Au surface sites of different coordination numbers (CN) and has been used successfully to model many catalytic reactions on the Au/TiO_2 system.[2,15–17,30]

The rutile TiO_2 (110) surface was modeled using (3 × 3) or (2 × 3) unit cells. Both unit cells have four O–Ti–O tri-layers with a total thickness of ~12.6 Å. The large (3 × 3) unit cell was used to calculate the binding of acetic acid at the perimeter of the Au/TiO_2 nanorod model surface, while all other simulations reported were carried out using the (2 × 3) unit cell. For both TiO_2 unit cells, the bottom two tri-layers were constrained to their lattice positions while the top two tri-layers were allowed to fully relax during the geometries optimization. All of the atoms in the Au nanorod were allowed to relax in the Z-direction to maintain lattice matching with the support. Allowing for the optimization of the Au atoms in the X and Y directions resulted in changes in reaction and activation energies of less than 0.1 eV.

All calculations reported herein were carried out using gradient corrected periodic plane wave DFT calculations as implemented in the Vienna *ab initio* Simulation Package (VASP).[31] Valence electrons were described with Kohn–Sham single-electron wave functions and expanded in a plane wave basis with an energy cutoff of 400 eV. The exchange-correlation energy was evaluated within the generalized gradient approximation using the PW91 functional.[32] The frozen core charges were represented with pseudopotentials constructed with the projector augmented-wave (PAW) method.[33,34] The DFT + U method was applied to correct the on-site Coulomb interactions in the oxide system.[35] The U value was chosen to be 4.0 eV to produce the band gap features observed experimentally.[36] Spin-polarization was considered for all calculations and was used when necessary. A vacuum gap of 10 Å was used in the z-direction between slabs. The Brillouin zone was sampled with (1 × 2 × 1) k-point mesh for the (3 × 3) unit cell and a (2 × 2 × 1) mesh for the (2 × 3) unit cell.[37] The electronic structures were converged self-consistently to within a tolerance of 1.0×10^{-5} eV. Geometries were considered as optimized when the force on each atom was less than 0.03 eV $Å^{-1}$.

The reaction paths and activation barriers were determined by using the climbing image nudged elastic band (NEB) method,[38,39] combined with the dimer method.[40] An eight-image NEB band was initially optimized along the minimal reaction path connecting the reactant and product states to the point where the forces normal to the reaction path on all of the images along the band were lower than 0.1 eV $Å^{-1}$. The transition state was subsequently isolated by the dimer method to the point where the force acting on the transition state dimer was lower than 0.03 eV $Å^{-1}$.

III Results and discussion

A FTIR study of acetic acid adsorption on different catalyst surfaces

Fig. 1 shows the IR spectra of the three catalyst surfaces of interest, namely (a) TiO_2, (b) Au/TiO_2, and (c) Au/SiO_2, interacting with acetic acid at 400 K. Black spectra show the catalyst surfaces in vacuum after pre-treatment, before acetic acid exposure; red spectra show the catalysts in contact with 0.5 Torr of acetic acid gas, followed by evacuation for 30 min at 400 K. The spectral features of acetic acid are similar comparing TiO_2 and Au/TiO_2, but very different on the Au/SiO_2 surface. These observations indicate that acetic acid absorbs as acetate species on TiO_2 and SiO_2 differently, and that Au nanoparticles result in very little contribution to the spectral features on Au/TiO_2.[41] Fig. 1d shows the gas phase IR spectrum of acetic acid (0.5 Torr) at 400 K.

When adsorbed at 400 K, acetic acid forms surface acetate species on all of the exposed TiO_2 surfaces.[42–50] According to other IR reports on acetate adsorption on TiO_2,[42–46] the main features labeled on Fig. 1a and b are assigned as follows: 2936 cm^{-1}, symmetric CH_3 stretching ($\nu_s(CH_3)$); 1532 cm^{-1}, asymmetric –COO stretching ($\nu_{as}(CO_2)$); 1454 cm^{-1}, symmetric –COO stretching ($\nu_s(CO_2)$); 1400 cm^{-1}, 1338 cm^{-1}, and 1324 cm^{-1}, CH_3 bending ($\delta(CH_3)$). The frequency difference between symmetric and asymmetric $\nu(CO_2)$ is 78 cm^{-1}, indicating the acetate is adsorbed on the TiO_2 surface in a symmetric bidentate orientation, as shown for chelating and bridging acetate species in the inset in Fig. 1a.[42–44] Clearly distinguishing between the chelating structure and the bridging structure by IR vibrational frequencies is difficult and debated in the literature.[42–44] DFT results for acetate adsorption on the TiO_2 rutile (110) surface (this work, Section IIIE), the anatase

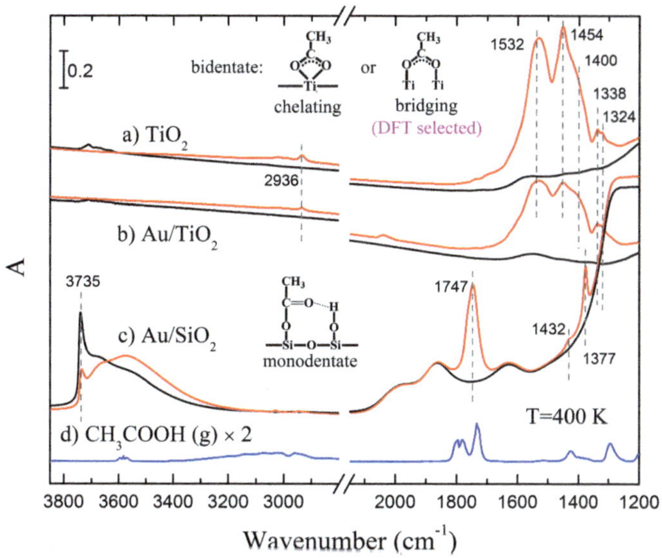

Fig. 1 IR spectra of acetic acid adsorption at 400 K on (a) TiO_2; (b) Au/TiO_2; and (c) Au/SiO_2. Black spectra—clean catalysts surface; red spetra—0.5 Torr acetic acid followed by evacuation for 30 min. Spectrum (d) shows 0.5 Torr gas phase acetic acid IR adsorption.

(010) surface,[51] and the anatase (101) surface,[52] however, all indicate that the bridging configuration is more stable than the chelating structure.

Acetic acid adsorbs on the SiO_2 surface as acetate species at 400 K as well.[47–50] However, the acetate/SiO_2 IR features are quite different from the acetate/TiO_2 features in the $\nu(CO_2)$ region: three bands at 1747 cm^{-1}, 1432 cm^{-1}, and 1377 cm^{-1} are observed. Young previously reported a careful IR comparison of the adsorption of acetic acid, acetyl chloride, and acetic anhydride on SiO_2 surfaces at 300 K and 423 K, and found the three bands described above in all cases.[49] He suggested that acetate adsorbs on SiO_2 *via* a monodentate configuration, where the 1747 cm^{-1}, 1432 cm^{-1}, and 1377 cm^{-1} bands corresponding to C=O stretching (ν(CO)), asymmetrical CH_3 bending ($\delta_{as}(CH_3)$), and symmetrical CH_3 bending ($\delta_s(CH_3)$), respectively. The higher frequency bands at 3030 cm^{-1} and 2946 cm^{-1} were identified as $\nu_{as}(CH_3)$ and $\nu_s(CH_3)$, respectively. The highest frequency band at 3735 cm^{-1}, which is ascribed to isolated SiO–H stretching on the SiO_2 surface, is heavily shifted and broadened to lower frequency by acetate adsorption, as strong hydrogen bonds form with the carbonyl moiety. One possible bidentate adsorption configuration of acetate on the SiO_2 surface is given in the inset to Fig. 1c.

B DFT calculated reaction paths for acetate formation and the most stable configuration of acetate on the Au/TiO_2 catalyst

Since the IR bands observed for acetate/TiO_2 species cannot discriminate between chelating and bridging acetate adsorption configurations (Fig. 1), DFT calculations were carried out to help isolate the most favorable site and structure for the acetic acid adsorption and acetate formation on the Au/TiO_2 catalyst. The calculated frequency difference between symmetric and asymmetric $\nu(CO_2)$ was found to be 91 cm^{-1}, which is consistent with the experimental value and within the accuracy of the method, allowing the acetate/TiO_2 structure to be assigned as a symmetrical bridging structure. The calculated results for adsorption of acetic acid on Au, indicate that it does not bind to the Au particle, even at the coordinatively unsaturated edge sites on Au with a coordination number (CN) of 7 (Fig. 2a). Gas phase acetic

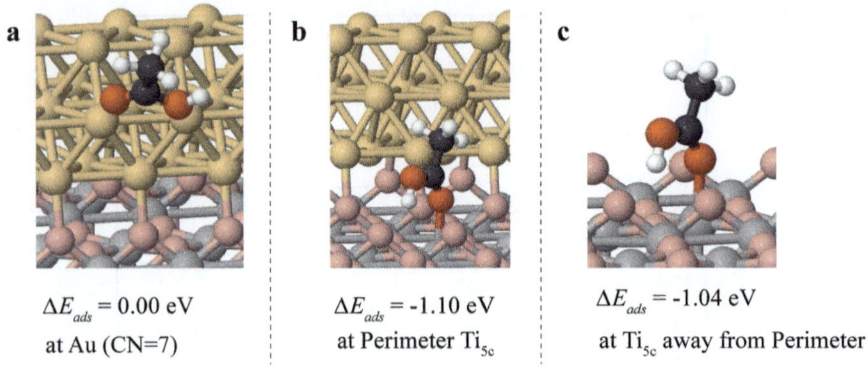

Fig. 2 Acetic acid adsorption at different sites. The Au atoms and Ti atoms are shown in yellow and grey respectively, whereas the O in the TiO_2 lattice, adsorbed O, C atoms and H atoms are shown in pink, red, black and white respectively. ΔE_{ads} refers to the adsorption energy.

acid molecules that approach the Au nanoparticle would need to overcome a 0.95 eV barrier in order to deprotonate the C(O)O–H group to form acetate, with an overall endothermicity of 0.04 eV, as shown in Fig. 3a. The lack of adsorption of acetic acid on Au and the high barrier required to activate it over Au to form acetate/Au exclude Au as the active surface for adsorption or reaction. This agrees well with the experimental observation shown in Fig. 1.

Unlike on Au nanoparticles, acetic acid binds strongly to the Ti_{5c} sites of the TiO_2 support, with adsorption energies ranging from −1.04 eV to −1.10 eV (Fig. 2b–c). It binds to the TiO_2 surface *via* a strong interaction between the negatively charged O of the carboxylate group and the Ti_{5c} sites together with a strong hydrogen bond between the proton of the acid and the bridging lattice oxygen on the surface of the TiO_2 support, as shown in Fig. 2b–c. The de-protonation of adsorbed acetic acid to form acetate on TiO_2 proceeds quite easily with a small barrier of only 0.09 eV and an exothermic heat of reaction at −1.05 eV, as shown in Fig. 3b.

We also compared the adsorption energy of the chelating and the bridging bidentate configuration of acetate/TiO_2 shown in Fig. 1 and found that the bridging

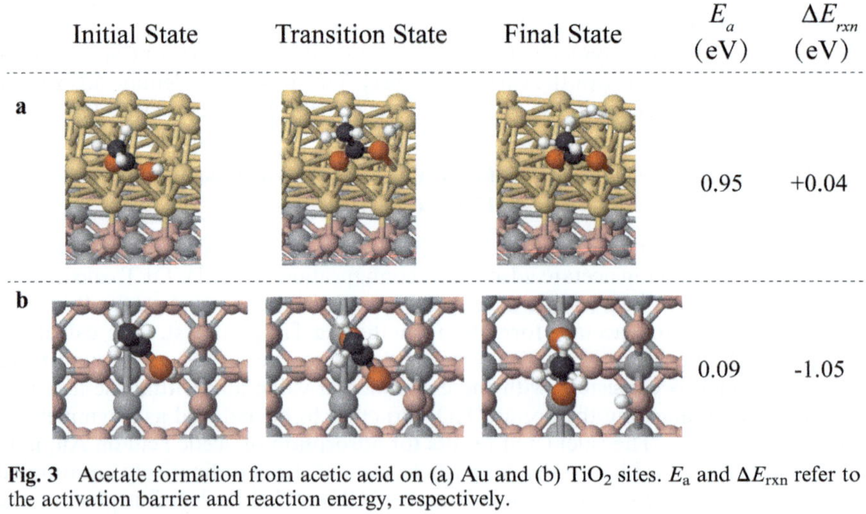

Fig. 3 Acetate formation from acetic acid on (a) Au and (b) TiO_2 sites. E_a and ΔE_{rxn} refer to the activation barrier and reaction energy, respectively.

configuration (Fig. 3b) is ~1.0 eV more stable than the chelating bidentate configuration. A recent STM study and DFT calculations of acetic acid adsorption on the anatase $TiO_2(101)$ surface also found the bridging configuration to be the stable adsorption state.[52] We therefore use the bridging bidentate configuration to simulate acetate/TiO_2.

C FTIR study of the interaction between adsorbed acetate and oxygen on different catalyst surfaces

The acetate-saturated surfaces shown in Fig. 1 were exposed to 1 Torr of O_2 at 400 K and monitored for 90 min under steady state. Fig. 4a shows the results for the oxidation of the acetate-saturated TiO_2 sample, where no significant changes occur in the IR spectra after 90 min. In contrast, Fig. 4b shows the development on the Au/TiO_2 catalyst, where the formation of the gold ketenylidene species (Au_2CCO) is observed by IR at 2040 cm^{-1}.[24] A small amount of Au_2CCO was already formed after acetic acid adsorption in Fig. 1b, showing that the high temperature O_2 pre-treatment (described in Section IIA) left a trace amount of adsorbed O species on the catalyst surface which is reactive at 400 K.[24] Little change in the absorbance of acetate features upon Au_2CCO formation was observed on the Au/TiO_2 sample in Fig. 4b, indicating that the partial oxidation of acetate, forming Au_2CCO, is a minor reaction causing the majority of the acetate species to remain unaffected.

The interaction between acetate/SiO_2 and 1 Torr of O_2 at 400 K was found to be minimal at 400 K, as shown in Fig. 4c. The negative results on TiO_2 and Au/SiO_2 indicate that the partial oxidation of acetate/TiO_2 to form Au_2CCO is catalyzed at the Au–Ti^{4+} perimeter sites at 400 K. Since Au/SiO_2 contains coordinatively-unsaturated Au atoms along the perimeter of the ~3 nm Au particles, the results from this experiment indicate that Au sites of this kind alone are insufficient for acetate partial oxidation and instead require the dual sites that reside at the interface of Au and TiO_2.

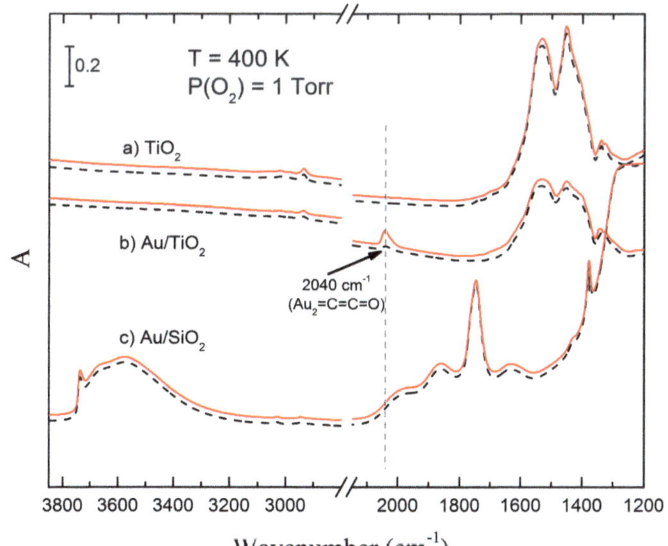

Fig. 4 IR spectra of adsorbed acetate reacting with 1 Torr of O_2 on three catalysts at 400 K: (a) on TiO_2; (b) on Au/TiO_2; and (c) on Au/SiO_2. Black dashed-line spectra—before O_2 exposure (reproduced from the red spectra of Fig. 1a); red spectra—in contact with 1 Torr of O_2 at 400 K for 90 min. The spectra have been shifted vertically by a small amount for easy comparison.

To further demonstrate the special properties of the Au–Ti^{4+} perimeter sites, a set of experiments investigating the interaction of acetate with O_2 at elevated temperatures on the three different surfaces were performed, as shown in Fig. 5. This experiment probes more extensive oxidation than experiments at 400 K. Since the Au/TiO_2 and TiO_2 catalysts were held on the same tungsten mesh, the reaction temperature was chosen to be 473 K to protect the Au/TiO_2 catalyst from sintering. As shown in Fig. 5a–b, after 210 min in 18 Torr of O_2, acetate intensity decreased only slightly on the TiO_2 sample, while most of the acetate disappeared on the Au/TiO_2 catalyst. Gas phase CO_2 was also detected in the cell, as highlighted in Fig. 5. The CO_2 is believed to be the product of the total oxidation of acetate and ketenylidene on the Au/TiO_2 catalyst. The other total oxidation product, H_2O, is not detected on the surface at 473 K, and is also not detected in the gas phase, possibly due to adsorption on the room temperature cell walls. The TiO_2 sample is held in the same experimental environment as the Au/TiO_2 catalyst, and thus gas phase CO_2 is observed above TiO_2 as well as above the Au/TiO_2 sample, as is observed in Fig. 5a.

More aggressive oxidation of acetate/SiO_2 using 11 Torr of O_2 at 550 K on the Au/SiO_2 sample is shown in Fig. 5c. Higher temperature is used here because no IR spectral change was detected at 473–500 K. At 550 K, during 60 min of reaction, most of the adsorbed acetate/SiO_2 species disappeared from the IR spectrum, as indicated by the loss of these spectral features and the reappearance of isolated SiO–H groups, but little oxidation product (no Au_2CCO and little CO_2(g)) was detected.

The more aggressive high temperature studies shown in Fig. 5 confirmed that without the Au–Ti^{4+} dual perimeter sites, no ketenylidene species is produced from acetate partial oxidation. It also demonstrated that the oxidation depth of the acetate/TiO_2 species on the Au/TiO_2 catalyst can be tuned by reaction temperature—partial oxidation at 400 K and total oxidation at 473 K.

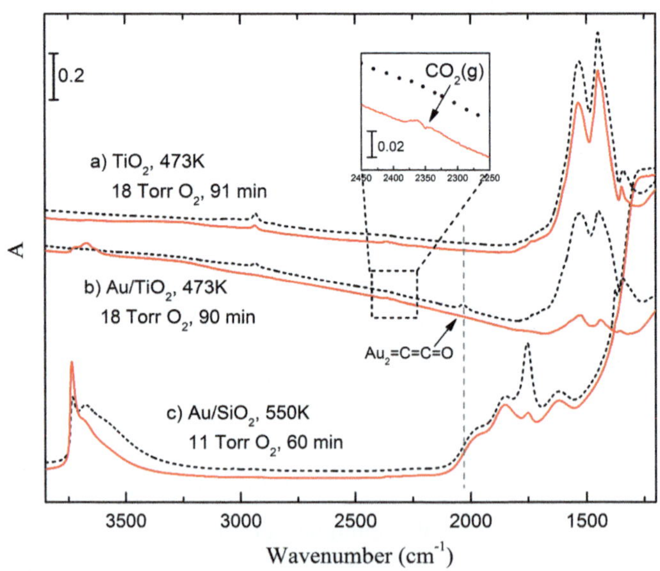

Fig. 5 IR spectra of adsorbed acetate reacting with O_2 on three catalysts at elevated temperatures: (a) on TiO_2 with 18 Torr of O_2 at 473 K; (b) on Au/TiO_2 with 18 Torr of O_2 at 473 K; and (c) on Au/SiO_2 with 11 Torr of O_2 at 550 K. Black dashed-line spectra—upon just reaching target temperature, before O_2 exposure; red spectra—after reaction in O_2.

D Partial oxidation of acetate with $^{18}O_2$ isotopic species on Au/TiO_2

The oxidative formation of Au_2CCO from acetate/TiO_2 on the Au/TiO_2 catalyst requires the synergistic action of dual TiO_2–Au sites that exist at the Au/TiO_2 interface. Here the acetate reactant is adsorbed on the TiO_2 support whereas the ketenylidene product that forms is bound to the Au nanoparticles, as shown in Scheme 1. The reaction requires multiple C–H bond activation steps as well as C–O bond scission in order for the overall reaction to proceed. The partial oxidation will not proceed without oxygen, indicating that the activation of the C–H and the C–O bond is assisted by the adsorption and dissociation of O_2. To identify the origin of the O atom in the final Au_2CCO product, an experiment involving $^{18}O_2$ was carried out.

As shown in Fig. 6, during the 90 min experiment with $^{18}O_2$ gas at 400 K, no ^{18}O-containing Au_2CCO species (characteristic $^{18}O\text{–}\nu(CO) = 2010\ cm^{-1}$)[24] is detected. This helps to confirm that the O in Au_2CCO is derived from the original acetate species.

E Kinetic studies of the formation of ketenylidene

Steady state partial oxidation of acetate on the Au/TiO_2 catalyst as shown in Fig. 7 was carried out over a temperature range of 400–420 K for kinetic studies, under 1 Torr of O_2. The integrated absorbance of the Au_2CCO band centered at 2040 cm^{-1} is used as a measure of the amount of ketenylidene product accumulated on the catalyst surface. The oxidation reaction becomes significantly retarded after 90 min at all temperatures, as indicated by the sluggish rate of growth of the Au_2CCO IR band when approaching the 90 min endpoint. The inhibition at longer times is likely the result of the inhibition of active surface sites by adsorbed oxygen, hydroxyl, and ketenylidene species. Normalization of the integrated absorbance of Au_2CCO species to the maximum measured value is carried out to make experiments executed on different days more comparable. Plots of the reaction progress at five different temperatures are shown in Fig. 7. The five experiments at various temperatures were carried out randomly in sequence.

Since the catalyst surface is saturated with acetate before reaction and remains nearly saturated during reaction, and since oxygen is abundant from the gas phase, the reaction is pseudo-first order in active site concentration located at the Au–Ti^{4+} perimeter, as shown in eqn (1).

$$\text{Rate} = \frac{d[Au_2CCO]}{dt} = k \times [\text{acetate}] \times [O_2] \times [\text{Active Site}] = k' \times [\text{Active Site}] \quad (1)$$

The solid lines under experimental data points shown in Fig. 7 display a mathematical fit to first-order kinetics. The normalization described earlier does not of course alter the reaction rate constant, k, obtained from the mathematical fit. The Arrhenius plot is constructed as shown in the inset of Fig. 7, yielding an apparent activation energy of 1.7 ± 0.2 eV.

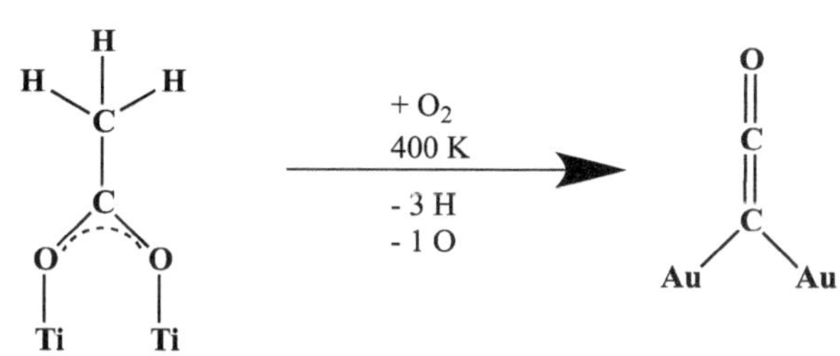

Scheme 1

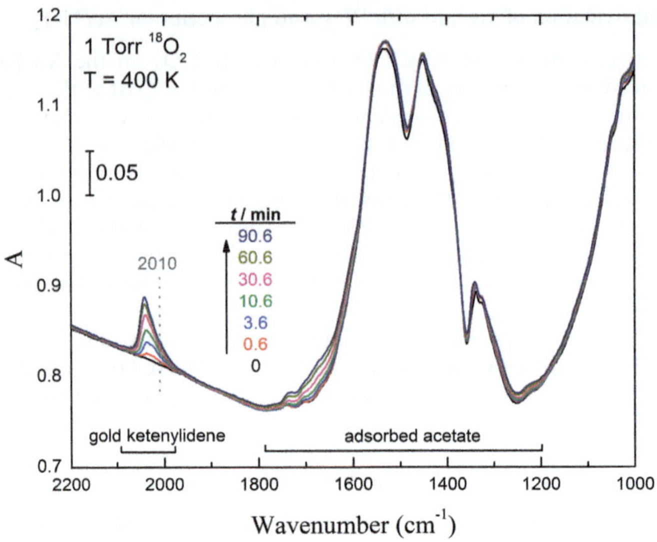

Fig. 6 IR spectra of the partial oxidation of acetate by $^{18}O_2$ on Au/TiO_2. Dotted line indicates the absorption frequency of $Au_2CC^{18}O$, which is not observed.

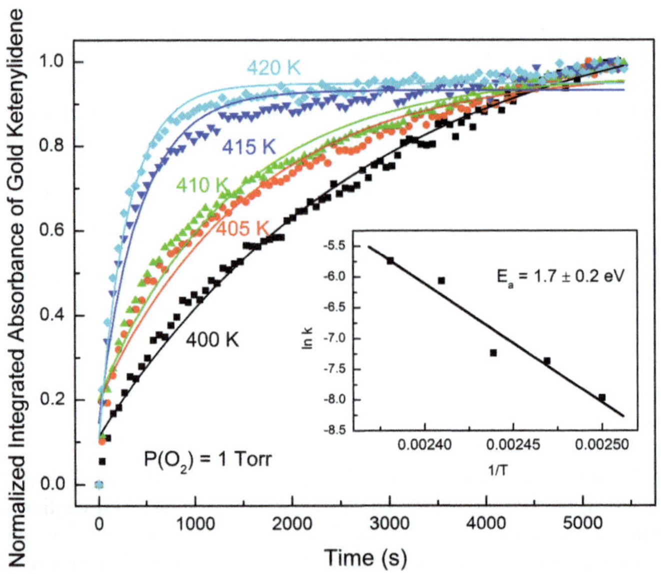

Fig. 7 Kinetic studies of the partial oxidation of acetate to ketenylidene on the Au/TiO_2 catalyst. Solid curves underneath experimental data points are mathematical fits to first-order kinetics. Inset shows the Arrhenius plot.

F Deuterium kinetic isotope effect (DKIE) in the partial oxidation of acetate species

As shown in Scheme 1, the partial oxidation of acetate/TiO_2 to form the observed Au_2CCO intermediate involves multiple C–H bond scission steps along with a single C–O bond scission step. In order to elucidate the nature of the transition state and follow the kinetics for the partial oxidation of acetate a DKIE study was designed, where a 1-to-1-ratio acetic acid-H_4 and acetic acid-D_4 mixture was pre-made in the gas-line and exposed to the Au/TiO_2 catalyst in a competitive experiment. ^{13}C

labeling of the reactant acetic acid-H_4 species ($CH_3{}^{13}COOH$) was carried out to determine whether the H- or the D-labeled acetate was responsible for the formation of the hydrogen-free gold ketenylidene product. As a result, observation of the formation of non-labeled Au_2CCO species at 2040 cm^{-1} allows the measurement of the rate of CD_3COO partial oxidation, while observation of the ^{13}C-isotopic labeled $Au_2C^{13}CO$ species at lower frequency (~1990 cm^{-1}) allows the rate of $CH_3{}^{13}COO$ partial oxidation to be simultaneously measured.

Fig. 8a shows the IR spectra development in the gold ketenylidene region during the DKIE study. As expected, the partial oxidation of the deuterium-labeled acetate proceeded at a slower rate than that of the non-labeled species as seen in the ketenylidene spectra shown in Fig. 8a. Fig. 8b shows a plot of the integrated absorbance of the two gold ketenylidene species as a function of reaction time. As shown in eqn (1) and Fig. 7, the reaction is pseudo-first order in the Au–Ti^{4+} dual perimeter site concentration, which decreases as the reaction moves forward. The rate of formation of the two ketenylidene species can be represented by eqn (2) and (3):

$$\frac{d[Au_2C^{13}CO]}{dt} = k_H \times [CH_3{}^{13}COO] \times [O_2] \times [\text{Active Site}] \tag{2}$$

$$\frac{d[Au_2CCO]}{dt} = k_D \times [CD_3COO] \times [O_2] \times [\text{Active Site}] \tag{3}$$

Since the concentration of the two acetate species, the O_2 pressure, and the concentration of the active sites are the same, dividing eqn (2) by eqn (3) yields eqn (4):

$$\frac{k_H}{k_D} = \frac{d[Au_2C^{13}CO]/dt}{d[Au_2CCO]/dt} \tag{4}$$

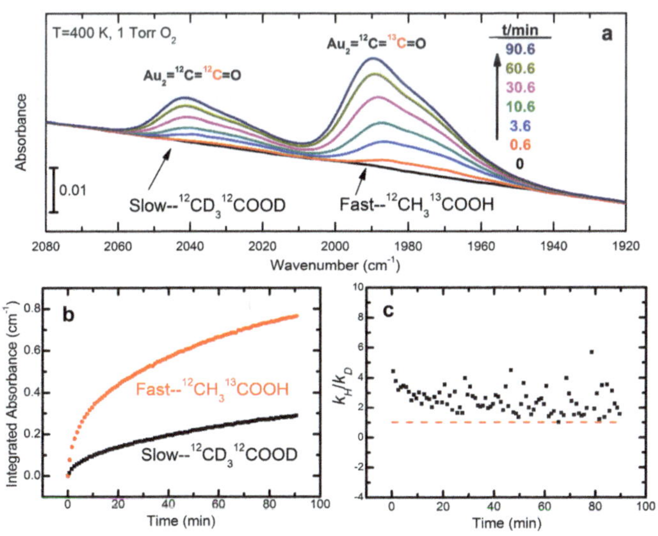

Fig. 8 Deuterium kinetic isotope study. (a) Partial oxidation of 1 : 1 ratio of adsorbed $CH_3{}^{13}COO/TiO_2$ and CD_3COO/TiO_2 at 400 K; (b) plots of the integrated absorbance of $Au_2C^{13}CO$ from acetic acid-H_4 oxidation and Au_2CCO from acetic acid-D_4 oxidation as a function of reaction time; (c) plots of k_H/k_D as a function of the reaction time and the extent of reaction.

Fig. 8c shows the DKIE, k_H/k_D, evaluated by point-by-point differentiation of the numerical data in Fig. 8b averaging for each point over three adjacent spectral scans. A linear least square fit of the DKIE data is shown in Fig. 8c. A DKIE in the range 4–2 is observed as the reaction proceeds. This observed DKIE is consistent with the results from Hutchings *et al.*, where DKIE values in the range of 4–2 were found for benzyl alcohol partial oxidation (forming benzaldehyde and toluene) over a Au–Pd/TiO_2 catalyst.[53] The observation of a DKIE indicates that C–H bond breaking in acetate/TiO_2 partial oxidation is of importance in the kinetics of formation of Au_2CCO species.

G DFT proposed mechanism of the acetate partial oxidation

Acetic acid appears to dissociate at the Ti_{5c} sites, as was reported above, to form a bidentate acetate intermediate bound to two Ti_{5c} sites and a bridging OH intermediate. We showed previously that molecular oxygen preferentially adsorbs and is activated at the Au–Ti^{4+} dual perimeter sites to form surface oxygen atoms. The barrier to activate O_2 at the Au–Ti^{4+} perimeter sites was calculated to be 0.4–0.6 eV depending upon the specific site of activation whereas the overall reaction energy is exothermic by 1.1 eV as is shown in Fig. 9a.[2] The oxygen adatoms that result can subsequently diffuse onto the Au nanoparticle with barriers that range from 0.23–0.72 eV depending upon the coordination number of the Au.[2,24] At the experimental temperature in the current study (≥400 K), it is expected that the O adatoms readily diffuse along the Au cluster and aid in the activation of the three C–H bonds of the bound acetate intermediates. It is also reasonable to assume that these O adatoms are readily formed at 1 Torr pressure which is consistent with the experimental results. The reaction which involves C–H as well as C–O activation appears only to occur in the presence of $O_2(g)$. This would suggest that the oxygen that is weakly coordinated to Au assists in the activation of the C–H bond. This is consistent with previous reports that weakly adsorbed oxygen on Au acts as a base and aids in the activation of weakly acidic C–H bonds.[54–56]

As shown in Scheme 1, the partial oxidation of acetate/TiO_2 forming Au_2CCO clearly involves both the Au surface sites and the TiO_2 support. We therefore examine a series of elementary steps that occur at the different Au/TiO_2 perimeter sites. The conversion of acetate (CH_3COO^-) to the product, ketenylidene (=C=C=O), requires a net activation of three C–H bonds and one C–O bond. After testing different sequences of steps, the most favorable (lowest energy barrier) reaction path is summarized eqn (5)–(10):

$$O_2(g) + Au^* + Ti^* \underset{k_{-1}}{\overset{k_1}{\rightleftarrows}} O^{Au} + O^{Ti} \tag{5}$$

$$H_3CCOO^{Ti} + O^{Au} \underset{k_{-2}}{\overset{k_2}{\rightleftarrows}} H_2CCOO^{TiAu} + OH^{Au} \tag{6}$$

$$H_2CCOO^{TiAu} + OH^{Au} \underset{k_{-3}}{\overset{k_3}{\rightleftarrows}} HCCOO^{TiAu} + H_2O^{Au} \tag{7}$$

$$H_2O^{Au} \xrightarrow{k_4} H_2O(g) + Au^* \tag{8}$$

$$HCCOO^{TiAu} + Au^* \xrightarrow{k_5} HCCO^{Au} + O^{Ti} \tag{9}$$

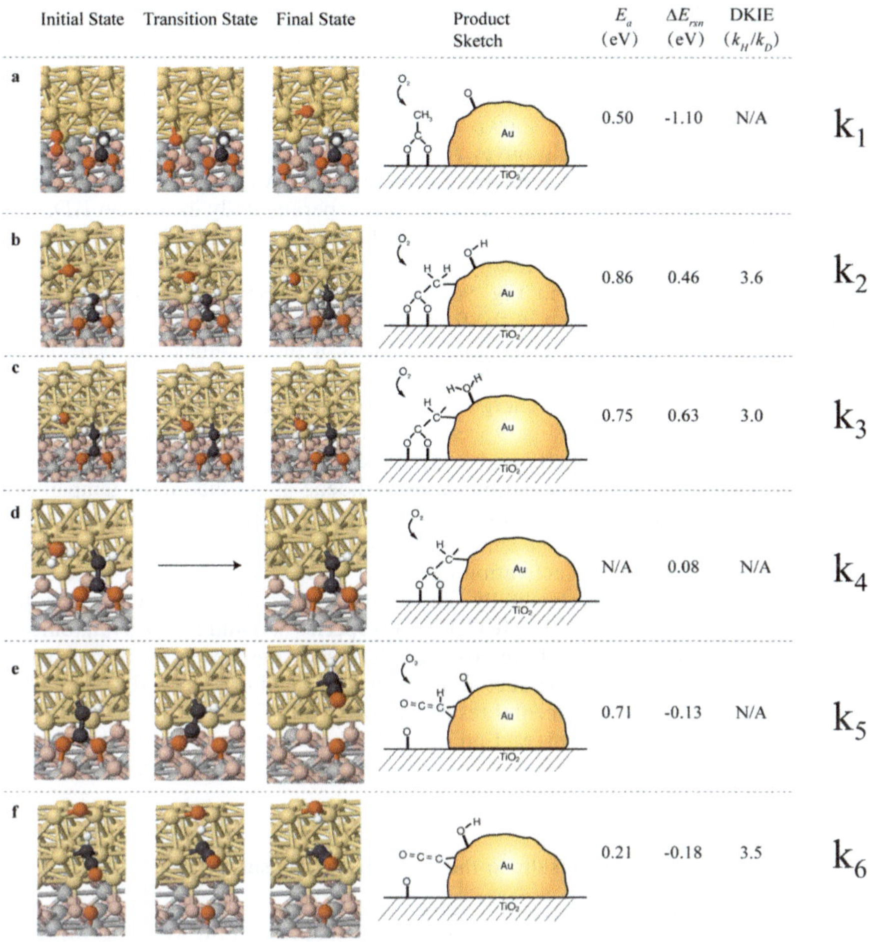

Fig. 9 Key steps for the ketenylidene formation at perimeter. a–d corresponding to eqn (6)–(9).

$$HCCO^{Au} + O^{Au} \xrightarrow{k_6} CCO^{Au} + OH^{Au} \quad (10)$$

where the superscripts Au, Ti and TiAu refer to the specific adsorption sites at the Au/TiO2 interface.

The detailed structures of the initial, transition and final states for each of these steps are shown in Fig. 9. The methyl moiety of the acetate/TiO_2 species that forms upon the de-protonation of bound acetic acid subsequently undergoes two sequential C–H bond activation steps by weakly coordinated oxygen atoms bound to Au with barriers of 0.86 and 0.75 eV and reaction energies of 0.46 and 0.63 eV, respectively as shown in Fig. 9b–c. The resulting H_2CCOO^{TiAu} intermediate that forms remains bound to the TiO_2 support. The subsequent activation of the second C–H bond, however, results in the formation of an unsaturated and unstable vinylic (CH) center which strongly binds to a neighboring Au site in order to stabilize the resulting $HCCOO^{TiAu}$ complex as is shown in Fig. 9c. These two C–H activation processes proceed *via* hydrogen abstraction from adsorbed oxygen and hydroxyl intermediates that are weakly bound to Au and go on to form H_2O which readily desorbs from Au as its binding energy is only −0.08 eV (Fig. 9d). After the activation of the two C–H bonds, the C–O bond of the carboxylate moiety breaks with an activation barrier of 0.7 eV, resulting in the formation of Au_2CHCO which is completely detached from the TiO_2 (Fig. 9e). The last C–H bond is ultimately activated by a weakly bound O on Au thus resulting in the formation of Au_2CCO and a weak Au-hydroxyl intermediate as shown in Fig. 9f. The final Au_2CCO configuration shown in Fig. 9f theoretically gives ν(CO) of 2041 cm^{-1}, consistent with the IR measurement (2040 cm^{-1}).[24] This reaction pathway to Au_2CCO also retains the O moiety from the original acetate ion, as shown experimentally in Section IID.

While the first dehydrogenation step has the highest intrinsic barrier (0.86 eV, Fig. 9b), the highest overall energy in the sequence of steps involves the C–O bond breaking step shown in Fig. 9e. The two C–H bond activation steps that precede C–O activation result in the formation of water which irreversibly desorbs from the surface (k_4) and readsorbs on the walls of the chamber. As such, the C–H activation steps are irreversible and non-equilibrated. The DFT-calculated adsorption energies, reaction energies and activation barriers for all of the steps considered in eqn 5–10 and Fig. 9 were used to construct a microkinetic model in order to simulate the chemistry and understand the kinetics. The simulations were used to model the rate of formation of ketenylidene as a function of temperature to determine the apparent activation barrier and compare with the experimental results presented in Fig. 7. The simulation results for the ln(rate) *vs* $1/T$ shown in Fig. 10 reveal an apparent activation energy of 1.72 eV which is very close to the value of 1.7 eV found experimentally. The results indicate that the rate is controlled by the C–O bond scission of the $HCCOO/TiO_2$ species (k_5) and by the availability of active sites. As the reaction proceeds the interfacial sites become covered by the hydroxyl and ketenylidene product along with unreacted oxygen and acetate. This is consistent with the experimental results shown in Fig. 7 which show a significant decrease in the rate at longer reaction times. High temperatures aid in the intrinsic activation of the C–O bond as well as in the generation of vacant sites by the associative desorption of oxygen.

H Connecting the DKIE results to the reaction mechanism

There are 3 C–H bond activation steps, involving the H transfer between the terminal methyl group of the acetate and adsorbed O or OH on Au in the proposed

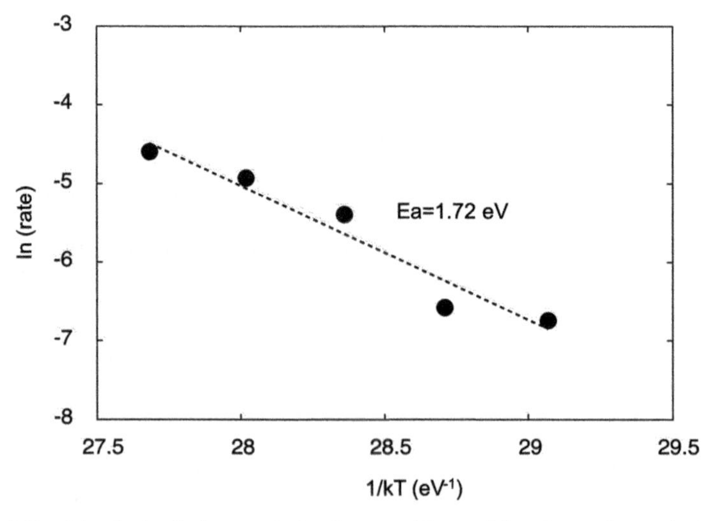

Fig. 10 Microkinetic simulation results for the rate of ketenylidene formation at the interfacial sites of Au/TiO_2 as a function of temperature. The calculated apparent activation energy was calculated here to be 1.72 eV.

reaction energy path presented in Fig. 9). Each of these steps has a calculated DKIE of 3–4 if they are considered individually and not coupled with other elementary steps. The last C–H activation step, labeled as k_6 in Fig. 9, takes place after the rate-limiting step, and hence does not contribute to the DKIE observed in the experiment. On the other hand, the first two C–H activation steps, which are labeled k_2 and k_3 in Fig. 9, can potentially contribute to the DKIE by resulting in the buildup of the more H-labeled intermediates than the D-labeled intermediates before the rate limiting step.

Both the C–H activation processes (reaction 2 and 3) are endothermic with low reverse activation barriers ($E_{-2}^* \sim 0.35$ eV and $E_{-3}^* \sim 0.15$ eV). If these steps were quasi-equilibrated the factor of 3–4 increase in switching from deuterium to hydrogen labeled acetic acid to form CH_2COO and CHCOO in k_2 and k_3, respectively, would cancel out, as the reverse rate constants k_{-2} and k_{-3} would be 3–4 times faster for the H-labeled intermediates over the D-labeled species and no significant DKIE would be observed. The postulated irreversible desorption of water in step k_4, however, makes the first two C–H activation steps irreversible thus generating an excess of the H-labeled CHCOO surface intermediate.

The microkinetic simulations reported above were also used to simulate the effects of the deuterium label on the surface concentrations for all of the oxygenate intermediates as well as on the overall rate. From the results, which are shown in Fig. 11, the first C–H activation step does not appear to accumulate more H-intermediate than the D-intermediate (Fig. 11b), indicating that this

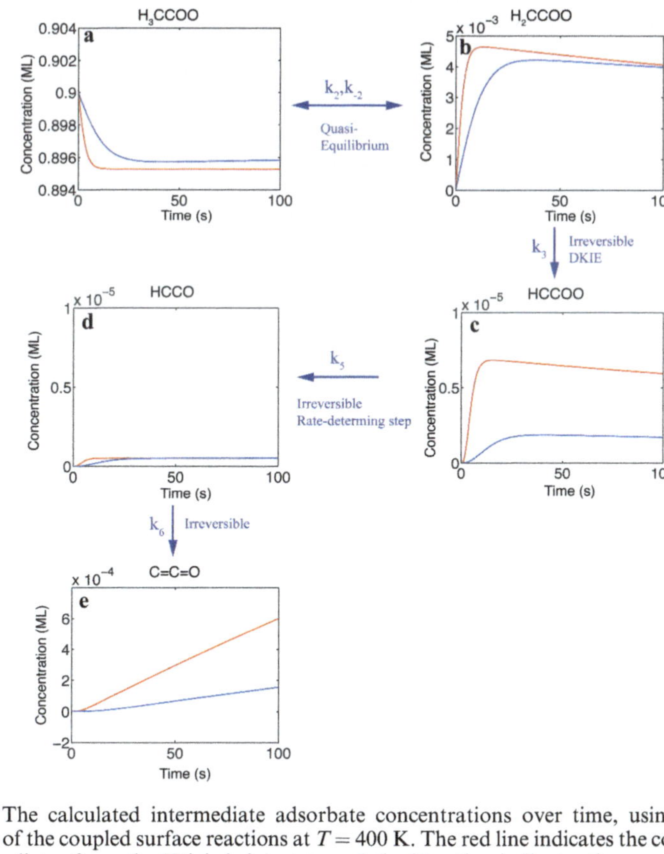

Fig. 11 The calculated intermediate adsorbate concentrations over time, using numerical solutions of the coupled surface reactions at $T = 400$ K. The red line indicates the concentration of intermediates from the H-labeled acetate and the blue line indicates the concentration of intermediates from the D-labeled acetate.

reaction is quasi-equilibrated. However, the k_3 step generates H_2O that readily desorbs with an energy of only 0.08 eV, thus eliminating the back reaction k_{-3}. This results in a buildup of HCCOO over DCCOO on the surface as shown in Fig. 11c. Adsorption of H_2O on the walls of the stainless steel chamber would act to remove H_2O from the catalyst surface.

The HCCOO and DCCOO intermediates that form subsequently react in the rate-limiting step that does not involve a DKIE to form the HCCO and DCCO intermediates which are quickly converted to the final ketenylidene products in the last rapid C–H activation step (Fig. 11e). The last C–H activation step is after the rate-limiting step and therefore its DKIE does not affect the total DKIE. The isotopic ratio of the final ketenylidene concentrations is therefore controlled by the intermediate concentrations that enter the rate-limiting step, k_5. As the concentration of the HCCOO intermediate is ~4 times higher than the DCCOO intermediate, the rate to generate ketenylidene from the HCCOO is ~4 times faster than the rate from the DCCOO, yielding an overall initial DKIE ~4 which is in good agreement with the calculations shown in Fig. 11c. This analysis confirms that the observed DKIE is the result of the C–H activation step that does not have a significant back reaction, step k_3.

I Energy barriers for ketenylidene formation on the TiO_2 surface

The results presented in section IIIA indicate that the initial exposure to acetic acid at the conditions of the experiment leads to a Au/TiO_2 catalyst that is covered with acetate and oxygen which bind strongly to the surface TiO_2 sites. As such, we also examined the possibility of ketenylidene formation on the TiO_2 support away from the Au–Ti^{4+} perimeter, following the same mechanism depicted in eqn (6)–(10) and shown as the red pathway in Fig. 10. The reaction starts from the di–σ bound acetate/TiO_2 configuration shown in Fig. 3b. Due to the upright adsorption configuration of the acetate along the Ti rows, the interaction and activation of the C–H bonds by oxygen atoms within the surface plane that connect to vicinal Ti_{5c} sites are excluded. C–H bond scission occurs instead on the $TiO_2(110)$ surface at the bridging lattice oxygen atoms that are outwardly-directed from the surface and can directly interact with the C–H bonds. The first three steps along the reaction path presented in eqn (6)–(8) were all calculated to have high activation barriers (>0.8 eV). The reactant, transition state and product structures along with their corresponding activation energies and overall reaction energies are shown in Fig. 12. The dehydrogenation steps shown in Fig. 12a–b generate very unstable $-CH_2$ and $-CH$ moieties, since there are no vicinal Au atoms to aid in their stabilization. As a result, high activation barriers of 1.19 eV and 2.28 eV are obtained consistent with the Brønsted–Evans–Polanyi relationship.[57,58] The subsequent activation of the C–O bond of the bound HCCOO intermediate results in the formation of HCCO and an O adatom adsorbed on the Ti_{5c} site of TiO_2 (Fig. 12c). The HCCO intermediate that forms is rather unstable and thus interacts with a vicinal Ti_{5c} site to partially stabilize its adsorption thus making the reaction somewhat exothermic ($\Delta E_{rxn} = -0.57$ eV). The stabilization of the HCCO intermediate at the Ti_{5c} sites, however, is not as strong as that provided by the Au at the Au/TiO_2 perimeter sites, thus resulting in a system that is still in a very high energy state, as shown by the red reaction pathway in Fig. 10. The C–O bond activation step at TiO_2 sites on the surface has a barrier of 0.83 eV, raising the overall barrier from the bound CH_3CO_2/TiO_2 state to 4.10 eV. This apparent activation barrier prohibits the reaction at ~400 K on TiO_2 sites alone.

The advantages of the Au–Ti^{4+} dual perimeter sites as compared to TiO_2 sites or Au sites alone are that Au can strongly interact with the dehydrogenated species,[24,59–61] while the Ti^{4+} sites stabilize the resulting O that forms as a result of C–O bond scission from the carboxylate end of the molecule.[2] These effects, together with the formation of reactive O adatoms generated at the Au–TiO_2 dual perimeter sites, help to promote the C–H and C–O bond cleavage. The combined activity of

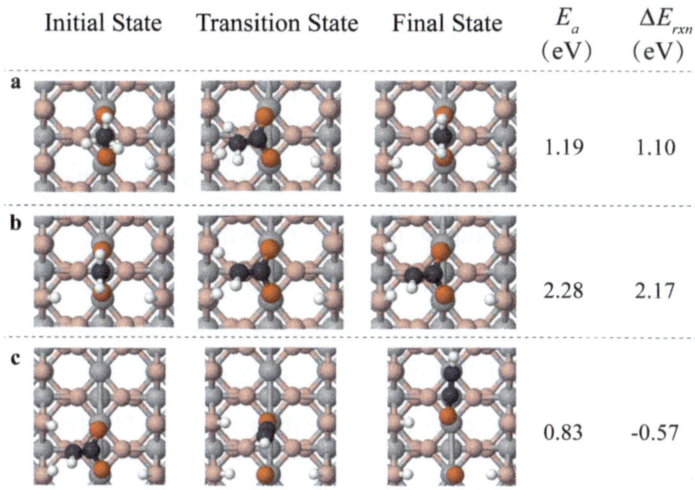

Fig. 12 Consecutive C–H bond and C–O bond scission (eqn (7)–(9)) of acetate/TiO_2 away from the Au/TiO_2 dual perimeter site.

Au and TiO_2 sites is therefore necessary for the formation of ketenylidene from acetate. The joint Au–Ti site readily activates molecular oxygen and supplies O atoms to the Au sites at the TiO_2 interface. Atomic oxygen is weakly bound to Au and acts as a Brønsted base, promoting the activation of the weakly acidic C–H bonds of the carboxylate intermediate to form the reactive HCCOO(a) intermediate on the vicinal oxide substrate. The dehydrogenated HCCOO(a) intermediate is sufficiently activated to enable the formation of strong Au–C and O–Ti bonds that stabilize the transition state at the Au–Ti^{4+} site and aid in HCC–O dissociation to form the gold ketenylidene product.

IV Summary and conclusions

Acetic acid is partially oxidized and undergoes de-oxygenation over nanometer Au particles supported on TiO_2. The dual Au–Ti sites that exist at the periphery of the Au/TiO_2 interface are able to activate molecular oxygen to form weakly held atomic oxygen intermediates on Au that readily catalyze the deprotonation of the adsorbed carboxylate intermediate. This activates the C–O bond to enable strong Au–C and Ti–O interactions which stabilize the transition state at the dual Au–Ti^{4+} site and promote its rupture and the formation of the unique surface ketenylidene intermediate. These reactions are important as they present possible routes to the de-oxygenation of acids and esters derived from bio-renewable intermediates. More specifically we find:

Acetic acid dissociatively adsorbs on the TiO_2 surface, to form bridging bidentate acetate anchored to the Ti_{5c} sites.

1. The partial oxidation of acetate/TiO_2 to form Au_2CCO occurs near 400 K, at the Au–Ti^{4+} dual perimeter sites present at ~3 nm Au nanoparticles. No ketenylidene formation occurs on Au/SiO_2 or TiO_2 under even more aggressive oxidation conditions, indicating that coordinative unsaturation in Au nanoparticles alone or TiO_2 alone is insufficient to support this reaction.

2. DFT simulations show that the Au–Ti^{4+} dual perimeter site is crucial for four processes: (1) activating O_2; (2) promoting the dehydrogenation steps; (3) breaking a C–O bond in the COO moiety; and (4) stabilizing the intermediates.

3. A $^{18}O_2$ isotope study revealed that the O moiety in the Au_2CCO intermediate species originated exclusively from the acetic acid, not from the gas phase oxygen.

4. The rate-limiting step for acetate partial oxidation was found to be the C–O bond scission of the HCCOO/TiO_2 intermediate. An initial deuterium kinetic isotope effect of ~4 was observed.

5. The overall calculated activation energy for ketenylidene formation on the Au/TiO_2 catalyst is 1.80 eV, matching with the 1.7 ± 0.2 eV apparent activation energy measured.

6. By raising the reaction temperature from ~400 K to 473 K, the partial oxidation process is converted to the total oxidation process for acetate/TiO_2 species adsorbed on the Au/TiO_2 catalyst, forming CO_2 and water.

Acknowledgements

We gratefully thank DOE-Office of Basic Energy Sciences under grant number DE-FG02-09ER16080, as well as the XSEDE computing resources from Texas Advanced Computing Center. We appreciate the generosity of Professor R. Zanella from UNAM who provided us the Au/SiO_2 sample. We also acknowledge a fellowship for Mrs. Isabel Green from AES Corporation through the AES Graduate Fellowships in Energy Research Program at the University of Virginia.

V References

1 I. X. Green, W. Tang, M. McEntee, M. Neurock and J. T. Yates, *J. Am. Chem. Soc.*, 2012, **134**, 12717–12723.
2 I. X. Green, W. Tang, M. Neurock and J. T. Yates, Jr., *Science*, 2011, **333**, 736–739.
3 M. Ojeda and E. Iglesia, *Chem. Commun.*, 2009, 352–354.
4 E. E. Stangland, K. B. Stavens, R. P. Andres and W. N. Delgass, *J. Catal.*, 2000, **191**, 332–347.
5 J. A. Rodriguez, S. Ma, P. Liu, J. Hrbek, J. Evans and M. Pérez, *Science*, 2007, **318**, 1757–1760.
6 G. Pattrick, E. v. d. Lingen, C. W. Corti, R. J. Holliday and D. T. Thompson, *Top. Catal.*, 2004, **30/31**, 273.
7 M. S. Chen and D. W. Goodman, *Science*, 2004, **306**, 252–255.
8 J. C. Fierro-Gonzalez, J. Guzman and B. C. Gates, *Top. Catal.*, 2007, **44**, 103–114.
9 Q. Fu, H. Saltsburg and M. Flytzani-Stephanopoulos, *Science*, 2003, **301**, 935–938.
10 J. G. Wang and B. Hammer, *Phys. Rev. Lett.*, 2006, **97**, 136107.
11 L. M. Molina and B. Hammer, *J. Catal.*, 2005, **233**, 399–404.
12 W. D. Williams, M. Shekhar, W. S. Lee, V. Kispersky, W. N. Delgass, F. H. Ribeiro, S. M. Kim, E. A. Stach, J. T. Miller and L. F. Allard, *J. Am. Chem. Soc.*, 2010, **132**, 14018–14020.
13 M. Shekhar, J. Wang, W.-S. Lee, W. D. Williams, S. M. Kim, E. A. Stach, J. T. Miller, W. N. Delgass and F. H. Ribeiro, *J. Am. Chem. Soc.*, 2012, **134**, 4700–4708.
14 N. Lopez, T. V. W. Janssens, B. S. Clausen, Y. Xu, M. Mavrikakis, T. Bligaard and J. K. Nørskov, *J. Catal.*, 2004, **223**, 232–235.
15 I. X. Green, W. Tang, M. Neurock and J. T. Yates, Jr., *Angew. Chem., Int. Ed.*, 2011, **50**, 10186–10189.
16 L. M. Molina, M. D. Rasmussen and B. Hammer, *J. Chem. Phys.*, 2004, **120**, 7673–7680.
17 S. Laursen and S. Linic, *Phys. Chem. Chem. Phys.*, 2009, **11**, 11006–11012.
18 L. M. Molina and B. Hammer, *Appl. Catal., A*, 2005, **291**, 21–31.
19 I. X. Green, W. Tang, M. Neurock and J. T. Yates, Jr., *J. Phys. Chem. C*, 2012, in preparation.
20 M. Haruta, T. Kobayashi, H. Sano and N. Yamada, *Chem. Lett.*, 1987, 405–408.
21 M. Valden, X. Lai and D. W. Goodman, *Science*, 1998, **281**, 1647–1650.
22 R. Meyer, C. Lemire, S. K. Shaikhutdinov and H.-J. Freund, *Gold Bull.*, 2004, **37**, 72.
23 J. Huang and M. Haruta, *Res. Chem. Intermed.*, 2011, **38**, 1–24.
24 I. X. Green, W. Tang, M. Neurock and J. T. Yates, *J. Am. Chem. Soc.*, 2012, **134**, 13569–13572.
25 R. Zanella, S. Giorgio, C. R. Henry and C. Louis, *J. Phys. Chem. B*, 2002, **106**, 7634–7642.
26 R. Zanella, A. Sandoval, P. Santiago, V. A. Basiuk and J. M. Saniger, *J. Phys. Chem. B*, 2006, **110**, 8559–8565.
27 I. X. Green and J. T. Yates, Jr., *J. Phys. Chem. C*, 2010, **114**, 11924–11930.

28 I. X. Green, C. Buda, Z. Zhang, M. Neurock and J. T. Yates, Jr., *J. Phys. Chem. C*, 2010, **114**, 16649–16659.
29 D. A. Panayotov, D. K. Paul and J. T. Yates, Jr., *J. Phys. Chem. B*, 2003, **107**, 10571–10575.
30 S. Laursen and S. Linic, *J. Phys. Chem. C*, 2009, **113**, 6689–6693.
31 G. Kresse, *Phys. Rev. B: Condens. Matter*, 2000, **62**, 8295.
32 J. P. Perdew and Y. Wang, *Phys. Rev. B: Condens. Matter*, 1992, **45**, 13244.
33 P. E. Blöchl, *Phys. Rev. B: Condens. Matter*, 1994, **50**, 17953.
34 G. Kresse and D. Joubert, *Phys. Rev. B: Condens. Matter Mater. Phys.*, 1999, **59**, 1758–1775.
35 S. L. Dudarev, G. A. Botton, S. Y. Savrasov, C. J. Humphreys and A. P. Sutton, *Phys. Rev. B: Condens. Matter Mater. Phys.*, 1998, **57**, 1505–1509.
36 B. J. Morgan and G. W. Watson, *Surf. Sci.*, 2007, **601**, 5034–5041.
37 H. J. Monkhorst and J. D. Pack, *Phys. Rev. B: Solid State*, 1976, **13**, 5188–5192.
38 G. Henkelman, B. P. Uberuaga and H. Jónsson, *J. Chem. Phys.*, 2000, **113**, 9901–9904.
39 G. Henkelman and H. Jónsson, *J. Chem. Phys.*, 2000, **113**, 9978–9985.
40 G. Henkelman and H. Jónsson, *J. Chem. Phys.*, 1999, **111**, 7010–7022.
41 S. Sato, K. Ueda, Y. Kawasaki and R. Nakamura, *J. Phys. Chem. B*, 2002, **106**, 9054–9058.
42 F. Gonzalez, G. Munuera and J. A. Prieto, *J. Chem. Soc., Faraday Trans. 1*, 1978, **74**, 1517–1529.
43 M. A. Hasan, M. I. Zaki and L. Pasupulety, *Appl. Catal., A*, 2003, **243**, 81–92.
44 Z. F. Pei and V. Ponec, *Appl. Surf. Sci.*, 1996, **103**, 171–182.
45 V. Sanchez Escribano, G. Busca and V. Lorenzelli, *J. Phys. Chem.*, 1990, **94**, 8945–8950.
46 M. J. Backes, A. C. Lukaski and D. S. Muggli, *Appl. Catal., B*, 2005, **61**, 21–35.
47 T. Fukushima, H. Arakawa and M. Ichikawa, *J. Chem. Soc., Chem. Commun.*, 1985, 729–731.
48 R. P. Eischens, *Science*, 1964, **146**, 486–493.
49 R. P. Young, *Can. J. Chem.*, 1969, **47**, 2237–2247.
50 A. G. Robietti and J. C. Thompson, *Spectrochim. Acta*, 1965, **21**, 2023–2030.
51 G. Capecchi, M. Faga, G. Martra, S. Coluccia, M. Iozzi and M. Cossi, *Res. Chem. Intermed.*, 2007, **33**, 269–284.
52 D. C. Grinter, M. Nicotra and G. Thornton, *J. Phys. Chem. C*, 2012, **116**, 11643–11650.
53 S. Meenakshisundaram, E. Nowicka, P. J. Miedziak, G. L. Brett, R. L. Jenkins, N. Dimitratos, S. H. Taylor, D. W. Knight, D. Bethell and G. J. Hutchings, *Faraday Discuss.*, 2010, **145**, 341–356.
54 B. J. Xu, X. Y. Liu, J. Haubrich, R. J. Madix and C. M. Friend, *Angew. Chem., Int. Ed.*, 2009, **48**, 4206–4209.
55 R. J. Madix and J. T. Roberts, *Surface Reactions*, Springer-Verlag, New York, 1994.
56 B. N. Zope, D. D. Hibbitts, M. Neurock and R. J. Davis, *Science*, 2010, **330**, 74–78.
57 J. N. Bronsted, *Chem. Rev.*, 1928, **5**, 231–338.
58 M. G. Evans and M. Polanyi, *Trans. Faraday Soc.*, 1938, **34**, 11–24.
59 P. P. Olivera, E. M. Patrito and H. Sellers, *Surf. Sci.*, 1995, **327**, 330–357.
60 A. V. Zeigarnik, R. E. Valdés-Pérez and O. N. Myatkovskaya, *J. Phys. Chem. B*, 2000, **104**, 10578–10587.
61 B. Xing, X.-Y. Pang and G.-C. Wang, *J. Catal.*, 2011, **282**, 74–82.

28 I. X. Green, C. Buda, Z. Zhang, M. Neurock and J. T. Yates Jr., *J. Phys. Chem. C*, 2010, 114, 16649–16659.
29 D. A. Panayotov, D. K. Paul and J. T. Yates Jr., *J. Phys. Chem. B*, 2003, 107, 10571–10575.
30 S. Laursen and S. Linic, *J. Phys. Chem. C*, 2009, 113, 6689–6693.
[illegible]

Catalyst nano-particle size dependence of the Fischer–Tropsch reaction

Rutger A. van Santen[*ab] **and Albert J. Markvoort**[ac]

Received 30th November 2012, Accepted 2nd January 2013
DOI: 10.1039/c2fd20135f

Computational catalytic studies indicate that the elementary reactions that constitute the Fischer–Tropsch reaction strongly dependent on the structure of the catalyst reaction center. Recent experimental evidence is available that, for metallic Fischer–Tropsch catalysts such as Co or Ru, the very small metallic particles show altered catalytic performance. To distinguish between changes in the relative concentration of reaction centres, changes in chemical reactivity, or rate controlling steps, transient SSITKA data are extremely useful. Here, we present kinetics simulations to extract molecular kinetic information from SSITKA data. We have applied such simulations to interpret published experimental SSITKA data on nano-particle size dependent Fischer–Tropsch (FT) kinetics. The FT catalytic cycle consists of four essential reaction steps. Their relative size determines activity as well as selectivity. The simulated SSITKA indicate three different regimes with different kinetic behaviour, where the two fundamental regimes to distinguish are the monomer-formation-limited and the chain-growth-limited regime. Particle size changes shift kinetics from one to the other regime. We note different effects of supports and choice of metal composition on changes in elementary rates or the relative number of reactive centres when the particle size is decreased in the nanometre regime.

Introduction

The Fischer–Tropsch (FT) reaction is a heterogeneous catalytic reaction that converts synthesis gas, *i.e.*, a mixture of carbon monoxide (CO) and hydrogen (H_2), into mainly linear hydrocarbons. Reaction is initiated by dissociative adsorption of CO to the catalyst surface. Adsorbed hydrogenated single carbon atom CH_x intermediates self-organize into adsorbed linear hydrocarbon chains, which desorb from the surface as alkanes or alkenes by reaction with hydrogen atoms or as aldehydes or alcohols by insertion of CO.[1–4] Fundamental interest in the workings of this reaction, which was discovered nearly a century ago,[5] has been renewed due to its current use in the conversion of natural gas, coal, and biomass into diesel quality liquid fuels.[6–8] A persistent challenge, of academic as well as industrial interest, is the design of a catalyst that maximizes longer hydrocarbon chain production while limiting the formation of undesired methane.

Two elements well-known for their activity and selectivity in the Fischer–Tropsch reaction are cobalt (Co) and ruthenium (Ru). To enhance the activity per unit weight of the metals, these metals are commonly dispersed on a support material, where the surface-to-volume can be further enhanced by decreasing the particle size. Recent

[a]*Institute for Complex Molecular Systems, Eindhoven University of Technology, Eindhoven, The Netherlands. E-mail: R.A.v.Santen@tue.nl; Fax: +31 40 245 5054; Tel: +31 40 247 3082*
[b]*Department of Chemistry and Chemical Engineering, Eindhoven University of Technology, Eindhoven, The Netherlands*
[c]*Computational Biology group, Eindhoven University of Technology, Eindhoven, The Netherlands*

experimental evidence is, however, available that for metallic Fischer–Tropsch catalysts as Co or Ru the very small metallic particles show altered catalytic performance. Particle size dependent studies for instance show that whereas turnover frequency is not a function of the cobalt or ruthenium particle size in the size range of 10 to 200 nm, the activity decreases for smaller particles, something that for cobalt nano-particles is accompanied by an increase in methane selectivity.[9,10] Other experiments have shown that metal oxides can be used for selectivity promotion. For such Co nano-particles the selectivity for longer chains can, for instance, be increased again by loading with MnO.

An explanation for this size dependent behaviour and the effect of promotion can be given by computational catalytic studies, which indicate that the elementary reactions that constitute the Fischer–Tropsch reaction strongly dependent on the structure of the catalyst reaction centre. But, to distinguish between changes in the relative concentration of reaction centres, changes in chemical reactivity, or rate controlling steps, experimental Steady-State Isotopic Transient Kinetic Analysis (SSITKA) data are extremely useful.[11] Such transient SSITKA experiments can provide surface residence times and coverages of reaction intermediates as a function of particle size. In such a SSITKA experiment, during the steady-state FT reaction isotopic switches (*e.g.*, from ^{12}CO to ^{13}CO) are performed, and subsequently the isotopically labeled products and reactants are monitored as a function of time. Thus it becomes possible to unravel overall rate into information on surface state of reactive adsorbed layer and information on the elementary rate constants. Hence, analysis of the experiments enables to distinguish between changes in the number of reactive centres from changes in the intrinsic kinetics.

From SSITKA experiments it for instance appeared that for carbon nanofiber supported cobalt catalysts the surface residence times of reversibly bonded CH_x intermediates increased, whereas that of CO decreased for small (<6 nm) Co particles.[12] Also Ru-based catalysts appear highly size dependent for particles smaller than 10 nm. Also here turnover frequency of CO consumption decreases as the particle size decreases, while now CO surface residence time increases as the Ru cluster size decreases.[11] MnO promoted cobalt nano-particles showed a decrease of CO residence time and an increase in the CH_x residence time with increasing MnO content.[13]

Here, we present kinetics simulations to extract molecular kinetic information from such SSITKA data.[11–13] These simulations are based on a complete scheme of Fischer–Tropsch that we recently introduced to predict steady state CO conversion and selectivity. We extend this model, in which we do not consider hydrogen transfer steps explicitly but lump those and other reaction steps into rate parameters for the carbon species, to include two isotopes. We analyse the influence of the individual lumped rate constants on the residence times of CO and CH_x on the surface. Finally, we apply this to the aforementioned published experimental data of supported Co and Ru catalytic nano-particles of varying size and are able to deduce explanations for the effect of promotion and the differences in particle size dependence.

Methods

Model

Our model, which we recently introduced[14] and which is based on the carbide hydrocarbon formation mechanism of the Fischer–Tropsch reaction that is currently generally accepted,[1] is schematically depicted in Fig. 1a. In the model, we do not consider hydrogen transfer steps explicitly, but lump those and other reaction steps into rate parameters for the carbon species. Surface occupancies are denoted by θ's, where sites can either be vacant (θ_v), or occupied by adsorbed CO (θ_{CO}) or by adsorbed chains of i carbons (θ_i), resulting in the constraint $\theta_v + \theta_{CO} + \sum_{i=1}^{\infty}\theta_i = 1$.

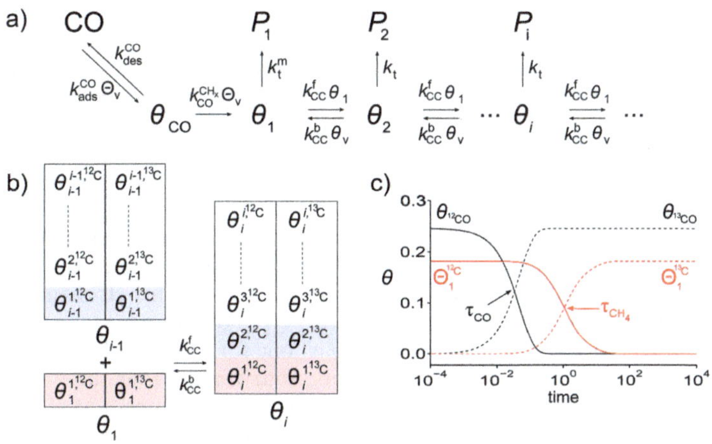

Fig. 1 Single reaction centre site lumped Fischer–Tropsch reaction model. (a) Schematic representation illustrating the lumped reaction steps, their rate constants, and their dependencies on reactants, where θ_{CO} represents adsorbed CO, θ_i adsorbed chains of i carbons, θ_v vacant surface sites and P_i desorbed alkanes of length i. (b) For transient simulations not just the surface coverage of θ_i is followed, but the coverage with ^{12}C and ^{13}C for each position in the chain, *i.e.*, $\theta_i^{j,C^{12}}$ and $\theta_i^{j,C^{13}}$, respectively. (c) A typical example of transient simulations following an isotope switch, illustrating how residence half-lives and surface coverages can be determined.

CO adsorbs to and desorbs from vacant surface sites with rate constants k_{ads}^{CO} and k_{des}^{CO}. Adsorbed CO can dissociate with rate constant $k_{CO}^{CH_x}$, resulting in an adsorbed C_1, where O_{ads} removal from the surface is considered to be fast. All chain growth reaction steps, converting a C_i chain plus a C_1 into a C_{i+1}, are considered reversible with rate constants k_{CC}^{f} and k_{CC}^{b}, independent of chain length. Finally, C_1 desorbs with rate k_t^m while chains C_i of at least length 2 desorb with, again chain length independent, rate k_t.

As illustrated in ref. 14, using this model, steady state surface concentrations can be calculated as a function of the individual lumped rate constants. This can be done, assuming mass-action kinetics, either by solving the mass balance equations or by integrating the reaction rate equations till a steady state is reached. As, in the steady state, the ratio $\alpha = \theta_{i+1}/\theta_i$ is chain length independent, which is a result of the rate constants being chain-length independent (for $i \geq 2$), the product distribution ($i > 1$) will have the characteristic logarithmic Schultz–Anderson–Flory dependence on chain length that is often found experimentally.[6,7] Because of this, all steady state surface concentrations for the chains can be written in terms of α and θ_1, *i.e.*, $\theta_i = \alpha^{i-1}\theta_1$. For the mass balance approach this has the advantage that all chain lengths can be taken into account as analytical expressions for $\theta_t = \sum_{i=1}^{\infty}\theta_i$ and $\sum_{i=1}^{\infty} i\theta_i$ can be derived.[15] Determination of the steady state is then reduced to solving a set of four equations with four unknowns, *i.e.*, α, θ_1, θ_{CO} and θ_v.

Drawback of the mass balance approach is, however, that just the steady state behaviour is obtained and thus no kinetic information can be deduced. Kinetics can be obtained by integrating the reaction rate equations, something that is however only feasible when truncating at a certain chain length N. This can be done either by making one chain growth step, *i.e.* from C_N to C_{N+1}, irreversible and lumping all longer chains together,[16] by lumping all chains longer than N together and estimate C_{N+1} from this as were all chains longer than N in steady state,[17] or by setting the forward chain growth rate for the step from C_N to C_{N+1} to zero. Here we use this latter approach which thus simply ignores all chains longer than N.

In the model so far just a single type of monomer C_1 is present. In order to model the transient kinetics of an isotope switch experiment (*e.g.* from ^{12}C carbon monoxide feed to ^{13}C carbon monoxide feed), however, two types of monomers are needed, *i.e.*, one monomer type for each isotope. Namely, in such an isotope switch experiment the system is, for instance, first brought to equilibrium using a carbon monoxide feed containing the ^{12}C isotope. Subsequently the feed is changed to ^{13}C carbon monoxide, after which the ^{13}C content of the formed reaction products are measured.

With two types of monomers, the number of different chains that can be formed explodes. Whereas for the case with one type of monomer, we needed 1 differential equation for the adsorbed CO and N differential equations for all chains up to the maximum chain length N considered, with two types of monomers two differential equations are needed for the adsorbed CO, *i.e.*, one for ^{12}CO and one for ^{13}CO, and two differential equations for the monomers, four for the four different possible dimers, eight for the trimers, *etc.* This yields in total $2 + \sum_{i=1}^{N} 2^i = 2^{N+1}$ differential equations for CO plus all possible chains up to length N, which is intractable for the typical maximal chain length $N = 50$ that is needed to prevent truncation effects.

One solution would be to use kinetic Monte Carlo simulations instead, where one does not work with concentrations but with numbers of molecules and only considers those chains that are actually present.[18] Such simulations are, however computationally demanding. Moreover, in order to compare with the transient kinetics experiments, knowledge about all individual chains is not needed. Rather, for the comparison with the experiments it suffices to know the fraction of the different isotopes in the formed product per chain length as a function of time. In the case that only forward chain growth reactions are considered, it then suffices to follow the fractions of the different isotopes per chain length. By knowing the fraction of the isotopes in the adsorbed chains and the fraction in the monomers that are added, the fractions in the resulting chains can be calculated. Keeping track on the overall fraction of the isotopes per chain length plus the surface concentrations of those chains would result in $2 + \sum_{i=1}^{N} 2 = 2(N + 1)$ differential equations.

But, in case the chain growth is reversible, as is evident from quantum chemical calculations,[19] apart from the overall isotope contents of a chain of length i, also the isotope fraction of the monomers that are cleaved off is needed. As monomers are added to and cleaved off exclusively from the end of the chain adsorbed to the surface, the fractions of the different isotopes vary per position in the chain, such that solely knowing the overall fraction of the isotopes per chain length no longer suffices. Because only the bottom carbon can be cleaved off a chain C_i, the fraction of the two isotopes at this bottom position is required to know the amounts of both isotopes that come off the chains and form new adsorbed C_1 species, as illustrated in Fig. 1b. However, once the bottom carbon is cleaved of, apart from a C_1, also a chain of length $i - 1$ is formed, where the second carbon of the original chain now is the carbon adsorbed to the surface that could be cleaved off in a subsequent reaction. To be able to keep track on the fraction of isotopes at the bottom of this chain C_{i-1} we thus also need to know the fraction at position 2 of the C_i chains. As a consequence, one needs to keep track on the fractions at all positions j in all chain lengths i, resulting in $i + 1$ differential equations for chains of length i, *i.e.*, one for the concentration of the chains and i for the fraction ^{13}C at all i positions in the chain. By introducing some redundancy, *i.e.*, not keeping one concentration plus i fractions ^{13}C but keeping the concentrations of both isotopes at all i positions, the number of differential equations for chains of length i almost doubles to $2i$, but will keep the equations rather similar to the 1 component case as described in ref. 14. When considering again chains up to length N, this thus yields in total $2 + \sum_{i=1}^{N} 2i = 2 + N(N + 1)$ differential equations, keeping it computationally tractable for the maximal required chain length $N = 50$.

Reaction rate equations describing transient kinetics of isotope switch

Here, in order to compare with transient kinetics experimentally measured using isotope labelled molecules, we thus extend the one component model[14] to be able to follow the concentrations $\theta_i^{j,^{12}\mathrm{C}}$ and $\theta_i^{j,^{13}\mathrm{C}}$ of both isotopes (^{12}C and ^{13}C) at each position j in the chains of length i, instead of just the concentration θ_i. For each $1 \le j \le i$ thus holds that $\theta_i^{j,^{12}\mathrm{C}} + \theta_i^{j,^{13}\mathrm{C}} = \theta_i$.

The differential equations for the CO coverages of the two isotopes become

$$\frac{\mathrm{d}\theta_{^{12}\mathrm{CO}}}{\mathrm{d}t} = k_{\mathrm{ads}}^{\mathrm{CO}} P_{^{12}\mathrm{CO}}\theta_{\mathrm{v}} - k_{\mathrm{des}}^{\mathrm{CO}}\theta_{^{12}\mathrm{CO}} - k_{\mathrm{CO}}^{\mathrm{CH}_x}\theta_{\mathrm{v}}\theta_{^{12}\mathrm{CO}} \tag{1}$$

and

$$\frac{\mathrm{d}\theta_{^{13}\mathrm{CO}}}{\mathrm{d}t} = k_{\mathrm{ads}}^{\mathrm{CO}} P_{^{13}\mathrm{CO}}\theta_{\mathrm{v}} - k_{\mathrm{des}}^{\mathrm{CO}}\theta_{^{13}\mathrm{CO}} - k_{\mathrm{CO}}^{\mathrm{CH}_x}\theta_{\mathrm{v}}\theta_{^{13}\mathrm{CO}} \tag{2}$$

respectively.

The differential equation for chains of length i are split to follow the concentrations $\theta_i^{j,^{12}\mathrm{C}}$ and $\theta_i^{j,^{13}\mathrm{C}}$ of both isotopes (^{12}C and ^{13}C) at each position j in the chains. Since monomers are added to and cleaved off exclusively from the side of the chain adsorbed to the surface, which we will denote as the bottom and having index 1, this yields

$$\begin{aligned}\frac{\mathrm{d}\theta_1^{1,^{12}\mathrm{C}}}{\mathrm{d}t} = {} & k_{\mathrm{CO}}^{\mathrm{CH}_x}\theta_{\mathrm{v}}\theta_{^{12}\mathrm{CO}} - k_{\mathrm{t}}^{\mathrm{m}}\theta_1^{1,^{12}\mathrm{C}} - k_{\mathrm{CC}}^{\mathrm{f}}\theta_1^{1,^{12}\mathrm{C}}\left(\theta_1^{1,^{12}\mathrm{C}} + \theta_1^{1,^{13}\mathrm{C}}\right) + k_{\mathrm{CC}}^{\mathrm{b}}\theta_{\mathrm{v}}\theta_2^{2,^{12}\mathrm{C}} \\ & + k_{\mathrm{CC}}^{\mathrm{b}}\theta_{\mathrm{v}}\sum_{i=2}^{\infty}\theta_i^{1,^{12}\mathrm{C}} - k_{\mathrm{CC}}^{\mathrm{f}}\theta_1^{1,^{12}\mathrm{C}}\sum_{i=1}^{\infty}\left(\theta_i^{1,^{12}\mathrm{C}} + \theta_i^{1,^{13}\mathrm{C}}\right)\end{aligned} \tag{3}$$

for the concentration of monomers containing a ^{12}C,

$$\begin{aligned}\frac{\mathrm{d}\theta_i^{1,^{12}\mathrm{C}}}{\mathrm{d}t} = {} & k_{\mathrm{CC}}^{\mathrm{b}}\theta_{\mathrm{v}}\left(\theta_{i+1}^{2,^{12}\mathrm{C}} - \theta_i^{1,^{12}\mathrm{C}}\right) - \mathrm{k}_{\mathrm{t}}\theta_i^{1,^{12}\mathrm{C}} - k_{\mathrm{CC}}^{\mathrm{f}}\left(\theta_1^{1,^{12}\mathrm{C}} + \theta_1^{1,^{13}\mathrm{C}}\right)\theta_i^{1,^{12}\mathrm{C}} \\ & + k_{\mathrm{CC}}^{\mathrm{f}}\theta_1^{1,^{12}\mathrm{C}}\left(\theta_{i-1}^{1,^{12}\mathrm{C}} + \theta_{i-1}^{1,^{13}\mathrm{C}}\right)\end{aligned} \tag{4}$$

for the concentration of chains of length i with a ^{12}C at the bottom position, and

$$\frac{\mathrm{d}\theta_i^{j,^{12}\mathrm{C}}}{\mathrm{d}t} = k_{\mathrm{CC}}^{\mathrm{b}}\theta_{\mathrm{v}}\left(\theta_{i+1}^{j+1,^{12}\mathrm{C}} - \theta_i^{j,^{12}\mathrm{C}}\right) - k_{\mathrm{t}}\theta_i^{j,^{12}\mathrm{C}} + k_{\mathrm{CC}}^{\mathrm{f}}\left(\theta_1^{1,^{12}\mathrm{C}} + \theta_1^{1,^{13}\mathrm{C}}\right)\left(\theta_{i-1}^{j-1,^{12}\mathrm{C}} - \theta_i^{j,^{12}\mathrm{C}}\right) \tag{5}$$

for the concentration of chains of length i with a ^{12}C at position j (with $j \ge 2$).

The analogous formulas for the ^{13}C isotope then are:

$$\begin{aligned}\frac{\mathrm{d}\theta_1^{1,^{13}\mathrm{C}}}{\mathrm{d}t} = {} & k_{\mathrm{CO}}^{\mathrm{CH}_x}\theta_{\mathrm{v}}\theta_{^{13}\mathrm{CO}} - k_{\mathrm{t}}^{\mathrm{m}}\theta_1^{1,^{13}\mathrm{C}} - k_{\mathrm{CC}}^{\mathrm{f}}\theta_1^{1,^{13}\mathrm{C}}\left(\theta_1^{1,^{12}\mathrm{C}} + \theta_1^{1,^{13}\mathrm{C}}\right) + k_{\mathrm{CC}}^{\mathrm{b}}\theta_{\mathrm{v}}\theta_2^{2,^{13}\mathrm{C}} \\ & + k_{\mathrm{CC}}^{\mathrm{b}}\theta_{\mathrm{v}}\sum_{i=2}^{\infty}\theta_i^{1,^{13}\mathrm{C}} - k_{\mathrm{CC}}^{\mathrm{f}}\theta_1^{1,^{13}\mathrm{C}}\sum_{i=1}^{\infty}\left(\theta_i^{1,^{12}\mathrm{C}} + \theta_i^{1,^{13}\mathrm{C}}\right)\end{aligned} \tag{6}$$

$$\begin{aligned}\frac{\mathrm{d}\theta_i^{1,^{13}\mathrm{C}}}{\mathrm{d}t} = {} & k_{\mathrm{CC}}^{\mathrm{b}}\theta_{\mathrm{v}}\left(\theta_{i+1}^{2,^{13}\mathrm{C}} - \theta_i^{1,^{13}\mathrm{C}}\right) - k_{\mathrm{t}}\theta_i^{1,^{13}\mathrm{C}} - k_{\mathrm{CC}}^{\mathrm{f}}\left(\theta_1^{1,^{12}\mathrm{C}} + \theta_1^{1,^{13}\mathrm{C}}\right)\theta_i^{1,^{13}\mathrm{C}} \\ & + k_{\mathrm{CC}}^{\mathrm{f}}\theta_1^{1,^{13}\mathrm{C}}\left(\theta_{i-1}^{1,^{12}\mathrm{C}} + \theta_{i-1}^{1,^{13}\mathrm{C}}\right)\end{aligned} \tag{7}$$

and

$$\frac{d\theta_i^{j,^{13}C}}{dt} = k_{CC}^{b}\theta_v\left(\theta_{i+1}^{j+1,^{13}C} - \theta_i^{j,^{13}C}\right) - k_t\theta_i^{j,^{13}C} + k_{CC}^{f}\left(\theta_1^{1,^{12}C} + \theta_1^{1,^{13}C}\right)\left(\theta_{i-1}^{j-1,^{13}C} - \theta_i^{j,^{13}C}\right). \tag{8}$$

Finally, the methane formation for both isotopes is given by

$$\frac{dP_1^{^{12}C}}{dt} = k_t^{m}\theta_1^{1,^{12}C} \quad \text{and} \quad \frac{dP_1^{^{13}C}}{dt} = k_t^{m}\theta_1^{1,^{13}C}, \tag{9}$$

respectively, whereas the amount of desorbing chains of at least length 2 is given by

$$\frac{dP_i}{dt} = k_t\left(\theta_i^{1,^{12}C} + \theta_i^{1,^{13}C}\right), \tag{10}$$

with the fraction of the ^{12}C isotope in desorbing chains of length i given by

$$f_i^{^{12}C} = \frac{\left\langle \theta_i^{j,^{12}C} \right\rangle_j}{\left\langle \theta_i^{j,^{12}C} \right\rangle_j + \left\langle \theta_i^{j,^{13}C} \right\rangle_j}, \tag{11}$$

Where $\langle \ldots \rangle_j$ denotes averaging over all positions j in the chain.

Isotope switch experiments are now performed as illustrated in Fig. 1c. First the system is brought to equilibrium with a ^{12}C carbon monoxide feed. Subsequently, the feed is changed to ^{13}CO. Then, the amount of ^{13}C in the CO as well as the methane that desorbs from the surface is followed as a function of time. From this residence times can be determined. Here τ is defined as the time in which half of the carbons in that species are replaced by the new isotope. Instead of only CO and methane in integrating the differential equations of course also the isotope fraction of all longer products can be obtained.

Brønsted–Evans–Polanyi relations

Using the model the effect of the individual lumped rate constants that constitute the Fischer–Tropsch process can be tested. Apart from that it is interesting to study the effect of correlated changes in the rate constants due to the surface reactivity,[20–23] that is identified with adsorption energies of C_{ads} and O_{ads}. The reactivity increases from right to left as for the metals in a row of the periodic system.[24] A well-developed procedure in computational catalysis uses Brønsted–Evans–Polanyi linear activation energy–reaction energy relations to relate activation energies of surface reactions with changes in the adsorption energies of the surface adatoms or reaction intermediates. In this way, the four controlling rate constants; $k_{CO}^{CH_x}$, k_t, k_{CC}^{f} and k_{CC}^{b} become correlated. Extensive literature exists on the selection of the Brønsted–Evans–Polanyi proportionality constants.[20–27] Brønsted–Evans–Polanyi relations

$$k_x = A_x e^{-\left(E_x^0 + \beta_x E_{ads(C)}\right)/RT} \tag{12}$$

are thus used to relate rate constants in the lumped model to surface reactivity, with temperature $T = 500$ K, R the gas constant, and parameters A_x, β_x and E_x^0 as given in Table 1. Reaction energies are related to the adsorption energy of carbon. In the case of CO dissociation the adsorption energies of atomic oxygen and carbon have been assumed to be proportional. Furthermore, $k_{ads}^{CO}P_{CO}$ and k_{des}^{CO} have been taken independent of $E_{ads(C)}$, with default values of 10^{10} s^{-1} and 2×10^9 s^{-1} respectively, such that the CO surface coverage in absence of reaction would be 0.83. Finally, for the steady state results based on the Brønsted–Evans–Polanyi relations k_{CC}^{b} is set to zero for simplicity. The influence of this back rate k_{CC}^{b} on the transients is however studied later on.

Table 1 Default parameters used in the Brønsted–Evans–Polanyi relations

Rate constant	A_x	β_x	E_x^0
k_t	$A_t = 10^{17}\ s^{-1}$	$\beta_t = -0.3$	$E_t^0 = 65\ kJ\ mol^{-1}$
$k_{CO}^{CH_x}$	$A_d = 10^{13}\ s^{-1}$	$\beta_d = 1.2$	$E_d^0 = 250\ kJ\ mol^{-1}$
k_{CC}^{f}	$A_f = 10^{13}\ s^{-1}$	$\beta_f = 0.0$	$E_f^0 = 35\ kJ\ mol^{-1}$

Results and discussion

Chain growth *versus* monomer formation limit

Fig. 2 shows the trends in α, CO conversion and surface coverage one obtains when using the Brønsted–Evans–Polanyi relations. Parameters for the black lines (see Table 1) are such that on metals as Ni and Cu the activation energies for dissociation are relatively high, but of product termination low. As one observes from these curves different regimes in relative reactivity can be distinguished. When one compares the reactivity curves from the right to the left, *i.e.* from low reactivity to high reactivity, initially the rate of CO dissociation is slow. Adsorbed CO dominates the surface, and upon dissociation of CO the C_1 species rapidly desorb as methane since the termination rate is relatively high. In this first regime, *i.e.*, the methanation regime where $k_{CO}^{CH_x} \ll k_t$, hardly any longer chains are being formed.

When going to the left, the concentration of monomers on the surface starts to increase when the rate of CO dissociation starts to compete with the rate of termination. Then also formation of chains starts to occur, as can be seen from the increase in chain growth parameter α, and CO consumption increases. In this region where the surface still contains plenty of adsorbed CO as well as vacant sites, which we denote the monomer formation limit regime, growth of the chains is limited by the dissociation rate of CO. The increase in α and R_{CO} continues till a surface reactivity is reached where the rate of CO dissociation becomes larger than the rate of C–C bond formation.

With the increased concentration of adsorbed chains the surface concentration of CO as well as that of vacant sites has now decreased so much that CO conversion even starts to decrease. For even more reactive surfaces CO is reacted away completely and replaced by growing hydrocarbon chains. In this regime, which we denote the chain growth limit regime, CO conversion is limited by the very low concentration of CO as the surface is poisoned with the growing chains.

Analytical expression for α and R_{CO} valid within the monomer formation limit and chain growth limit can be deduced from the kinetic equations corresponding to the lumped model shown in Fig. 1a. In the monomer formation model limit $\left(k_{CO}^{CH_x}\theta_{CO} \ll k_{CC}^{f}\theta_1\right)$:

$$R_{CO} = k_{CO}^{CH_x}\theta_{CO}(1-\theta_{CO}) \quad \text{and} \quad \alpha = \left(1 + \left(\frac{k_t^{\,2}}{k_{CC}^{f}k_{CO}^{CH_x}\theta_{CO}(1-\theta_{CO})}\right)^{1/3}\right)^{-1} \tag{13}$$

whereas in the chain growth model limit[14] $(k_{CO}^{CH_x}\theta_{CO} \gg k_{CC}^{f}\theta_1)$:

$$R_{CO} = \sqrt{k_t k_{CC}^{f}(1-\theta_{CO})^3} \quad \text{and} \quad \alpha = \left(1 + \sqrt{\frac{k_t}{k_{CC}^{f}(1-\theta_{CO})}}\right)^{-1} \tag{14}$$

where the expressions for α have been deduced for the case that it has a high value $(k_t \ll k_{CC}^{f}\theta_1)$ and the rate of methane formation (k_t^{m}) is considered equal to the rate of termination (k_t). These limits are also drawn in Fig. 2.

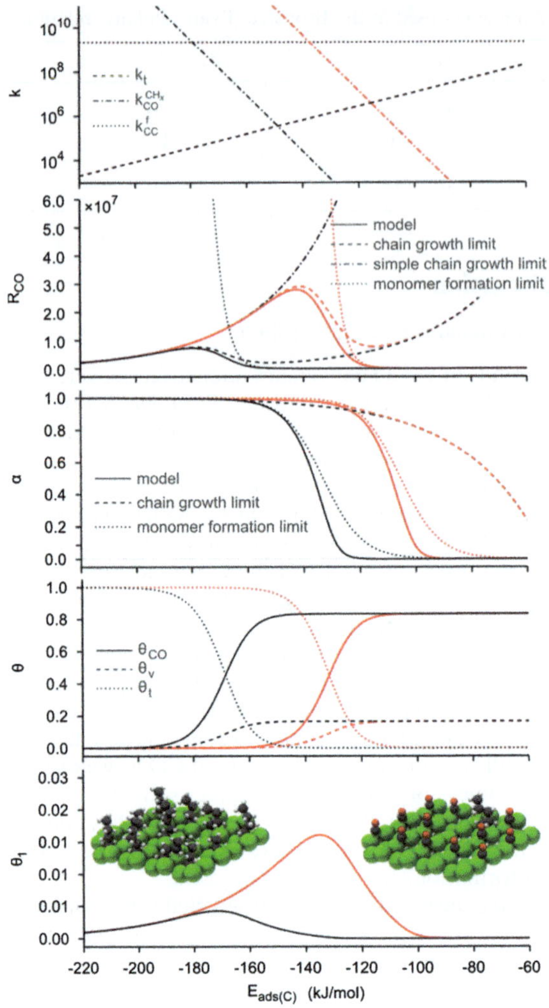

Fig. 2 Influence of the CO activation energy $E_d{}^0$. (a) The rate constants resulting from Brønsted–Evans–Polanyi relations, (b) CO consumption R_{CO}, (c) chain growth parameter α, (d–e) surface coverage, where $\theta_t = \sum_{i=1}^{\infty}\theta_i$, as a function of the adsorption energy of carbon $E_{ads(C)}$. Solid lines represent the simulated values, the dotted lines the monomer formation limit from eqn (13), dashed lines the chain growth limit from eqn (14), and the dashed-dotted line $R_{CO} = \sqrt{(k_t k_{CC}^{f})}$, which we denote the simple chain growth limit. The black lines correspond to an initial value (*i.e.* at $E_c = 0$) of the CO activation energy of 250 kJ mol^{-1} and red lines to an initial value of 200 kJ mol^{-1}.

As is shown by the red lines in Fig. 2, for a surface with a different structure, that causes the relative rate of the CO to CH_x transformation to increase compared to k_{CC}^{f} and k_t, the maximum of R_{CO} shifts to lower values of $E_{ads(C)}$, with a higher value of k_t. Whereas the chemistry is essentially similar, the CO consumption rate maximum is now substantially higher. High α and maximum of CO consumption now do not coincide as well as for the lower reactivity surface. Maximum C_{2+} yield is favoured by a value of k_t high enough that poisoning by growing hydrocarbon chains does not yet occur, but low compared to k_{CC}^{f}. Then $k_{CO}^{CH_x}$ has to be fast compared to k_t as well for overall rate of chain growth to be consistent with the chain growth model limit.

Residence time dependence on individual rate constants

It is also possible to study the dependence on individual rate constants (k_{des}^{CO}, $k_{CO}^{CH_x}$, k_{CC}^{f}, k_{CC}^{b}, and k_t) by varying one rate constant a time. In Fig. 3 the dependence on the dissociation rate constant $k_{CO}^{CH_x}$ is shown. Apart from the surface concentrations and the performance parameters α and CO conversion, here also the half lives from the isotope switch simulations are shown as a function of $k_{CO}^{CH_x}$ for various values of $k_{CC}^{f} = k_{CC}^{b}$. The solid lines correspond to $k_{CC}^{f} = k_{CC}^{b} = 0$, *i.e.*, no chain growth such that only methane is produced. In the plot of the residence times, we recognize three different regimes. In the first, *i.e.* the low dissociation regime $k_{CO}^{CH_x} < k_t$, the rate of termination is fast compared to the rate of C_1 formation. Consequently, concentration of C_1 as well as α is very small, while the CO coverage is high. The residence time of CH_4 (τ_{CH_4}) is equal to k_t^{-1}, that of CO equal to $k_{des}^{CO^{-1}}$.

In the second regime, *i.e.*, $k_t < k_{CO}^{CH_x} < k_{CC}^{f}$, CO on the surface becomes gradually displaced by growing hydrocarbon chains. $k_{CO}^{CH_x}$ remains rate controlling, resulting in an intermediate value of α. This is thus the monomer formation limit. The residence time of methane becomes controlled not just by k_t but also by the rate of incorporation of C_1 into growing hydrocarbon chains, reducing τ_{CH_4} to $(k_{CC}^{f} = k_t)^{-1}$.

In the third regime, *i.e.*, $k_{CO}^{CH_x} > k_{CC}^{f}$, k_t is rate controlling, α is high and the CO coverage low. In this chain growth limit, τ_{CH_4} remains constant at $(k_{CC}^{f} + k_t)^{-1}$, while τ_{CO} decreases with increasing $k_{CO}^{CH_x}$ following $(k_{des}^{CO} + k_{CO}^{CH_x}\theta_v)^{-1}$ as it is consumed rapidly.

Fig. 4 shows residence times, steady state production, surface coverage, and reference CO concentration (θ_{CO}^{ref}) as a function of CO surface coverage, where $\theta_{CO}^{ref} = k_{ads}^{CO}P_{CO}/(k_{ads}^{CO}P_{CO} + k_{des}^{CO})$ would be the CO coverage in absence of CO dissociation and where the CO coverage is changed by varying the rate of CO desorption.

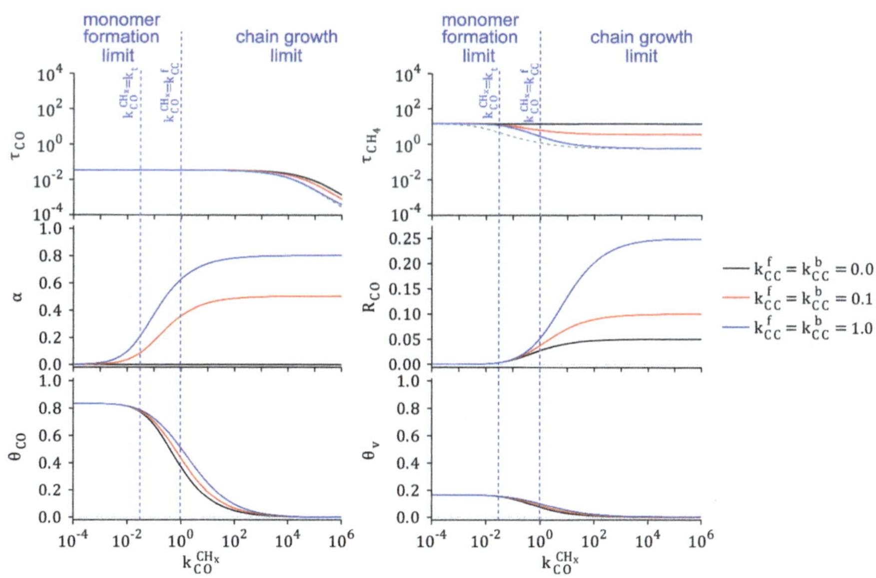

Fig. 3 Dependence of half lives for CO (τ_{CO}) and methane (τ_{CH_4}), steady state production (R_{CO}), chain growth rate α, and surface coverage on the dissociation rate constant $k_{CO}^{CH_x}$. Values of not mentioned rate constants are $k_{ads}^{CO}P_{CO} = 100$, $k_{des}^{CO} = 20$, and $k_t = 0.05$. Three regimes are discerned, which are separated by the dashed vertical lines at $k_{CO}^{CH_x} = k_t$ and $k_{CO}^{CH_x} = k_{CC}^{f}$ for the blue case, *i.e.*, $k_{CC}^{f} = k_{CC}^{b} = 1$. For this case also the approximations $\ln(2)/\left(k_{des}^{CO} + k_{CO}^{CH_x}\theta_v\right)$ for τ_{CO} and $\ln(2)/(k_t + k_{CC}^{f}(\theta_1 + \theta_t))$ for τ_{CH_4} are shown.

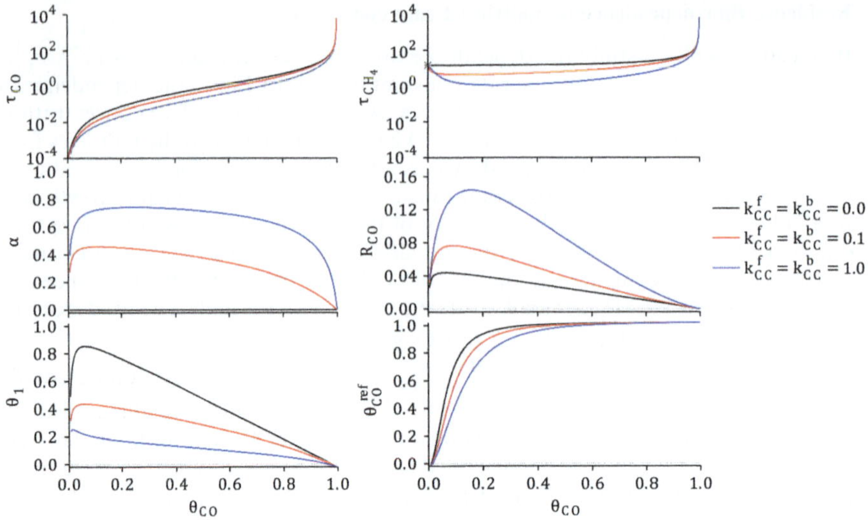

Fig. 4 Dependence of half lives for CO (τ_{CO}) and methane (τ_{CH_4}), steady state production (R_{CO}), chain growth rate α, and surface coverage on the CO surface coverage, where this CO surface coverage is varied by changing CO desorption rate k_{des}^{CO}. Also the reference CO surface coverage is shown, *i.e.*, the CO coverage that would be present in the absence of dissociation. Values of not-mentioned rate constants are $k_{ads}^{CO}P_{CO} = 100$, $k_{CO}^{CH_x} = 11$, and $k_t = 0.05$. For the blue case, *i.e.*, $k_{CC}^f = k_{CC}^b = 1$ also the approximation $\ln(2)/\left(k_{des}^{CO} + k_{CO}^{CH_x}\theta_v\right)$ for τ_{CO} is shown, as well as the point $\ln(2)/k_t$ for τ_{CH_4}.

The right bottom part of the Figure shows that, starting from very low desorption rate, *i.e.*, $\theta_{CO}^{ref} \approx 1$, the surface concentration CO decreases more strongly than the reference CO coverage. As also observed experimentally[28] the surface becomes initially covered with C_1 species that replace CO. At lower coverage of CO it becomes replaced increasingly by the growing hydrocarbon chains, covering a large extend of the surface. The coverage of C_1 gradually has a maximum with decreasing θ_{CO} as for very high desorption rates the surface finally becomes almost empty.

The conversion of CO shows a maximum as a function of CO coverage as well. This also holds for α. Although it is almost constant over a large range, it vanishes when the CO coverage is either small or close to 1. At high CO coverage the surface becomes non-reactive as there are hardly any surface sites available for chain growth, while at low θ_{CO} the surface becomes non-reactive because the surface is completely covered with growing chains.

The top part of the Figure illustrates the changes in residence time. The behaviour of the residence time of CO is consistent with the decreasing residence time expectation when CO desorption rate increases, *i.e.*, τ_{CO} equals $\left(k_{des}^{CO} + k_{CO}^{CH_x}\theta_v\right)^{-1}$. For very low CO coverage, τ_{CH_4} equals k_t^{-1}. At the intermediate CO coverage regime, τ_{CH_4} decreases due to the consumption of C_1 into growing chains, such that as in the previously discussed case of Fig. 3, the residence time of CH_4 is low when α is high. The increase in τ_{CH_4} at the high CO coverage boundary stems from the fact that there τ_{CH_4} is dominated by the CO conversion rate.

Fig. 5 shows the dependence of residence time, chain growth rate α, rate of CO conversion, and surface coverage as a function of k_{CC}^b for three values of k_t. We observe the expected decrease in α upon increasing the rate of termination. Then the rate of product formation is so high that chains no longer get the chance to grow and the decrease in concentration of growing chains increases the concentration of surface vacancies. One observes that the value of α is not affected by k_{CC}^b

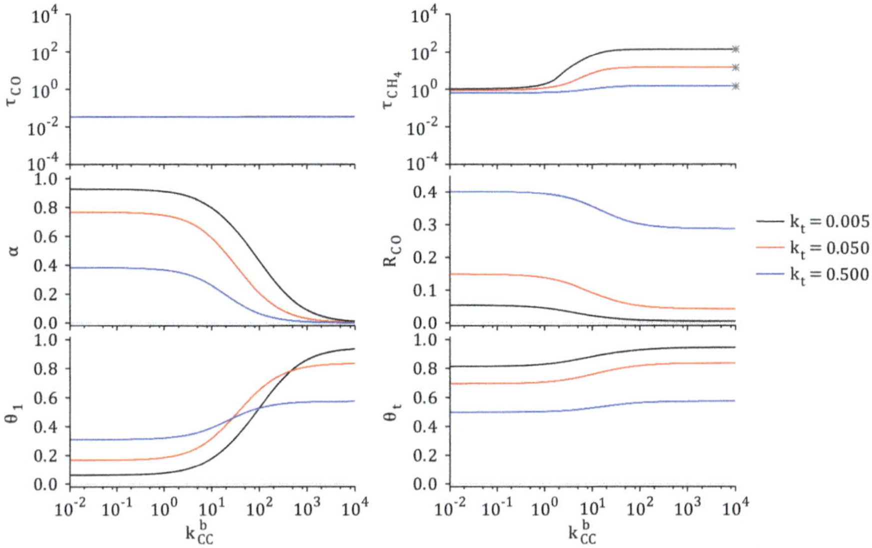

Fig. 5 Dependence of half lives for CO (τ_{CO}) and methane (τ_{CH_4}), steady state production (R_{CO}), chain growth rate α, and surface coverage on the backwards rate constant k^b_{CC}. Values of not mentioned rate constants are $k^{CO}_{ads}P_{CO} = 100$, $k^{CO}_{des} = 20$, $k^{CH_x}_{CO} = 11$, and $k^f_{CC} = 1$. Also the approximation $\ln(2)/k^{CO}_{des}$ for τ_{CO} is shown, as well as the points $\ln(2)/k_t$ for τ_{CH_4}.

until k^b_{CC}/k^f_{CC} equals 1. When k^b_{CC} increases further α decreases initially slowly. When k^b_{CC}/k^f_{CC} is 10 it has decreased by barely ten percent. When this ratio increases further α starts to decline steeply. This is very important as it appears from DFT calculations that the ratio k^b_{CC}/k^f_{CC} may strongly depend on surface structure and metal.[19] As long as k^b_{CC}/k^f_{CC} is smaller than 10 the effect of the reverse reaction can be essentially ignored and the classical expression $\alpha = k^f_{CC}\theta_1/(k^f_{CC}\theta_1 + k_t)$ can be used. For large k^b_{CC}, τ_{CH_4} is equal to k_t^{-1} and it decreases for smaller k^b_{CC}. τ_{CO} is constant, *i.e.*, approximately ${k^{CO}_{des}}^{-1}$.

Discussion and conclusion

Based on the residence time dependencies on the individual rate constants as observed in the previous section, we will now analyse transient isotope switch experiments in order to decompose overall catalyst kinetics into differences between rates of elementary reactions and changes in numbers of reactive centres. Especially the group of Holmen[11–13] has recently produced transient kinetic information for cobalt (Co) and ruthenium (Ru) as a function of particle size changes in the nano-size regime, which changes the presence of catalytic sites per particle. Changes for each metal are quite specific and also depend on catalyst support or promoting reducing oxide used. Table 2 compares reported data and interpretation based on our simulations.

In case 1, *i.e.*, carbon nanofiber supported cobalt catalyst, the decrease in α with particle size implies an increase in k_t or increase in apparent rate of C–C bond formation. The longer τ_{CH_4} is not consistent with an increase of k_t. The longer τ_{CH_4}, decrease of α and decrease of R_{CO} are consistent with a decreased $k^{CH_x}_{CO}$ (see eqn (13)). The shorter τ_{CO} of reversible CO implies a lower CO coverage. These data are indicative for the validity of the monomer formation kinetics model.

The data of case 2, *i.e.*, silica-supported cobalt catalyst, are contradictory to the previous case, which may be due to altered properties of CO due to interaction with the silica support. The decrease in α and shorter τ_{CH_4} are consistent with

Table 2 Experimentally observed effects of catalyst particle size and promotion on CO and methane residence times, conversion, and α, together with the explanation according to our model

Effect of decrease in particle size					
	Experimental observation				**Model explanation**
	τ_{CH_4}	τ_{CO}	R_{CO}	α	
1. Co (carbon support)[12]	longer	shorter	decreases	decreases	$k_{CO}^{CH_x}$ decreases
2. Co (silica support)[13]	shorter	shorter	decreases	decreases	k_t increases decrease in number of sites
3. Ru (alumina support)[11]	constant	longer	decreases	constant	decrease in reactivity by increased CO coverage
Effect of promotion					
	Experimental observation				**Model explanation**
4. Co (MnO promoted)[13]	longer	longer	increases	increases	k_t decreases, increase in number of sites

increased rate of termination, k_t. The decrease in R_{CO} has to be due to a decrease in number of reactive centers.

In case 3, *i.e.*, alumina-supported ruthenium catalyst, only the rate of desorption changes, but none of the reactivity parameter. One concludes that in this case there is simply a reduction in reactive centres with decreasing particle size.

In case 4, *i.e.*, MnO promoted cobalt catalyst, a longer τ_{CH_4} implies a decrease in CO dissociation rate or decrease in k_t. As α increases, k_t has to decrease, unless the rate of C–C bond formation sharply increases. The increase in rate of CO consumption is consistent with an increase in number of reactive centres. A decrease of k_t effects CO conversion negatively, unless CO dissociation is slow compared to the rate of C–C bond formation (monomer formation model limit). However, a faster rate of CO dissociation counteracts the increase of τ_{CH_4}. The increase in CO consumption and decrease of k_t are consistent with dual reaction centre site model.[14] The addition of promoting reducible oxides may lead to the presence of dual reaction centre sites, resolving the problem of suppression of CO conversion by the growing hydrocarbon chains; on the metal atoms chain growth occurs, while the oxide cation assists CO dissociation[29] by exclusively interacting with the oxygen atom.

To conclude, we have presented an extension of the lumped Fisher–Tropsch model from ref. 14 to include two isotopes, in order to analyse transient SSITKA data with a complete Fischer–Tropsch model. We analysed the influence of the individual lumped rate constants on the residence times of CO and CH_x on the surface. This, for instance, showed a decrease in the residence time for CH_x follow by a decrease in residence time for CO for increasing CO dissociation rate $k_{CO}^{CH_x}$, *i.e.*, in going from the monomer formation limit to the chain growth limit; a minimum in the CH_x residence time as a function of the CO surface coverage on varying the CO desorption rate; and an increase in the CH_x residence time with increasing backwards rate constant k_{CC}^{b}. Finally, these insights were used to give explanations for the experimentally observed effect of promotion and the differences in particle size dependence as observed in published experimental SSITKA data of supported Co and Ru catalytic nano-particles.

References

1 R. A. van Santen, I. M. Ciobica, E. van Steen and M. M. Ghouri, *Adv. Catal.*, 2011, **54**, 127.
2 P. Biloen and W. M. H. Sachtler, *Adv. Catal.*, 1981, **30**, 165.
3 A. T. Bell, *Catal. Rev. Sci. Eng.*, 1981, **23**, 203.
4 R. A. van Santen, M. M. Ghouri, S. Shetty and E. J. M. Hensen, *Catal. Sci. Technol.*, 2011, **1**, 891.

5 F. Fischer and H. Tropsch, *Brennst.-Chem.*, 1926, **7**, 97.
6 H. H. Storch, N. Golumbic and R. B. Anderson, *The Fischer–Tropsch and Related Syntheses*, 1951, Wiley, New York.
7 R. B. Anderson, H. Kölbel and M. Rálekv, *The Fischer–Tropsch synthesis*, 1984, Academic Press, Orlando.
8 A. P. Steynberg and M. E. Dry, *Fischer–Tropsch Technology*, 2004, Elsevier.
9 G. L. Bezemer, J. H. Bitter, H. P. C. E. Kuipers, H. Oosterbeek, J. E. Holewijn, X. Xu, F. Kapteijn, A. J. van Dillen and K. P. de Jong, *J. Am. Chem. Soc.*, 2006, **128**, 3956.
10 J. M. G. Carballo, J. Yang, A. Holmen, S. Garcia-Rodriguez, S. Rojas, M. Ojeda and J. L. G. Fierro, *J. Catal.*, 2011, **284**, 102.
11 S. L. Shannon and J. G. Goodwin, *Chem. Rev.*, 1995, **95**, 677.
12 J. P. den Breejen, P. B. Radstake, G. L. Bezemer, J. H. Bitter, V. Froseth, A. Holmen and K. P. de Jong, *J. Am. Chem. Soc.*, 2009, **131**, 7197.
13 J. P. den Breejen, A. M. Frey, J. Yang, A. Holmen, M. M. van Schooneveld, F. M. F. de Groot, O. Stephan, J. H. Bitter and K. P. de Jong, *Top. Catal.*, 2011, **54**, 768.
14 A. J. Markvoort, R. A. van Santen, P. A. J. Hilbers and E. J. M. Hensen, *Angew. Chem., Int. Ed.*, 2012, **51**, 9015.
15 H. M. M. ten Eikelder, A. J. Markvoort, T. F. A. de Greef and P. A. J. Hilbers, *J. Phys. Chem. B*, 2012, **116**, 5291.
16 E. T. Powers and D. L. Powers, *Biophys. J.*, 2006, **91**, 122.
17 P. A. Korevaar, S. J. George, A. J. Markvoort, M. M. J. Smulders, P. A. J. Hilbers, A. P. H. J. Schenning, T. F. A. de Greef and E. W. Meijer, *Nature*, 2012, **481**, 492.
18 A. J. Markvoort, H. M. M. ten Eikelder, P. A. J. Hilbers, T. F. A. de Greef and E. W. Meijer, *Nat. Commun.*, 2011, **2**, 509.
19 I. M. Ciobica, G. J. Kramer, Q. Ge, M. Neurock and R. A. van Santen, *J. Catal.*, 2002, **212**, 136.
20 A. Logadottir, T. H. Rod, J. K. Norskov, B. Hammer, S. Dahl and C. J. H. Jacobsen, *J. Catal.*, 2001, **197**, 229.
21 V. Pallassana and M. Neurock, *J. Catal.*, 2000, **191**, 301.
22 H. Toulhoat and P. Raybaud, *J. Catal.*, 2003, **216**, 63.
23 R. A. van Santen and M. Neurock, *Molecular Heterogeneous Catalysis*, 2006, Wiley.
24 T. Bligaard and J. K. Norskov, in *Chemical bonding at surfaces and interfaces*, ed. A. Nilsson, L. G. M. Pettersson and J. K. Norskov, 2008, Elsevier, Amsterdam, ch. 4.
25 A. Nilsson, L. G. M. Pettersson and J. K. Nørskov, *Chemical bonding at Surfaces and Interfaces*, 2008, Elsevier, Amsterdam.
26 R. A. van Santen, *Acc. Chem. Res.*, 2009, **42**, 57.
27 R. A. van Santen, M. Neurock and S. G. Shetty, *Chem. Rev.*, 2010, **110**, 2005.
28 G. A. Beitel, C. P. M. de Groot, H. Oosterbeek and J. H. Wilson, *J. Phys. Chem. B*, 1997, **101**, 4035.
29 W. M. H. Sachtler and M. Ichikawa, *J. Phys. Chem.*, 1986, **90**, 4752.

8 F. Fischer and H. Tropsch, *Brennst. Chem.*, 1926, 7, 97.

9 H. H. Storch, N. Golumbic and R. B. Anderson, *The Fischer–Tropsch and Related Syntheses*, 1951, Wiley, New York.

10 R. B. Anderson, H. Kölbel and M. Rálek, *The Fischer–Tropsch Synthesis*, 1984, Academic Press, Orlando.

11 [illegible] *Fischer–Tropsch Technology*, 2004, Elsevier.

12 [illegible]

13 [illegible]

14 [illegible]

15 [illegible]

16 [illegible]

17 [illegible]

18 [illegible]

A computational study of the influence of the ceria surface termination on the mechanism of CO oxidation of isolated Rh atoms

Weiyu Song, Antonius P. J. Jansen and Emiel J. M. Hensen*

Received 15th November 2012, Accepted 25th January 2013
DOI: 10.1039/c3fd20129e

The reaction mechanism for CO oxidation by isolated Rh atoms stabilized on $CeO_2(111)$, $CeO_2(110)$ and $CeO_2(100)$ surfaces is investigated by a combination of Density Functional Theory and kinetic Monte Carlo calculations. On Rh/$CeO_2(111)$, one adsorbed CO molecule on Rh was found to form a stable intermediate structure with surface O. The reaction cycle cannot be closed because of the strong adsorption of the CO_2 complex. The presence of a second adsorbed CO significantly decreases the desorption energy, thus opening a possible reaction path. Formation of the oxygen vacancy is accompanied by reduction of surface cerium. On Rh/$CeO_2(110)$, adsorbed CO can easily react with a ceria surface O atom due to the lower Ce–O bond energy. Since surface O atom migration is much more facile on Rh/$CeO_2(110)$ than on Rh/$CeO_2(111)$, CO_2 desorption is also more easy for the former surface. Molecular oxygen will adsorb on the resulting vacancy. After desorption of the second CO_2 product molecule by reaction of adsorbed CO with another surface O atom, the adsorbed oxygen molecule migrates spontaneously to the vacancy site and dissociates with negligible barrier. The role of molecular oxygen is to heal the oxygen vacancy rather than being involved in a direct reaction with adsorbed CO. The Rh/$CeO_2(100)$ model was found to be inactive for CO oxidation, mainly because of the geometric constraints for the adsorbed CO molecule to react with one of the surface O atoms, despite the low Ce–O bond energy of the $CeO_2(100)$ surface. The main reason is the large distance between the C of adsorbed CO and the ceria O surface atoms. The particularities of the CO oxidation mechanism for isolated Rh atoms on these ceria surfaces are in agreement with the experimental activity trends.

1 Introduction

Transition metal-based catalysts are indispensable in modern chemical industry and also serve in many environmental catalysis applications. In recent decades, much attention has been devoted to the understanding of the catalytic reactivity of transition metal nanoparticles. With the advance of computational chemistry, significant insights have been obtained in the structure sensitivity of these nanoparticles for particular elementary reaction steps.[1] Usually, transition metal nanoparticles are supported on an oxide carrier material that serves the purpose to maintain a high dispersion of the metal. Recently, the attention in catalysis is shifting from nanoparticles to sub-nanometer and atomically dispersed transition metals stabilized on oxide supports.[2,3] This is especially useful in order to understand the interaction

Laboratory of Inorganic Materials Chemistry, Schuit Institute of Catalysis, Eindhoven University of Technology, P.O. Box 513, 5600 MB Eindhoven, The Netherlands. E-mail: e.j.m.hensen@tue.nl

of the transition metal with the oxide support. There are many indications that the support may play an active role in the catalytic cycle.[2] In this respect, the transition metal-catalyzed reaction between CO and O_2 has been investigated in great detail, not only as a model reaction but also because of its industrial relevance. Ceria is a widely used oxide support for transition metals, especially for CO oxidation.[4–12] Recently, evidence is growing that the active phase of some transition metals when supported on ceria during CO oxidation is not metallic but oxidic.[4,5,7] Toyota researchers found indications for a reversible structural rearrangement of Pt–O_x–Ce species, reducing (oxidizing) conditions destabilizing (stabilizing) the Pt–O_x–Ce ensemble.[7] Raman spectroscopy also shows that rhodium, iridium, palladium, and platinum form surface-phase metal oxide (M–O) structures that interact strongly with the surface of CeO_2.[12]

Recently, Ligthart *et al.* reported that the composition of the active phase in Rh/CeO_2 catalysts strongly depends on the reaction conditions and the particle size. Whilst typically under conditions of steam reforming the Rh nanoparticles are metallic, very small particles that tend to become slowly oxidized leading to deactivation.[13] Applied in CO oxidation, Rh nanoparticles smaller than 2 nm become fully oxidized and are significantly more active than larger particles which remain metallic. There are indications that the oxidation of small Rh nanoparticles on ceria leads to a strong dispersion of Rh over the support.[4] The formation of a surface oxide film of Rh commensurate with the CeO_2 surface under oxidative conditions is confirmed by quantum-chemical DFT calculations.[14] In essence, under reducing conditions Rh tends to form three-dimensional metallic clusters on CeO_2(111), whereas a two-dimensional film is more stable in the presence of oxygen. It was found that a Rh dimer will disperse into atomic Rh species on ceria under these conditions. This suggests that a Rh-oxide film may disperse into isolated Rh species. This is supported by an experimental model study of Rh/CeO_2 showing that oxygen adsorption at edges of Rh clusters results in isolated Rh species.[15] All these studies point to the possibility that atomically dispersed Rh species supported on ceria may play a role in CO oxidation. Therefore, the first motivation of the current contribution is to investigate whether single site Rh atoms stabilized on ceria can be active in the CO oxidation reaction.

Another related topic of interest is the influence of the support surface structure on the catalytic activity. It is now possible to control the crystallographic surface planes of ceria nanocrystals.[16,17] Used as a catalyst itself, ceria nanocrystals exposing (110) and (100) planes are more active in CO oxidation than crystals predominantly exposing (111) facets. This structure sensitivity has been related to the oxygen vacancy formation energy, which increases in the order (110) < (100) < (111).[18,19] Recently, Chen *et al.* systematically studied by DFT calculations the possible reaction paths for CO oxidation *via* ceria lattice oxygen on CeO_2(111) and (110) surfaces.[20] Indeed, lattice oxygen atoms are more reactive on CeO_2(110) than on CeO_2(111) surfaces in CO oxidation. Several recent reports confirm the influence of nanosized ceria on the catalytic activity of metal nanoparticles. Guzman *et al.* showed that smaller CeO_2 crystals increase the CO oxidation activity of Au.[5] By combination of computational and experimental methods, the Illas group found a lower vacancy formation energy for nanosized CeO_2.[8] They also showed how oxygen atoms of the ceria support are supplied to supported Pt nanoparticles. Ligthart *et al.* reported that the catalytic activity of Rh supported on ceria nanocrystals decreases in the order: Rh/CeO_2(110) > Rh/CeO_2(111) > Rh/CeO_2(100) planes.[4] A recent theoretical paper proposed a mechanism of CO oxidation with lattice oxygen atoms *via* a Mars–van Krevelen mechanism on a stepped-CeO_2(111) supported Au nanocluster.[21] Such mechanism was found to be impossible on a planar CeO_2(111) surface.[22] These result shows the importance of the structure of the ceria surface in CO oxidation catalysis. Therefore, the second motivation of the current contribution is to study the influence of the ceria surface plane on the CO oxidation activity of isolated Rh atoms.

The mechanism for CO oxidation on ceria-supported transition metals has been extensively studied.[21–24] Previous theoretical works proposed a molecular reaction pathway: CO reacts with adsorbed O_2,[22] similar to what has been suggested for CO oxidation on Pt-oxide[25] and other oxide-supported metals.[26] The main role of the ceria support is believed to maintain a particular particle size and geometry of the metal conducive to adsorption of CO and O_2. We consider this mechanism to be insufficient to account for the influence of the support structure. Firstly, CO_2 can be detected when ceria-supported catalysts are exposed to CO in the absence of O_2, evidencing the involvement of oxygen atoms of the ceria.[5,9] Secondly, the density of Ce^{3+} increases with the CO oxidation temperature.[9] These experiments provide strong evidence that ceria provides reactive oxygen atoms partaking in CO oxidation reaction. This is in accordance with the notion that ceria has a high oxygen storage capacity. Theoretically, this mechanism was first examined by Camellone and Fabris employing substitutional and supported Au on a $CeO_2(111)$ surface.[23] Their results show that both supported and substitutional Au atoms promote CO oxidation *via* O vacancy formation. These vacancies can be efficiently healed by molecular oxygen for the substitutional case, completing the reaction cycle. However, the stability of the substitutional system is still an issue of debate.[27] For the supported case, the O vacancy formed during the reaction attracts positively charged Au, turning them into negative Au species that prevent further CO and O_2 adsorption. It leads to catalyst deactivation. The authors then suggested that the Au–Au cohesive energy in a supported cluster of larger size can prevent the deactivating diffusion process. This mechanism was recently examined by Kim *et al.*[22] in their Mars–van Krevelen mechanism. However, the relatively low desorption energy of CO_2 from the Au–COO* surface complex is caused by the diffusion of the gold atom in the oxygen vacancy site.[23] For the larger cluster model employed by Kim *et al.*, in which the Au diffusion process is indeed hindered, the energy of CO_2 desorption is the same as the sum of the energies for the removal of a single oxygen atom from CeO_2 and the oxidation of CO to CO_2. Very recently, the same authors proposed a possible Mars–van Krevelen mechanism based on the on a stepped-$CeO_2(111)$ supported Au nanocluster.[21] The study shows a rather promising way to explain the role of the ceria support in CO oxidation. Our previous contribution on ceria-supported Rh_3O_x cluster for CO oxidation suggests a new mechanism involving oxygen spillover from the ceria surface to the Rh cluster.[24] The mechanism takes advantage of the well known oxygen storage properties of ceria and explains the experimental observation that CO can react with O from the ceria surface, generating Ce^{3+}. Here, we investigate the mechanism of CO oxidation by isolated Rh atoms supported on different CeO_2 surfaces. The main focus is to understand the role of surface O atoms from the ceria support in the reaction mechanism and the reactivity differences between Rh atoms dispersed on different terminations of ceria.

2 Calculation details

Density functional theory (DFT) with the PBE (Perdew–Burke–Ernzerhof) functional[28] as implemented in the Vienna *Ab initio* Simulation Package (VASP)[29–31] was employed. A Hubbard U term was added to the PBE functional (DFT + U) employing the rotationally invariant formalism by Dudarev *et al.*,[32] in which only the difference ($U_{eff} = U - J$) between the Coulomb U and exchange J parameters enters. Spin-polarized calculations were performed. The projector augmented wave method (PAW)[33–35] was used to describe the interaction between the ions and the electrons with the frozen-core approximation.[34] The Ce (4f, 5s, 5p, 5d, 6s), O (2s, 2p), and Rh (4p, 4d, 5s) electrons were treated as valence states using a plane-wave basis set with a kinetic energy cut off of 400 eV. For Ce, a value of $U_{eff} = 4.5$ eV was used, which was calculated self-consistently by Fabris *et al.*[36] using the linear response approach of Cococcioni and de Gironcoli.[37] This value is within the 3.0–5.5 eV range reported to provide localization of the electrons left upon oxygen removal from CeO_2.[38] The $CeO_2(111)$, $CeO_2(110)$ and $CeO_2(100)$ surfaces were modeled as periodic slabs with three, five and four CeO_2 trilayers, respectively, and a vacuum space between the slabs larger than 10 Å. In Fig. 1, we provide the bulk truncated structures of the three surfaces, respectively. Bulk CeO_2 contains four-fold coordinated O and eight-fold coordinated Ce. The $CeO_2(111)$ surface exposes three-fold O and seven-fold Ce atoms. The $CeO_2(110)$ surface also exposes three-fold O as well as six-fold Ce atoms. The $CeO_2(100)$ surface exposes two-fold O and six-fold Ce atoms. It is noted that $CeO_2(111)$ has a compact structure consisting of an uninterrupted Ce–O bonding network all over the surface, whereas the $CeO_2(110)$ and (100) surfaces contain separated rows of CeO_2 units. These structural differences are consistent with the fact that the (111) facet is the most stable one, followed by the (110) and (100) surfaces. For $CeO_2(111)$ and (100), a 3×3 surface cell was utilized and a 2×3 cell for $CeO_2(110)$. Due to the large size of the surface cells ($\sim$121 Å^2 for each surface cell), only a $1 \times 1 \times 1$ k-point mesh was used for the Brillouin zone integration. The use of a higher $2 \times 2 \times 1$ k-point mesh lead to negligible energy differences. As an example, the difference for CO adsorption on Rh/$CeO_2(111)$ using the different k-point setting is 0.014 eV. The bulk equilibrium lattice constant (5.49 Å) previously calculated by PBE + U ($U_{\rm eff} = 4.5$ eV) was used.[39,40] The bottom layers were kept fixed to their bulk position. Atoms were relaxed until forces were smaller than 0.05 eV Å^{-1}. The location and energy of transition states were calculated with the climbing-image nudged elastic band method.[41]

It can be shown from first principles that an exact description of reaction kinetics is given by the master equation[42]

$$\frac{dP_\alpha}{dt} = \sum_\beta \left[W_{\alpha\beta} P_\alpha - W_{\beta\alpha} P_\beta \right].$$

Here, α and β are all possible configurations of the system, the P's are their probabilities, and the W's are rate constants. We have solved this equation using kinetic Monte Carlo (kMC).[43] Usually, the number of configurations for surface reactions is very large, but in the present case there are only limited configurations for a single Rh adatom plus adsorbates and a small part of the ceria support. This means that it is also possible to solve the master equation by use of a numerical ODE-solver. Because the rate constants differ by many orders of magnitude, there are many reaction conditions at which numerical problems are encountered. We have therefore used kMC as it is more robust.

3 Results and discussion

3.1 Rh/CeO_2 (111)

The potential energy diagram and structures of the surface intermediates and the transition states for the reaction on this model are shown in Fig. 2 and 3, respectively. Our previous work showed that a single Rh atom prefers to adsorb on the three-fold hollow site formed by the surface oxygen atoms of $CeO_2(111)$.[14] The Rh atom is oxidized by surface cerium with the resulting reduced Ce^{3+} ion being in the next nearest neighbor position to the Rh ion. CO adsorbs on the ionic Rh

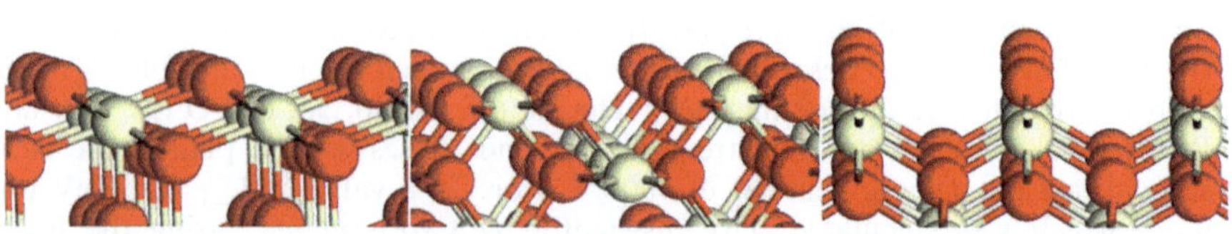

Fig. 1 Bulk truncated structures (side view) of CeO_2 (a) (111), (b) (110) and (c) (100) surfaces.

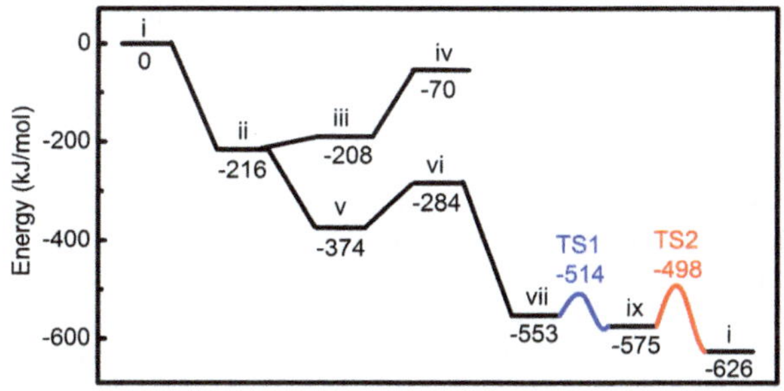

Fig. 2 Reaction energy diagram for CO oxidation on $Rh/CeO_2(111)$.

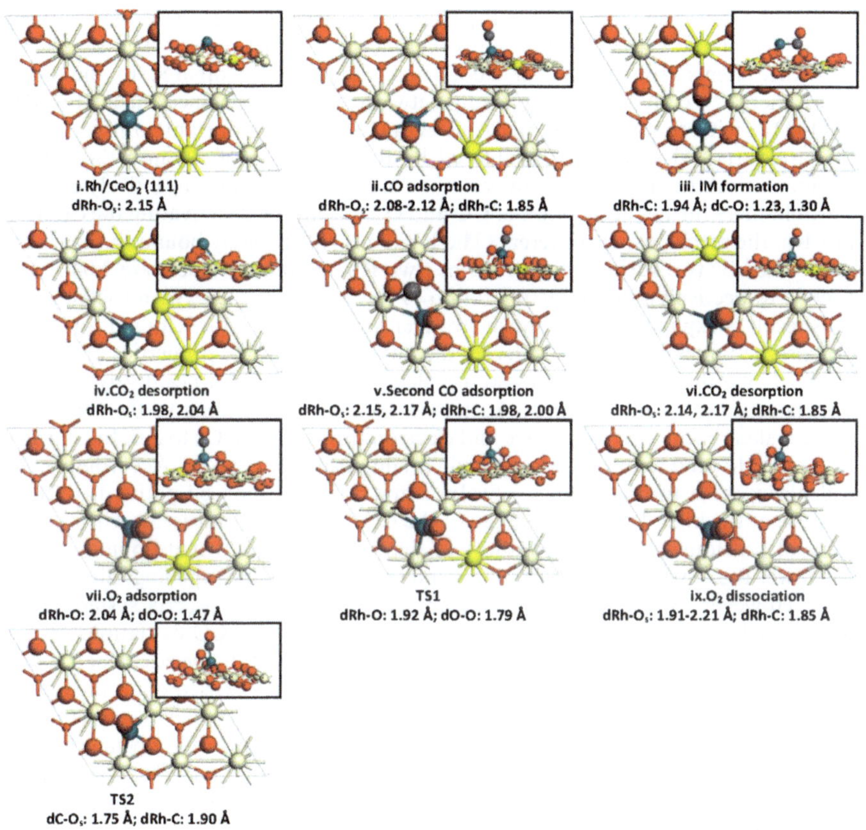

Fig. 3 The structures of the intermediate and transition state structures on $Rh/CeO_2(111)$ with the top slab layer and the cluster shown. Color scheme: red sphere (surface O), white sphere (Ce^{4+}), blue sphere (Rh), black sphere (C) and yellow sphere (Ce^{3+}). The subsurface O atoms are represented by small spheres.

site with an adsorption energy of −216 kJ mol^{-1}. During CO adsorption, Rh migrates from its initial three-fold hollow site to the neighbouring bridge site (state ii, Fig. 3). Rh is not only coordinated to two surface O atoms with bond distances of 2.08 and 2.13 Å, but also to one subsurface O atom present at a distance of 2.13 Å from Rh. The distance of the C atom to Rh is 1.85 Å. The bond distance of C–O is 1.16 Å, only slightly affected with respect to the computed gas-phase value (1.14 Å).

The subsurface O anion coordinating to the Rh atom is slightly elevated compared to its original position. The adsorbed CO then migrates to form an intermediate structure with the C also connected to a surface O (state iii, Fig. 3). As will be described later, desorption of this CO_2* complex is difficult, hindering the likelihood of reaction path I. Thus, we did not try to locate a transition state for this CO migration process. For the adsorbed CO_2* complex, the Rh–C bond distance was elongated to 1.94 Å. The Rh was pulled out slightly being only coordinated to two bridge O atoms at a distance of 2.13 Å, while the bond with the subsurface O is broken. It takes 138 kJ mol^{-1} to desorb CO_2. Similar to findings in previous theoretical work on Au_{13}/CeO_2,[22] the desorption energy is equivalent to the sum of the energies for the removal of a single oxygen atom from CeO_2 and the oxidation of CO to CO_2. Even when we consider the contribution of the entropy for CO_2 desorption ($T\Delta S = -64$ kJ mol^{-1} at 298 K), the rate of this elementary step will be very low at temperatures typically employed for CO oxidation. Recently, the corresponding mechanism was also explored for Pt_1/FeO_x, in which CO_2 desorption is relatively easy (76 kJ mol^{-1}); the reaction is exothermic by -19 kJ mol^{-1}.[3] The significant difference stems from the lower Fe–O bond energy as compared to the Ce–O bond energy. To make a comparison between the reaction mechanism for Rh/CeO_2(111) with that proposed for Au/CeO_2(111),[23] it is important to note that in the latter study desorption is facilitated by the migration of Au into the oxygen vacancy site. After migration, however, the Au atom will become negatively charged which hinders further adsorption of CO and O_2, essentially deactivating the catalyst. For Rh, the mechanism is different. The Rh ion is still strongly bound to the bridge site formed by two surface oxygen atoms (two Rh–O bond distances are 2.04 Å and 1.99 Å, respectively), inhibiting migration of the Rh ion to the oxygen vacancy site. This result is explained by the much stronger metal–O bond for Rh than for Au. It can also be understood by considering the different electronic properties of Rh and Au. For Au, the 6s orbital can take up electrons, making Au negatively charged. This is causing the migration of Au to the vacancy site, leading to the formation of a covalent bond with surface Ce and electron transfer from Ce to Au. In contrast, Rh prefers to bind to oxygen anions and remain positively charged. Accordingly, Rh will not migrate to the vacancy site. In summary, the proposed path I is difficult because of the high desorption energy of the CO_2* complex, consistent with conclusions of previous work.[22] The authors of that work suggested that doping of ceria will lower the oxygen vacancy formation energy, thus possibly opening this reaction path. However, considering that the stability of substitutional ceria systems is an issue of debate,[2,7] we explored here alternative pathways for CO oxidation.

In this study we also considered the formation of $Rh(CO)_2$ species, which is well documented in the experimental literature.[15,44] Abbet *et al.* carried out a theoretical study of CO oxidation by isolated $Pd(CO)_2$ supported on MgO.[45] Thus, it would be interesting to examine if two adsorbed CO molecules on Rh would open a reaction path for CO oxidation. A second CO molecule can adsorb on Rh/CeO_2 with an adsorption energy -158 kJ mol^{-1}, which is much lower than the adsorption energy of the first CO (-216 kJ mol^{-1}). This can be understood from the bond order conservation principle. After optimization, the C from the second adsorbed CO is connected to a surface O (1.31 Å), with the surface O slightly pulled out of the surface (state v, Fig. 3). The distance of C to Rh is 2.01 Å. It takes only 90 kJ mol^{-1} to desorb CO_2, much less than CO_2 desorption in path I (138 kJ mol^{-1}). After desorption, one O vacancy on the ceria surface is generated. At the same time, one more surface Ce ion is reduced. The reaction further proceeds by adsorption of molecular O_2 from the gas phase with one O atom going to the vacancy site, the other O being connected to both Rh and a surface Ce. The bond distance of the adsorbed O_2 is elongated to 1.47 Å, suggesting the bond activation of adsorbed O_2. Upon adsorption of O_2, one of the surface Ce^{3+} is re-oxidized to Ce^{4+}. The barrier for O_2 dissociation is only 40 kJ mol^{-1}. For the transition state structure, the O–O bond distance is elongated to 1.90 Å. The final state structure is 20 kJ mol^{-1} more stable than the initial state. One of the dissociated O atoms sits on the bridge site between Ce and Rh, with bond distances of 2.25 Å and 1.91 Å, respectively, while the other O atom heals the surface vacancy site. The remaining surface Ce^{3+} is re-oxidized to Ce^{4+}. The O atom on the interface between Rh and Ce then reacts with adsorbed CO to form CO_2, which desorbs into the gas phase. The barrier for this process is 77 kJ mol^{-1} and releases 51 kJ mol^{-1}. The single Rh ion migrates back to the three-fold hollow site of the ceria surface, completing the reaction cycle. The adsorption of a second CO on Rh plays an important role in the whole reaction cycle for the $Rh/CeO_2(111)$ model. By comparing the structures of the $Rh(CO)/CeO_2$ (state iii, Fig. 3) and $Rh(CO)_2/CeO_2$ (state v, Fig. 3), we can identify two factors influencing the desorption energy of CO_2: (i) the bond strength of O from CO_2 to surface Ce and (ii) the bond strength of C from CO_2 to Rh. For $Rh(CO)/CeO_2$, the O from CO_2 is connected to three surface O, and Rh connects only to one C atom. Accordingly, the bond strengths of O to Ce and Rh to C are larger than that of $Rh(CO)_2/CeO_2$.

3.2 Rh/CeO_2 (110)

The potential energy diagram and structures of the surface intermediates and the transition states for the reaction on $CeO_2(110)$ are shown in Fig. 4 and 5, respectively. The preferred adsorption site for Rh on this surface is the four-fold hollow site formed by the surface O atoms. The binding energy of Rh is much stronger than on the (111) surface (140 kJ mol^{-1} more stable). This corresponds to a much shorter Rh–O bond (2.01 Å) as compared to the $Rh/CeO_2(111)$ surface (2.15 Å). Two electrons are transferred to the ceria surface, generating two Ce^{3+} ions. A CO molecule adsorbs on the top site of Rh with an adsorption energy of 103 kJ mol^{-1}, which is much lower than on $Rh/CeO_2(111)$ (−216 kJ mol^{-1}). This can be understood from the bond order conservation principle: the coordination number of Rh to surface O are three and four for $Rh/CeO_2(111)$ and $Rh/CeO_2(110)$, respectively. The distance of the C atom of CO to Rh is 1.87 Å. The bond distance of CO* is 1.17 Å. These parameters are consistent with weak adsorption of CO compared with that on $Rh/CeO_2(111)$. The adsorbed CO then tilts to a surface O atom with negligible barrier, forming an intermediate structure (state iii, Fig. 5), which is more stable by 54 kJ mol^{-1}. For this intermediate structure, one of the surface O atoms is slightly shifted to the Rh, and one subsurface O atom below the Rh adatom is lifted to the Rh (Rh–O: 2.15 Å). In this structure, Rh is five-fold coordinated, namely to four surface O atoms (Rh–O: 2.07–2.12 Å) and a subsurface O atom (Rh–O: 2.14 Å). Subsequently, CO* reacts with the slightly pulled out O atom to form a CO_2* complex. The barrier for this process is 47 kJ mol^{-1} and releases 55 kJ mol^{-1}. One more surface cerium is reduced to Ce^{3+} in this process, which arises from the generated oxygen vacancy. For the TS, the bond length of C to the O atom is 1.94 Å. The CO_2 desorption energy is 76 kJ mol^{-1}, much smaller than the corresponding value for the (111) surface (138 kJ mol^{-1}). During the process of the CO_2

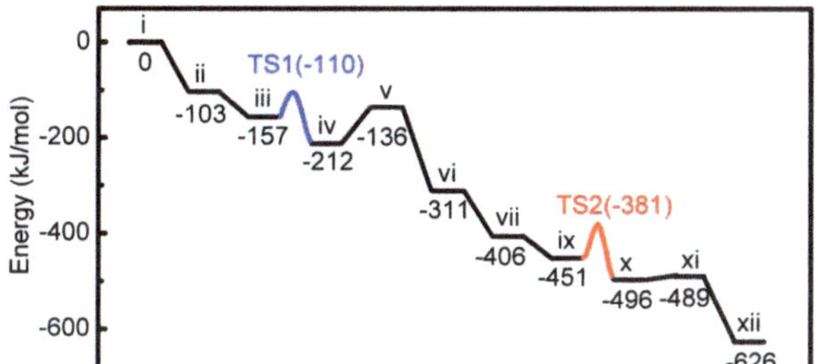

Fig. 4 Reaction energy diagram for CO oxidation on $Rh/CeO_2(110)$.

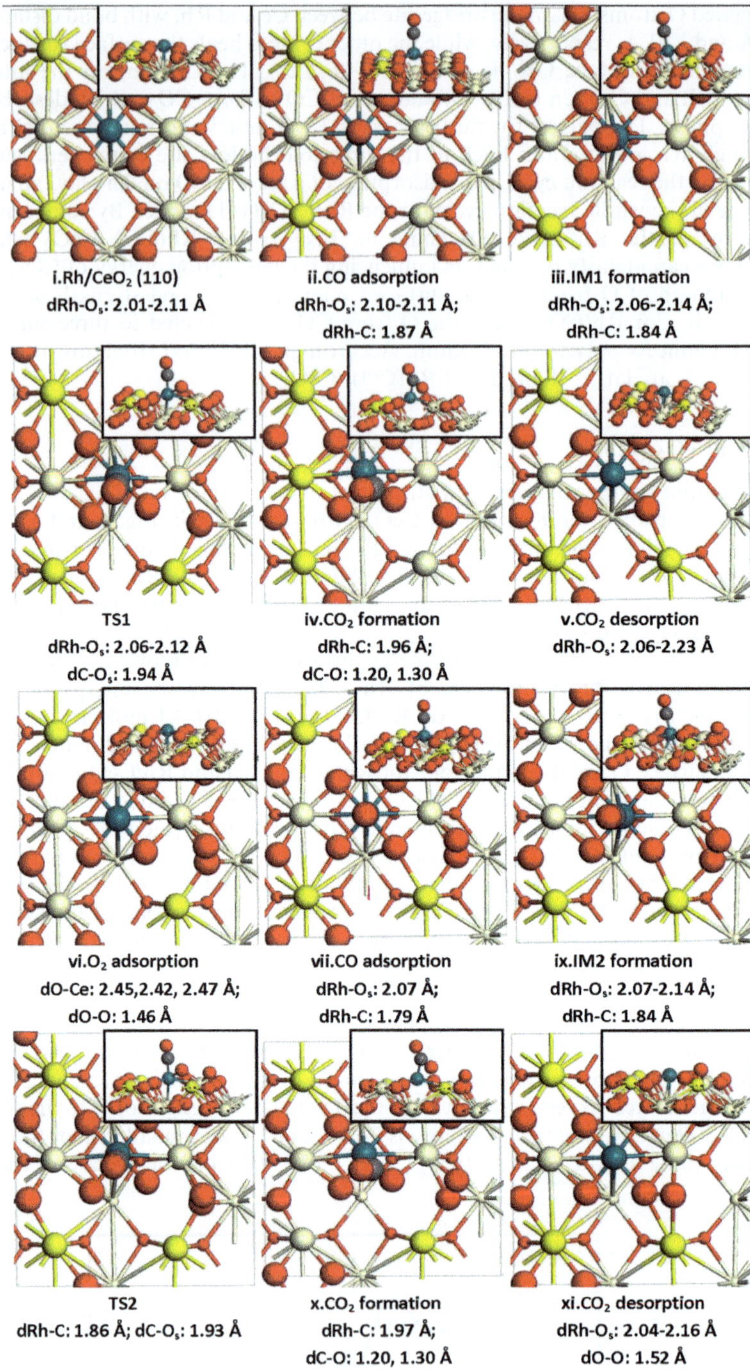

Fig. 5 Structures of the intermediates and transition state structures for CO oxidation on Rh/CeO_2(110).

desorption, the nearby surface O spontaneously migrates to the generated oxygen vacancy site, forming a new bond with Rh (state v, Fig. 5). Accordingly, the generated vacancy migrates to the next nearest neighbor of the Rh adatom.

The oxygen vacancy migration mechanism has been mentioned as one of the main reasons for the high catalytic activity of ceria. However, Esch *et al.* reported that diffusion of a single oxygen vacancy on $CeO_2(111)$ is very slow at room temperature.[46] A recent theoretical work by the group of Hu predicts a high barrier for the direct vacancy hopping mechanism (1.6 eV),[47] consistent with the experimental findings. Our results show that oxygen migration is much more easy for $CeO_2(110)$. The facile migration of the surface O to fill the generated vacancy site close to the Rh site makes CO_2 desorption possible. To provide evidence for this conclusion, we also computed the desorption energy without the migration of the O atom (based on state v, Fig. 5). In this case, adsorbed CO further reduces the additional O atom and no surface O migration is available. The desorption energy is now much higher (167 kJ mol^{-1}). This confirms that the migration of the surface O on the $CeO_2(110)$ surface is crucial to facilitate CO_2 desorption and to explain the high reactivity.

After CO_2 desorption and oxygen vacancy generation, an O_2 molecule adsorbs to the vacancy with an adsorption energy of 175 kJ mol^{-1}. The O–O bond distance is elongated slightly to 1.46 Å. This is because of the electron transfer from the ceria support to the anti-bonding orbital of O_2. Accordingly, one of the Ce^{3+} ions was oxidized to Ce^{4+}. CO adsorbs on the Rh site with an adsorption energy of 95 kJ mol^{-1}. The adsorbed CO tilts to the surface O atom, forming intermediate structure state ix in Fig. 5. The Rh adatom is coordinated to four surface O and one subsurface O (Rh–O distance: 2.13 Å). Oxidation of CO occurs with a barrier of 70 kJ mol^{-1}. In the transition state, the distance of C to O is 1.93 Å. It only takes 7 kJ mol^{-1} to desorb CO_2*, much smaller than the desorption energy of the first CO_2 molecule. During desorption of CO_2, the adsorbed O_2* spontaneously migrates to the generated vacancy site close to the Rh, forming a new bond with Rh (Rh–O: 2.16 Å) (state xi, Fig. 5). The bond distance of O–O is further elongated to 1.52 Å. Subsequently, O_2* dissociates with negligible barrier and releases 137 kJ mol^{-1}, with one of the O atoms filling the vacancy site and completing the reaction cycle.

3.3 Rh/CeO_2(100)

The most stable adsorption site for Rh on $CeO_2(100)$ is the bridge site between the surface O atoms. One surface Ce^{4+} is reduced to Ce^{3+} *via* an electron transferred from Rh. CO adsorption occurs on the top site of Rh with an adsorption energy of 159 kJ mol^{-1}. Various possible reaction paths were explored to react the adsorbed CO with surface O atoms, namely the surface O atoms attached to Rh and neighboring to Rh (Fig. 6). Exploration of a reaction pathway involving the adsorbed CO molecule with the O atoms bound to Rh always led to the initial CO adsorption state. The distance of the other (four) surface O atoms to CO is too large (4.65–5.60 Å, Fig. 6) for the reaction with CO. The adsorption of a second CO molecule does not change these findings. The conclusion is therefore that the structure of the Rh/$CeO_2(100)$ model is such that CO adsorbed to Rh cannot be oxidized. This provides a tentative explanation for the low activity of Rh supported on ceria nanocubes, exposing mainly (100) planes.[4]

3.4 Kinetic Monte Carlo simulations

Kinetic Monte Carlo calculations were performed for the Rh/$CeO_2(111)$ and Rh/$CeO_2(110)$ model systems. The Rh/$CeO_2(100)$ system was not further considered, because no reasonable reaction pathway has been identified. Under experimental conditions ($P_{CO} = P_{O2} = 0.045$ atm, $T = 333$ K), the kMC simulation predicts much higher activity in CO oxidation for Rh/$CeO_2(110)$ (TOF = 210 s^{-1}) than Rh/$CeO_2(111)$ (17 s^{-1}). The higher reactivity of the former system is in accordance with experimentally observed differences.[4] The much higher activity of

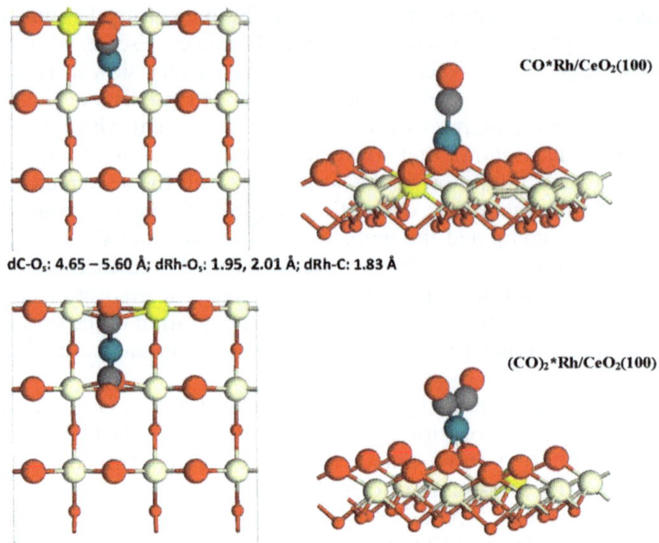

Fig. 6 The structures of CO adsorbed on $Rh/CeO_2(100)$.

$Rh/CeO_2(110)$ arises from more facile migration of surface oxygen as discussed above. Note that these simulations predict reaction orders with respect to CO and O_2 of zero for both structural models. As discussed in detail in our previous work,[24] this is at odds with experimental results showing oxygen reaction orders of unity. This discrepancy is due to the low sticking coefficient for oxygen adsorption, which is difficult to predict.[24]

Instead of directly participating in the CO oxidation, molecular oxygen adsorbs at the interface between the oxygen vacancy and the Rh atom followed by facile dissociation. This conclusion is similar to the one at the basis of a previously proposed mechanism involving Au-doped ceria[22] and the mechanism for a Rh_3O_x cluster.[24] This new mechanism can successfully explain why experimentally CO_2 is detected even without addition of O_2,[5,9] in contrast to earlier models.[22] The oxygen vacancy is generated *via* desorption of the CO_2 from the support surface, accompanied by reduction of Ce^{4+} to Ce^{3+}. This is in accordance with results from Raman spectroscopy that show that the Ce^{3+} density increases with the CO oxidation reaction.[9] Ce^{3+} ions were suggested to play a critical role for various reactions employing ceria as the support.[48,49] Kim *et al.* assumed a mechanism involving molecular oxygen adsorption on a Ce^{3+} site close to the metal.[22] In their mechanism, adsorbed molecular oxygen reacts with adsorbed CO. In our mechanism, in contrast, Ce^{3+} formation is due to the generation of an oxygen vacancy during the CO oxidation process. Our mechanism and the previous proposed mechanism for Au-doped ceria[23] suggest that Ce^{3+} ions are formed during the CO oxidation reaction. That is, Ce^{3+} formation is the result instead of the reason of the CO oxidation reaction. In line with this, Raman spectroscopic experiments show that the density of Ce^{3+} correlates to the rate of CO oxidation reaction.[9]

4 Conclusions

In this contribution, possible reaction paths of CO oxidation were studied for isolated Rh atoms supported on three CeO_2 surfaces with (111), (110) and (100) terminations. On $Rh/CeO_2(111)$, the reaction only proceeds when two CO molecules are adsorbed on the Rh ion. The presence of a second CO molecule facilitates the desorption of CO_2. Gas phase oxygen adsorbs on the vacancy site followed by dissociation. The other CO molecule then reacts with one of the O atoms leading to CO_2 desorption, completing the reaction cycle. On Rh/CeO_2(110), adsorbed CO can easily react with a ceria surface O atom due to the lower Ce–O bond energy. The facile migration of surface O atoms facilitates the CO_2 desorption process, which is much easier for Rh/CeO_2(110) than for Rh/CeO_2(111). Molecular oxygen will adsorb on the resulting vacancy. After desorption of the second CO_2 product molecule by reaction of adsorbed CO with another surface O atom, the adsorbed oxygen molecule migrates spontaneously to the vacancy site and dissociates with negligible barrier. The role of molecular oxygen is to heal the oxygen vacancy rather than being involved in a direct reaction with adsorbed CO. The Rh/CeO_2(100) model was found to be inactive for CO oxidation, mainly because of the geometric constraints for the adsorbed CO molecule to react with one of the surface O atoms, despite the low Ce–O bond energy of the CeO_2(100) surface. The main reason is the large distance between the C of adsorbed CO and the ceria O surface atoms. The particularities of the CO oxidation mechanism for isolated Rh atoms on these ceria surfaces are in agreement with the experimental activity trends.

Acknowledgements

Advanced Dutch Energy Materials Innovation Lab (ADEM) is acknowledged for financial support. Supercomputing facilities were provided by the Netherlands Organization for Scientific Research (NWO). The authors appreciate useful discussions with Prof. Rutger van Santen, Dr. Cristina Popa, Dr. Michel Ligthart and Dr. Volkan Degirmenci.

References

1 R. A. van Santen, M. Neurock and S. G. Shetty, *Chem. Rev.*, 2010, **110**, 2005–2048.
2 M. Flytzani-Stephanopoulos and B. C. Gates, *Annu. Rev. Chem. Biomol. Eng.*, 2012, **3**, 545–574.
3 B. Qiao, A. Wang, X. Yang, L. F. Allard, Z. Jiang, Y. Cui, J. Liu, J. Li and T. Zhang, *Nat. Chem.*, 2011, **3**, 634–641.
4 D. A. J. M. Ligthart, R. A. van Santen and E. J. M. Hensen, *Angew. Chem., Int. Ed.*, 2011, **50**, 5306–5310.
5 J. Guzman, S. Carrettin and A. Corma, *J. Am. Chem. Soc.*, 2005, **127**, 3286.
6 J. A. Farmer and C. T. Campbell, *Science*, 2010, **329**, 933–936.
7 (*a*) M. Hatanaka, N. Takahashi, T. Tanabe, Y. Nagai, K. Dohmae, Y. Aoki, T. Yoshida and H. Shinjoh, *Appl. Catal., B*, 2010, **99**, 336–342; (*b*) M. Hatanake, N. Takahashi, T. Tanabe, Y. Gagai, A. Suda and H. Shinjoh, *J. Catal.*, 2009, **266**, 182–190.
8 G. N. Vayssilov, Y. Lykhach, A. Migani, T. Staudt, G. P. Petrova, N. Tsud, T. Skála, A. Bruix, F. Illas, K. C. Prince, V. Matolín, K. M. Neyman and J. Libuda, *Nat. Mater.*, 2011, **10**, 310–315.
9 Y. Lee, G. He, A. J. Akey, R. Si, M. Flytzani-Stephanopoulous and I. P. Herman, *J. Am. Chem. Soc.*, 2011, **133**, 12952–12955.
10 H. Yoshida, Y. Kuwauchi, J. R. Jinschek, K. Sun, S. Tanaka, M. Kohyama, S. Shimada, M. Haruta and S. Takeda, *Science*, 2012, **335**, 317–319.
11 T. Uchiyama, H. Yoshida, Y. Kuwauchi, S. Ichikawa, S. Shimada, M. Haruta and S. Takeda, *Angew. Chem., Int. Ed.*, 2011, **50**, 10157–10160.
12 (*a*) L. L. Murrell, S. J. Tauster and D. R. Anderson, *Stud. Surf. Sci. Catal.*, 1991, **71**, 275–289; (*b*) L. L. Murrell, S. J. Tauster and D. R. Anderson, *Stud. Surf. Sci. Catal.*, 1993, **75**, 681–690.
13 D. A. J. M. Ligthart, R. A. van Santen and E. J. M. Hensen, *J. Catal.*, 2011, **280**, 206–220.
14 W. Song, C. Popa, A. P. J. Jansen and E. J. M. Hensen, *J. Phys. Chem. C*, 2012, **116**, 22904–22915.
15 J. Xu, D. R. Mullins and S. Overbury, *J. Catal.*, 2006, **243**, 158–164.
16 X. W. Liu, K. B. Zhou, L. Wang, B. Y. Wang and Y. D. Li, *J. Am. Chem. Soc.*, 2009, **131**, 3140–3141.
17 Z. Wu, M. Li and S. H. Overbury, *J. Catal.*, 2012, **285**, 61–73.
18 M. Nolan, J. E. Fearon and G. W. Watson, *Solid State Ionics*, 2006, **177**, 3069–3074.

19 M. Nolan, S. C. Parker and G. W. Watson, *Surf. Sci.*, 2005, **595**, 223–232.
20 F. Chen, D. Liu, J. Zhang, P. Hu, X. Q. Gong and G. Lu, *Phys. Chem. Chem. Phys.*, 2012, **14**, 16573–16580.
21 H. Y. Kim and G. Henkelman, *J. Phys. Chem. Lett.*, 2013, **4**, 216–221.
22 H. Y. Kim, H. M. Lee and G. Henkelman, *J. Am. Chem. Soc.*, 2012, **134**, 1560–1570.
23 M. F. Carmellone and S. Fabris, *J. Am. Chem. Soc.*, 2009, **131**, 10473–10483.
24 W. Song, A. P. J. Jansen, V. Degirmenci, D. A. J. M. Ligthart and E. J. M. Hensen, *Chem. Commun.*, 2013, **49**, 3851–3853.
25 X. Q. Gong, R. Raval and P. Hu, *Phys. Rev. Lett.*, 2004, **93**, 106104.
26 L. M. Molina and B. Hammer, *Appl. Catal., A*, 2005, **291**, 21–31.
27 (*a*) C. J. Zhang, A. Michaelides, D. A. King and S. J. Jenkis, *J. Chem. Phys.*, 2008, **129**, 194708; (*b*) C. J. Zhang, A. Michaelides, D. A. King and S. J. Jenkis, *J. Phys. Chem. C*, 2009, **113**, 6411–6417.
28 J. P. Perdew, K. Burke and M. Ernzerhof, *Phys. Rev. Lett.*, 1996, **77**, 3865–3868.
29 G. Kresse and J. Furthmuller, *Comput. Mater. Sci.*, 1996, **6**, 15–50.
30 G. Kresse and J. Furthmuller, *Phys. Rev. B*, 1996, **54**, 11169–11186.
31 http://www.vasp.at/.
32 S. L. Dudarev, G. A. Botton, S. Y. Savrasov, C. J. Humphreys and A. P. Sutton, *Phys. Rev. B*, 1998, **57**, 1505–1509.
33 P. E. Blöchl, *Phys. Rev. B*, 1994, **50**, 17953–17979.
34 G. Kresse and D. Jouber, *Phys. Rev. B*, 1999, **59**, 1758–1775.
35 O. Bengone, M. Alouani, P. E. Blöchl and J. Hugel, *Phys. Rev. B*, 2000, **62**, 16392–16401.
36 S. Fabris, S. de Gironcoli, S. Baroni, G. Vicario and G. Balducci, *Phys. Rev. B*, 2005, **72**, 237102.
37 M. Cococcioni and S. de Gironcoli, *Phys. Rev. B*, 2005, **71**, 035105.
38 C. W. Castleton, J. Kullgren and K. Hermansson, *J. Chem. Phys.*, 2007, **127**, 244704.
39 J. Da Silva, M. Ganduglia-Pirovano, J. Sauer, V. Bayer and G. Kresse, *Phys. Rev. B*, 2007, **75**, 045121.
40 C. Popa, M. Ganduglia-Pirovano and J. Sauer, *J. Phys. Chem. C*, 2011, **115**, 7399–7410.
41 G. Henkelman, B. P. Uberuaga and H. Jonsson, *J. Chem. Phys.*, 2000, **113**, 9901.
42 X. Q. Zhang and A. P. J. Jansen, *Phys. Rev. E*, 2010, **82**, 046704.
43 A. P. J. Jansen, *An Introduction to Kinetic Monte Carlo Simulations of Surface Reactions* (Springer, Berlin, 2012).
44 P. Basu, D. Panayotov and J. T. Yates, *J. Am. Chem. Soc.*, 1988, **110**, 2074–2081.
45 S. Abbet, U. Heiz, H. Hakkinen and U. Landman, *Phys. Rev. Lett.*, 2001, **86**, 5950–5953.
46 E. Esch, S. Fabris, L. Zhou, T. Montini, C. Africh, P. Fornasiero, G. Comelli and R. Rosei, *Science*, 2005, **309**, 752.
47 H. Y. Li, H. F. Wang, Y. L. Guo, G. Z. Lu and P. Hu, *Chem. Commun.*, 2011, **47**, 6105–6107.
48 J. B. Park, J. Graciani, J. Evans, D. Stacchiola, S. D. Senanayake, L. Barrio, P. Liu, J. F. Sanz, J. Hrbek and J. A. Rodriguez, *J. Am. Chem. Soc.*, 2010, **132**, 356.
49 J. A. Rodriguez, J. Graciani, J. Evans, J. B. Park, F. Yang, D. Stacchiola, S. D. Senanayake, S. G. Ma, M. Perez, P. Liu, J. F. Sanz and J. Hrbek, *Angew. Chem., Int. Ed.*, 2009, **48**, 8047.

Morphology, dimension, and composition dependence of thermodynamically preferred atomic arrangements in Ag–Pt nanoalloys

Lei Deng,[ab] Huiqiu Deng,*[b] Shifang Xiao,[b] Jianfeng Tang[ab] and Wangyu Hu*[b]

Received 4th December 2012, Accepted 16th January 2013
DOI: 10.1039/c3fd20138d

The present article is on Metropolis Monte Carlo simulations coupled with semi-empirical potentials to obtain the thermodynamically preferred configurations of Ag–Pt nanoalloys. The effects of particle size, morphology or alloy composition on the surface segregation and the chemical ordering patterns were investigated. Surface segregation of Ag is observed in all Ag–Pt nanoalloys. Such segregation develops quickly as the increase of particle sizes or global Ag composition. Generally, Ag surface enrichment is more apparent for more open particles except for large sized icosahedron (ICO) nanoalloys. The most energetically favorable chemical ordering patterns gradually evolve from Pt-core/Ag-shell to onion-like structures when the global Ag composition increases. Due to the site preference of Ag segregation, the presence of partly alloyed facets and Ag blocked vertices or edges at low global Ag compositions can modify the electronic and geometric structures on the nanoalloys' surface. The coupling between Pt and Ag sites is a topic of particular interest for catalysis. The detailed atomistic understanding of atomic arrangements in Ag–Pt nanoalloys is essential to intelligently design robust and active nanocatalysts with a low cost.

1 Introduction

Noble metal nanoparticles (NPs) have received enormous attention in scientific researches and already been utilized in many technologically important reactions such as in the chemical and petrochemical industry, electrochemical fuel cells, and automobile exhaust purification.[1–13] All the applications result from the NPs excellent catalytic activity, selectivity and stability. Different from their single-metallic counterparts, bimetallic NPs, now referred to as "nanoalloys" (NAs), display new properties and capabilities that make them promising candidates for designing heterogeneous nanocatalysts.[1–13] The heterogeneous catalysis is a typical surface process and the surface atoms of catalysts mainly determine the kinetics of the reactions. Compared to monometallic NPs, supported NAs usually have composition-dependent surface geometric and electronic structures. For a NA system, there exists an additional degree of freedom due to the possibility of one of the species segregating to the surface, which reduces the free energy of the NA.[1,7] Because of the surface segregation and nanoscale effect, the surface structures and chemical ordering patterns of NAs may be myriad and different from those of the ideal bulk alloys, which further tunes the catalytic activity or selectivity by so-called "ensemble and ligand effect".[14–19] It is well known that the atomic arrangement of

[a]*Department of Applied Physics, Hunan Agricultural University, Changsha 410128, China*
[b]*Department of Applied Physics, Hunan University, Changsha 410082, China. E-mail: hqdeng@hnu.edu.cn; wangyuhu2001cn@yahoo.com.cn*

nanocatalysts depends strongly on the alloy composition, particle size or shape, and operating conditions.[1,10–13,20,21] Therefore, systematic studies of the atomic arrangements of thermodynamically preferred NAs play an essential role in intelligently designing durable and active nanocatalysts with low costs.

Alloying of Pt nanocatalysts has been found to be able to reduce Pt loading, to improve catalytic activity and selectivity,[1–7] and to increase the stability of NAs against sintering or dissolution during electrochemical cycle.[9,22] Particular attention has been paid to the Ag–Pt nanosystem as an electrocatalyst for fuel cell relevant processes. Considerable experimental researches have been dedicated to understanding the thermodynamics and kinetics of Ag–Pt NAs, where Ag–Pt NAs were produced with structures between core–shell and homogeneous alloyed structures depending on the synthesis conditions, especially the presence of a suitable stabilizer.[23–32] The performance of Ag–Pt nanocatalysts depends greatly on the degree of surface Ag–Pt alloying which is determined by the nature of the Pt precursor. For example, the mass specific activity of Pt is one order of magnitude higher in methanol electro-oxidation and several times higher in formic acid electro-oxidation than that with no Ag–Pt alloying.[23–25] An unique composition dependence is also found for both electro-oxidation of methanol and catalytic oxidation of *o*-phenylenediamine by hydrogen peroxide.[26,27] In addition, proper alloying with Ag not only improves the CO poisoning of the Pt catalyst but also enhances the catalytic activity greatly.[27] Furthermore, recent studies demonstrate that Ag–Pt NAs with submonolayer coverage of Ag could simultaneously achieve high activity and selectivity for C_2H_2 hydrogenation over large temperature operating windows.[31,32] That is to say, the common trade-off encountered for hydrogenation catalysts, increasing activity at the expense of selectivity, did not occur for Ag–Pt NAs. Therefore, the knowledge of the precise surface structure and chemical ordering pattern of Ag–Pt nanocatalyst is of tremendous interest and importance.

Unfortunately, it is still a very complicated task to obtain the atomic-scale structural and compositional information of NAs at the experimental level.[1,33] Furthermore, it is well-known that the surface structure and atomic arrangement of NAs are influenced by many factors, including the NAs' size, morphology and composition. The complex interplay between these factors represents a complication, due to the lack of reliable experimental protocols for controlling the spatial distribution of the chemical species in a given NA. This is less problematic at the theoretical level, as effective methods are available to explore the atomic-scale structural and compositional information of NAs.[34–48] The first-principles approach has recently evolved into a powerful method to provide insight into the surface chemistry of nanoclusters.[34–36] However, practical nanocatalysts usually consist of over several hundred atoms and the variety of structures in NAs is much richer than that in monometallic NPs, where accurate first-principles approaches become very computationally intensive. As an option, molecular dynamics (MD) and Monte Carlo (MC) simulations with semi-empirical potentials offer effective tools to study the properties of NAs and complement ongoing experimental efforts. For instance, understanding the dynamics of formation, evolution of morphology, surface segregation and atomic scale mixing of monometallic NPs and NAs by MD and MC simulation is still an active area of research.[37–48]

The above considerations motivate systematic studies of the atomic arrangements of Ag–Pt NAs as a function of overall composition, shape, and size, which represent the most interesting elements of NAs with respect to catalytic activity and selectivity. The paper is structured as follows: in Section 2 we briefly describe the structure model of NAs, atomic interaction potentials, and simulation methods; in Section 3.1 the surface segregation, thermodynamic stability and chemical ordering patterns of Ag–Pt NAs are given; in Section 3.2 we present the preference segregation and atomic mixing on the surface of Ag–Pt NAs and discuss their impact on catalytic activity and selectivity; Section 4 contains the conclusions.

2 Model and method

2.1 Structure and magic number of NAs

In the extensive structural survey, guided by actual observations as well as energy minimization of atomistic models with many-body interactions by past experience,[37–42] we constructed a broad range of structural motifs and morphologies as shown in Fig. 1. These include (i) face-centered cubic (FCC) structures with various morphologies, such as octahedron (OCT) and its variants, cube-octahedron (CUB) and truncated octahedron (TOC) obtained by truncating symmetrically the six vertices, thus having square and hexagonal (triangular in the case of CUB) facets; (ii) icosahedron (ICO) structures having a quasi-spherical shape and a close-packed surface with 20 distorted (111)-like facets; (iii) pentagonal decahedral (DEC) arrangements and their variants, Ino-decahedron (IDEC)[49] and Marks-decahedron (MDEC),[50] where reentrant (111) facets are introduced *via* a modified Wulff construction to optimize the energetic stability of NPs. The ICO, CUB and IDEC morphologies are structured in perfect shells, while TOC and MDEC are in imperfect ones. Because of their discrete nature, atomistic models of various structural motifs and morphologies correspond to specific size sequences (magic numbers). According to the magic numbers, a series of model NPs with 10^1–10^4 atoms were considered in the present paper.[50–54] For each composition considered, a starting configuration of Ag–Pt NA was generated by designating randomly the proper numbers of atoms (12.5%, 25.0%, 37.5%, 50.0%, 62.5%, 75.0% and 87.5%, respectively) as Ag atoms, with the remaining designated as Pt atoms.

2.2 Interaction potentials

In the present simulation, the modified analytic embedded atom method (MAEAM) potentials are employed to describe the energetics in metal systems.[42–46] These potentials generally provide good description for many-body interactions and have already been successfully applied for the study of bulk, surface, and NPs of metals and alloys.[42–46] The MAEAM potential functions and the parameters for Ag and Pt were given in refs. 43 and 44. The parameters were determined by fitting to a number of bulk equilibrium properties of metals such as lattice constants, cohesive energy, vacancy formation energy, elastic constants, *etc.* To determine the cross interaction potential for a binary alloy, two adjustable parameters had been used and fitted to

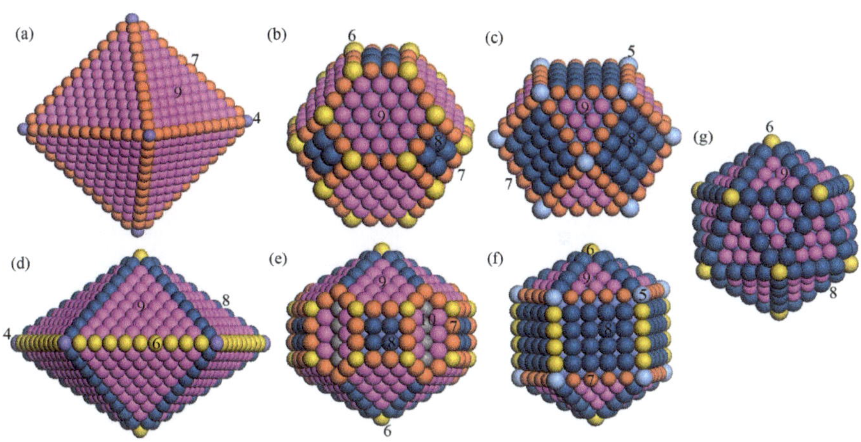

Fig. 1 Atomistic models representing different geometric motifs: (a) OCT, (b) TOC, (c) CUB, (d) DEC, (e) MDEC, (f) IDEC, (g) ICO. The numbers represent the coordination number of different surface sites.

the lattice constants and heats of formation values available for disordered solid solutions and imaginary intermetallics from the first-principles calculations and experimental data. Then various physical properties of Ag–Pt alloys have been tested carefully by means of the interaction potentials.

Fig. 2 shows the heats of formation ΔH for disordered solid solutions and imaginary intermetallics $AgPt_3$ ($L1_2$), AgPt ($L1_0$), and Ag_3Pt ($L1_2$), respectively. The present heats of formation are reasonably consistent with the semi-empirical potentials and first-principles calculations as well as experimental data.[55–60] Therefore, the present Ag–Pt cross interaction potential can be used for a wide range of components with reasonable accuracy. It is clear that the heats of formation of solid solutions and intermetallics are positive, which implies that the chemical ordering between Ag and Pt is unfavorable and these imaginary intermetallics do not exist.[60] Furthermore, the unfavorable interaction between Ag and Pt would result in phase separation in Ag–Pt bulk alloys.

2.3 Metropolis MC simulations

The MC method is particular advantageous in studying the NAs' configurations under thermodynamic equilibrium as it does not require the knowledge of exact diffusion pathways, which may be complex and cooperative in nature. Thus, MC simulations were carried out to probe the global equilibrium structures and segregation properties of Ag–Pt NAs, which involved calculating the initial potential energy using MAEAM potentials, randomly selecting 2 atoms, swapping the 2 atomic positions, and calculating the new potential energy.[20,21,45,46] The acceptance or rejection of the trial is based on the standard Metropolis algorithm: if the change of total energy ΔE is negative, the configuration would be accepted; if not, the configuration is accepted only if $\exp(-\Delta E/k_BT)$ is greater than a random number selected from a uniform distribution.[21] Then, the required statistical quantities of NAs configurations belonging to the NPT ensemble were obtained by averaging over 20 000 MC steps per atom after the system achieving equilibration state. The NAs' equilibrium configurations were obtained under a range of temperatures (T = 50, 300, 600 and 900 K).

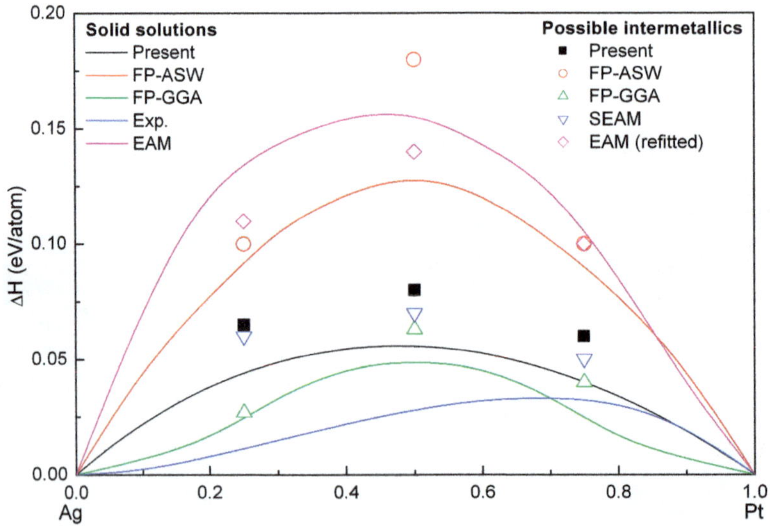

Fig. 2 The heats of formation ΔH for disordered solid solutions (lines) and imaginary intermetallics (symbols). The results of semi-empirical potentials calculations and first-principles calculations as well as experimental data can be found in refs. 55–60.

3 Results and discussion

3.1 Surface segregation, chemical ordering patterns and thermodynamic stabilities of Ag–Pt nanoalloys

In the Ag–Pt bimetallic system, the surface energy of Ag is much less than that of Pt, which favors surface enrichment of Ag.[61–69] The lattice constant of fcc Ag (4.09 Å) is much larger than that of Pt (3.91 Å), thus the segregating of Ag atoms from the bulk to the surface will decrease the elastic strain energy.[66–69] In addition, the positive heat of formation of bulk Ag–Pt alloys also facilitates atomic segregation.[66–69] Therefore, from the point of view of thermodynamics, a strong segregation of Ag atoms onto the surface of Ag–Pt NAs is expected to minimize the free energy.[1,20]

The size dependence of surface segregation in Ag–Pt NAs with fixed morphology, TOC, is shown in Fig. 3. For small NAs, all available Ag atoms segregate to the surface as indicated by the "Max Segregation" lines. In all cases, MC simulations show that Ag is significantly enriched on the surfaces as compared to its bulk composition ratio, serving as an *a posteriori* confirmation of above thermodynamic considerations. It is seen that present surface Ag composition is in good agreement with available MC simulation results using surface modified pair potential (SMPP).[20] The EXAFS results showed a lower average coordination number (CN) for Ag, also indicating partial surface Ag segregation.[30] From Fig. 3, the following general behaviors are observed: Ag atoms strongly segregate to the surface and surface Ag compositions increase with the NAs' sizes.

The composition dependence of surface segregation in Ag–Pt NAs is shown in Fig. 3 and Fig. 4. It is found that Ag atoms strongly segregate again to the surface and surface Ag compositions increase with global Ag compositions. Fig. 3 and Fig. 4 indicate that a nearly pure Ag surface layer will form when there are enough Ag atoms in the NAs. For fixed particle size or global Ag concentration, the formation of a nearly pure Ag surface is limited by the nanoscale effect, which means the supply of surface sites or Ag atoms is finite for NAs. Apparently, this nanoscale effect could be circumvented by increasing the particle sizes or global Ag concentrations, as shown in Fig. 3 and Fig. 4.

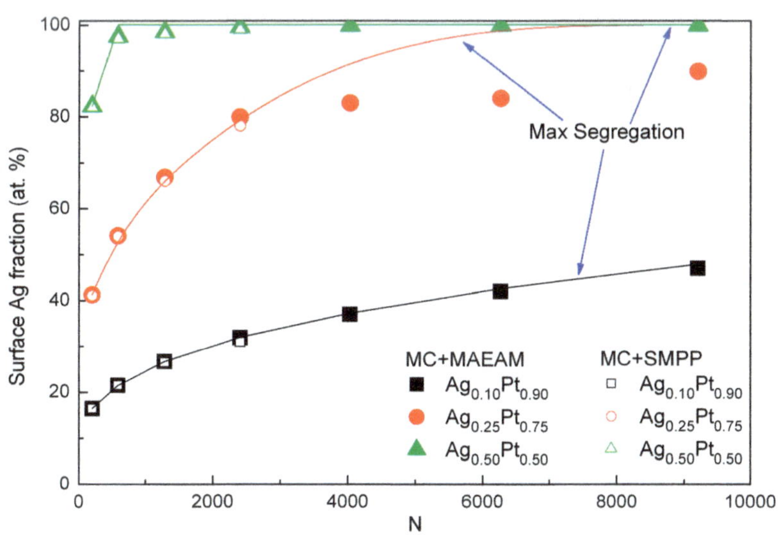

Fig. 3 The segregation in TOC structured Ag–Pt NAs with different composition ratios as a function of particle sizes. The lines represent theoretical max segregation. The full and open symbols refer to the results of present MAEAM potentials and that of SMPP, respectively.

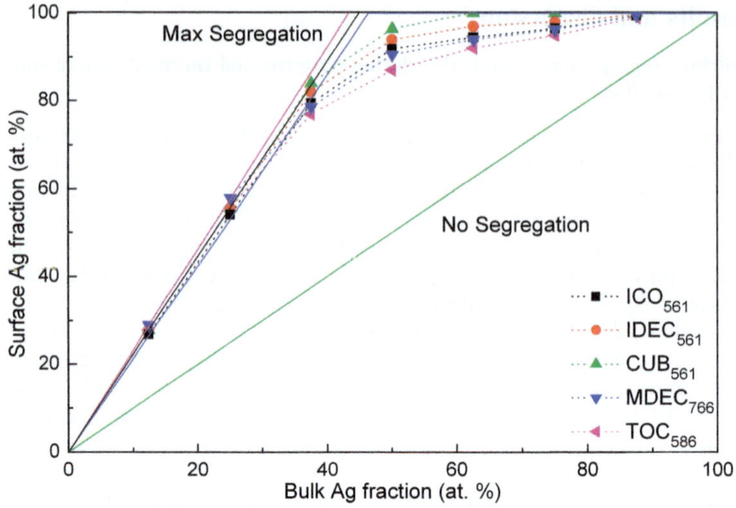

Fig. 4 The composition dependence of surface segregation in Ag–Pt NAs with different morphologies but almost fixed sizes.

Because of the strong surface segregation and the nanoscale effect, the chemical ordering patterns of thermodynamically stable Ag–Pt NAs will be myriad and different from those of the ideal bulk alloys. As an example, the detailed configurations of 561-atom CUB and 766-atom MDEC NAs are shown in Fig. 5 and Fig. 6, respectively. At small global Ag concentration, the lowering of surface free energy overwhelms any bulk free energy changes such that all the available Ag atoms segregate to the surface, while the bulk remains pure Pt. Due to a nanoscale effect, there are never enough Ag atoms to form a Ag monolayer. Therefore, the energetically favorable configurations comprise Pt-core/mixed-shell structure. While a complete surface segregation may give the lowest surface free energy, a Ag monolayer can be obtained when the global Ag concentration is increased. For example, the number of surface atoms constitutes 44.9% of total atoms for 561-atom NAs, then a Pt-core/Ag-shell structure was obtained at global Ag content of 0.45, where the surface energy and mixing energy are both minimized. When the global Ag concentration is further increased, the excess Ag atoms are pushed to the center of NAs as shown in Fig. 5 and Fig. 6. It is of interest to note that the Ag shell did not populate the subsurface layer for higher global Ag concentration, and the subsurface layer was enriched by Pt atoms; then a three-shell onion-like structure appeared. This kind of subsurface arrangement of diluted impurities Pt, Ni, Cu, and Pd has been found in Au–Pt, Ag–Ni, Ag–Cu, and Ag–Pd NAs.[46,64,65] Both

Fig. 5 The detailed atomic configurations (top) and cross sections (bottom) of 561-atom CUB NAs as a function of global Ag content: (a) 0.125, (b) 0.250, (c) 0.375, (d) 0.500, (e) 0.625, (f) 0.750, and (g) 0.875. Ag is in gray and Pt in blue.

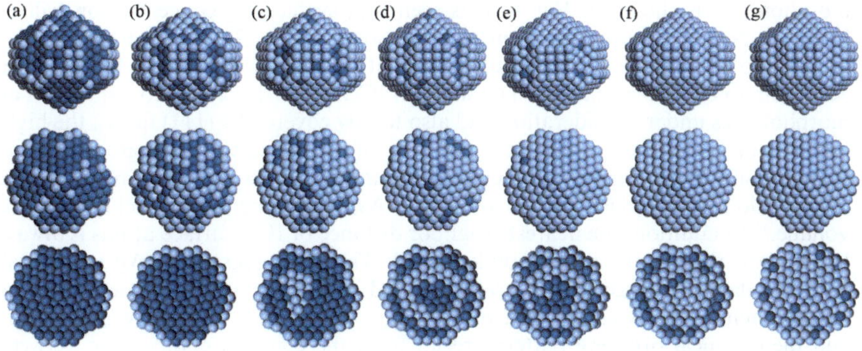

Fig. 6 The side view (top) and top view (middle) of atomic configurations and cross sections (bottom) of 766-atom MDEC NAs as a function of global Ag content: (a) 0.125, (b) 0.250, (c) 0.375, (d) 0.500, (e) 0.625, (f) 0.750, and (g) 0.875. Ag is in gray and Pt in blue.

the onion-like structures and Pt-core/mixing-shell ones have been realized by different synthesis routes.[70–72] While it is apparent that alloy composition has significantly impacted on the chemical ordering patterns of Ag–Pt NAs, it is also likely that these patterns are little dependent on the particle sizes (not shown here).

The morphology and size dependence of surface Ag concentrations for $Ag_{0.25}Pt_{0.75}$ NAs is shown in Fig. 7. It can be seen that surface Ag concentrations of ICO, IDEC, and CUB NAs are very similar, as are those of MDEC and TOC NAs. The surface Ag concentration is promoted by increasing the particle sizes. However, the completed surface segregation of Ag atoms in $Ag_{0.25}Pt_{0.75}$ NAs is lost at a range of particle sizes, which will be discussed in detail later. These results indicate that surface segregation of Ag is a result of a subtle balance between energetic factors such as the surface energy and the strain energy. From the viewpoint of surface energy, the (111) facets are the least favorable surface sites for Ag segregation.[20,46] As a result, these morphology-dependent trends may result from the contribution of (111) facets. The contribution of (111) facets to the surface of considered

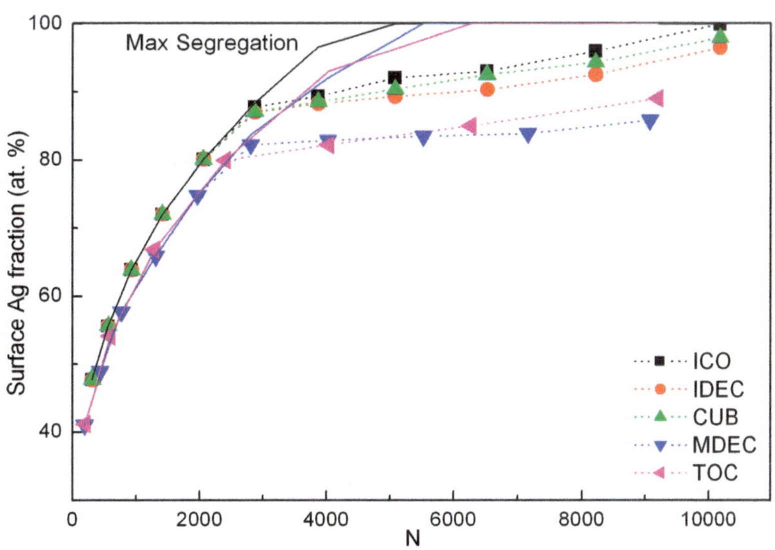

Fig. 7 The segregation *versus* particle sizes of $Ag_{0.25}Pt_{0.75}$ NAs with different morphologies.

morphologies is shown in Fig. 8. It is seen that surface Ag concentration will be higher in NAs with more open surface (as shown in Fig. 7). However, a question arises as to how the highest surface Ag concentration of ICO NAs can be explained according to the above analysis. The shape of ICO NA is the closest to a sphere of all morphologies under consideration and also totally covered by (111) facets; the highest ratio of (111) facets will have a negative effect on the surface segregation and should result in the lowest surface Ag concentration. The answer comes from the contribution of the elastic strain of ICO NAs, which increases quickly with the volume.[41] In addition, the nearest-neighbor distance of ICO surface atoms is always larger than that in the perfect bulk crystal. Thus, the segregated Ag atoms will accommodate this large tensile strain imposed on surface atoms. These results led us to conclude that the size and morphology dependence of surface segregation and the chemical ordering patterns result from the complex competition between lowering the surface energy and the strain energy. The former is always present for all NAs and is dominant for small ones, while the latter is dominant for high strained particles, especially for large-sized ICO NAs. Revisiting Fig. 5 and Fig. 6, it appears that the chemical ordering patterns of Ag–Pt NAs are almost independent of their morphologies.

As mentioned above, the nanoscale effect can be circumvented by increasing the particle sizes or global Ag concentration. As shown in Fig. 3 and Fig. 7, the surface Ag concentration is indeed promoted by increasing the particle sizes. However, the completed surface segregation of Ag atoms in $Ag_{0.25}Pt_{0.75}$ NAs is lost and a partial surface alloy is formed at a range of particle sizes, regardless of particle morphology. The detailed atomic configurations of $Ag_{0.25}Pt_{0.75}$ NAs with large sizes are shown in Fig. 9. It can be seen that surface Pt atoms mainly locate at (111) facets. It is interesting to note that all of the (100) facet Pt atoms are monomers and most of the (111) facet Pt are in the form of monomer, dimer and trimer. We propose that not only the chemical ordering patterns but also the size mismatch between Ag and Pt atoms play a crucial role in the incomplete surface segregation of Ag. The pronounced size mismatch of the two components is responsible for a miscibility gap that promotes surface segregation,[20,68] but at the same time acts as driving force for surface alloy formation at specific Ag concentrations.[58] As mentioned above, the chemical ordering patterns of Ag–Pt NAs are strongly dependent on alloy compositions. At higher or lower Ag global composition, such as $Ag_{0.10}Pt_{0.90}$ or $Ag_{0.50}Pt_{0.50}$ NAs shown in Fig. 3, a Pt-core/mixed-shell or Ag-rich-core/Ag-shell structure is formed,

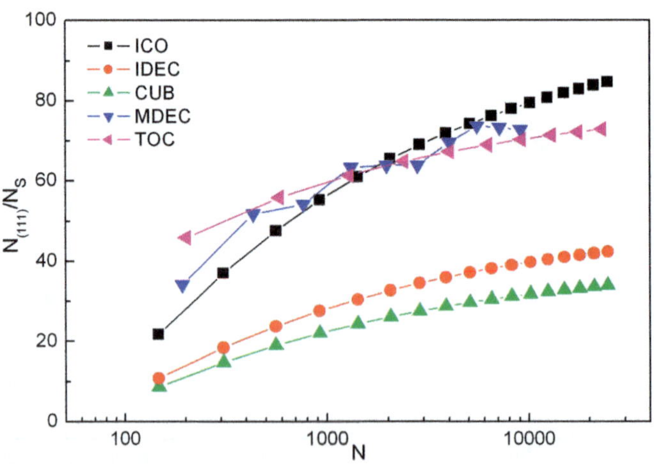

Fig. 8 The contribution of (111) facets to the surface sites *versus* particle sizes for NPs with different morphologies.

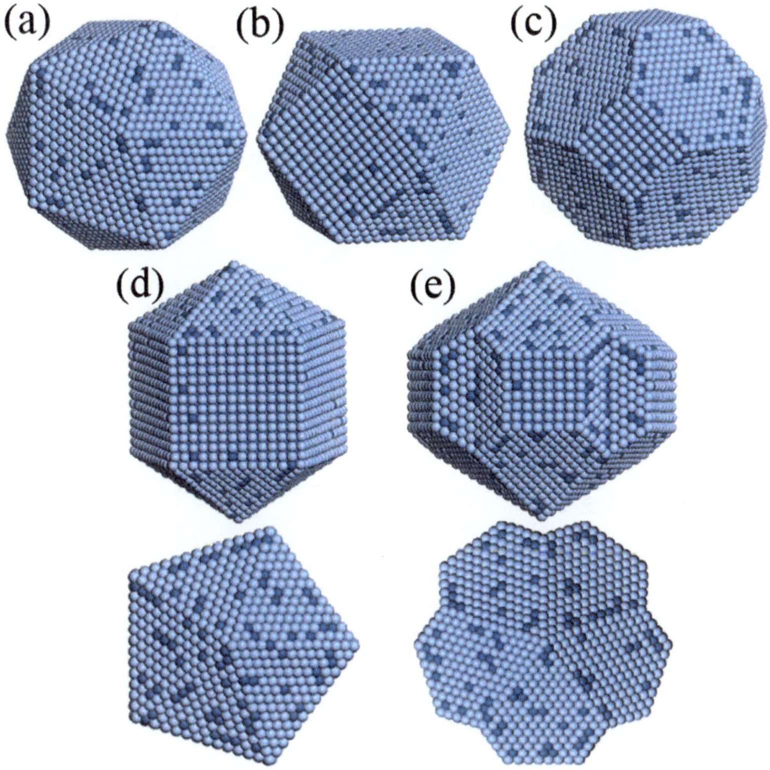

Fig. 9 The detailed atomic configurations of $Ag_{0.25}Pt_{0.75}$ NAs: (a) 8217-atom ICO, (b) 8217-atom CUB, (c) 9201-atom TOC, (d) 8217-atom IDEC, (e) 9075-atom MDEC. Ag is in gray and Pt in blue.

where the strain energy between the shell and core are relative small and can be released completely by surface relaxation. For $Ag_{0.25}Pt_{0.75}$ NAs, if a perfect Pt-core/Ag-shell structure is formed, the strain energy between the Ag shell and Pt core becomes comparable to the segregation energy and cannot be released completely by surface relaxation. Therefore, it is not surprising that completed surface segregation of available Ag in $Ag_{0.25}Pt_{0.75}$ NAs is prohibited and a partial surface alloy is formed.

3.2 Site preferences on NAs surface and their impacts on catalytic performance

The precise atom arrangement on the surface of Ag–Pt NAs is a topic of particular interest for catalysis. Even small changes in the atomic arrangement can result in significant variations in the catalytic activity and selectivity. The NAs' surfaces are composed of low-coordinated vertices, edges and facets with different orientations, which usually have their particular segregation behavior. To examine this particular segregation, we further analyzed the site preferences of Ag atoms on the surface of Ag–Pt NAs. Fig. 10 shows the normalized distribution of Ag on different coordinated sites as a function of global Ag content. For very low Ag content, we observe that Ag prefers CN 5–6 which is suggestive of low coordination sites such as vertices or edges as shown in Fig. 5 and Fig. 6. As the Ag content increases, we observe higher average CN 7–8 which is representative of edges or (100) facets. Further increasing in Ag content results in average CN 9–10, which corresponds to Ag rich (111) facets. The site preference of Ag segregation decreases

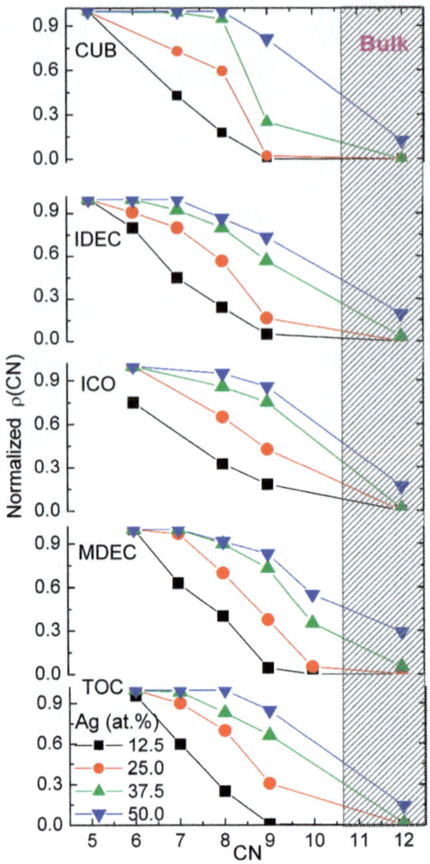

Fig. 10 The normalized distribution of Ag on different coordinated sites as a function of global Ag content for 561-atom CUB.

in the following order: vertex > edge > (100) facet > (111) facet > bulk. This site preference segregation results from CN-dependent surface free energies and is in agreement with the behavior observed for Ag–Pt, Au–Pt, and Au–Pd NAs.[31,32,45,73] In a real system, this site preference of Au or Ag segregation indicates the possibility to synthesize thermodynamically stable NPs with vertices or edges decorated by Au or Ag and Pt or Pd atoms located mainly on the facets.[31,32,73]

It is reasonable to anticipate that the catalytic performance of nanocatalyst would be enhanced, modified or suppressed by alloying that modifies surface electronic and geometric structures. No doubt, the surface atom distribution, the type and proportion of exposed facets as well as the presence of vertices and edges on Ag–Pt NAs will make the catalysis and reaction mechanism highly complex. Thus, it is still a challenge to relate the surface structure and atomic distribution of Ag–Pt NAs to observed catalytic activity and selectivity. In the following, we will focus primarily on selected catalytic reactions of practical interest, including oxygen reduction reaction (ORR), CO electro-oxidation in alkaline media and selective hydrogenation of the C=O group in α,β-unsaturated aldehydes. We believe that these discussion can facilitate identification of possible mechanisms and allow prediction of a good candidate for specific reactions.

In general, most electrocatalytic processes should be more facile in alkaline solutions than in acidic solutions. At the same time, a much wider range of materials are stable in alkaline environments, including much less expensive materials such as Ag, while very few electrode materials are stable under strongly acidic conditions, especially under strongly oxidizing conditions. As an example, Ag–Pt nanocatalysts have been utilized for the ORR in alkaline media especially in chloralkali electrolysis, exhibiting comparable catalytic activities to pure Pt catalysts but showing better long-term stabilities at industrial NaOH concentrations and elevated temperatures.[74–76] It is well known that OH adsorbs at low-coordination vertex and edge sites of Pt NPs preferentially over (111) facet sites.[77–79] Because the ORR rate is limited by O and OH desorption on Ag–Pt surfaces, the less reactive high-coordination (111) facets are the most conducive to the ORR. For low Ag loading as shown in Fig. 5, 6 and 10, the Ag atoms also occupy high-coordination (111) facets. The occurrence of such a coupling between Pt sites and Ag sites on the NAs' surface plays a key role in ORR in alkaline media. From the viewpoint of thermodynamics, the adsorbed O and OH will be strongly bound to Pt and Ag sites, respectively.[77–80] From the viewpoint of kinetics, it was found that Ag has low O_2 dissociation barriers (0.00 eV with solvation due to the strong solvent stabilization of O_{ad}) and O hydration barriers (0.12 eV), but they suffer from a huge H_2O formation barrier (0.98 eV) as compared with Pt (0.30 eV).[80] Thus, O_2 can first dissociate and form OH on Ag, and then the OH concentration gradient produced by the different activities of Pt and Ag sites can drive OH migrate to the Pt sites and finish the H_2O formation. It is expected that present Ag–Pt NAs with such coupling between surface Pt and Ag sites might be promising candidates for ORR in alkaline media.

Apart from ORR, this strategy may be applied to the CO electro-oxidation on Ag–Pt NAs in alkaline media. It is generally assumed that the active OH species adsorb at low-coordination "defect" sites preferentially over facet sites and react with adsorbed CO.[77–79] At low CO coverage, however, the Pt–CO bond is too strong for oxidation by OH at vertex and edge.[77] Upon decreasing the particle size, therefore, the restricted supply of CO to the active sites may become rate limiting for CO electro-oxidation. As shown in Fig. 5, 6 and 10, the high-coordination (111) facets of Ag–Pt NAs are decorated by Ag atoms. Considering the electronic effect, CO adsorption on Pt sites adjacent to Ag atoms is strengthened relative to Pt (111) facets, thus Ag–Pt NAs would be less favorable for CO electro-oxidation processes.[77–79] From the viewpoint of geometric effects, however, each Ag atom appears to block the adsorption of CO by physical site blocking and would inhibit CO poisoning. Furthermore, based on the possible bifunctional mechanism, oxidation of the strongly bound CO on Pt requires reaction with OH adsorbed on (111) Ag, where it is more weakly bound than on vertices and edges, and thus more reactive. It is not surprising that alloying with Ag can not only improve the CO poisoning of the Pt nanocatalyst but also greatly enhance its catalytic activity, which is in agreement with experimental observations.[77,81] Therefore, it is possible to predict that Ag–Pt NAs might be a good candidate for the methanol oxidation. In fact, higher activities of the methanol and formic acid oxidation were reported with Ag–Pt NAs than with Pt NPs.[23–26]

A possibility to tune the catalytic selectivity of Pt NPs is to use inactive Ag atoms preferentially segregating to specific sites (like vertices and edges) to avoid an unwanted reaction. For instance, allylic alcohols obtained by preferred hydrogenation of the C=O group of α,β-unsaturated aldehydes are valuable intermediates for the production of perfumes, flavorings and pharmaceuticals. However, the hydrogenation of the C=C group of α,β-unsaturated aldehydes is more thermodynamically favorable than the hydrogenation of their C=O group.[82] Therefore, selective hydrogenation of the C=O group in α,β-unsaturated aldehydes is very difficult. For typical catalysts, such as Pt and Ru, both the C=C and C=O groups are hydrogenated to yield the saturated alcohol or aldehydes. While Ag catalysts were able to hydrogenate selectively the C=O group and produce allylic alcohols as main product under the gas phase.[82] Since low-coordination step and kink of Pt surface (edge and vertex for Pt NPs) sites facilitate hydrogenation of both C=C and C=O bonds,[83] Ag–Pt nanocatalysts with these low-coordination sites blocked by Ag (as shown in Fig. 5, 6 and 10) will control the intramolecular selectivity by hydrogenation preferentially at the C=O group while keeping the C=C bonds intact. While direct experimental verification of our prediction is currently not available, Pt nanocatalysts with step sites blocked by Ag have been found to simultaneously achieve high activity and selectivity for C_2H_2 hydrogenation over large temperature operating windows.[31,32] Further research such as DFT calculations can be used to fully understand the influence of the atomic arrangement of the Ag–Pt surface on catalytic activity and selectivity, in conjunction with thermodynamic or micro-kinetic models.[84]

4 Conclusions

In this work, we have used Metropolis MC simulations coupled with semi-empirical potentials to study the atomic arrangements of Ag–Pt NAs and their dependence on particle size, morphology and alloy composition. MC simulations show that Ag is significantly enriched on the surfaces to reduce their free energy, and increasing the NA sizes or global Ag composition leads to higher surface Ag concentrations. Because of the strong surface segregation and the nanoscale effects in Ag–Pt NAs, the chemical ordering patterns are significantly dependent on the global Ag composition. When the global Ag composition is increased, the most favorable patterns gradually evolve from Pt-core/alloying shell to perfect Pt-core/Ag-shell structures, and then to onion-like structures with Ag shell and Pt-rich subsurface, independent of particle sizes and morphologies. Furthermore, the surface segregation of NAs depends on particle morphology, especially the (111) facets-to-surface ratios as (111) facets are the least favorable surface sites for Ag segregation. The lowering of surface energy is always present for all NAs and is dominant for small NAs, thus, surface Ag concentration is higher in more open particles. For large sized ICO NAs, the lowering of strain energy is preferred, which promotes surface segregation of Ag. Thus, they have the closest surface but the highest surface Ag concentration. At low global Ag compositions, the preference of Ag segregation decreases in the following order: vertex > edge > (100) facet > (111) facet > bulk. As a result, the presence of partially alloyed facets along with vertices and edges blocked by Ag is a topic of particular interest for catalysis. Considering that the surface electronic and geometric structures is modified by this coupling between Pt sites and Ag sites, their ORR activity in alkaline media can be enhanced from the viewpoint of thermodynamics and kinetics. Apart from ORR, this coupling will not only greatly enhance the catalytic activity of CO electro-oxidation in alkaline media but also improve the CO poisoning with respect to Pt nanocatalysts. Furthermore, it is suggested that Ag–Pt NAs might be promising candidates for selective hydrogenation of the C=O group in α,β-unsaturated aldehydes. These detailed understanding of atomic arrangements of Ag–Pt NAs is essential to intelligently design robust and active nanocatalysts with a low cost.

Acknowledgements

This work was financially supported by the National Natural Science Foundation of China (50801024, 51071063 and NSFC-NSAF 10976009), the Science and Technology Program of Hunan Province (No. 2012FJ3054), and the Science Foundation for the Talents of Hunan Agricultural University (No. 12YJ04). It was also partly supported by the Fundamental Research Funds for the Central Universities, Hunan University. The authors thank the National Supercomputing Center in Changsha for the computing resources provided.

References

1 R. Ferrando, J. Jellinek and R. L. Johnston, *Chem. Rev.*, 2008, **108**, 845.
2 Z. M. Peng and H. Yang, *Nano Today*, 2009, **4**, 143.
3 M. Sankar, N. Dimitratos, P. J. Miedziak, P. P. Wells, C. J. Kiely and G. J. Hutchings, *Chem. Soc. Rev.*, 2012, **41**, 8099.
4 A. C. Chen and P. Holt-Hindle, *Chem. Rev.*, 2010, **110**, 3767.
5 D. S. Wang and Y. D. Li, *Adv. Mater.*, 2011, **23**, 1044.
6 J. Gu, Y. W. Zhang and F. Tao, *Chem. Soc. Rev.*, 2012, **41**, 8050.
7 X. W. Liu, D. S. Wang and Y. D. Li, *Nano Today*, 2012, **7**, 448.
8 V. R. Stamenkovic, B. S. Mun, M. Arenz, K. J. J. Mayrhofer, C. A. Lucas, G. F. Wang, P. N. Ross and N. M. Markovic, *Nat. Mater.*, 2007, **6**, 241.
9 J. Zhang, K. Sasaki, E. Sutter and R. R. Adzic, *Science*, 2007, **315**, 220.
10 J. X. Wang, H. Inada, L. J. Wu, Y. M. Zhu, Y. M. Choi, P. Liu, W. P. Zhou and R. R. Adzic, *J. Am. Chem. Soc.*, 2009, **131**, 17298.
11 J. X. Wang, C. Ma, Y. M. Choi, D. Su, Y. M. Zhu, P. Liu, R. Si, M. B. Vukmirovic, Y. Zhang and R. R. Adzic, *J. Am. Chem. Soc.*, 2011, **133**, 13551.
12 M. E. Grass, Y. Yue, S. E. Habas, R. M. Rioux, C. I. Teall and G. A. Somorjai, *J. Phys. Chem. C*, 2008, **112**, 4797.
13 F. Tao, M. E. Grass, Y. Zhang, D. R. Butcher, J. R. Renzas, Z. Liu, J. Y. Chung, B. S. Mun, M. Salmeron and G. A. Somorjai, *Science*, 2008, **322**, 932–934.
14 W. M. H. Sachtler, *Faraday Discuss. Chem. Soc.*, 1981, **72**, 7.
15 J. W. A. Sachtler and G. A. Somorjai, *J. Catal.*, 1983, **81**, 77.
16 J. A. Rodriguez, *Surf. Sci. Rep.*, 1996, **24**, 223.
17 F. Maroun, F. Ozanam, O. M. Magnussen and R. J. Behm, *Science*, 2001, **293**, 1811.
18 J. R. Kitchin, J. K. Norskov, M. A. Barteau and J. G. Chen, *Phys. Rev. Lett.*, 2004, **93**, 156801.
19 D. A. Hansgen, D. G. Vlachos and J. G. Chen, *Nat. Chem.*, 2010, **2**, 484.
20 J. K. Strohl and T. S. King, *J. Catal.*, 1989, **116**, 540.
21 G. F. Wang, M. A. Van Hove, P. N. Ross and M. I. Baskes, *Prog. Surf. Sci.*, 2005, **79**, 28.
22 J. B. Park, S. F. Conner and D. A. Chen, *J. Phys. Chem. C*, 2008, **112**, 5490.
23 D. Zhao, Y. H. Wang, B. Yan and B. Q. Xu, *J. Phys. Chem. C*, 2009, **113**, 1242.
24 Y. Y. Feng, J. H. Ma, G. R. Zhang, G. Liu and B. Q. Xu, *Electrochem. Commun.*, 2010, **12**, 1191.
25 Z. M. Peng, H. J. You and H. Yang, *Adv. Funct. Mater.*, 2010, **20**, 3734.
26 W. W. He, X. C. Wu, J. B. Liu, K. Zhang, W. G. Chu, L. L. Feng, X. N. Hu, W. Y. Zhou and S. S. Xie, *J. Phys. Chem. C*, 2009, **113**, 10505.
27 W. W. He, X. C. Wu, J. B. Liu, K. Zhang, W. G. Chu, L. L. Feng, X. N. Hu, W. Y. Zhou and S. S. Xie, *Langmuir*, 2010, **26**, 4443.
28 Y. Y. Feng, G. R. Zhang, J. H. Ma, G. Liu and B. Q. Xu, *Phys. Chem. Chem. Phys.*, 2011, **13**, 3863.
29 J. B. Xu, T. S. Zhao and Z. X. Liang, *J. Phys. Chem. C*, 2008, **112**, 17362.
30 B. J. Hwang, S. M. S. Kumar, C. H. Chen, R. W. Chang, D. G. Liu and J. F. Lee, *J. Phys. Chem. C*, 2008, **112**, 2370.
31 L. J. Jones, Z. Buras and M. J. Gordon, *J. Phys. Chem. C*, 2012, **116**, 12982.
32 L. J. Jones and M. J. Gordon, *J. Phys. Chem. C*, 2012, **116**, 23472.
33 A. I. Frenkel, *Chem. Soc. Rev.*, 2012, **41**, 8163.
34 D. Liu, Y. F. Zhu and Q. Jiang, *J. Phys. Chem. C*, 2009, **113**, 10907.
35 Z. D. Pozun, K. Tran, A. Shi, R. H. Smith and G. Henkelman, *J. Phys. Chem. C*, 2011, **115**, 1811.
36 C. Y. Lu and G. Henkelman, *J. Phys. Chem. Lett.*, 2011, **2**, 1237.
37 C. L. Cleveland and U. Landman, *J. Chem. Phys.*, 1991, **94**, 7376.
38 C. L. Cleveland, U. Landman, T. G. Schaaff, M. N. Shafigullin, P. W. Stephans and R. L. Whetten, *Phys. Rev. Lett.*, 1997, **79**, 1873.
39 Y. G. Chushak and L. S. Bartell, *J. Phys. Chem. B*, 2001, **105**, 11605.
40 F. Baletto, R. Ferrando, A. Fortunelli, F. Montalenti and C. Mottet, *J. Chem. Phys.*, 2002, **116**, 3856.
41 F. Baletto and R. Ferrando, *Rev. Mod. Phys.*, 2005, **77**, 371.
42 Z. Zhang, W. Y. Hu and S. F. Xiao, *Phys. Rev. B*, 2006, **73**, 125443.
43 S. F. Xiao, W. Y. Hu, W. H. Luo, Y. R. Wu, X. F. Li and H. Q. Deng, *Eur. Phys. J. B*, 2007, **54**, 479.
44 W. H. Luo, W. Y. Hu and S. F. Xiao, *J. Phys. Chem. C*, 2008, **112**, 2359.
45 L. Deng, W. Y. Hu, H. Q. Deng and S. F. Xiao, *J. Phys. Chem. C*, 2010, **114**, 11026.
46 L. Deng, W. Y. Hu, H. Q. Deng, S. F. Xiao and J. F. Tang, *J. Phys. Chem. C*, 2011, **115**, 11355.

47 R. Huang, Y. H. Wen, Z. Z. Zhu and S. G. Sun, *J. Mater. Chem.*, 2011, **21**, 11578.
48 B. Y. Wang, M. X. Liu, Y. T. Wang and X. S. Chen, *J. Phys. Chem. C*, 2011, **115**, 11374.
49 S. Ino, *J. Phys. Soc. Jpn.*, 1969, **27**, 941.
50 L. D. Marks, *Rep. Prog. Phys.*, 1994, **57**, 603.
51 T. P. Martin, *Phys. Rep.*, 1996, **273**, 199.
52 M. José-Yacamán, J. A. Ascencio, H. B. Liu and J. Gardea-Torresdey, *J. Vac. Sci. Technol., B*, 2001, **19**, 1091.
53 A. S. Barnard, *Rep. Prog. Phys.*, 2010, **73**, 086502.
54 J. M. Montejano-Carrizales, F. Aguilera-Granja and J. L. Moran-Lopez, *Nanostruct. Mater.*, 1997, **8**, 269.
55 S. Takizawa, K. Terakura and T. Mohri, *Phys. Rev. B*, 1989, **39**, 5792.
56 M. H. F. Sluiter, C. Colinet and A. Pasturel, *Phys. Rev. B*, 2006, **73**, 174204.
57 R. A. Johnson, *Phys. Rev. B*, 1990, **41**, 9717.
58 G. Grochola, I. K. Snook and S. P. Russo, *J. Phys. Chem. C*, 2009, **113**, 7541.
59 J. Cai and Y. Y. Ye, *Phys. Rev. B*, 1996, **54**, 8398.
60 I. Karakaya and W. T. Thompson, *Bull. Alloy Phase Diagrams*, 1987, **8**, 334.
61 Y. G. Ma and P. B. Balbuena, *Surf. Sci.*, 2008, **602**, 107.
62 L. L. Wang and D. D. Johnson, *J. Am. Chem. Soc.*, 2009, **131**, 14023.
63 Y. H. Zhang, Z. Y. Duan, C. Xiao and G. F. Wang, *Surf. Sci.*, 2011, **605**, 1577.
64 F. Baletto, C. Mottet and R. Ferrando, *Phys. Rev. Lett.*, 2003, **90**, 135504.
65 S. R. Calvo and P. B. Balbuena, *Surf. Sci.*, 2005, **581**, 213.
66 L. O. Paz-Borbón, R. L. Johnston, G. Barcaro and A. Fortunelli, *J. Chem. Phys.*, 2008, **128**, 134517.
67 F. Calvo, E. Cottancin and M. Broyer, *Phys. Rev. B*, 2008, **77**, 121406.
68 K. Yun, Y. H. Cho, P. R. Cha, J. Lee, H. S. Nam, J. S. Oh, J. H. Choi and S. C. Lee, *Acta Mater.*, 2012, **60**, 4908.
69 F. R. Negreiros, F. Taherkhani, G. Parsafar, A. Caro and A. Fortunelli, *J. Chem. Phys.*, 2012, **137**, 194302.
70 S. Schwamborn, L. Stoica and W. Schuhmann, *ChemPhysChem*, 2011, **12**, 1741.
71 K. Kim, K. L. Kim and K. S. Shin, *J. Phys. Chem. C*, 2011, **115**, 23374.
72 M. Guerra-Balcázar, F. M. Cuevas-Muniz, L. Álvarez-Contreras, L. G. Arriaga and J. Ledesma-García, *J. Power Sources*, 2012, **197**, 121.
73 H. J. Zhang, T. Watanabe, M. Okumura, M. Haruta and N. Toshima, *Nat. Mater.*, 2011, **11**, 49.
74 F. H. B. Lima, C. D. Sanches and E. A. Ticianelli, *J. Electrochem. Soc.*, 2005, **152**, A1466.
75 M. Chatenet, L. Genies-Bultel, M. Aurousseau, R. Durand and F. Andolfatto, *J. Appl. Electrochem.*, 2002, **32**, 1131.
76 M. Chatenet, M. Aurousseau, R. Durand and F. Andolfatto, *J. Electrochem. Soc.*, 2003, **150**, D47.
77 J. S. Spendelow, J. D. Goodpaster, P. J. A. Kenis and A. Wieckowski, *J. Phys. Chem. B*, 2006, **110**, 9545.
78 M. T. M. Koper, *Nanoscale*, 2011, **3**, 2054.
79 G. Garcia and M. T. M. Koper, *ChemPhysChem*, 2011, **12**, 2064.
80 Y. Sha, Ph.D. Thesis, California Institute of Technology, 2011, p. 76.
81 C. X. Xu, Y. Y. Li, F. Tian and Y. Ding, *ChemPhysChem*, 2010, **11**, 3320.
82 P. Claus, *Top. Catal.*, 1998, **5**, 51.
83 R. M. Rioux, H. Song, M. Grass, S. Habas, K. Niesz, J. D. Hoefelmeyer, P. Yang and G. A. Somorjai, *Top. Catal.*, 2006, **39**, 167.
84 F. Yang, Q. F. Zhang, Y. W. Liu and S. L. Chen, *J. Phys. Chem. C*, 2011, **115**, 19311.

General discussion

Professor Asakura opened the discussion of the paper by Professor Sauer: During the diffusion of VO_x, there is a charge transfer between V and Ce. What is the effect of the charge transfer on the diffusion process? Does it hinder or accelerate the diffusion of VO_x?

Dr Paier replied on behalf of Professor Sauer: Our findings indicate that the migration of VO_x on the ceria surface is not necessarily accompanied by a charge transfer process, *i.e.* reduction of the (for minima) preferred oxidation state +5 to the oxidation state +4 of the V ion. This is seen in Fig. 5 and 6 of our paper. In path 1, transition state 4 (TS4) contains V(+4) and corresponds to an energy barrier of 1.94 eV, which is a large barrier. In path 2, V preserves its oxidation state +5 during migration, and hence no charge transfer is involved. However, both pathways require a change in the coordination number of V (*e.g.* from 4 to 3) and therefore the role of the charge transfer is difficult to assess. Note that TS5, which is connected to the highest barrier found, has a higher number of coordinatively unsaturated atoms (compared to TS4), it appears that the number of "cus" atoms is most likely the reason for the large barriers. It appears that the charge transfer plays a minor role.

Mr Song asked: Regarding the phase diagram (Fig. 3), you stated that the reference system is pure bulk V. How do you change the chemical potential of V? Because the reference is bulk V, you can not change the composition. And the influence from temperature to chemical potential of solid V should be also small.

Dr Paier replied: In order to avoid questions related to the dependence of the chemical potential of V on activity and temperature, the phase diagram you are referring to is based on the difference of the chemical potential to the DFT total energy (*i.e.* 0 temperature, 0 activity limit).

Professor Bowker commented: There are some parallels here with the paper in Session 2 on MoO_3 deposited on iron oxide where it was found that Mo_3O_9 trimers appeared to dominate the surface by STM imaging. In that case it may be that the Mo was deposited mainly as trimers, so the kinetic blocking you suggest for VO_x may have been avoided.

Professor Freund followed up Professor Bowker's comment: Concerning transition metal oxide trimers in general and the comparison between experiments with Mo_3O_9, in particular, I would like to point out that in the case of VO_x on ceria, model catalyst preparation of metal was dosed onto the surface in an oxygen ambient as opposed to metal oxide the vapor in the former case.

Professor Hermann enquired: Why did you restrict your theoretical studies to VO_x particles up to trimeric vanadium?

Professor Sauer responded: We did not go beyond trimers because we did not expect a new effect for larger ring sizes. The trimer is the first species in which each monomer is connected by two oxygen bridges with its neighbor. What we are currently doing however, are calculations on a vanadia monolayer on ceria as a limiting case.

Professor Hermann asked: Have you performed V_xO_y calculations on free particles with tetrameric or higher vanadium components?

Professor Sauer replied: Yes we did and the results are published. In the gas phase, rings rather than cages are the most stable species.[1]

1 M. Sierka, J. Döbler, J. Sauer, G. Santambrogio, M. Brümmer, L. Wöste, E. Janssens, G. Meijer and K. R. Asmis, *Angew. Chem., Int. Ed.*, 2007, **46**, 3372–3375.

Professor Hermann then remarked: How do you connect your studies for small VO_x particles up to trimers with that for VO_x monolayers on CeO_2 substrate?

Professor Sauer answered: We will check if the trend we observe going from monomers to trimers will persist for the limiting case of the monolayer. When speaking about reactivity, we should be aware of the fact that for deposited clusters the uncovered $CeO_2(111)$ surface may also be involved. This is impossible for the monolayer since there is no exposed CeO_2.

Professor Hermann further enquired: In your diffusion and migration studies you seem to have considered migration paths where only twisting motions were taken into account. Along the same lines, we have recently studied NH_4 diffusion at V_2O_5 surfaces and find diffusion paths with combinations of tumble and twist motion to yield lowest barriers.[1,2] Did you also test such combinations?

1 M. Gruber and K. Hermann, *J. Chem. Phys.*, 2013, **138**, 194701.
2 M. Gruber and K. Hermann, *J. Chem. Phys.*, 2013, **138**, 094704.

Dr Paier responded on behalf of Professor Sauer: The twisting motion or mode you mentioned connects M1 with M3 (*cf.* Fig. 6 in our paper). If we agree that twisting refers to a rotation about an axis perpendicular to the surface, this was seen to be the case for M1 → M3, which could be assigned to a clear and pronounced twisting motion. The other modes we found for the migration could be assigned to rocking-type of modes, which should relate to the "tumble motion" in question. Hence we can answer in the affirmative. On the other hand, I would caution against usage of the notion "tumbling" in this special case. The VO_2 species studied is tightly bound on the $CeO_2(111)$ surface and what one observes are vibrational modes of a strongly bound vanadia cluster on the surface. Therefore, we prefer to classify modes as twisting, rocking *etc.*

Professor Campbell opened the discussion of the paper by Professor Neurock: The fast-to-equilibrium assumption you made for the first two reaction steps in Fig. 10 to get the measured activation energy to equal the DFT energies is inconsistent with your explanation of the factor of 4 kinetic isotope effect observed. Please clarify.

Professor Neurock replied: You are correct in that there was an inconsistency in the previous results. We have corrected this in the revised manuscript. We had assumed that the initial activation of the C–H bonds were quasi-equilibrated which was inconsistent with the later indication that the 2nd C–H activation step is irreversible due to the loss of water. We have replaced the previous potential energy surface which assumed quasi-equilibrium with a plot of ln (rate) *vs.* $1/T$ from the simulations which reveals an apparent barrier of 1.72 eV. The results which are the direct outcome from the simulation are consistent with the experimental results. The apparent barrier is difficult to decompose into distinct elementary steps as it is comprised of the rate-determining C–O activation step together with other elementary steps that control the surface coverage. Both the experiments as well as the simulations indicate that the high apparent activation barrier is the result of a high intrinsic C–O activation energy as well as the energy cost of generating vacancy sites to carry out the reaction. Higher temperatures help to aid in the recombinative desorption of O which generates the vacant sites needed carry out the reaction. The simulations also reveal the kinetic isotope effect of 4.0 which is also consistent with the experimental results.

Dr Philip Davies enquired: In your paper you indicate that both the acetic acid and the dioxygen adsorb at the gold/oxide interface sites. Is there any evidence that competition between the two adsorbates affects the kinetics of the reaction and could this be exploited to provide further clues to the mechanism?

Professor Neurock replied: Yes, our kinetic simulations show that there is significant site competition between the acetate and the oxygen at the start of the reaction. At longer times, the sites are covered with acetate, oxygen, hydroxyl and the ketenylidene intermediates which inhibit oxidation of acetate. This can also be seen from the experimental kinetic data, where the reaction becomes slower as time goes on. The kinetics might exploited by examining other oxide supports which might have strong enough interactions with oxygen and oxygenate intermediates to activate the C–O and O_2 at the interface but weak enough to allow for the rapid formation of vacancy sites.

Professor Bowker commented: Some of these data show similarity with work from Madix *et al.* concerning acetic acid adsorption on copper where ketene was the major product in gas phase during TPD.[1] For silver, it cannot be done without oxygen.[2] Is it possible that in the work you've described that acetate transfers from titania onto gold, where the ketene is then produced?

1 M. Bowker and R. J. Madix, *Appl. Surf. Sci.*, 1981, **8**, 299.
2 M. Barteau, M. Bowker and R. J. Madix, *J. Catal*, 1981, **67**, 118.

Professor Neurock responded: Indeed, our results reported here are quite similar to the elegant previous work by Madix on Ag and Cu. We presented more detailed spectroscopic analyses along with a direct comparison with the results from Madix in a previously published paper.[1] As you nicely point out, Cu can directly activate the C–H bond to form the ketene surface intermediate. Oxygen, however, which is necessary to activate the C–H bonds on Ag and Au, is more aggressive and subsequently activates the remaining C–H bonds to form the ketenylidene. Our results show the presence of the characteristic C=C=O stretching mode of the ketenylidene species at 2040 cm^{-1} which is consistent with previous results for Sim and King[2] who show the formation of ketenylidene in the oxidation of acetone on a O'-precovered Ag(111) 2004–2040 cm^{-1}. Detailed ^{13}C and ^{18}O isotopic labeling studies helped to further confirm the formation of the ketenylidene. The spectroscopic data indicates that the active acetate species resides on the TiO_2. It is clear, however, that the resulting ketenylidene that forms is bound to Au through its terminal C. Our theoretical results would suggest that the transition occurs after the two C–H bonds are activated. In addition, the activation of the C–O bond does not proceed on Au alone but instead occurs at the Ti^{4+} sites at the Au interface.

1 I. Green, W. Tang, M. Neurock and J. Yates, *J. Am. Chem. Soc.*, 2012, **134**, 13569–13572.
2 W. S. Sim, and D.A. King, *J. Phys. Chem.*, 1996, **100**, 14794–14802.

Dr Willock noted: In the mechanism you put forward the rate-limiting step is for the cleavage of a C–O bond that leaves an O atom over a five-coordinate Ti centre on the TiO_2 surface. It seems that the leaving group that this process implies will be quite unstable and lead to the large barrier that is calculated. A possible alternative is that the acid group is protonated prior to the C–O(H) bond being broken which would create a hydroxyl on the surface rather than a lone O atom.

Professor Neurock responded: The intrinsic barrier to activate the C–O bond was 0.71 eV with an overall reaction energy that is exothermic. This suggests that the Ti^{4+} sites at the Au interface help to stabilize adsorbed O' and aids in the reaction. The barrier to activate the C–OH bond will likely be a little lower than that to activate the C–O bond but would require that the O–H bond remain intact throughout the conversion of acetic acid or that we transfer a proton back to the adsorbed oxygenate intermediate at some point. The IR results in Fig. 1–2 clearly indicate that the O–H bond of the acid is broken. In addition, we have shown in recent calculations[1] that the barrier to transfer a proton back to the O of the bound oxygenate is greater than 0.7 eV which would suggest that the activation proceeds *via* C–O activation rather than C–OH activation.

1 I. Green, M. McEntee, W. Tang, M. Neurock and J. Yates, *Catal. Lett.*, 2013.

Professor Behm asked two questions: First, in view of my previous comment on the stability of the active oxygen species for CO oxidation on supported Au catalysts (the active oxygen species for reaction at room temperature and above is stable adsorbed up to 400 °C) I would like to ask about the stability of the molecular adsorbed O_2 state (against desorption). Is that species really sufficiently stable, to fulfil this, or are we talking about a reaction mechanism which may be dominant at low temperatures, but not at realistic reaction temperatures?

Second, it looks like that in the comparison of reaction barrier between experiment and theory, two different things are compared. While calculations use reaction barriers in the rate-determining step, experiments determine apparent activation energies, which include the contribution from competing desorption. Hence, for similar values of the reaction barrier in the rate-determining step one would possibly obtain very different numbers in calculation and experiment.

Professor Neurock responded: In the present work on the partial oxidation of acetic acid, the kinetic experiments were carried out between 400–420 K. Molecular oxygen readily dissociates and the active oxygen intermediates are thought to be atomic oxygen and hydroxyl intermediates bound to Au and Ti^{4+}–Au sites at the Au/TiO_2 interface which are stable at reaction conditions. In our previous studies for CO oxidation on Au/TiO_2, these experiments were carried out at 120 K. Both molecular as well as atomic oxygen were thought to be the active forms at the low temperatures of the experiment. At the higher temperatures where CO oxidation is typically carried out (~300 K), the more likely intermediates would be atomic oxygen or surface hydroxyl intermediates (in the presence of water).

To answer your second question, we have used the intrinsic activation barriers along with the adsorption and desorption energies as input to microkinetic simulations. The simulations were carried out at different temperatures in order to compare with the apparent experimental activation energy. As such, the barriers do take into account the competing desorption steps. The simulations result in an apparent activation barrier of 1.72 eV which is in good agreement with the experimental value of 1.7 ± 0.2 eV.

Professor Hermann asked: Did you assume an ideal rutile or anatase crystal structure for the TiO_2 support and did you include reconstruction/relaxation of the support surface in your modelling?

Professor Neurock answered: In the present work, we assumed that the TiO_2 support was in the form of the ideal rutile structure. In more recent studies we have examined in more detail the influence of the surface structure (anatase *versus* rutile), exposed facets and the surface termination on the reactions of oxygenates. This will be the topic of a future communication. Our focus herein was to provide understanding how the ideal Au–Ti interfacial sites might carry out elementary steps involved in the selective oxidation of acetic acid.

Mr Song queried: Regarding your Au nanorod model, only z of the Au nanorod was allowed to relax, while x and y directions are fixed. I am wondering to what extent can this model represent the realistic situation, relaxation in x and y directions. Will this constrained calculation cause a large error?

Professor Neurock responded: The partially constrained Au rod has been used successfully in the past to model the reactions at the Au/support interface.[1–3] The lateral x and y direction were held fixed in order to maintain an appropriate match with the underlying oxide lattice. This approximation was tested by allowing for the full optimization of all of the Au atoms. The changes in the reaction energies and activation barriers were found to be less than 0.1 eV, which indicates that fixing the x and y positions doesn't significantly affect the results.

1 L. M. Molina, M. D. Rasmussen and B. Hammer, *J. Chem. Phys.*, 2004, **120**, 7673–7680.
2 S. Laursen and S. Linic, *Phys. Chem. Chem. Phys.*, 2009, **11**, 11006–11012.
3 I. Green, W. Tang, M. Neurock and J. T. Yates, *Science*, 2011, **333**, 736–739.

Professor Hutchings said: In your mechanism the support is integral to the observed catalysis. Are there other supports that could have the same attributes? By modifying the support would it be possible to stabilize the ketene species? In this way maybe improved supports could be designed.

Professor Neurock replied: This is an excellent point. We suspect that other supports with moderate Lewis acid–base site pairs or redox-active oxides such as CeO_2 and ZrO_2 would aid in catalyzing the reaction. These substrates are known to aid in catalyzing the ketonization of carboxylic acids.[1–7] Au/CeO_2 is already known to be a good oxidation catalyst.[8,9]

1 C. A. Gaertner, J. C. Serrano-Ruiz, D. J. Braden and J. A. Dumesic, *J. Catal.*, 2009, **266**, 71.
2 C. A. Gaertner, J. C. Serrano-Ruiz, D. J. Braden and J. A. Dumesic, *Ind. Eng. Chem. Res.*, 2010, **49**, 6027.
3 A. D. Murkute, J. E. Jackson and D. J. Miller, *J. Catal.*, 2011, **278**, 189.
4 R. W. Snell and B. H. Shanks, *Appl. Catal., A*, 2013, **451**, 86.
5 Yamada *et al.*, *Mol. Catal., A*, 2011, **346**, 79–86.
6 Glinski *et al.*, *Appl. Catal., A*, 2007, **323**, 77–85.
7 Jackson *et al.*, *Appl. Catal., A*, 2012, **431**, 157–163.
8 Flytzani-Stephanopoulos *et al.*, *Science*, 2003, **301**, 935–938.
9 Neurock, Vajda *et al.*, *ACS Catal.*, 2012, **2**, 2409–2423.

Professor Hutchings opened the discussion of the paper by Professor van Santen: This is a fascinating study and it is clear that using this approach you have developed a very elegant model that can explain a lot the observations concerning the product distributions observed in Fischer–Tropsch synthesis. However, a main assumption is that the rate of chain termination is independent of the carbon number, apart from C_1. Typically C_2 would be distinctly different from C_3 and higher chain lengths. However, the model works very well, can you comment on this?

Professor van Santen answered: In our current work we ignored for instance the difference between rate of C_2 formation and that of the higher hydrocarbons. This obviously is not correct. C_2 formation is unique since it is the recombination of two C_1 species. However since C_2 is a small fraction of the total amount of hydrocarbon when chain growth parameter α is large it is not part of the simulations discussed in the paper. Microkinetics simulations can be set up to incorporate fully the effect of chain dependent C_1 insertion. Also the current model ignores readsorption of olefins and their incorporation, which also should be part of follow-up studies.

Professor Hutchings remarked: When considering other metals with Fe for example one would have to consider CO_2 formation. Can this be accommodated in the model?

Professor van Santen replied: Yes. This would be in particular important for Fe. In order to do so properly also O adsorption should be considered in more detail. This can in principle be done, but is not part of the current model.

Professor Bowker said: The scheme in Fig. 1a shows the reaction pathway you considered. Can you tell us a bit more about the calculations – are some steps considered as equilibrium – is there only one rate-determining step or are forward and back steps considered independently?

Professor van Santen answered: We solve the corresponding partial differential equations rigorously without assuming rate-limiting steps. This is a necessity since the rate-limiting step may vary when particle size or surface reactivity changes. The latter is explicitly discussed in the paper. It distinguishes monomer formation limit from chain growth kinetics limit.

Professor Bowker then commented: From all the efforts in surface science over the last few decades, we know that generally there are strong coverage-dependent effects on the magnitude of kinetic barriers. Have you taken into account lateral interactions between adsorbates in your microkinetic modelling?

Professor van Santen responded: The most significant lateral interactions to take into account is that which controls the adsorption energy of CO. We adjusted the CO adsorption energy to its value at high coverage.

Professor Hermann enquired: Your studies you seem to have chosen rate constants by matching with experimental data. I gathered this from the qualitative table (without numbers) that you showed. Could you also use theoretical results from DFT studies to determine the rate constants?

Professor van Santen answered: Yes. We have demonstrated this in a recent paper.[1]

1 R. A. van Santen *et al.*, *J. Phys. Chem C*, 2013, **117**, 4488.

Dr Whiston noted: The rate constants calculated by quantum mechanical calculation in the lumped kinetic model can be validated by predicting the product distribution observed in Fischer–Tropsch over the catalyst. This is a very useful validation of whether the model has some basis in reality for a real catalytic system. Does the model in fact accurately predict both product distribution (molecular weight distribution of hydrocarbon products) and variation with temperature and feed composition over both Co and Ru based catalysts?

Professor van Santen answered: This paper only describes lumped microkinetics results in which we did not vary the temperature and pressure. Parameters are based on microkinetics simulations that properly predict temperature dependence of chain growth parameter as well as methane production.

Professor Campbell said: It is known that Ni makes mostly methane while Fe/FeC makes longer HCs. What is the explanation for this in your model? (*i.e.*, Which rate constant differences are most critical for this?)

Professor van Santen responded: The rate of methane formation from adsorbed CH_x species has to be slow compared to the rate CH_x formation from CO. The rates of CH_x insertion into the growing chain do not vary strongly for the different metals. On the other hand Ni has a high activation energy for CO activation, even when hydrogen assisted and the rate of methane formation from adsorbed CH_x is relatively fast (M–C bond to surface of Ni is relatively weak). This is very different for the Fe catalysis. Now the rate of C–O bond cleavage is faster and rate of methane formation from adsorbed CH_x slow, because of the strong M–C bond interaction. In this case the surface concentration of CH_x can build up with high enough concentration to insert into the growing hydrocarbon chain.

Professor Campbell then commented: If my memory serves me well, the rate is negative order in CO pressure and near 1st order in hydrogen pressure. Is this consistent with CO activation being rate-controlling?

Professor van Santen replied: Yes. Our microkinetics simulations indicate that under this condition the CO surface concentration remains high. This has an inhibiting effect on CO dissociation because CO vacancies are needed for its activation. In a paper to in preparation we will show explicitly the corresponding negative order in CO in microkinetics simulations. The positive order in hydrogen that we also reproduce relates to the low H_2 coverage (it is suppressed by CO) and to the need to hydrogenate the carbonaceous intermediates so that reaction will proceed. The rate of chain growth termination, in which also hydrogen transfer reaction are important, is also a slow step!

Professor Hutchings asked: In Fischer–Tropsch synthesis the critical process is the activation of CO. Can you use the insight from your approach to aid catalyst design? As you suggest experimental approaches, *e.g.* the addition of a second metal, that could be tested experimentally?

Professor van Santen replied: One of the important parts of the paper is the suggestion that excess methane formation occurs on sites where CO activation is slow, but hydrogenation of adsorbed C fast. Terrace sites are candidates for this. Step-edge sites would be the ideally structured sites that have increased M–C interactions as well as enhanced CO activation rates. These would occur on the edges of not too small particles. Promotion with reducible oxides in particular can be helpful, since they will assist CO dissociation on highly CO covered surfaces and also enhance the interaction of C atoms with the surface.

Professor Bowker queried: Is it possible to hydrogenate the CO (*e.g.* to HCO) *before* CO dissociation?

Professor van Santen answered: Hydrogen-activated CO dissociation competes with direct CO dissociation on surfaces with a high activation barrier for direct CO activation. On Co and Ru the formation of intermediate formyl is endothermic by 100 kJ mol^{-1}. The activation energy for its formation is of the order of 130 kJ mol^{-1}. So when the activation energy for direct CO dissociation is higher than 130 kJ mol^{-1} the hydrogen-activated CO path takes over. This will typically happen on the terraces of Co and Ru and also on more reactive surfaces of less reactive metals. An important question relates to the issue of which activation energy of C–O bond cleavage and CH_x formation is acceptable on a catalyst that is a good Fischer–Tropsch catalyst. It should be so low that its rate becomes fast compared to the rate of CH_x transformation to methane.

Professor Hermann opened a general discussion of Professor Sauer's, Professor Neurock's and Professor van Santen's papers: Before this theory session we have heard a fair number of exciting experimental results on nanoparticles but only very few comparisons with data from theoretical studies, simulations or modeling, were made. Can we identify reasons for this situation? Ideally, studies on complex microscopic systems such as nanoparticles should combine experimental and theoretical efforts in a collaboration to arrive at conclusive answers where both sides can profit from each other.

Professor Sauer commented: We should not only ask all our experimental colleagues in the room why they don't use theory – we should also ask how can you afford not to use it?

Professor Freund added: I find the direct collaboration between experiment and theory exceedingly important. This, however, should be based on an interplay between detailed calculations addressing as close as possible the experimental situation, and detailed spectroscopic and other experiments attempting to collect as much information on a given system as possible.

In the literature I see more and more cases where, for example, an activity or selectivity measurement is presented on a dispersed metal catalyst and the results are explained on the basis of periodic slab calculations on a metal single crystal, where neither the support–metal interface nor the finite size of the metal particle is taken into account or is even critically discussed. This appears to me like looking for a lost key where the light is, irrespective of where the key has been lost.

Professor Sauer responded to Professor Freund's remark: For surface problems we can use either periodic models or (embedded) cluster models and both involve approximations. In the last decade the size of both periodic cells and embedded clusters that can be studied with a given accuracy has increased substantially. In professional applications, convergence studies with the model size are made and for many problems, there is now an overlap of the application area of both methods. The approach that I favor is a joined computational–experimental approach. In numerical quantum chemistry we know exactly what we calculate, but we are often limited in accuracy and our models may not be realistic. In experiments, we often do not know precisely the structure and composition of the sample we measure. Assigning measured spectroscopic signatures (IR, NMR) or STM images to specific atomic structures is therefore a major way that theory and experiments work together. In the more general perspective, theory should calculate what is not easily measured and experiments should also be done on well-defined model systems which are accessible to computations on representative models. In our computational study in collaboration with the experiments in Professor Freund's laboratory we have shown how fruitful such an approach can be.

Professor Neurock commented: As we attempt to compare theory and experiment, it is important that the models of the active catalyst appropriately mimic the active surface structures and compositions that result under experimental conditions. Many catalytic reactions carried out over supported metal particles (with the exception of Au) are carried out under conditions where the surfaces are nearly fully covered with adsorbed intermediates. Many of the calculations reported in the literature, however, are carried out at low coverage on single crystal. This unfortunately doesn't tell us much about catalysis on supported particles under working conditions. The simulations should appropriately treat the adsorbate–adsorbate interactions that are important at the higher coverages at working conditions. The calculations should also consider the differences between single crystal surfaces and supported nanoparticles. At very high coverages, single crystal surfaces can typically only accommodate surface coverages below 1 ML before they begin to reconstruct due to the lattice strain. Supported particles on the other hand can accommodate coverages up to and exceeding 1 ML. They also do not have to reconstruct but can more easily relieve stain by expanding. This is well established in the literature on the oxidation of metal particles. We have seen such expansions for alkane oxidation reactions over Pt[1] as well as for CO oxidation and Fischer–Tropsch synthesis reactions[2] over CO-covered metal surfaces.

1 A. Allian, *et al.*, *J. Am. Chem. Soc.*, 2011, **133**, 4498–4517.
2 B. T. Loveless, C. Buda, M. Neurock and E. Iglesia, *J. Am. Chem. Soc.*, 2013, **135**, 6107–6121.

Professor Bowker continued this discussion: We must remember that an aim of science is to understand the details of how reaction processes take place. Through surface science and fundamental catalysis we gain rules for how reactions take place and also we can quantify kinetic barriers and coverage effects upon them. Surely the ultimate aim of our area of science must be to have an expert system in place, designed from these rules, trends and quantification that we have worked hard to define, which will mean that we will only need theory to predict yields and selectivity? We can then confirm these predictions by experiment.

Professor Park agreed that there is gap between theory and experiment: The experimental system involves additional complexity that is not often reflected by the theory. For example, on changing the size of nanoparticles, the change of oxidation states occurs. Typically the smaller Rh, Ru, and Co nanoparticles exhibit higher oxidation states, which in turn influences the catalytic activity and selectivity significantly. I think this point might deserve more discussion.

Professor Attard commented: I wonder how well quantum mechanics can treat "weak" interactions related to catalysis such as precursor states, van der Waals and hydrogen bonding and coming back to Professor Somorjai's comments relating to enzyme catalysis, subtle interactions such as those involved in enantioselectivity? I would also as a surface electrochemist be interested in the question of theory and charge localisation.

Professor Neurock closed the general discussion in response to Professor Attard: The theory community has made great progress in appropriately treating the effects of water (hydrogen bonding) and van der Waals interactions. There are now a number of studies in which the water is explicitly treated in calculations in metal surfaces,[1] oxides and zeolites and helps to confirm experimental results on the role of hydrogen bonding as well as stabilization of charged intermediates. In addition, we have seen the development of new functionals within density functional theory that can treat van der Waals interactions. Professor Sauer has carried out very elegant work in combining high level MP2 calculations and couple cluster theory to correct standard periodic DFT calculations to provide very accurate treatments of van der Waals interactions for zeolite catalyzed reactions.[2] In addition, he has used this approach to test the accuracy and appropriateness of lower level models.

1 C. D. Taylor and M. Neurock, *Curr. Opin. Solid State Mater. Sci.*, 2005, **9**, 49–65.
2 S. Svelle, C. Tuma, X. Rozanska, T. Kerber and J. Sauer, *J. Am. Chem. Soc.*, 2009, **131**, 816–825.

Professor Asakura opened the discussion of the paper by Professor Hensen†: Is the role of Rh adsorption site for CO only? Is oxygen activated in defect sites? Is there any role of Rh to activate the oxygen?

Mr Song responded: Thank you very much for your question. Actually, the O_2 molecule adsorbs at the interface between Rh and generated vacancy site followed by the dissociation. This actually suggests that Rh also plays role in the O_2 activation process.

Professor Behm enquired: Are the numbers given for the CO adsorption energies (1 CO, 2 CO) differential adsorption energies (per CO) or integral (average) adsorption energies per CO?

† Prof. Hensen's paper was presented by Mr Song, Laboratory of Inorganic Materials Chemistry, Schuit Institute of Catalysis, Eindhoven University of Technology, The Netherlands.

Mr Song replied: They are given for differential adsorption energies.

Professor Hermann remarked: Cerium dioxide can, depending on the (*hkl*) direction of its surface, exhibit one to three different surface terminations. You seem to have focused on oxygen-terminated surfaces only. What was the reason and did you also consider Ce metal termination?

Mr Song responded: The relative stability of oxygen-terminated surfaces *vs.* Ce-terminated surfaces has been studied before.[1] It was shown that the oxygen-terminated surface has the lowest free energy for a wide range of oxygen partial pressures up to 1 atm. and only at ultra-high vacuum does the Ce-terminated 111 become the more stable. Thus here in our current study, we focus on oxygen-terminated surfaces.

1 Y. Jiang, J. B. Adams and M. van Schilfgaarde, *J. Chem. Phys.*, 2005, **123**, 064701.

Professor Hermann then asked: Did you take into account relaxation effects of the topmost oxygen layer in your calculations of the oxygen terminated CeO_2 surfaces?

Mr Song replied: In our calculation, we fixed the bottom tri-layer (O–Ce–O) and relax the top two tri-layers (O–Ce–O).

Professor Freund enquired: When you looked at adsorption of a single Rh atoms on CeO_2(111), did you also consider adsorption on steps of the CeO_2(111) surface, whose structure is based on the removal of a O–Ce–O-tri-layer? I ask because when we deposit metals on a CeO_2(111) film we see in STM the metal nucleating at the steps!

Mr Song thanked Professor Freund for the comment: In the current study, we only focus on the plane surface of ceria, namely (111), (110) and (100). We fully agree with you that the steps sides play an important role. One very recent theoretical study based on Au cluster supported stepped CeO_2(111) model indeed also suggests a more reactive reaction path for CO oxidation compared with CeO_2(111) surface.[1] Whereas, here we intend to understand the difference of catalytic activity of experimentally observed Rh supported on different shapes of ceria nanoparticles (nanocube, nanorod, *etc.*) for CO oxidation. And it was shown that these different nanoshapes mainly expose (111), (110) and (100) surfaces. The stepped sides could not be identified. Our theoretical studies show that the (110) surface is more reactive, which suggests the important role of surface morphology in catalysis. We thought this final conclusion is in accordance with your comments.

1 H.-Y. Kim and G. Henkelman, *J. Phys. Chem. Lett.*, 2013, **4**, 216–221.

Professor Bowker noted: Rh doesn't work well on the (100) surface, but is it possible for Rh to be below surface plane of O atoms, rather than above them?

Mr Song replied: The situation you are referring to is the doped system. Indeed, there are lots of theoretical studies working on this doped ceria for CO oxidation. However, the stability issue of the doped ceria still remains in debate. Thus, here we only focus on the adsorbed Rh on ceria for CO oxidation.

Professor Bowker then asked: Do you know what the charge state of the Rh is in this modelling?

Mr Song responded: It is very difficult to give an accurate value of the charge state of Rh. We can only have a relatively qualitative description of the electronic state of

Rh. When Rh is adsorbed at the ceria surface, some of surface cerium atoms were reduced from +4 to +3 (which could be concluded from the +1 spin magnetic moment and one integer electron on f orbital). From this, we can conclude that some of the electrons are transferred to the ceria surface. So Rh should show a positive charge state.

Professor Campbell opened the discussion of the paper by Dr Deng: Terry King or Andy DePristo did some calculations very similar to yours using EAM for bimetallic clusters back in the 1980's, but it is not cited here. Can you please summarize their results.

Dr Deng communicated in reply: Thank you very much for the constructive comments. King *et al.* performed excellent and pioneering work in this field.[1] Their most relevant paper for Ag–Pt bimetallic nanoparticles has been cited as ref. 20 in our paper. Their results of TOC structured clusters have also been summarized and compared with our results in Fig. 3. It can be seen that present surface Ag composition is in good agreement with their MC simulation results using surface modified pair potential (SMPP).

1 J. K. Strohl and T. S. King, *J. Catal.*, 1989, **116**, 540.

Professor Campbell commented: The embedded atom method (EAM or MEAM) includes multi-body interactions, not just two-body interactions, and it works very well without angular corrections for FCC and HCP metals, which do not have significant directional preferences for bonding (unlike BCC metals).

Professor Hermann remarked: In your work you seem to have considered only particles which are confined by high density surfaces. However, many metal nanoparticles can also include sizable surface areas which have steps and kinks. Examples are quasi-spherical Cu nanoparticles prepared in the Inorganic Chemistry Department at the Fritz-Haber Institute, Berlin, and observed by atom-resolved electron microscopy. In view of these results, could your particle shapes be a consequence of the specific potential models that you are using?

Dr Deng replied: Thank you very much for the constructive comments. Industrial catalysts are often supported nanoparticles of varying shapes and sizes, whose structures and electronic states depend on the support and the conditions of synthesis or operating, and are too complex to be completely captured at the theoretical level. Thus, reasonable simplified model systems are necessary. For example, although quasi-spherical Cu nanoparticles prepared at the Fritz-Haber Institute include sizable surface areas which have steps and kinks,[1] a model system within DFT calculations was used to confirm the effect of surface defects at the Cu surface on the catalytic properties, where a flat Cu(111) surface represents the ideal defect-free catalyst, whereas a stepped Cu(211) surface was used to include the effect of surface defects. Therefore, in the present work only the nanoparticles consisting of atomically smooth terraces and edges were used to formulate the simplified model system. Such terraces and edges were based on a modified Wulff construction and guided by actual observations as well as atomistic models with many-body interactions by past experience, rather than a consequence of the specific potential models that we used. While these model nanoparticles may not capture all the aspects of a complex catalyst, this convenient model system contains the main common features of nanoparticles and allows for the study of the morphology, dimension, and composition dependence of surface segregation and chemical ordering patterns. In addition, the trends of surface segregation and chemical ordering patterns determined for model Ag–Pt nanoparticles should potentially be able to be extrapolated to the trends in real nanoparticle with non-magic numbers and surface defects such as steps, kinks, adatoms, *etc.*

1 M. Behrens *et al.*, *Science*, 2012, **336**, 893.

Professor Freund queried: We heard in the paper by Dr Lu in Session 1 about modeling of EXAFS spectra of bimetallic particles. Would it be possible to use the results of your calculations to produce theoretical EXAFS spectra which may be averaged over the various configurations in order to compare them with experimental spectra, and to evaluate how sensitive the calculated spectra are to variations of the various parameters?

Dr Deng thanked Professor Freund for the constructive comment: EXAFS is the method of choice for the determination of the arrangement of atoms for bimetallic nanoparticles. For example, with the help of coordination number extracted from MC simulation results or EXAFS analysis, it is possible to calculate the alloying extent and predict the structural model of bimetallic nanoparticles. As we have shown before, the calculated coordination number and alloying extents of Au–Ag nanoparticles with different sizes and compositions based on MC simulations are consistent with experimental EXAFS analysis.[1] Moreover, EXAFS can give information about the electronic state of the respective metal atoms, which can not be directly obtain due to the limited accuracy of EAM potentials and most of the other empirical potentials.

1 L. Deng *et al.*, *J. Phys. Chem. C*, 2011, **115**, 11355–11363.

Professor Asakura added in response to Professor Freund: EXAFS can be analyzed by real model structures. But if the size was large (more than 2 nm) it is quite difficult to express the real structure only from EXAFS.

Professor Bowker asked: The surface morphology is dictated by the surface energy for the monometallics, but do you have fixed shape configurations for added silver? Isn't it likely that different levels of silver may change the most stable morphology of the particle?

Dr Deng replied: One of the classic models to describe particle shape (whether it is small, large, free-standing, or as precipitates) is the Wulff construction, which dictates a particle shape given its orientation-dependent surface free energy. While there have been extensions, in all cases the deviations from the Wulff shape for monometallic nanoparticles are relatively minor. For bimetallic nanoparticles, however, there is the additional degree of freedom of surface segregation of one or more species. Thus, the orientation-dependent surface free energy would be changed by different levels of surface composition, which means that the shape and composition profiles of the bimetallic nanoparticle are generally dependent on the particle size and global composition.[1] For Ag–Pt particles, however, the surfaces are almost covered by Ag atoms due to their strong surface segregation except for very low global Ag composition, which means that the shapes of Ag–Pt particles is mainly determined by the orientation-dependent surface free energy of Ag. Therefore, based on a modified Wulff construction,[2] only the nanoparticles consisting of atomically smooth terraces and edges (such as TOC, CUB, ICO, IDEC, and MDEC) were used to formulate the simplified model system in the present paper. Then a starting configuration was generated by designating randomly the proper numbers of atoms as Ag atoms, with the remaining designating as Pt atoms.

1 E. Ringe *et al.*, *Nano Lett.*, 2011, **11**, 3399–3403.
2 C. L. Cleveland and U. Landman, *J. Chem. Phys.*, 1991, **94**, 7376.

Dr Boscoboinik remarked: This work is quite interesting. However, it would useful to compare it with experimental data. In fact, this has been done before for Monte Carlo simulations of surface bimetallic alloys in which the distributions of atom ensembles on the surface was compared with temperature-programmed desorption experiments of CO desorbing from these ensembles for real surface alloys.[1,2] By comparison with the experiments the parametrization of the simulations was optimized to achieve the best fit between ensemble distribution and integrated areas in the TPD peaks. This is one way to do it, but in principle other techniques can also be used for the comparison between simulation and experiment, for example infrared spectra of CO on the bimetallic alloys will give a series of peaks for which the frequency depends on the atom ensemble in which the CO molecule is adsorbed (single Pt atoms surrounded by Ag should for example give a frequency corresponding to atop sites, while pairs of Pt atoms surrounded by Ag can provide bridge sites).

1 J. A. Boscoboinik, F. C. Calaza, M. T. Garvey and W. T. Tysoe, *J. Phys. Chem. C*, 2010, **114**, 1875–1880.
2 M. Garvey, J. A. Boscoboinik, L. Burkholder, J. Walker, C. Plaisance, M. Neurock and W. T. Tysoe, *J. Phys. Chem. C*, 2012, **116**, 4692–4697.

Dr Deng thanked Dr Boscoboinik for the constructive comments: It is true that the distributions of atom ensembles on Au–Pd (see your references and ref. 1) and Ag–Pd alloy surface[2,3] have been studied using MC simulations and compared with TPD experiments. For bimetallic nanoparticles, the similar size and shape distributions of atom ensembles on different coordinated sites of Au–Pd[4] and Au–Ag nanoparticles[5] were analyzed quantitatively, which show the varied composition- and temperature-dependent effects. For Ag–Pt nanoparticles, as a result of the strong surface Ag segregation, Pt ensembles mainly occupy the (111) facets, in the form of Pt monomers, dimers, and trimers. We will analyze quantitatively the surface Pt ensembles and compare it with experimental data (if any or when they are available).

1 J. A. Boscoboinik *et al.*, *Phys. Rev. B*, 2008, **77**, 045422.
2 Y. Ma *et al.*, *Phys. Chem. Chem. Phys.*, 2011, **13**, 10741–10754.
3 A. K. Engstfeld, *et al.*, *Phys. Chem. Chem. Phys.*, 2012, **14**, 10754–10761.
4 B. Shan *et al.*, *Phys. Rev. B*, 2009, **80**, 035404.
5 L. Deng *et al.*, *J. Phys. Chem. C*, 2011, **115**, 11355–11363.

Dr Beaumont commented: The use of this particular theoretical technique to calculate/find the most thermodynamically stable configurations of nanoparticles in vacuum is interesting. Coming from the perspective of using nanoparticles as catalysts (the underlying motivation of work from your paper) it is not at all clear that we always reach the thermodynamically most stable nanoparticle in experiments where the nanoparticles undergo restructuring. There is only a limited range of temperatures accessible before particle sintering occurs and so only kinetically accessible structures (at temperatures below the sintering temperature) can be obtained on particle restructuring.

The generation of nanoparticles through colloidal synthesis methods very often results in a kinetically determined structure permitted by the synthesis conditions, for instance in our work presented in Session 1 of this meeting. To understand the structure of a working catalyst it is therefore necessary to consider kinetically accessible nanoparticle structures from the available starting point and it would be very interesting to calculate the barriers to accessing different types of restructured particles (including the thermodynamically most stable form you've already calculated) when starting with structures that can be achieved synthetically. I suspect MC simulations may not be the technique for this, but it would nevertheless be an interesting thing to try with another method. (The issue of real catalyst particles also interacting with both the reactants and the support can't really be ignored here either and would of course be very valuable to include if there is a way to do so).

Dr Deng thanked Dr Beaumont for the insightful suggestions: The stability of bimetallic nanoparticles is very interesting from both the practical and theoretical points of view. The present article is to obtain thermodynamically preferred configurations of Ag–Pt nanoalloys in vacuum. To further understand the structure of nanocatalysts under different environments, it is helpful to consider both the effects of reactants and support on particle restructuring. Further work on the adsorption- and support-induced surface segregation of Pt-based bimetallic nanoparticles is being prepared in our group, where MC simulations are used to obtain the thermodynamically stable configurations (similar to ref. 1). From the kinetic point of view, only in a limited range of temperatures (at high enough temperatures but below the sintering temperature), the thermodynamically most stable structures found here can be directly compared with the synthetically structures kinetically obtained by nanoparticle restructuring. However, the thermodynamically preferred configurations predicted by MC studies would be helpful to understand the ageing of nanoparticles prepared under non-equilibrium conditions.[2] On the other hand, it is true that MC simulations are not suitable to study the kinetic evolution of different types of nanoparticle structures (free or supported, clean or adsorbed). To this end, Molecular Dynamics simulation with well-designed annealing cycles[3] might be a better choice, but the results strongly depend on the metal–metal, metal–adsorption or metal–support potentials (if any). Another suitable method would be Kinetic MC simulation.[4] A KMC code for nanoparticle restructuring (morphology evolution, surface segregation and dealloying) in vacuum or under electrochemical environment is under development in our group.

1 N. Savargaonkar *et al.*, *J. Catal.*, 1996, **162**, 277–283.
2 F. Yin, Z. W. Wang and R. E. Palmer, *J. Exp. Nanosci.*, 2012, **7**, 703–710.
3 E. Favry *et al.*, *Phys. Chem. Chem. Phys.*, 2011, **13**, 9201–9208.
4 M. Müller and K. Albe, *Beilstein J. Nanotechnol.*, 2011, **2**, 40–46.

Professor Hermann opened a general discussion of Professor Hensen's and Dr Deng's papers: I would like to address the problem of atom charges and charge transfer. In physics we learn that the charge of a specific atom inside a molecule (or solid) cannot be defined rigorously due to the quantum mechanical nature of the system. On the other hand, atom charges and their differences caused by charge transfer are widely used in chemistry (and physics) to describe electronic behavior and we have heard examples during this meeting. Sometimes, even quantitative values from quantum chemical calculations (valence charges, populations, Bader, Hirshfeld charges, *etc.*) are attached to atoms where, depending on the evaluation method, quite different results may be obtained. In view of this conceptual variety what can we learn from charge analyses?

Professor Campbell noted: It is important to clearly distinguish between atom charge and formal oxidation state. Formal oxidation state can be extracted from quantum mechanical calculations using spin population analysis instead of the much more common total charge population analysis. It clearly explains spin-related measurements like EPR and Mossbauer spectroscopy which are used to measure unit differences in oxidation states. These formal oxidation states are also used with great predictive success by organometallic chemists, *via* their famous electron-counting rules.

Professor Sauer agreed with Professor Campbell: Atomic charges have a meaning only within the concept in which they are defined. Oxidation numbers, in transition metal or rare earth compounds, are defined by the occupation of d- and f-shells, respectively. The safest way to determine them from calculations is from natural orbital populations or spin density distribution. Experimentally, one uses spectroscopy to determine the occupation of d- or f-shells, which are not always obvious. For example, ligands are not always "redox-innocent" (Wieghardt)[1] and in our presentation we look at systems which can have Ce^{3+}/V^{5+} or Ce^{4+}/V^{4+} populations.

1 P. Ghosh *et al.*, *J. Am. Chem. Soc.*, 2003, **125**, 1293–1308.

Professor Somorjai made a general comment: Five theory papers very well embedded with strong experimental work is a good sign showing major theory contribution. However, charge flow leading to acid–base chemistry and covalent chemistry leading to neutral atom adsorption are both present on catalyst surfaces. Is there theory treatments of combined effects of both types of chemistry?

Professor Neurock closed the general discussion in response to Professor Somorjai: Our paper on the partial oxidation of acetic acid demonstrates both covalent as well as acid–base chemistry as the elementary reactions appear to occur on Au–Ti, Ti–O and Au–O site pairs at the Au/TiO_2 interface. The Ti^{4+}–O^{2-} sites on TiO_2 are Lewis acid–base pairs that are known to heterolytically deprotonate acids to form carboxylate anions and protons that bind to the surface Ti^{4+} and O^{2-} sites, respectively. This is consistent with our spectroscopic and theoretical results which reflect the clear heterolytic activation of the acetic acid reactant. Our theoretical results also show that there is a charge transfer from the Au to the $Ti4^+$ sites at the Au/TiO_2 interface and that these the interfacial Au–Ti^{4+} sites can activate O_2 at higher temperatures to generate negatively charged oxygen atoms that migrates onto Au. The atomic oxygen can be considered a "charge carrier" that is generated at the Au/TiO_2 support and "spills" over onto Au. Atomic oxygen on Au behaves as a base which subsequently aids in the activation of weakly acidic C–H bonds of the bound acetate and CH_xCO_2* intermediates. These reactions proceed over Au–O site pairs where the Au atom inserts into the C–H bond to form a covalent Au–CH_x bond while the negatively charged oxygen atom on Au abstracts the resulting hydrogen atom as a proton. The Au–O site pair is very similar to the metal–ligand pairs that carry out σ-bond metathesis steps in organometallic chemistry. C–H activation proceeds in such organometallic systems by a classic oxidative addition at the metal center together with a proton transfer to a basic ligand. As such, the C–H activation over the Au–O site pair involves covalent and acid–base character. More generally the partial oxidation of the acetic acid to ketenylidene has both covalent as well as acid–base chemistry.

Alumina support and Pd_n cluster size effects on activity of Pd_n for catalytic oxidation of CO

Matthew D. Kane, F. Sloan Roberts and Scott L. Anderson*

Received 14th December 2012, Accepted 15th January 2013
DOI: 10.1039/c3fd20151a

A series of model catalysts were prepared by depositing different size Pd_n clusters on alumina films grown to variable thickness on a Ta(110) support. Samples were characterized by a combination of X-ray photoelectron spectroscopy, low energy He^+ scattering, and temperature-programmed reaction and desorption (TPR/TPD). For the activity studies, the samples were first exposed to $^{18}O_2$ at T_{ox}, and then to ^{13}CO at 180 K, where CO sticks to Pd, but not to the alumina support. CO oxidation activity increased with increasing thickness of the alumina support up to ~4.5 nm, but was constant for greater thicknesses. Activity increased, with T_{ox} up to 400 K, but then declined for T_{ox} = 500 K. Activity was also found to be non-monotonically dependent on deposited cluster size, with Pd_n ($n \leq 6$) being generally more reactive than the larger clusters studied. Activity was only weakly correlated with exposed Pd binding sites, which decreased with increasing cluster size, however, there does appear to be a correlation between activity and electronic structure, as probed *via* the Pd 3d binding energy. Unlike previous systems we have studied, the activity of small Pd_n on these alumina films was quite stable, with essentially no changes observed in up to eight successive TPR experiments.

1 Introduction

Metal-catalysed CO oxidation has been extensively studied, both because it has practical importance, and because it is relatively simple and amenable for study using surface science techniques.[1–12] The rich literature on CO oxidation over oxide-supported Pd catalysts is particularly attractive from the perspective of size-selected cluster measurements, providing mechanistic insight to aid interpretation. Another motivation is that we recently observed non-monotonic cluster size dependence for CO oxidation over $Pd_n/TiO_2(110)$, and found that activity was clearly correlated with variations in Pd_n electronic properties.[9] Here we report a study under similar conditions, of CO oxidation catalysed by Pd_n deposited on variable thickness alumina films grown on Ta(110).

The literature for CO and O_2 interactions with Pd is enormous, but it is worth mentioning a few relevant results. This reaction has been studied extensively for both Pd single crystals[1,6,13–16] and Pd nanoparticles supported on metal oxides.[2,17–27] It was shown by Ertl and others that CO oxidation on noble metals occurs by a Langmuir–Hinshelwood mechanism, where both reactants must adsorb on the surface before combining to form CO_2, and that kinetics are controlled by competition between CO and oxygen for binding sites.[1,6,13,14] The Freund and Matolin groups studied CO on alumina-supported Pd, and noted an increase in low temperature CO desorption with decreasing Pd particle size, attributed to CO weakly bound atop small Pd particles, as opposed to in high coordination binding sites on larger

Department of Chemistry, The University of Utah, 315 S. 1400 E, Salt Lake City, UT, USA. E-mail: anderson@chem.utah.edu; Fax: +801 5816681; Tel: +801 585 7289

particles.[3,24,28,29] The Campbell group studied Pd particles grown on both alumina films and single crystal α-Al_2O_3(0001),[30,31] examining both Pd sintering behaviour and the nature of O_2 binding. Bowker and co-workers reported studies of TiO_2-supported Pd, probing CO reverse spillover, oxygen spillover, and strong metal support interaction effects.[32–34] Goodman and co-workers also did important work in this and related systems, exploring pressures bridging the gap between UHV and practical working catalyst conditions.[35–40] They showed that even at elevated pressures, mechanistic insight from UHV studies can be applied.[6,41,42]

There are also a number of studies of cluster size effects on CO oxidation catalysed by Pd_n, in addition to the Pd_n/TiO_2 study mentioned above. Heiz, Landman, together with many co-workers and collaborators, have studied cluster size effects of Pd_n supported on MgO/Mo(100).[4,19,20,43–48] This work suggests there is an increased activation barrier for O_2 dissociation with decreasing Pd_n cluster size ($n \leq 30$), resulting in higher temperatures required for efficient O_2 activation. It was shown that small amounts of co-adsorbed CO can actually enhance O_2 dissociation. Related work on Au_n/MgO/Mo(100) showed that MgO film thickness could be used to tune the activity of the Au_n clusters for CO oxidation.[49,50]

We previously attempted a study of CO oxidation over Pd_n/alumina,[8] using the process reported by Shaikhutdinov *et al.*[3] to grow a 0.5 nm thick alumina film on NiAl(110). At 100 K, O_2 was observed to oxidize all Pd_n studied, with the exception of Pd_4 – an exception that was subsequently explained by theory from the Khanna group.[51] Attempts to study CO oxidation failed, however, because oxygen initially bound to the Pd clusters, readily spills over to the thin alumina support, and reacts with the underlying NiAl, as discussed by Shaikhutdinov *et al.*[3] In fact, Shaikhutdinov *et al.* used this effect to thicken the alumina layer beneath their Pd particles, generating an active catalyst. One of the motivations of the present work is to examine both cluster size and alumina film thickness effects in the Pd/alumina system.

2 Experimental methodology

Apparatus

The experiments in this paper were conducted in an UHV system with a base pressure $\leq 2 \times 10^{-10}$ Torr as described previously.[52,53] The vacuum system is made up of three main sections: the cluster deposition beamline, the main deposition/analysis chamber, and the load lock/high-pressure chamber. The differentially-pumped cluster deposition beamline drops the operating pressure from 20 mTorr in the cluster source chamber down to 2×10^{-10} Torr in the final chamber leading into the analysis/deposition chamber. A laser vaporization source is used to create Pd_n^+ clusters ($n \leq 25$) that are guided by a series of quadrupole ion guides, mass selected by a quadrupole mass filter, and then guided into the deposition/analysis chamber, where they are deposited on the sample surface at approximately 1 eV atom^{-1}. Deposition occurs through a mask with 2 mm orifice, located ~0.1 mm in front of the sample surface. The main deposition/analysis chamber contains facilities for *in situ* characterization using X-ray photoelectron spectroscopy (XPS), ultraviolet photoelectron spectroscopy (UPS), and ion scattering spectroscopy (ISS). In the experiments described here, the XPS was recorded using an Al Kα source, and binding energies were corrected to the Ta $4d_{5/2}$ binding energy of the Ta(110) single crystal used as the substrate for alumina film growth. The main chamber also contains a residual gas analyzer (RGA) to monitor background partial pressures and gas doses, and a differentially pumped section that houses a second mass spectrometer that views the sample through a 2.5 mm diameter skimmer orifice. This mass spectrometer is used to monitor species desorbing from the deposited cluster spot, with minimal sensitivity to species desorbing from outer regions of the sample or from other surfaces in the vacuum system. When the sample is positioned in front of the skimmer orifice, it can be dosed with gases using a set of seven dosing tubes that are directed toward the sample position, while monitoring gasses in the chamber background and desorbing from the surface, using the two mass spectrometers. The load lock/high-pressure chamber is a small antechamber attached to the bottom of the main chamber, with base pressure $\leq 5 \times 10^{-9}$ Torr. The sample is inserted into the antechamber through a gate valve and triple differential seal, and the antechamber can be pressurized without affecting the main chamber pressure. For the present experiments the antechamber was used for growth of alumina support films under high O_2 pressures.

Model catalyst preparation

Typical samples consisted of Pd_n clusters deposited on an alumina film grown on a 5 $\times$ 10 $\times$ 1 mm Ta(110) single crystal (Surface Preparation Laboratory). The Ta crystal was spot welded to two tantalum heating wires suspended from a pair of tungsten rods attached to a liquid nitrogen-cooled probe, all supported by a manipulator system. The sample temperature was monitored by a K-type thermocouple, spot welded to the back of the Ta(110) single crystal. The sample temperature was controlled between 115 K and 1200 K using resistive heating, or up to 1500 K (limited by the thermocouple) by electron bombardment of the back side of the Ta crystal. Alumina films in the thickness range used here, are conductive enough to allow the Pd_n^+ to neutralize upon deposition. By integrating the sample current during deposition, the number of Pd_n^+ deposited can be determined, allowing production of cluster spots with reproducible Pd coverage. All the samples discussed here had Pd_n deposition densities corresponding to 1.53×10^{14} Pd atoms cm^{-2}, *i.e.*, ~0.1 of a close-packed Pd monolayer. Deposition was done at room temperature, with beam energy of 1 eV atom^{-1}.

The Ta(110) crystal was cleaned by Ar^+ sputtering while the temperature was held at 900 K, followed by annealing at 1400 K for 1 hour. XPS of the annealed sample was used to verify that no Al or Pd remained from previous experiments. The sample was periodically annealed for longer time periods (up to 48 hours) to help anneal out any build-up of bulk defects over experimental cycles. The alumina film was then grown by evaporation of Al onto the Ta(110) single crystal in the antechamber, while holding the surface at 600 K in a background of 5×10^{-6} Torr of ultra-high purity $^{16}O_2$, that was further purified by passing the gas through a frozen ethanol bath (160 K). After growth, the samples were annealed at 700 K for 5 minutes in a background of 5×10^{-6} Torr of $^{16}O_2$, to ensure that the surface was fully oxidized. This alumina growth procedure was adapted from one described by Goodman and co-workers,[54] with the changes being a lower sample temperature during the Al deposition and addition of the post-growth oxygen annealing. We adopted the lower growth temperature because ISS showed that films grown at higher sample temperatures had small (~1%) concentrations of Ta atoms in the surface layer, even for films with >6 nm thickness. ISS showed a sharpening of the aluminum and oxygen peaks after the post-growth annealing, indicating a more homogeneous surface environment in the film. Ratios of the Ta and Al XPS signals were used to determine the thickness of the thin alumina film, using literature values for the photoemission cross sections and β parameters[55] and effective attenuation lengths calculated using the program of Powell *et al.*[56] The estimated absolute uncertainty in film thickness is ±20%, however, the relative uncertainty in comparing thicknesses of different films is estimated to be less than 5%.

As discussed below, tests were made of the effects of both alumina film thickness and growth rate on the activity of supported Pd clusters for CO oxidation. Activity was found to be independent of film growth rate for rates below ~0.8 nm min^{-1}, but faster rates led to decreased activity of the supported Pd, presumably due to increased roughness/disorder in the alumina film. Unless stated otherwise, the samples discussed below were prepared with growth rates in the 0.3–0.5 nm min^{-1} range – well below the range where activity begins to depend on growth rate. As shown below, the alumina film, in absence of deposited Pd, is inert under our conditions.

Characterization and reactivity measurements

After the alumina film was grown and Pd_n was deposited, the sample was characterized by XPS at room temperature in order to measure the core level binding energies of both the Pd_n clusters and the support. The sample was then cooled to 135 K for reactivity studies. After initial cooling, and prior to dosing with reactants, the sample was flashed to 560 K to desorb any adventitious background gases that might have adsorbed during the time required for XPS and sample cooling. The sample was then dosed with 10 L $^{18}O_2$ with the sample held at T_{ox}, followed by a 10 L exposure of ^{13}CO at a sample temperature of 180 K. During dosing, the sample was positioned ~1 cm from the O_2 and CO dosing tubes, and 2 mm from the mass spectrometer skimmer cone. We calibrated the doses by comparing CO adsorption resulting from variable doses delivered through the dose tubes with doses from the chamber background, using TPD from Pd(111) and Ni(100) to calibrate both the amount of adsorbed CO and the mass spectrometer absolute sensitivity (by measuring saturated monolayer desorption).[57] Every day, we measured the mass spectrometer signal for a fixed background pressure of Ar, allowing us to correct for any day-to-day variations in mass spectrometer sensitivity due to factors such as multiplier gain.

For most experiments, $^{18}O_2$ exposure was done at $T_{ox} = 400$ K to allow comparison with previous results for $Pd_n/TiO_2(110)$ where strong dependence of CO oxidation activity on Pd_n size was found.[9] The CO dose temperature of 180 K was chosen to be high enough to minimize CO binding to the alumina support, but low enough to result in minimal reaction of the impinging ^{13}CO with oxygen pre-adsorbed during the $^{18}O_2$ dose. To test this latter point, the mass spectral signals for all CO_2 isotopologs were monitored during the ^{13}CO dose. A small pulse of $^{13}C^{16}O^{18}O$ was observed just at the start of the ^{13}CO dose, but the integrated signal in this $^{13}C^{16}O^{18}O$ pulse was only ~2–3% of the integrated $^{13}C^{16}O^{18}O$ production observed during the TPR, confirming that 180 K is, indeed, cold enough to largely suppress reaction.

After dosing, the sample was cooled to the 135 K TPR starting temperature, and moved to a position ~1 mm in front of the differentially pumped mass spectrometer orifice. The sample temperature was then ramped at 3 K s^{-1} to 560 K, while the mass spectrometer rapidly cycled between all the masses of interest, to measure species desorbing from the surface. This TPR process (including 400 K $^{18}O_2$ and 180 K ^{13}CO dosing) was repeated two more times to examine how the activity of the samples might change with gas adsorption and heating. Finally, the sample was probed by a ^{13}CO TPD experiment where the $^{18}O_2$ dose was omitted, and only ^{13}CO was dosed (180 K). This final CO TPD provides a measure of the density of CO binding sites remaining on the surface after the three TPR experiments. In addition to the expected $^{13}C^{16}O^{18}O$ product, the signals for all other CO_2 isotopologs were measured, as were both ^{13}CO and ^{12}CO. We conclude, therefore, that under our conditions, ^{16}O in the alumina film does not participate in the chemistry significantly, and that any adventitious ^{12}CO that might have adsorbed on the samples during deposition, XPS characterization, and cooling, was removed by the combination of the 560 K flash and 400 K O_2 exposure.

Post-reaction sample characterization was performed using XPS and ISS. Comparison of pre- and post-reaction XPS probes changes in the Pd chemical environment. ISS is sensitive only to the top-most layer of atoms, and because our samples were all prepared with the same total number of Pd atoms deposited, variations in the post-reaction Pd ISS intensity provide insight into the fraction of Pd in the surface layer. The Pd ISS signal could be affected both by morphology changes (*i.e.*, single layer *vs.* multilayer clusters) and by adsorbates remaining on the surface of the Pd clusters after the series of TPR experiments.

3 Results and discussion

As reported previously, Pd_n supported on a 0.5 nm alumina film grown on NiAl(110) readily oxidize upon O_2 exposure, but are ineffective for CO oxidation.[8] As discussed by Shaikhutdinov *et al.*,[3] the problem is that the 0.5 nm film is too thin to serve as an effective barrier to oxygen diffusion, so that oxygen activated on the Pd clusters tends to spill over and react with the underlying NiAl substrate, rather than oxidizing co-adsorbed CO. Therefore, before examining the dependence of Pd_n activity on cluster size, it is important to understand how alumina film properties vary with thickness, and how thickness influences the activity of Pd_n supported on alumina/Ta(110).

ISS and XPS characterization Pd_{20} on alumina films of varying thickness

Alumina films grown epitaxially on Ta(110), Mo(110) and Re(0001) have been studied by Goodman[54,58,59] and shown to have well defined long range crystal order when prepared with a sample temperature at 900 K. As described above, we used lower annealing temperatures in this experiment, because He^+ ion scattering (ISS) shows that even quite thick films grown or annealed above 900 K have low level Ta contamination in the surface layer, presumably due to Ta diffusion through the alumina thin film. Madey and co-workers used ISS to show that the growth mode of alumina films on Re(0001) and Ru(0001) is dependent on growth conditions. Al deposited on Re(0001) and Ru(0001) grows layer by layer and can then be oxidized to form stoichiometric alumina, however this method does not lead to good long range order.[60] Good long range order was obtained by growing alumina at elevated temperatures (1173 K) in a background of O_2, the growth mode under those conditions was initial formation of 3D alumina islands that coalesced into a continuous film with increasing coverage.[61] In particular, at least 1.5 nm of alumina was required to achieve a continuous film. Fig. 1 compares ISS data for 0.1 ML Pd_{20} as-deposited on alumina films with 1.4 nm and 4.7 nm thicknesses, grown as discussed above. Note that for the thinner film (1.4 nm ≈ 3 ML), a shoulder is observed on the high E/E_0 side of the Pd peak, indicating that there is Ta in the surface layer. Our relative ISS detection sensitivities for Al, Pd, and Ta, determined by measurements on thick alumina films, freshly sputtered/annealed Ta(110), and 0.1 ML Pd on alumina films, are roughly equal for Ta and Pd, but ten times lower for Al. Therefore this Ta shoulder corresponds to ~3.5% of Ta exposed in the surface layer, indicating the presence of some combination of exposed Ta substrate and/or Ta in the alumina surface layer. For the 4.7 nm film, the Ta shoulder is absent, indicating that a continuous alumina film is present, with no Ta in the surface layer. In addition, note that the background in the 0.5 to 0.8 E/E_0 range is substantially reduced in the thicker alumina film. This continuous background is due to multiple and sub-surface scattering, and is sensitive to Ta in the near-surface region, due to the much larger cross section for He^+ scattering from Ta, compared to Al.

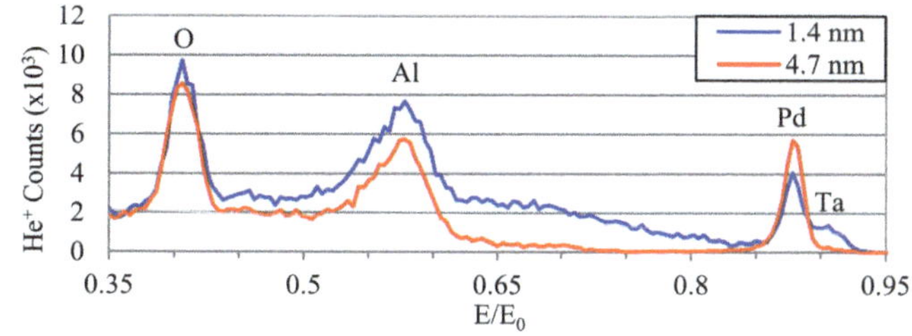

Fig. 1 He^+ ISS for as-deposited 0.1 ML Pd_{20} on 1.4 nm (blue) and 4.7 nm (red) alumina films.

The XPS binding energies for Pd on alumina/Ta can vary both due to charging of the alumina film, and possible film thickness-dependent screening/image charge effects on the photoemission final state. To illustrate the magnitude of these effects, Fig. 2 plots the Al 2s and O 1s binding energies as a function of film thickness, with the binding energy scale shifted so that the Ta $4d_{5/2}$ peak in the Ta(110) substrate is at the literature value of 226.4 eV for metallic Ta.[62] Note that the Al 2s peak was used, rather than the more commonly tabulated Al 2p, because of less interference from Ta photoelectrons. With this approach to setting the binding energy scale, it can be seen that the Al and O binding energies are constant within the experimental uncertainty (~±0.15 eV) for thicknesses up to ~5 nm, but then shift to higher binding energies for thicker films. Nonetheless, they remain within the ranges of binding energies reported in the NIST database for Al 2s (116.1–121.2) and O 1s (528.27–533.1 eV).[63] It is unclear whether the shift to higher energies for the thicker films simply results from charging, or from formation of a Schottky barrier causing band bending, as suggested by Madey,[61] or some combination of the two. It has previously been reported that thin alumina films (≈ 3 nm or less) grown on a conductive metal substrate do not show significant charging.[64] If the binding energy scale were set using some reference value for either the O 1s or Al 2s peaks, rather than Ta 4d from the conductive support, this might better correct for charging of the film, however it is not obvious what value should be chosen, or how Schottky barriers might affect this calibration.

From our perspective, the important point is how the Pd 3d binding energies shift both with cluster size and with thickness of the alumina support film. Fig. 3 shows how the Pd $3d_{5/2}$ binding energy changes for Pd_{20} deposited on alumina films of varying thickness, compared with the analogous changes observed in the Al 2s and O 1s peaks from the films. To allow easy comparison of all three binding energies, they are plotted as shifts relative to the binding energies measured for the thinnest alumina film (1.4 nm). Note that the binding energy scale is inverted to allow better comparison with the CO oxidation activity, also plotted in Fig. 3, and discussed below.

In contrast to the roughly constant binding energies for Al 2s and O 1s, the Pd 3d binding energy initially shifts to lower binding energy as alumina thickness is increased from 1.4 to ~5 nm. For thicker films, where the Al 2s and O 1s binding energies shift gradually to higher binding energies, the Pd 3d binding energy increases only slightly for 6.3 nm, and then increases dramatically for the thickest film. As already noted, a somewhat different pattern would be obtained if some other method were used to reference the XPS binding energy scale, however, the important point is that the dependence of Pd 3d binding energy on film thickness is qualitatively different from those for Al 2s and O 1s, indicating that the chemical environment of the supported Pd changes with film thickness.

It is not clear why the Pd 3d binding energy for the 9.1 nm film is shifted so far to higher binding energy compared to the thinner films. There is some evidence that this

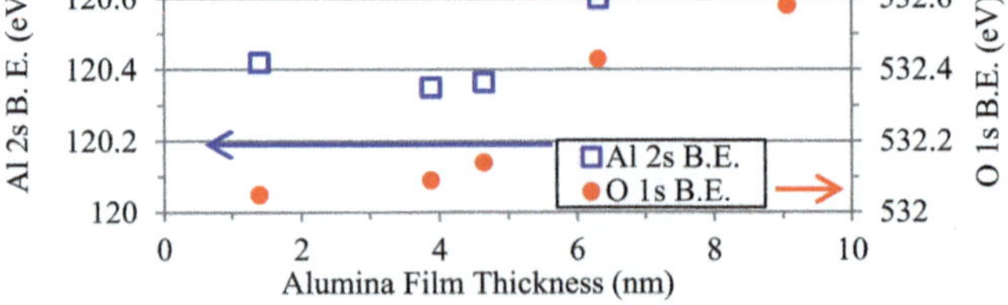

Fig. 2 Al 2s binding energy (blue square) and O 1s binding energy (red circle) as a function of alumina film thickness.

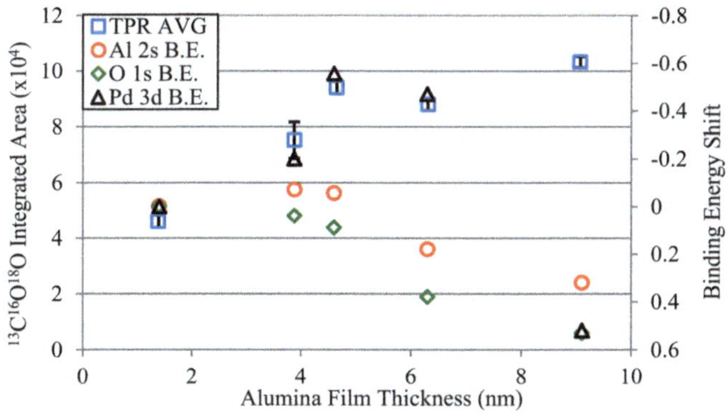

Fig. 3 Shift in peak positions from values measured on a 1.4 nm thick film for Al 2s (red circle), O 1s (green diamond) from a freshly grown alumina film and as-deposited Pd 3d binding energies for 0.1 ML Pd_{20}/alumina/Ta(110) as a function of alumina thickness (black triangle). Average $^{13}C^{16}O^{18}O$ production (blue square) in three consecutive TPR runs for the same samples.

thick film charges substantially more than the thinner films, and it may be that the supported Pd clusters charge differently than the underlying support. We note that the Al 2s, O 1s, and Pd 3d peak widths are constant and close to the resolution limit of our (non-monochromatized) XPS system, however, for the 9.1 nm film, the Pd 3d peak width increases by 0.5 eV as compared to that for the thinner films. Interestingly, the peak widths for the Al 2s and O 1s remain unchanged. In the cluster size-dependent reactivity studies, we therefore kept the alumina support thickness in a narrow range around 5 nm.

The effects of alumina film thickness on the CO oxidation activity of Pd_{20}

Fig. 4 compares CO_2 signals observed during the first TPR run for a 6.0 nm alumina film grown on Ta(110) (dotted black line), and from Pd_{20} deposited on various

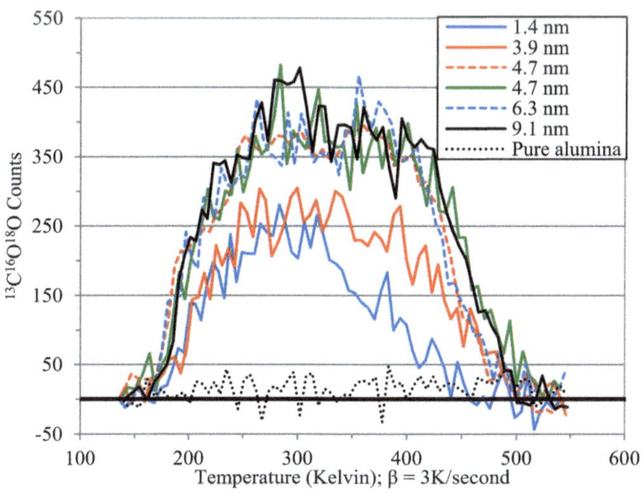

Fig. 4 $^{13}C^{16}O^{18}O$ desorption from 0.1 ML Pd_{20}/alumina samples with variable alumina thickness, following 10 L $^{18}O_2$ exposure at 400 K and 10 L ^{13}CO exposure at 180 K. Alumina thickness (nm): 1.4 (solid blue), 3.9 (solid red), 4.7 (dashed red), 4.7 (green), 6.3 (dashed blue), 9.1 (solid black). Dotted black data are for a 6.0 nm pure alumina film with no Pd deposited.

thickness alumina films grown on Ta(110). The reaction conditions included O_2 exposure at 400 K, CO exposure at 180 K, and ramping temperature from 135 K to 560 K at 3 K s^{-1}, as described above. The first point to note is that the Pd-free alumina/Ta(110) support generates no significant CO_2, indicating that the CO_2 signal generated from Pd_{20}-containing samples can reasonably be attributed to chemistry associated with the Pd_{20}. For the Pd_{20}-containing samples, it can be seen that for films thinner than ~4.5 nm, the CO oxidation activity increases with increasing alumina thickness.

Fig. 5 shows the integrated CO_2 production observed during three consecutive TPR runs for each Pd_{20}/alumina sample. It can be seen that the integrated CO_2 production is rather consistent from run to run, and this is true as well of the temperature dependence (Fig. 4). Evidently, the state of the samples does not change significantly from run to run, at least for the rather modest maximum temperature used here. Because the CO_2 production has good run-to-run stability, Fig. 5 also shows the average and standard error of the integrated CO_2 production for the three runs.

The average CO_2 production is also plotted in Fig. 3, for comparison with the Pd 3d binding energy. With the exception of the 9.1 nm film, where charging may have been a problem in the XPS, there appears to be a correlation between Pd 3d binding energy and activity. The nature of the correlation – that samples with high Pd 3d binding energies are relatively inactive, and *vice versa* – is the same as what we observed as a cluster-size-dependent correlation for Pd_n/TiO_2.[9] This similarly suggests that perhaps part of the increase in activity with increasing thickness may be due to changing electronic environment for the supported Pd. As noted, we have no definitive explanation for the large shift to higher binding energy of the Pd binding energy for the 9.1 nm film. As discussed above, there is some evidence that Pd on this thick film is simply charging due to electron emission during XPS, resulting in a spuriously large apparent binding energy. Of course, we cannot rule out other explanations, such as a change in the initial chemical state of the Pd in this sample, although this scenario seems unlikely given that there is virtually no difference in chemical activity of the 9.1 nm film, compared to those in the 4–6 nm range.

In addition to the increase in integrated CO_2 production with increasing thickness, the CO_2 desorption temperature dependence is also sensitive to alumina thickness, with relatively less CO_2 desorption at higher temperatures for the thinnest alumina film. Clearly, if sub-4.5 nm alumina films were used for cluster deposition, it would be difficult to separate the effects of cluster size from those from run-to-run

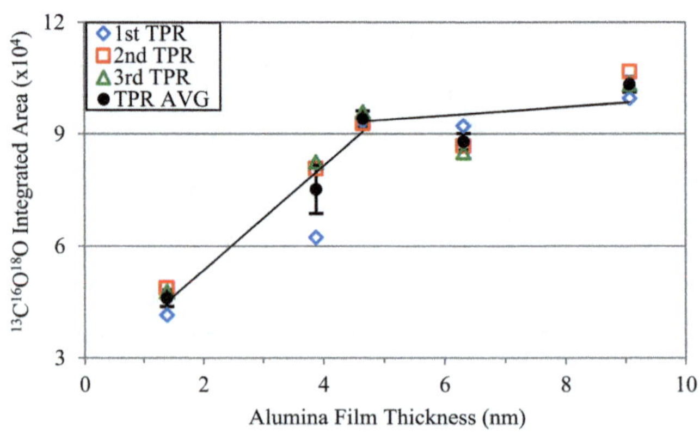

Fig. 5 $^{13}C^{16}O^{18}O$ desorption from three consecutive TPR runs on 0.1 ML Pd_{20}/alumina samples, as a function of the alumina film thickness. TPR conditions: 10 L $^{18}O_2$ oxidation at 400 K, 10 L ^{13}CO exposure at 180 K, 3 K s^{-1} heating ramp. The black circles show the average of the three TPR runs (TPR AVG).

variations in alumina thickness. Fortunately, for films thicker than ~4.5 nm, the CO_2 production activity and temperature dependences become nearly independent of alumina thickness. Therefore, the size-dependent experiments described below were all done using films with thicknesses of 5.0 ± 0.3 nm, where sensitivity to thickness is small, but the film is still thin enough that charging is minimal during Pd_n^+ deposition and XPS.

As noted above, for the thinnest alumina film, ISS shows some Ta in the surface layer, which could indicate the presence of exposed areas of the Ta(110) substrate. As a control, we did an experiment where 0.1 ML of Pd_{20} was deposited directly onto a lightly oxidized Ta(110) surface, with no alumina deposited. The deposited Pd_{20} appears in XPS with a Pd 3d binding energy of 336.5 eV (corresponds to a positive 0.34 eV shift in Fig. 3), which is about what would be expected by extrapolating the binding energy trend in Fig. 3 to zero alumina thickness. This $Pd_{20}/TaO_x/$ Ta(110) sample showed no CO_2 product under TPR conditions. Unreactive CO was observed to desorb in a feature peaking around 150 K.

In addition to the CO_2 product shown in Fig. 4, desorption of unreacted ^{13}CO is also observed during TPR. Fig. 6A shows the ^{13}CO signal recorded during the first TPR run for each sample. Fig. 6B shows ^{13}CO desorption during a final run, for which the sample was dosed with 10 L ^{13}CO at 180 K as usual, but without the prior

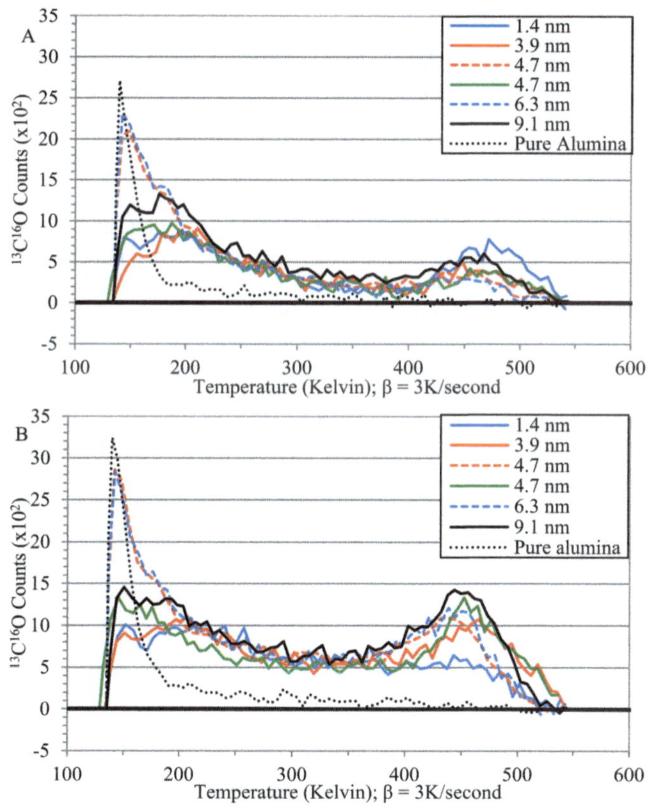

Fig. 6 (A) Unreacted $^{13}C^{16}O$ desorbing during the first TPR run on samples with 0.1 ML of Pd_{20}/alumina with variable alumina thickness after 10 L $^{18}O_2$ oxidation at 400 K and 10 L ^{13}CO exposure at 180 K. (B) $^{13}C^{16}O$ desorption after 10 L ^{13}CO exposure at 180 K with no O_2 exposure. Alumina film thicknesses (nm): 1.4 (blue), 3.9 (red), 4.7 (dashed red), 4.7 (green), 6.3 (dashed blue), 9.1 (black). Dotted black data are for a 6.0 nm pure alumina film with no Pd deposited.

$^{18}O_2$ exposure. No CO_2 products are observed during this final TPD run, indicating that no active oxygen is left on the sample surfaces at the end of the preceding TPR runs.

10 L ^{13}CO exposure at 180 K should saturate all CO binding sites that are stable at 180 K (mostly associated with Pd), but it should not fill more weakly-bound sites, mostly associated with alumina. As shown in Fig. 6, however, there was significant ^{13}CO desorption below 180 K, resulting from adsorption of background ^{13}CO remaining after the CO dose, as the sample cooled to the 135 K TPD starting temperature. The magnitude of this sub-180 K ^{13}CO exposure was roughly 0.3 L, however, during the set of experiments shown in Fig. 4, 5, and 6, the sample mounting was repaired, which resulted in longer cool-down times, and higher sub-180 K ^{13}CO exposures. The desorption curves shown as solid lines were measured before the sample repair, and those measured afterward are shown as dotted or dashed lines. Comparison of the data sets suggests that the "extra" weakly-bound CO on the post-repair samples has little or no effect on the amount or temperature dependence of the CO oxidation (Fig. 4), and also does not seem to affect the CO desorption for $T > 180$ K (Fig. 6). Nonetheless, we note that all the cluster-size-dependent data reported below were taken on the repaired sample, with good run-to-run consistency in the amount of sub-180 K CO adsorbed.

For the 6.0 nm thick alumina film with no Pd (dotted black line), only a sharp desorption feature was observed, beginning as soon as heating started, and with little CO desorbing above ~180 K. For the Pd_{20}/alumina samples there are two new CO desorption features. All the Pd_{20} samples have a low temperature feature that peaks below 200 K, with a tail extending to ~400 K. For the post-repair samples (dashed curves), this Pd-associated desorption feature merges into the sharp feature attributed to desorption from alumina, below ~180 K. In addition to the low temperature feature, the Pd_{20} samples show a clear desorption peak at around 450 K. These two CO desorption features associated with Pd are qualitatively similar to those observed for CO bound to $Pd_n/TiO_2(110)$ under identical conditions. In that system, it was shown *via* temperature-dependent ISS experiments that the low temperature binding site (approximately 200 K for Pd/TiO_2) corresponds to CO bound at the periphery of the Pd clusters, while the high temperature peak (approximately 430 K for Pd/TiO_2) corresponds to CO bound on top of the Pd clusters, such that it prevents He^+ scattering from the underlying Pd.[57]

CO desorption during the final TPD experiments performed after the three sequential TPR runs on each sample, are shown in Fig. 6B. It can be seen that the same three desorption features are observed as in the TPR runs (Fig. 6A), however, the intensity is higher, particularly for temperatures above 250 K. This increase is attributed to two factors. In TPR, some fraction of the CO binding sites are presumably blocked by pre-adsorbed ^{18}O, resulting in less ^{13}CO adsorption during the 180 K dose. Furthermore, a significant amount of CO_2 is generated during TPR (compare Fig. 4 and 6A), mostly between 220 K and 450 K, consuming a substantial fraction of the CO adsorbed on the samples. It can be seen that the thinnest films showed the greatest amount of unreacted CO desorbing at high temperatures (Fig. 6A), whereas in the final TPD (Fig. 6B), the amount of high temperature CO desorption increases with alumina film thickness. This trend is consistent with the CO oxidation activity of these samples (Fig. 4). The thinnest alumina films result in relatively low CO oxidation activity, which leaves a larger fraction of CO remaining to desorb unreacted (compare Fig. 6A and 6B).

One important feature of these model catalysts is that they appear to be quite stable, at least under the conditions explored. Fig. 7 shows that both the amount and temperature dependence of CO_2 production for a Pd_{20}/4.7 nm alumina sample were essentially identical in three sequential TPR runs, and also that no significant CO_2 was observed in the final TPD run, where the $^{18}O_2$ exposure was omitted. One

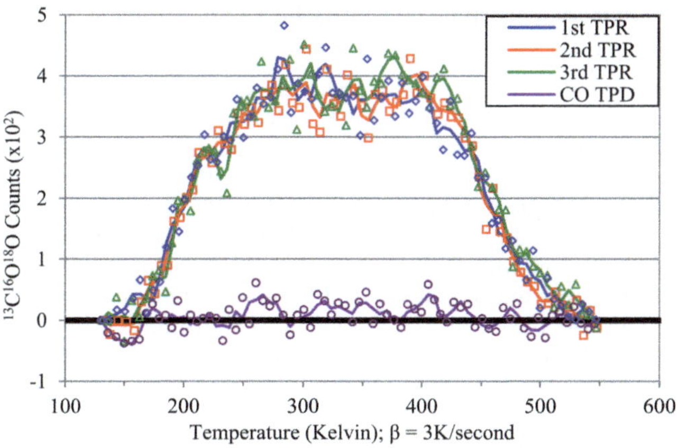

Fig. 7 $^{13}C^{16}O^{18}O$ desorption from 0.1 ML Pd_{20} on 4.7 nm thick alumina/Ta(110) in three consecutive TPR runs, each preceded by 10 L 400 K $^{18}O_2$ and 10 L 180 K ^{13}CO exposure. ^{13}CO desorption during a sequent CO TPD run, with 10 L ^{13}CO exposure at 180 K with no O_2 exposure.

could argue that this highly reproducible activity might mean that even before the first TPR run, the Pd_{20} had diffused and sintered into some morphology that is stable under these conditions. As shown below, however, samples prepared with identical Pd coverage, differing only in the size of Pd_n deposited, have significantly different CO oxidation activities, and these cluster size differences are stable in repeated TPR runs. Stable, but size-dependent behaviour is inconsistent with facile diffusion/sintering, thus the results suggest that these samples are, indeed, thermally stable, at least for the rather mild reaction conditions studied. Selected tests were run on samples where up to eight TPR runs were performed, with no significant change from first to last. This high stability is quite different from what we observed under similar conditions for $Pd_n/TiO_2(110)$, where activity dropped rapidly in sequential TPR runs.[65]

ISS was also run on all samples after the series of TPR and TPD experiments. As-deposited ISS was done only for a few, separately prepared samples, to avoid ISS damage effects on the TPR/TPD results. For the few film thicknesses where we have both as-deposited and post-TPR/TPD ISS data, there is no change in Pd ISS intensity, within experimental error (~5%), indicating that the amount of Pd in the surface layer is not significantly affected by TPR/TPD under the conditions used. This observation is consistent with the stability of the TPR results in multiple runs.

Finally, we note that no ^{18}O signal was observed in the post-TPR/TPD ISS data, indicating that no ^{18}O is left on or in the surface layer, either bound to Pd or to the alumina support. If ^{18}O was left on the surface after TPR, it could be reacted away in the final CO TPD, however, in that case CO_2 should be observed during this final TPD, contrary to observation. Taken together, the ISS and TPD results show that oxygen is the limiting reactant, and is quantitatively reacted away in TPR under our conditions. Furthermore, we can conclude that ^{18}O used in TPR does not spill over to, or exchange with the alumina support, which was grown with normal $^{16}O_2$. After a similar set of TPR and TPD runs for $Pd_n/TiO_2(110)$, 20–30% of the surface oxygen is ^{18}O, and it was shown that most of this ^{18}O dissociated on the Pd clusters, and spilled over to the support where it is not active for CO oxidation.[65] The lack of spillover in the Pd/alumina system is presumably one reason why the CO oxidation activity is substantially higher, compared to $Pd/TiO_2(110)$ under identical conditions.

The effects of oxidation condition on activity for CO oxidation by Pd_{20}

In a previous study of CO oxidation over Pd_n/TiO_2 using TPR conditions similar to those here, we found that activity was highly dependent on oxidation temperature. No cluster sizes were active for $T_{ox} \leq 200$ K, select sizes were active for $T_{ox} = 300$K, and all but Pd_1 showed (size-dependent) activity for $T_{ox} = 400$ K.[10] Fig. 8A shows CO_2 production for a sample prepared by depositing Pd_{20} on a 1.4 nm alumina film, and then probed by sequential TPR runs with 10 L of $^{18}O_2$ at T_{ox}, followed by 10 L ^{13}CO at 180 K. The first three runs had $T_{ox} = 180$, 400, and 500 K. The final run had $T_{ox} = 400$ K again, but with 50 L $^{18}O_2$ exposure. Fig. 8B shows desorption of unreacted ^{13}CO measured during the same TPR runs. The maximum CO_2 production was observed in the runs with $T_{ox} = 400$ K. Oxidizing at 180 K, where $Pd_{20}/TiO_2(110)$ is completely inactive, resulted in almost no change in the low temperature portion of the CO_2 desorption feature, but substantially less CO_2 production above room temperature. Consistent with the decreased overall conversion of CO to CO_2, there was substantially more desorption of unreacted CO in the $T_{ox} =$ 180 K run, mostly in the high temperature peak. It is interesting that increasing T_{ox} to 500 K also resulted in significantly lower CO_2 production, but without the large increase in desorption of unreacted CO. Given that CO and oxygen are known to compete for binding sites (see introduction), the decrease in CO_2 production may simply indicate that higher T_{ox} resulted in a greater-than-optimal coverage of the Pd

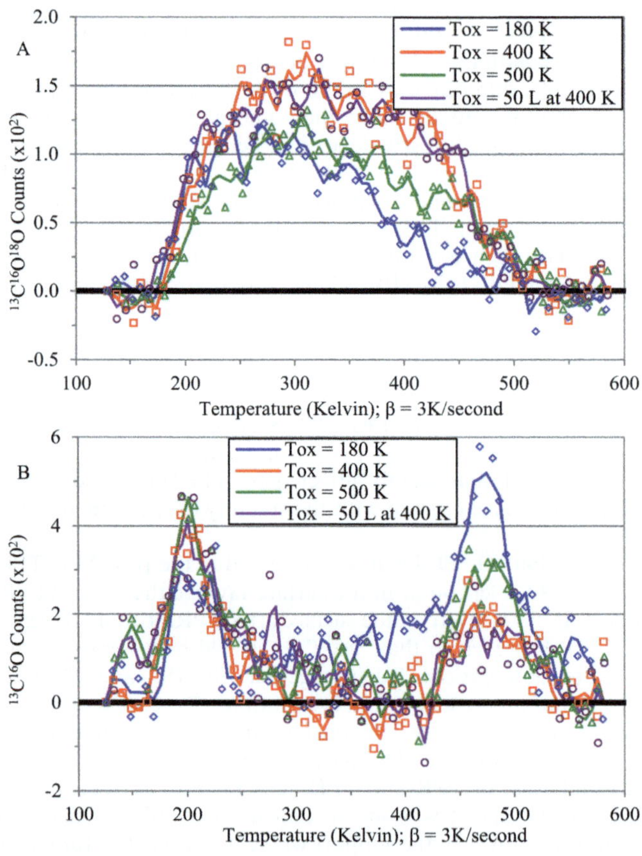

Fig. 8 (A) $^{13}C^{16}O^{18}O$ desorption and (B) unreacted $^{13}C^{16}O$ desorption from 0.1 ML Pd_{20}/1.4 nm alumina/Ta(110) during consecutive TPR runs with different $^{18}O_2$ exposures: 10 L at 180 K (blue), 10 L at 400 K (red), 10 L at 500 K (green), 50 L at 400 K (purple).

clusters by oxygen. Note, however, that in the second TPR with $T_{ox} = 400$ K, the $^{18}O_2$ exposure was increased by a factor of five, which might have been expected to result in similar "over oxidation" of the Pd sites. Instead, the $T_{ox} = 400$ K TPR 50 L O_2 exposure resulted in essentially identical CO_2 production to the $T_{ox} = 400$ K TPR.

The effects of Pd_n size (n = 1, 2, 3, 4, 5, 6, 7, 10, 16, 20, and 25) on CO oxidation activity

CO oxidation and CO TPD were studied for a series of Pd_n/alumina/Ta(110) samples, using TPR/TPD conditions identical to those used to generate the data reported in Fig. 1–7. To avoid confusing the cluster size dependence of interest with effects of alumina film thickness, these samples were all deposited on 5.0 ± 0.3 nm thick films – in the range where activity is insensitive to thickness. For the same reason, the growth rate was kept in the 0.3 nm min^{-1} range, and all experiments reported were done with essentially identical sub-180 K CO exposures. The CO_2 desorption temperature dependence for all cluster sizes were found to be qualitatively similar to those shown for Pd_{20} in Fig. 7, and desorption of unreactive CO during TPR, and of CO in the final TPD showed bi-modal temperature dependence similar to those shown in Fig. 6. Also as in Fig. 7, CO_2 production in successive TPRs was found to be constant within the experimental uncertainty, for all cluster sizes.

The integrated CO_2 production, *i.e.*, the CO oxidation activity, is modestly dependent on cluster size, as shown by the blue squares in Fig. 9, which show the integrated CO_2 production averaged over the three TPR runs for each sample. The error bars show the standard error in the three runs. The fact that activity is size dependent in a non-monotonic fashion implies that the samples retain some memory of the size of the deposited clusters, *i.e.*, that diffusion and sintering are not facile enough under the conditions studied to completely average out the effects of deposited size. Furthermore, the fact that activity does not vary significantly in sequential TPR runs, shows whatever the state of the deposited clusters is, it is reasonably stable.

It is interesting to compare the effects of cluster size in this system with those we observed for $Pd_n/TiO_2(110)$, and those observed by Heiz and co-workers for Pd_n/MgO/Mo(100). Under conditions identical to those used for Fig. 9, CO oxidation over Pd_n/TiO_2 has a stronger size dependence with Pd_1/TiO_2 being essentially

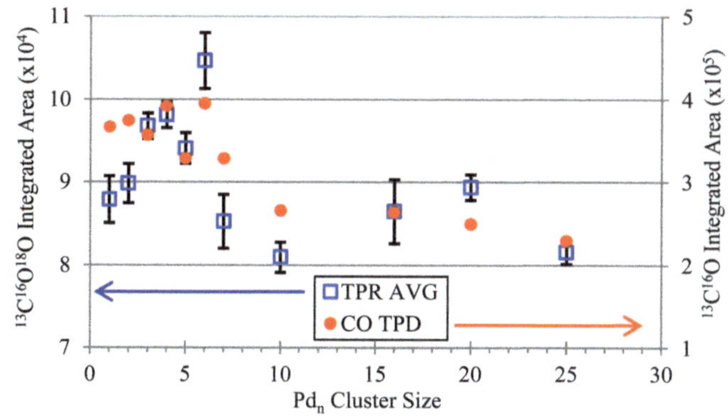

Fig. 9 Average $^{13}C^{16}O^{18}O$ production (blue squares) in three TPR runs for samples with 0.1 ML of Pd deposited as different Pd_n, on a 5.0 ± 0.3 nm thick alumina film grown on Ta(110). TPR conditions: 10 L $^{18}O_2$ exposure at 400 K, 10 L ^{13}CO exposure at 180 K. $^{13}C^{16}O$ desorption from a subsequent CO TPD (red dots) with 10 L ^{13}CO exposure at 180 K, but no O_2 exposure.

inactive, and the other Pd_n/TiO_2 having activities that vary by a factor of ~2.[9] The highest activity in that system was for Pd_{20}/TiO_2, with lower activities observed both for larger and smaller clusters. Even for Pd_{20}/TiO_2, only ~5% of the absorbed CO reacted to form CO_2, and no activity was observed for T_{ox} below room temperature.[10] In contrast, all Pd_n/alumina/Ta(110) samples are found to be quite active, with ~45% of the adsorbed CO reacting to form CO_2, and substantial activity even for $T_{ox} = 180$ K. As might be expected for a system where reactivity is high for all Pd_n, the dependence on activity is much weaker, amounting to only ~25% variation between least and most reactive. There does appear to be a significant trend of activity increasing from Pd_1 to Pd_6, then dropping abruptly for larger clusters. As shown in Fig. 10, the increased activity for Pd_n with $n \leq 6$, is correlated with increased CO_2 production at low temperatures, suggesting that the activation energy for the reaction of adsorbed CO and oxygen is smaller for the small clusters.

Also shown in Fig. 9 is the CO desorption measured in the final CO TPD run, without pre-oxidation of the samples. As noted, no CO_2 is observed in this run, therefore the CO integrated desorption should be a measure of the number of (mostly Pd-associated) CO binding sites on the samples. It can be seen that for Pd_n, $n \leq 6$, the CO intensity is roughly constant, and that it drops by a factor of 30% for Pd_{10-25}. This may indicate that there is a structural transition, such as from single layer to partially two layer structures, with increasing cluster size, and of course this decrease in availability of Pd sites might be expected to be reflected in the CO oxidation activity.

The non-zero activity observed here for deposited Pd_1 might be an indication that atoms deposited on alumina/Ta(110) sinter to form clusters which are reactive. We note, however, that Heiz and co-workers have observed significant activity for model catalysts prepared by deposition of Pd atomic ions on MgO/Mo(100).[4,44,46] In particular, they examined CO oxidation over Pd_n/MgO/Mo(100), and found that Pd_1 were active in the temperature range of 260–500 K and the activity is increased when Pd is deposited on electron-rich O-vacancies in the MgO oxide layer.[66,67] It was also shown that while there was a strong Pd_n size dependence for deposited clusters, the defect-rich film was not shown to enhance the activity for the larger clusters.[20,48] In future experiments we plan to use ISS on the as-deposited clusters to probe morphology.

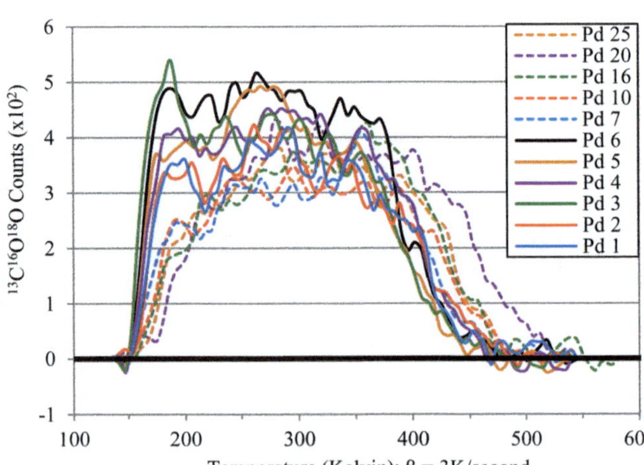

Fig. 10 $^{13}C^{16}O^{18}O$ production from the first of three consecutive TPR runs with samples containing 0.1 ML of Pd deposited as different Pd_n, on a 5.0 ± 0.3 nm thick alumina film grown on Ta(110). TPR conditions: 10 L $^{18}O_2$ exposure at 400 K followed by 10 L ^{13}CO exposure at 180 K.

The apparent lack of sintering of Pd_n/alumina/Ta(110) observed here is in contrast to studies by other authors of Pd on alumina surfaces. For example, Campbell and co-workers studied Pd particle growth from 0.8 ML equivalents of Pd evaporated onto α-Al_2O_3(0001), and found particles with average diameter of 5 nm, increasing to ~6 nm after heating to 680 K, and to 8 nm after heating to 1000 K.[30] Similarly, Freund and co-workers have reported a number of studies of Pd particles grown by Pd evaporation and annealing on alumina/NiAl(110) with Pd loadings as low as 0.3ML.[3,68,69] 1 ML of Pd evaporated onto alumina films grown on a Ta support were studied by the Goodman group and showed good thermal stability.[37,69,70]

Four factors may increase the thermal stability of our samples. Our alumina films are relatively thick, and grown and annealed at relatively low temperature. Both factors are likely to result in films that are rough and/or defective relative to α-Al_2O_3(0001) or the crystalline 0.5 nm film grown on NiAl(110), and defects tend to provide anchoring sites for atoms and clusters. Another factor is that pre-formed clusters deposited on surfaces are expected to diffuse and sinter less than deposited atoms, because more bonds need to be broken, leading to increased diffusion barrier. This effect was demonstrated nicely by Buratto and co-workers for Au_n/TiO_2(110). Au_1 was observed to sinter to large multi-layer particles at room temperature, while clusters, even as small as Au_2, were stable for hours.[71] Furthermore, the Pd loading for our samples is relatively low which increases inter-cluster distances, forcing diffusion over longer length scales for sintering to occur. Finally, we were careful in these studies to keep the maximum temperature of the TPR and TPD scans below 575 K, and to minimize the time spent at high temperature.

Fig. 11 compares the CO oxidation activity of Pd_n/alumina/Ta(110) with the measured Pd 3d binding energy for as-deposited clusters. It can be seen that the two are generally anti-correlated (note inverted binding energy scale), suggesting that this system is another example where electronic structure of the supported clusters, as probed by the shifting core level, is a controlling factor for activity as an oxidation catalyst. Similar anti-correlations have been seen for CO oxidation over Pd_n/TiO_2(110)[9] and carbon oxidation by water in electrocatalysis by Pt_n on glassy carbon.[72]

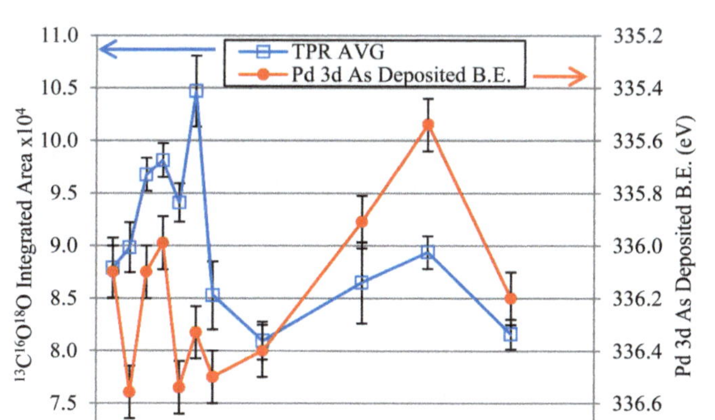

Fig. 11 Average $^{13}C^{16}O^{18}O$ production (blue squares) in three TPR runs for samples with 0.1 ML of Pd deposited as different Pd_n, on a 5.0 $\pm$ 0.3 nm thick alumina film grown on Ta(110). TPR conditions: 10 L $^{18}O_2$ exposure at 400 K, 10 L ^{13}CO exposure at 180 K. As deposited Pd 3d binding energy (red dots) for as-deposited Pd_n.

4 Conclusions

We have used CO_2 production and CO desorption from Pd_{20} deposited on a series of variable thickness alumina films grown on Ta(110) to observe how behaviour of supported Pd changes with support thickness, and to find a thickness range where properties are insensitive to thickness. The alumina film provides a very stable environment for the Pd clusters, such that the size-dependent CO oxidation activity is stable in repeated TPR cycles. There is significant, non-monotonic variation in activity with size, which is weakly correlated with the density of Pd sites on the surface, but also must depend on properties such as the cluster electronic properties or availability of particular binding sites, which vary with size.

There is more work to be done with this system, including complete characterization of the as-deposited samples by XPS, UPS, ISS, and CO TPD, temperature-dependent ISS and UPS experiments to probe the nature of oxygen and CO binding, and studies of the effects of Ta incorporation into the alumina surface layer on activity and stability of the catalysts. Nonetheless, it is clear already the Pd_n/alumina/Ta(110) is substantially more active and more thermally stable than Pd_n/TiO_2, and provides an interesting system in which to explore the effects of cluster size on activity and physical properties of model catalysts.

Acknowledgements

This work was supported by the U.S. Department of Energy, Condensed Phase and Interfacial Molecular Science program Grant No. DEFG03-99ER15003.

References

1 H. Conrad, G. Ertl and J. Kueppers, *Surf. Sci.*, 1978, **76**, 323–342.
2 T. Schalow, B. Brandt, D. E. Starr, M. Laurin, S. K. Shaikhutdinov, S. Schauermann, J. Libuda and H.-J. Freund, *Phys. Chem. Chem. Phys.*, 2007, **9**, 1347–1361.
3 S. Shaikhutdinov, M. Heemeier, J. Hoffmann, I. Meusel, B. Richter, M. Bäumer, H. Kuhlenbeck, J. Libuda, H.-J. Freund, R. Oldman, S. D. Jackson, C. Konvicka, M. Schmid and P. Varga, *Surf. Sci.*, 2002, **501**, 270–281.
4 S. Kunz, F. F. Schweinberger, V. Habibpour, M. Rottgen, C. Harding, M. Arenz and U. Heiz, *J. Phys. Chem. C*, 2010, **114**, 1651–1654.
5 S. Kunz, K. Hartl, M. Nesselberger, F. F. Schweinberger, G. Kwon, M. Hanzlik, K. J. J. Mayrhofer, U. Heiz and M. Arenz, *Phys. Chem. Chem. Phys.*, 2010, **12**, 10288–10291.
6 S. M. McClure and D. W. Goodman, *Chem. Phys. Lett.*, 2009, **469**, 1–13.
7 W. E. Kaden, W. A. Kunkel and S. L. Anderson, *J. Chem. Phys.*, 2009, **131**(11), 114701–114715.
8 T. Wu, W. E. Kaden, W. A. Kunkel and S. L. Anderson, *Surf. Sci.*, 2009, **603**, 2764–2770.
9 W. E. Kaden, T. Wu, W. A. Kunkel and S. L. Anderson, *Science*, 2009, **326**, 826–829.
10 W. E. Kaden, W. A. Kunkel, M. D. Kane, F. S. Roberts and S. L. Anderson, *J. Am. Chem. Soc.*, 2010, **132**, 13097–13099.
11 M. M. Gadgil, R. Sasikala and S. K. Kulshreshtha, *J. Mol. Catal.*, 1994, **87**, 297–309.
12 M. M. Gadgil and S. K. Kulshreshtha, *J. Mol. Catal. A: Chem.*, 1995, **95**, 211–222.
13 T. Engel and G. Ertl, *Chem. Phys. Lett.*, 1978, **54**, 95–98.
14 T. Engel and G. Ertl, *J. Chem. Phys.*, 1978, **69**, 1267–1281.
15 I. Z. Jones, R. A. Bennett and M. Bowker, *Surf. Sci.*, 1999, **439**, 235–248.
16 G. A. Kok, A. Noordermeer and B. E. Nieuwenhuys, *Surf. Sci.*, 1983, **135**, 65–80.
17 B. Brandt, T. Schalow, M. Laurin, S. Schauermann, J. Libuda and H.-J. Freund, *J. Phys. Chem. C*, 2007, **111**, 938–949.
18 C. J. Harding, S. Kunz, V. Habibpour and U. Heiz, *Phys. Chem. Chem. Phys.*, 2008, **10**, 5875–5881.
19 C. J. Harding, S. Kunz, V. Habibpour and U. Heiz, *Chem. Phys. Lett.*, 2008, **461**, 235–237.
20 C. J. Harding, S. Kunz, V. Habibpour, V. Teslenko, M. Arenz and U. Heiz, *J. Catal.*, 2008, **255**, 234–240.
21 M. Bowker, P. Stone, R. Bennett and N. Perkins, *Surf. Sci.*, 2002, **497**, 155–165.
22 V. Matolin, I. Jungwirthova and E. Tomkova, *Prog. Surf. Sci.*, 1990, **35**, 175–178.
23 E. Gillet and V. Matolin, *Z. Phys. D: At., Mol. Clusters*, 1991, **19**, 361–365.

24 I. Stara and V. Matolin, *Surf. Sci.*, 1994, **313**, 99–106.
25 V. Matolin, I. Stara, N. Tsud and V. Johanek, *Prog. Surf. Sci.*, 2001, **67**, 167–181.
26 V. Matolin, V. Johanek, I. Stara, N. Tsud and K. Veltruska, *Surf. Sci.*, 2002, **507–510**, 803–807.
27 J.-H. Fischer-Wolfarth, J. A. Farmer, J. M. Flores-Camacho, A. Genest, I. V. Yudanov, N. Rösch, C. T. Campbell, S. Schauermann and H.-J. Freund, *Phys. Rev. B: Condens. Matter Mater. Phys.*, 2010, **81**, 241416.
28 I. Jungwirthovk, I. Star and V. Matolin, *Surf. Sci.*, 1997, **377–379**, 644–649.
29 N. Tsud, V. Johánek, I. Stará, K. Veltruská and V. Matolín, *Surf. Sci.*, 2000, **467**, 169–176.
30 S. L. Tait, L. T. Ngo, Q. Yu, S. C. Fain Jr and C. T. Campbell, *J. Chem. Phys.*, 2005, **122**, 064712.
31 S. Penner, P. Bera, S. Pederson, L. T. Ngo, J. J. W. Harris and C. T. Campbell, *J. Phys. Chem. B*, 2006, **110**, 24577–24584.
32 M. Bowker and E. Fourre, *Appl. Surf. Sci.*, 2008, **254**, 4225–4229.
33 R. A. Bennett, C. L. Pang, N. Perkins, R. D. Smith, P. Morrall, R. I. Kvon and M. Bowker, *J. Phys. Chem. B*, 2002, **106**, 4688–4696.
34 M. Bowker, P. Stone, R. Bennett and N. Perkins, *Surf. Sci.*, 2002, **497**, 155–165.
35 M. S. Chen, Y. Cai, Z. Yan, K. K. Gath, S. Axnanda and D. W. Goodman, *Surf. Sci.*, 2007, **601**, 5326–5331.
36 D. C. Meier and D. W. Goodman, *J. Am. Chem. Soc.*, 2004, **126**, 1892–1899.
37 E. Ozensoy and D. W. Goodman, *Phys. Chem. Chem. Phys.*, 2004, **6**, 3765–3778.
38 Y. Q. Cai, A. M. Bradshaw, Q. Guo and D. W. Goodman, *Surf. Sci.*, 1998, **399**, L357–L363.
39 D. R. Rainer, M. Koranne, S. M. Vesecky and D. W. Goodman, *J. Phys. Chem. B*, 1997, **101**, 10769–10774.
40 D. R. Rainer, M. C. Wu, D. I. Mahon and D. W. Goodman, *J. Vac. Sci. Technol., A*, 1996, **14**, 1184–1188.
41 S. M. McClure, M. Lundwall, F. Yang, Z. Zhou and D. W. Goodman, *J. Phys.: Condens. Matter*, 2009, **21**, 474223–474230.
42 S. M. McClure, M. Lundwall, Z. Zhou, F. Yang and D. W. Goodman, *Catal. Lett.*, 2009, **133**, 298–306.
43 B. Yoon, U. Landman, V. Habibpour, C. Harding, S. Kunz, U. Heiz, M. Moseler and M. Walter, *J. Phys. Chem. C*, 2012, **116**, 9594–9607.
44 M. Moseler, M. Walter, B. Yoon, U. Landman, V. Habibpour, C. Harding, S. Kunz and U. Heiz, *J. Am. Chem. Soc.*, 2012, **134**, 7690–7699.
45 V. Habibpour, Z. W. Wang, R. E. Palmer and U. Heiz, *J. Appl. Sci.*, 2011, **11**, 1164–1170.
46 C. J. Harding, S. Kunz, V. Habibpour, V. Teslenko, M. Arenz and U. Heiz, *J. Catal.*, 2008, **255**, 234–240.
47 M. Arenz, S. Gilb and U. Heiz, *Chem. Phys. Solid Surf.*, 2007, **12**, 1–47.
48 M. A. Roettgen, S. Abbet, K. Judai, J.-M. Antonietti, A. S. Woerz, M. Arenz, C. R. Henry and U. Heiz, *J. Am. Chem. Soc.*, 2007, **129**, 9635–9639.
49 M. E. Vaida, T. M. Bernhardt, C. Barth, F. Esch, U. Heiz and U. Landman, *Phys. Status Solidi B*, 2010, **247**, 1001–1015.
50 C. Harding, V. Habibpour, S. Kunz, A. N.-S. Farnbacher, U. Heiz, B. Yoon and U. Landman, *J. Am. Chem. Soc.*, 2009, **131**, 538–548.
51 R. Robles and S. N. Khanna, *Phys. Rev. B: Condens. Matter Mater. Phys.*, 2010, **82**, 085428.
52 K. J. Boyd, A. Lapicki, M. Aizawa and S. L. Anderson, *Rev. Sci. Instrum.*, 1998, **69**, 4106–4115.
53 S. Lee, C. Fan, T. Wu and S. L. Anderson, *J. Chem. Phys.*, 2005, **123**, 124710.
54 P. J. Chen and D. W. Goodman, *Surf. Sci.*, 1994, **312**, L767–L773.
55 J. J. Yeh and I. Lindau, *At. Data Nucl. Data Tables*, 1985, **32**, 1–155.
56 C. J. Powell and A. Jablonski, *NIST Electron Effective-Attenuation-Length Database*, NIST, Gaithersburg, MD, 2011.
57 W. E. Kaden, W. A. Kunkel, F. S. Roberts, M. Kane and S. L. Anderson, *J. Chem. Phys.*, 2012, **136**, 204705.
58 C. Xu and D. W. Goodman, *Chem. Phys. Lett.*, 1996, **263**, 13–18.
59 M.-C. Wu and D. W. Goodman, *J. Phys. Chem.*, 1994, **98**, 9874–9881.
60 Y. Wu, E. Garfunkel and T. E. Madey, *Surf. Sci.*, 1996, **365**, 337–352.
61 Y. Wu, E. Garfunkel and T. E. Madey, *J. Vac. Sci. Technol., A*, 1996, **14**, 2554–2563.
62 C. J. Powell, *Appl. Surf. Sci.*, 1995, **89**, 141–149.
63 *Nist X-ray Photoelectron Spectroscopy Database, Version 4.1* (National Institute of Standards and Technology, Gaithersburg, 2012), http://srdata.nist.gov/xps/.
64 T. L. Barr, *Thin Solid Films*, 1994, **253**, 277–284.
65 W. E. Kaden, W. A. Kunkel, M. D. Kane, F. S. Roberts and S. L. Anderson, *Unpublished results*.

66 S. Abbet, A. M. Ferrari, L. Giordano, G. Pacchioni, H. Hakkinen, U. Landman and U. Heiz, *Surf. Sci.*, 2002, **514**, 249–255.
67 S. Abbet, U. Heiz, H. Häkkinen and U. Landman, *Phys. Rev. Lett.*, 2001, **86**, 5950–5953.
68 M. Bäumer, J. Libuda, A. Sandell, H. J. Freund, G. Graw, T. Bertrams and H. Neddermeyer, *Ber. Bunsen-Ges.*, 1995, **99**, 1381–1386.
69 M. Heemeier, S. Stempel, S. K. Shaikhutdinov, J. Libuda, M. Bäumer, R. J. Oldman, S. D. Jackson and H.-J. Freund, *Surf. Sci.*, 2003, **523**, 103–110.
70 D. R. Rainer, S. M. Vesecky, M. Koranne, W. S. Oh and D. W. Goodman, *J. Catal.*, 1997, **167**, 234–241.
71 X. Tong, L. Benz, P. Kemper, H. Metiu, M. T. Bowers and S. K. Buratto, *J. Am. Chem. Soc.*, 2005, **127**, 13516–13518.
72 S. Proch, M. Wirth, H. S. White and S. L. Anderson, *J. Am. Chem. Soc.*, 2013, **135**, 3073–3086.

Energetics of elementary reaction steps relevant for CO oxidation: CO and O_2 adsorption on model Pd nanoparticles and Pd(111)

Matthias Peter, Sergey Adamovsky, Jose Manuel Flores Camacho and Swetlana Schauermann*

Received 2nd January 2013, Accepted 31st January 2013
DOI: 10.1039/c3fd00001j

The energetics of elementary surface processes relevant for CO oxidation, particularly CO and O_2 adsorption, were investigated by a direct calorimetric method on model Pd nanoparticles and on the extended Pd(111) single crystal surface. The focus of this study lies on a detailed understanding of how a nanometer scale confinement of matter affects the binding strength of gaseous adsorbates. We report adsorption energies and sticking coefficients of CO and O_2 measured as a function of the adsorbate surface coverage both on pristine and O-covered Pd surfaces. The reduced dimensions of the Pd substrate were found to affect the binding strength of the adsorbates in two principle ways: (i) *via* the change of the local adsorption environment that can result *e.g.* in stronger adsorbate bonding at the particle's low coordinated surface sites and (ii) *via* the contraction of the Pd lattice in small clusters and a concomitant weakening of chemisorptive interaction. Particularly for O_2 adsorption, the change of the adsorption site from a three-fold hollow on Pd(111) to the edge site on Pd nanoparticles (~4 nm sized on average) was found to result in a strong increase of the Pd–O bond strength. In contrast, CO adsorbs weaker on Pd nanoparticles as compared to the extended Pd(111) surface. In total, the binding energies of adsorbates on Pd and with this their surface coverages turn out to depend in a non-monotonic way on the particular structure of Pd surfaces, including the local structure of the adsorption site as well as the global properties of the small clusters arising *e.g.* from the lattice contraction.

I. Introduction

Confinement of matter at a nanoscopic scale is well known to result in new specific properties of a material. These new properties inherent to nanoparticles are particularly important for a number of practical applications of heterogeneous catalysis, such as *e.g.* CO oxidation over transition metals.[1,2] This reaction has attracted much attention in recent decades both in surface science and ambient pressure studies and many phenomena have already been well understood.[3–7] Despite the comprehensive understanding and general agreement on the microscopic mechanisms of CO oxidation over many transition metals, less information is available on the thermodynamics of the underlying surface processes. Particularly for metallic nanoparticles, the quantitative information on the adsorption and reaction enthalpies of elementary reaction steps, such as dissociative oxygen adsorption, CO adsorption on clean and oxygen-containing surfaces, CO and recombination

Department of Chemical Physics, Fritz-Haber-Institut der Max-Planck-Gesellschaft, Faradayweg 4-6, 14195 Berlin, Germany. E-mail: schauermann@fhi-berlin.mpg.de

with an adsorbed oxygen atom, is missing. It is a current scientific challenge to obtain detailed quantitative information on the binding strength of the gaseous adsorbates and on the reaction enthalpies as a function of the particle size and the particle surface composition, since these parameters essentially determine the surface coverages under the operation conditions and with this strongly affect the absolute reaction rates. The information would allow a deeper fundamental understanding of how the surface binds reactants and guides them through various elementary steps to the products, and provide a basis for a rational design of new catalytic and functional materials.

In previous studies related to CO oxidation, the energetics of surface processes were probed mainly by desorption-based methods such as temperature programmed desorption (TPD) and equilibrium adsorption isotherm measurements.[7–13] However, these methods can be applied correctly only for fully reversible adsorption–desorption processes such as *e.g.* CO adsorption and desorption on clean metals. In contrast, this condition does not apply for CO recombination with an oxygen atom or oxygen dissociative adsorption, which can be additionally accompanied by some side processes such as *e.g.* oxygen diffusion into the subsurface,[12–15] refaceting,[11,14] particle reconstruction,[7,16,17] as well as surface[17–19] and bulk oxide formation.[18,20] These obstacles and also the need for assumptions on the pre-exponential factors, which have to be made for the kinetic modeling of the desorption process, strongly limit the quantitative determination of binding and reaction energies related to CO oxidation by traditional desorption-based methods. A strategy to overcome these shortcomings is a direct calorimetric measurement of adsorption and reaction enthalpies under isothermal conditions.[21–24]

In this work, we report on the first direct calorimetric measurement of binding energies of gas phase molecules relevant to CO oxidation on model Pd nanoparticles and compare them to Pd(111) single crystal. The energetics of surface processes were probed on well-defined Pd nanoparticles supported on a thin Fe_3O_4 crystalline oxide film prepared *in situ* under ultra-high vacuum (UHV) conditions. We apply a recently developed UHV single crystal adsorption calorimeter (SCAC) based on molecular beam techniques[24] to investigate the energetics of dissociative oxygen adsorption and CO adsorption on clean and O-precovered Pd nanoparticles and on Pd(111), as a reference. With this approach we address the question: how do the reduced dimensions of a metallic substrate (extended laterally stiff single crystal *vs.* small atomically flexible nanoparticles) and its surface composition affect the adsorption strength of the reactive species?

II. Experimental procedure

The adsorption experiments were performed at the Fritz-Haber-Institute (Berlin) in an ultra-high vacuum (UHV) single crystal adsorption calorimetry (SCAC) apparatus based on molecular beam techniques described in detail elsewhere.[24] An effusive doubly differentially pumped multi-channel array source was employed to produce an oxygen (Linde, purity 99.999%) beam at the intensity 1×10^{14} molecules cm^{-2} s^{-1} (5.3×10^{-7} mbar on the sample surface) or CO beam (Linde, purity 99.995%) at the intensity 1×10^{14} molecules cm^{-2} s^{-1} (5.3×10^{-7} mbar on the sample surface). The beam was cut into pulses of 266 ms length by a remote-controlled chopper, after that the gas impinged on a sample prepared on an ultrathin (1 μm) single crystal (Pd(111) or model supported catalysts grown on Pt(111)). The small change in the temperature of the sample caused by adsorption of oxygen was measured by a heat detector, consisting of a 9 μm-thick β-polyvinylidene fluoride (β-PVDF) pyroelectric ribbon coated with Au on both sides. The energy calibration was performed by applying pulses of laser light (HeNe, Linos, wavelength 632.8 nm, 5 mW), which passes through the same path as the molecular beam and is chopped in a way identical to the molecular flux. Simultaneously, the fraction of the molecules adsorbed in a single beam pulse (*i.e.*, the sticking coefficient) was defined by the modified King–Wells method[25] with a quadrupole mass spectrometer (QMS Hiden, HAL 301/3F) in a non-line-of-sight geometry.

The $Pd/Fe_3O_4/Pt(111)$ model catalyst was prepared by growing a thin (~50 Å) $Fe_3O_4(111)$ film on a Pt(111) single crystal by repeated cycles of Fe (>99.99%, Goodfellow) physical vapor deposition at a substrate temperature of ~120 K and subsequent oxidation at 875 K (see refs. 26 and 27 for details). The cleanliness and the quality of the oxide film were checked by AES and LEED. Pd particles were grown *in situ* by physical vapor deposition from a Pd rod (Goodfellow, >99.9%) using a commercial evaporator (Focus, EFM 3) with a deposition rate of 0.3 Å min^{-1} (2.04×10^{14} atoms cm^{-2} min^{-1}). The nominal deposition thickness value amounts to 4.0 Å, or to 2.8×10^{15} atoms cm^{-2}. Directly after Pd deposition, the sample was annealed to 600 K and cooled to 300 K, and a microcalorimetric measurement of oxygen or CO adsorption heats was performed. The evaporation rates of Fe and Pd were calibrated by a quartz-crystal microbalance (QCM, Sigma instruments).

III. Results and discussion

III.A Combined energy and sticking coefficient measurement: CO adsorption on supported Pd nanoparticles and Pd(111)

The model surface employed in this study consisted of Pd nanoparticles supported on a well-ordered thin Fe_3O_4 film grown on a Pt(111) single crystal (Fig. 1; for details of the preparation procedure and structural characterization by scanning tunneling microscopy (STM) see refs. 26 and 27). On average, a Pd particle on this surface is about 4 nm in diameter and contains ~720 Pd atoms; the particle density on the support amounts to 3.8×10^{12} particles cm^{-2}. According to STM data, the particles are crystalline and exhibit a nearly hexagonal shape with a flat top facet, which indicates that the particles grow in (111) orientation and their sides are terminated with (111) and (100) facets.

In the calorimetric experiment, two parameters were measured simultaneously and independently: the sticking coefficient and the corresponding adsorption energy (differential adsorption heat). The details of the calorimetric measurement were comprehensively discussed elsewhere.[24] Briefly, we prepare a model catalyst on an ultrathin (1 μm) Pt(111) single crystal and transfer it *in situ* to the calorimeter. In

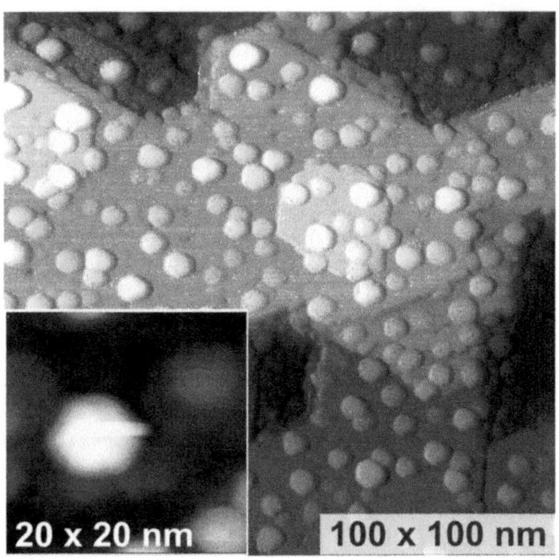

Fig. 1 STM image of the $Pd/Fe_3O_4/Pt(111)$ model catalyst used in this study, nominal Pd deposition thickness 4 Å, from ref. 27.

a calorimetric experiment a pulse of gas from a chopped molecular beam impinges on an ultrathin single crystal surface. A fraction of the molecules striking the surface adsorb and this heat input causes a transient temperature rise (~10 mK), which is measured by a pyroelectric heat detector. The absolute energy calibration is performed by applying pulses of laser light of known energy, which passes through the molecular beam and is chopped in a way identical to the molecular flux. Two parameters, critically influencing the calibration – the absolute power of the laser light and the reflectivity of the sample – are measured *in situ*. Simultaneously, the absolute fraction of the molecules adsorbed in a single beam pulse is measured in a sticking coefficient measurement by the King–Wells method[25] in order to determine the absolute value of deposited energy per number of adsorbed molecules.

Fig. 2 shows the complete calorimetric data set obtained at 300 K: the sticking coefficients (Fig. 2a and 2b) and the differential adsorption heats (Fig. 2c and 2d) plotted as a function of CO surface coverage for Pd nanoparticles and Pd(111) single crystal. Note that at this temperature CO does not permanently chemisorb on the Fe_3O_4 support[28] but on the Pd particles only and the experimental time scale is long enough in order to let the molecules diffuse over the surface and find the strongest adsorption site.[29]

The initial CO sticking coefficient on Pd(111) (Fig. 2b) reaches the value 0.72 ± 0.03 and remains nearly constant until the CO surface coverage of $\sim 2.7 \times 10^{14}$ CO molecules cm^{-2} (or 0.18 CO molecules per surface Pd atom) is achieved. After that it decreases to a quasi steady state value[21] of about 0.3 for a CO coverage of 0.76×10^{15} molecules cm^{-2} (or 0.5 CO molecules per surface Pd atom) in saturation. The apparent non-zero sticking coefficient close to saturation arises from the partial CO desorption between the beam pulses and additional adsorption of CO when the next pulse hits the surface (see full discussion in refs. 21,23,24). The CO saturation coverage on Pd(111) is in excellent agreement with the previously reported value 0.5 with reference to the number of surface Pd atoms, which was obtained by a number of independent surface-sensitive techniques (see *e.g.* refs. 30–33). The value of the initial sticking coefficient measured in this study is somewhat lower than the previously reported values of 0.8–1.0[34] and 0.95 ± 0.05.[35] Qualitatively similar dependence of the sticking coefficient on surface CO coverage was observed on Pd

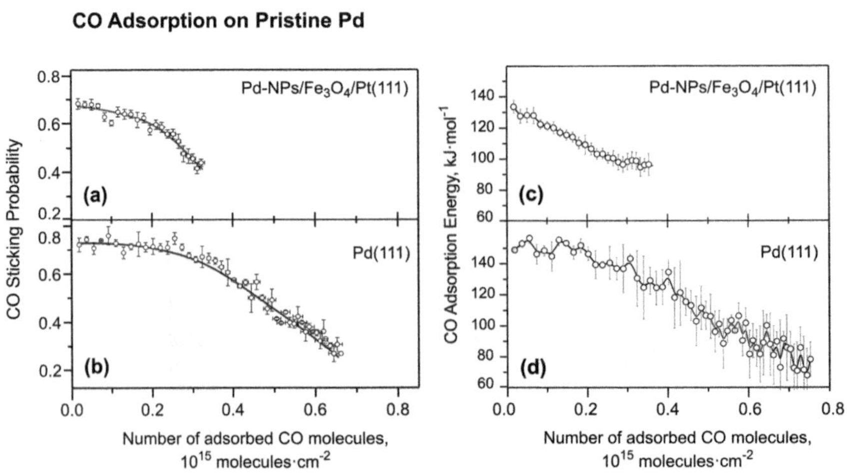

Fig. 2 Sticking probability plotted as a function of number of adsorbed CO molecules for Pd nanoparticles (a) and for Pd(111) (b). Differential heats of adsorption plotted as a function of number of adsorbed CO molecules for Pd nanoparticles (c) and Pd(111) (d). All data were obtained at 300 K. The data are shown as an average of four to six independent measurements on freshly prepared surfaces. The error bars correspond to the error of the mean.

nanoparticles, with the initial sticking coefficient amounting to 0.68 ± 0.02 and the saturation coverage of 0.4×10^{15} molecules cm^{-2} (Fig. 2a). The final CO surface coverage expressed as a number of adsorbed CO molecules per surface Pd atom is $\sim$0.48, which is in good agreement with the value 0.5 measured on Pd(111)[30–33] and Pd(100)[36] single crystal surfaces at room temperature. This result shows that the absolute adsorption capacity of investigated Pd nanoparticles is close to that of the single crystal surfaces and indicates that no dramatic changes in the adsorption mechanism occur upon reduction of the particle sizes down to 4 nm. It is apparent that the general adsorption mechanism follows a typical precursor type behavior, which was previously observed on the Pd single crystal surfaces (compare *e.g.* refs .34 and 35).

The sticking coefficient data can be rationalized on a basis of the so-called "capture zone" effect.[37–41] This phenomenon is based on the existence of two adsorption channels: the adsorbate molecules can either directly impinge on the metal particles and chemisorb or they can be trapped in a weakly bound state on the support and reach the particles *via* surface diffusion. The capture zone is defined as an area from which impinging adsorbates can be collected by the particles. This effect was predicted and later experimentally confirmed to considerably enhance the adsorption rate on the particles.[5,7] In the present experiments, the capture zone effect was found to play an important role in the overall adsorption rate. The magnitude of this effect can be evaluated from the comparison of the measured initial sticking probability data on the supported particles and the estimate of the highest possible initial sticking probability on the particles of the same size in absence of the contribution from the support. For the Pd particles investigated in this study (average particle size of $\sim$4 nm and the island density 3.8×10^{12} particles cm^{-2}), about 48% of the Fe_3O_4 surface area is covered by Pd particles. This means that only $\sim$48% of the impinging molecules arrive directly on the metal surface and about 52% interact first with the support. If one assumes the highest possible sticking coefficient on metallic Pd to be 1.0 and fully neglects the contribution from the capture zone effect, an initial sticking coefficient close to 0.48 can be expected (corresponding to the scenario when 48% of the molecules completely stick to the metal particles). However, the experimentally measured initial sticking coefficient 0.68 is a factor of 1.4 higher than the estimated upper limit for the direct sticking on the metal particles. This observation can be rationalized only in the scope of the "capture zone" model, where the additional 20% of the impinging molecules become first weakly trapped on the support and reach the Pd particles, where they strongly chemisorb, within their residence time. Thus, the capture zone effect leads to an enhancement of the adsorption rate by a factor of at least 1.4. If one assumes the experimentally measured value of the sticking coefficient 0.74 on Pd(111), the enhancement factor increases to $\sim$1.9. Note that CO does not chemisorb on Fe_3O_4 at the experimental temperature of 300 K,[28] so that the enhanced adsorption cannot be explained simply by CO chemisorption on the support. Similar enhancement of the initial sticking probability due to the "capture zone effect" has been reported for a number of model surfaces consisting of Pd nanoparticles supported on various oxides.[26,40,41]

The differential adsorption heats obtained on Pd nanoparticles and Pd(111) are plotted as a function of CO surface coverage in Fig. 2c and 2d. Each curve is an average of 4 to 6 independent measurements on freshly prepared model systems. On Pd(111), the heat of CO adsorption initially amounts to 149 ± 4 kJ mol^{-1}. After adsorption of about 0.76×10^{15} CO molecules cm^{-2} the surface of the particles reaches saturation (with the corresponding CO surface coverage of 0.5 ML with respect to the number of surface Pd atoms) and the adsorption heat decreases to the final non-zero value 75 ± 9 kJ mol^{-1}, which is related to the adsorption–desorption equilibrium established in quasi steady state regime.[21] Note that the final heat value is subject to large statistical errors since a very small absolute number of CO molecules is added in a single CO pulse, which is difficult to measure. Two reasons can account for the decreasing adsorption energy with growing CO coverage: intermolecular repulsion of neighboring CO molecules and increasing competition for the d-electrons of Pd nanoclusters participating in the CO–Pd bonding. For the Pd(111) single crystal surface, both the initial heat of adsorption as well as the saturation coverage at 300 K were found to be in a very good quantitative agreement with the data reported earlier in the literature.[42–45]

A qualitatively similar dependence of the adsorption heat on CO coverage was observed for Pd nanoparticles. On this surface, the initially high adsorption heat 134 ± 4 kJ mol^{-1} continuously decreases with growing CO coverage and reaches a lower steady state value ~90 kJ mol^{-1} after adsorption of ~0.4 × 10^{15} molecules cm^{-2}. Remarkably, the initial adsorption energy is 15 kJ mol^{-1} lower than the corresponding value on the Pd(111) surface despite the fact that Pd nanoparticles contain a large fraction of low-coordinated surface sites, such as edges and corners, which are often discussed to be strong binding sites. Previously, we comprehensively investigated the dependence of the CO initial adsorption energy on Pd as a function of the cluster size in the broader size range from 2 to 8 nm (or from 120 to 4900 Pd atoms per particle).[46] The initial adsorption heats were compared in the lowest coverage limit, where the adsorption heat is determined by the interaction of individual CO molecules with metal, and is not perturbed by the CO intermolecular repulsion or by the competition for the d-electrons. In the whole investigated size region the adsorption heat was observed to strongly decrease with decreasing particle size: from 126 ± 3 kJ mol^{-1} on 8 nm-sized Pd particles to 106 ± 1 kJ mol^{-1} on the smallest ~2 nm clusters. Additionally, all investigated particles showed smaller initial adsorption heat as compared to the single crystal surface (149 ± 3 kJ mol^{-1}).

It has to be pointed out that the surface of the Pd particles investigated in this work exposes mainly (111) terraces along with a smaller fraction of (100) facets and low-coordinated defect sites such as edges and corners. According to theoretical calculations, such irregular sites exhibit higher CO adsorption energies than Pd(111).[47] Experimentally, there is a general agreement that the CO adsorption energy on the (100) plane is about 10–15 kJ mol^{-1} higher than on the (111) plane (see *e.g.* ref. 42). However, the dramatic effect of low-coordination defect sites (like step edges) on the CO adsorption energy, such as observed for Pt(111),[48] is *not* seen on Pd(111).[42,44] Ramsier, Lee and Yates[44] even concluded from comparing CO TPD and adsorption energies for different stepped Pd facets that "structure of the Pd surface is only of minor importance in CO adsorption/desorption kinetics." These experimental results suggest that the degree of coordination of the Pd surface atoms on the stepped surfaces is not as important in determining the CO adsorption energy as on surfaces of other metals like Pt. Therefore, the increasing fraction of low-coordination Pd atoms with decreasing particle size may not be the dominant effect in determining how the CO adsorption energy varies with Pd particle size. Instead, other size-dependent properties of a Pd nanocluster have to be responsible for the obtained decrease of the initial adsorption enthalpy of carbon monoxide with decreasing the particle size.

First, the decrease of the chemisorption energy on Pd nanoparticles can result from the contraction of the lattice parameter of a small metal nanoparticle. Previously, it has been demonstrated experimentally[49,50] that the interatomic bond length in small metal particles decreases with decreasing particle size. It is now well established also by computational studies of various metals including Pd[51–53] that relaxed structures of metal clusters exhibit shorter interatomic distances than bulk crystals. This phenomenon has been rationalized as a result of decreasing average coordination number of the atoms with the higher surface-to-volume ratio of the cluster.[51–53] Particularly for Pd, it was shown that the average calculated Pd–Pd nearest-neighbor distance decreases linearly with decreasing linear dimension of a nanoparticle in the range of 55–260 atoms per cluster.[54] This lattice contraction in the small metal clusters was theoretically shown to result in a reduction of the adsorbate binding energy. Particularly for CO adsorption on the Pd clusters consisting of 55–260 atoms, it was demonstrated that the clusters with optimized structures show systematically lower adsorption energies by about 10–15 kJ mol^{-1} than the clusters with a bulk terminated geometry, *i.e.* with all interatomic distances *d*(Pd–Pd) = 275 pm.[55] This finding agrees with the principle of bond order conservation:[56,57] in the contracted clusters, one expects weaker adsorption bonds and stronger binding within the adsorbate as a result of better saturated valences of the substrate atoms. The decrease of the adsorbate binding energy with the lattice constant can be also rationalized in terms of the strain effect, which is associated with a downward shift of the valance d-band with decreasing lattice parameter.[55,58] Such effects were found not to be restricted to 3-fold hollow sites on (111) facets, but reflect a general trend that also holds for adsorption at other sites of the particles such as bridge sites and edges.[55]

A second reason for the decrease of adsorption heat of a gas-phase molecule on the small metal clusters is a feasible weakening of the dispersion force (van der Waals interaction) that is induced by dynamic response of bulk electrons of the metal to charge density fluctuations in an adsorbed molecule. Previously, it has been shown that the electron population at the Fermi edge, which is mainly relevant for this interaction, drastically changes with the cluster size in the range of a few nanometers.[59] The gradual decrease of the initial adsorption heat with the decreasing cluster size can be reasonably understood, since smaller clusters contain less electrons available for dynamic response. Similar behavior has been previously observed in TPD studies for methane on Pd[60] and NO on Ag nanoclusters[61] and was also attributed to the reduced polarizability of the small metal nanoparticles.

III.B Combined energy and sticking coefficient measurement: O_2 dissociative adsorption on supported Pd nanoparticles and Pd(111)

Dissociative oxygen adsorption was investigated on the same two Pd surfaces at 300 K, at which oxygen adsorbs dissociatively and does not stick to the oxide support. Fig. 3 shows the obtained sticking coefficients (Fig. 3a and 3b) and the adsorption energies (Fig. 3c and 3d) plotted as a function of O surface coverage.

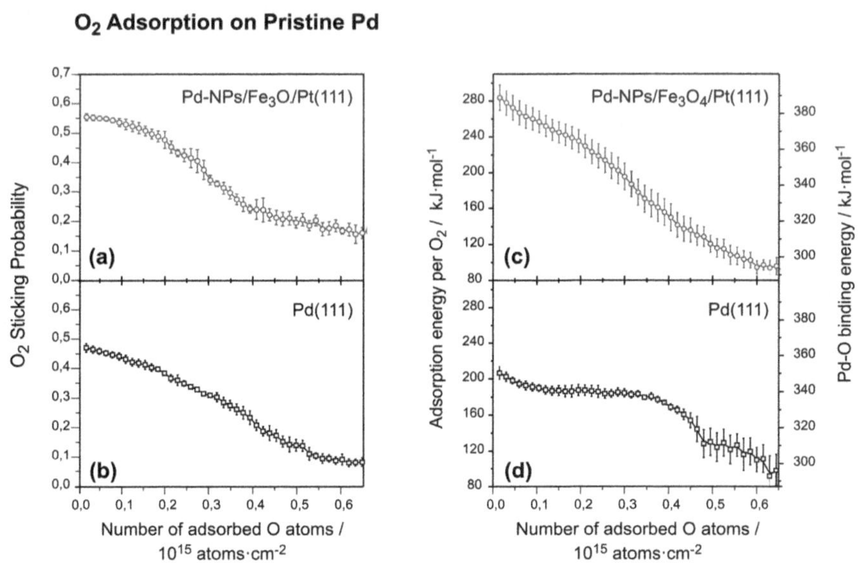

Fig. 3 Sticking probability of oxygen plotted as a function of number of adsorbed O atoms for Pd nanoparticles (a) and for Pd(111) (b). Differential adsorption heat per mole of O_2 (left axis) and Pd–O binding energy (right axis) plotted as a function of the number of adsorbed O atoms. The sticking coefficients and adsorption heats were obtained at 300 K and are shown as an average of four to six independent measurements on freshly prepared surfaces. The error bars correspond to the error of the mean.

On Pd(111), the initial O_2 sticking probability reaches 0.47 ± 0.01 and steadily decreases until the O surface coverage approaches the value of ~0.5 × 10^{15} O atoms cm^{-2} (or ~0.3 O atoms per surface Pd atom, Fig. 3b). The measured saturation coverage is in agreement with the literature data ranging from 0.25 to 0.4 observed for different O_2 exposures on this surface.[9,10,12,13,15] Previously, formation of O coverages larger than 0.25 was observed for high O_2 exposures and a number of reasons were suggested to be responsible for this phenomenon, such as *e.g.* slow formation of compressed O overlayers or O diffusion into the subsurface region.[12–15] Apparently, similar processes might take place also under our reaction conditions and they appear to be more pronounced on Pd nanoparticles. The detailed analysis of oxygen adsorption in the high coverage regime both on Pd(111) and Pd nanoparticles will be the subject of a forthcoming publication.[62] Here, we will mainly focus on oxygen coverages up to 0.25, at which these side processes do not play any significant role.

On Pd nanoparticles, a qualitatively similar dependence of the sticking coefficient on surface O coverage was observed, with the initial sticking coefficient amounting to 0.55 ± 0.01 and the saturation coverage lying between 0.4 × 10^{15} and 0.5 × 10^{15} O atoms cm^{-2} (Fig. 3a). Note that the initial sticking coefficient of O_2 on the Pd clusters is even higher than on Pd(111) despite the fact that only ~48% of the surface is covered by Pd. To rationalize this high dissociative sticking coefficient, we estimated the capture zone effects on this surface. If one assumes the same value of the initial sticking coefficient as on Pd(111) (0.47), the fraction of the molecules that impinge directly on the Pd nanoparticles and remain adsorbed amounts to 0.23. The rest of the adsorbed molecules (relative fraction ~0.32) must first be trapped on the Fe_3O_4 support and then diffuse to the metal particles within their residence time and dissociate on it (note that O_2 neither dissociatively nor molecularly adsorbs on the oxide support under our reaction conditions). This scenario does not seem to be very likely though, since it implies that a molecule impinging on the surface has a higher probability to be trapped on the Fe_3O_4 support and then undergo the dissociative adsorption on Pd particles than directly stick and dissociate on metallic nanoparticles (0.62 *vs.* 0.47, correspondingly). More realistically, O_2 molecules have higher initial probability to dissociate on Pd nanoparticles than on Pd(111) as the first type of surface contains a large number of potentially highly reactive low-coordinated adsorption sites. To evaluate the feasibility of this hypothesis, we estimated the magnitude of the capture zone effect for the boundary case when the initial sticking coefficient of O_2 on Pd nanoparticles is equal to unity. In this case, the fraction of the molecules impinging directly on the particles and dissociating there should amount to 0.48 and only the remaining fraction of O_2 amounting to 0.07 has to arrive from the support and dissociate on the metal, which might be a realistic scenario. Based on the available data set, the exact quantification of the initial sticking coefficient on Pd nanoparticles is not possible, since the experimentally measured value is a convolution of the intrinsic initial sticking coefficient with the capture zone effects. The strategy to tackle this problem would be *e.g.* a systematic variation of the particles sizes and density on the support, thus allowing an estimation of the relative contribution of the capture zone effect. The corresponding experiments are currently under way. Based on the available data, we can only conclude that the intrinsic initial sticking coefficient on Pd nanoparticles lies in the range between 0.47 and 1.0.

Fig. 3c and 3d show the dependence of the differential O_2 adsorption heat (left axis) and the calculated Pd–O bonding strength (right axis) plotted as a function of O atoms surface coverage for Pd nanoparticles and Pd(111).

Both model surfaces show a very strong coverage dependence of oxygen binding energies. For Pd(111), the initial adsorption energy on the clean surface amounts to ~205 ± 7 kJ mol^{-1}, which drops to about 100 kJ mol^{-1} close to saturation. As discussed above, two reasons might potentially account for the decreasing adsorption energy with increasing O coverage: interatomic repulsion of neighboring O atoms, and increasing competition for the d-electrons of Pd atoms participating in the O–Pd bonding. A qualitatively similar dependence of the adsorption energy on the O coverage was observed for Pd nanoparticles. However, there is a very pronounced difference in the initial adsorption energy between the Pd clusters and Pd(111). While O binds with ~205 kJ mol^{-1} on the pristine Pd(111) surface, the initial binding energy rises to ~280 ± 13 kJ mol^{-1} on the pristine Pd nanoparticles (both values are differential adsorption heats for O_2 molecule). Note that on average only about 7 O atoms per Pd nanoparticle (comprising ~320 surface Pd atoms) contribute to the measured initial adsorption energy, which allows us to consider this energy value as a limiting case for the interaction of an individual O atom with a pristine Pd nanoparticle. Also note that neither subsurface O diffusion nor surface oxide formation can occur at the experimental low temperature conditions (300 K) and low oxygen coverages, which could otherwise affect the measured values.

The energy value obtained by SCAC for Pd(111) in the lowest coverage limit (205 kJ mol^{-1} at the surface coverage 0.01×10^{15} O atoms cm^{-2}) is in good agreement with literature values of O adsorption energies at low coverages situated in the range of 210 to 230 kJ mol^{-1}.[10,13,15] In contrast, the very high initial energy value obtained for Pd nanoparticles (275 kJ mol^{-1} at 0.01×10^{15} O atoms cm^{-2}) cannot be compared to any value reported in TPD studies either for low Miller index Pd surfaces, or for stepped Pd crystals, which typically span the range between 190 and 230 kJ mol^{-1}.[9,13,15,63] Thus, this high initial adsorption energy must be related to some structural feature inherent to Pd nanoparticles, which cannot be reproduced even by the steps of the high Miller index Pd surfaces. Such adsorption sites on Pd nanoparticles might be the low coordinated surface sites such as particle edges and corners. In previous STM studies on supported Pd nanoparticles, oxygen was shown to modify such sites first, indicating thus that they are the strongest adsorption sites on a Pd nanoparticle.[64] To check the feasibility of this explanation, we carried out CO titration experiments by infrared reflection absorption spectroscopy (IRAS) to identify the preferential adsorption sites of oxygen on Pd nanoparticles. In previous IRAS studies, the CO molecule was shown to be a suitable probe molecule for probing different adsorption sites on Pd nanoparticles, allowing independent identification of the regular adsorption sites on the (111) terraces and the low-coordinated adsorption sites such as edges and corners.[65] With this spectroscopic approach, it was experimentally confirmed that oxygen adsorption first modifies the low coordinated surface sites such as edges and corners. For more details, the reader is referred to ref. 66. From these observations we can unambiguously attribute the unexpectedly high initial adsorption energy (280 kJ mol^{-1}), measured in a calorimetric experiment on supported Pd nanoparticles, to oxygen adsorption at edge and corner sites. It remains, however, an open question as to why the edge sites of Pd nanoclusters interact so strongly with adsorbed oxygen. Further theoretical work is needed in order to rationalize the microscopic nature of this phenomenon, which can presumably originate from the reconstruction or refaceting of the particles edges that was previously reported in the literature.[64,67] It is also feasible that the edge sites exhibit a higher activity in dissociating of O_2 molecules, which can account for the experimentally observed high values of the initial sticking coefficient on Pd nanoparticles.

III.C CO adsorption on supported Pd nanoparticles and Pd(111) precovered by O

Under the reaction conditions, both CO and oxygen interact with the surface, which is partially or nearly fully covered by co-reactants. The presence of co-adsorbates can significantly change the binding energies and consequently strongly affect the surface coverages under the operation conditions. For the reactions involving a dissociative step, such as CO oxidation, even small changes in the surface coverage might result in switching to a "poisoning" regime, where high concentration of one reactant (CO) can nearly completely suppress dissociation of the other one (O_2).[5,7] Therefore it is a very important issue and a current scientific challenge to accurately determine the binding energies in the systems of two co-adsorbates and understand how this binding strength depends on the structure of the underlying metal substrate. To approach this goal, we measured the binding energies of CO on both Pd surfaces pre-saturated with oxygen. Oxygen deposition was carried out at 300 K to enable dissociation of molecular species, and immediately afterwards CO adsorption energies were measured at 100 K, at which CO does not react with O to produce CO_2.

The results of these experiments are displayed in Fig. 4a and 4b for Pd nanoparticles and Pd(111), respectively. They show that both surfaces precovered with oxygen exhibit similar adsorption capacity for CO to the clean Pd surfaces. The measured saturation values are $\sim 0.75 \times 10^{15}$ molecules cm^{-2} on Pd(111) and $\sim 0.50 \times 10^{15}$ molecules cm^{-2} on O/Pd nanoparticles (compare to the values 0.76 $\times 10^{15}$ molecules cm^{-2} and 0.4×10^{15} molecules cm^{-2} on clean Pd surfaces, correspondingly). However, it should be noted that CO adsorption on the clean surfaces was investigated at 300 K, while the experiments on the O-precovered surfaces were carried out at 100 K to avoid surface reaction between adsorbed CO and O. According to previous experimental and theoretical studies on Pd(111), CO occupies two hollow sites (hcp and fcc) at 300 K forming a $c(2 \times 4)_{2CO}$ overlayer with the saturation coverage of 0.5 CO molecules per surface Pd atom.[68,69] At 100 K, the CO saturation coverage increases on the clean Pd(111) surface to 0.75 due to occupation of an additional on-top adsorption site per unit cell.[70,71] On Pd(111) precovered with oxygen, CO forms a new mixed $p(2 \times 2)_{CO+O}$ phase in the temperature range 60–110 K.[72] STM results in combination with DFT calculations suggest that this overlayer structure is consistent with the occupation of a (fcc) hollow site by one oxygen

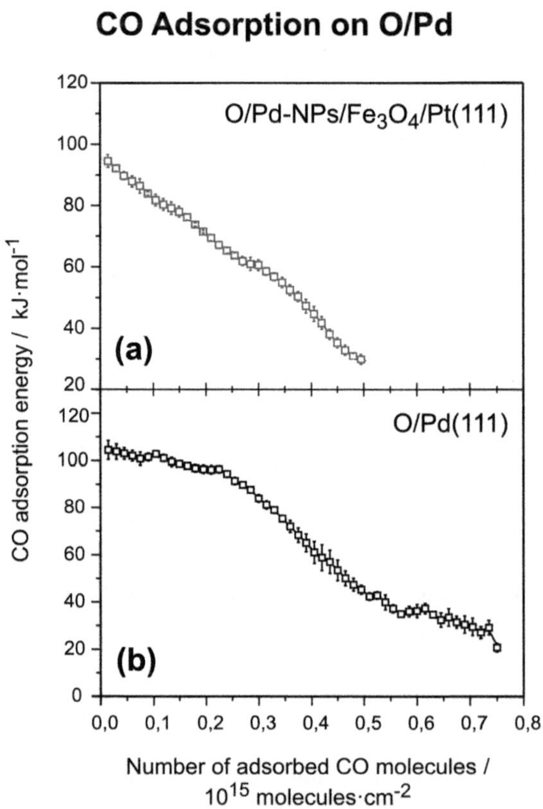

Fig. 4 Differential heats of CO adsorption plotted as a function of number of adsorbed CO molecules for Pd nanoparticles (a) and Pd(111) (b) pre-covered with a monolayer of O atoms. Oxygen was deposited at 300 K, CO adsorption experiments were carried out at 100 K.

atom per unit cell and by either one or two CO molecules per unit cell.[72] In the latter case, for which the unit cell contains one oxygen atom and two CO molecules, the CO saturation coverage should amount to 0.5 per surface Pd atom, which is in good agreements with the experimentally measured value ~0.51.

On O/Pd(111), the initial heat of CO adsorption amounts to 104 ± 4 kJ mol^{-1}, which is ~45 kJ mol^{-1} lower than the CO adsorption energy on the pristine Pd(111) surface. The initial CO adsorption heat on Pd nanoparticles precovered with oxygen reaches 95 ± 2 kJ mol^{-1}. This value is ~39 kJ mol^{-1} lower than the initial heat of CO adsorption on the clean Pd nanoparticles. Interestingly, both O-presaturated surfaces show a similar decrease of the CO adsorption energy by ~40–45 kJ mol^{-1}. Moreover, the same energy difference (~40 kJ mol^{-1}) in CO adsorption heat was observed over the whole range of CO surface coverages on both surfaces (data not show).

Two major observations can be made from the comparison of clean *vs.* O-precovered Pd surfaces with respect to the CO adsorption energies. First, the presence of a monolayer of co-adsorbed oxygen atoms lowers the CO binding energy by about 40–45 kJ mol^{-1} both on the single crystal surface and on Pd nanoparticles. This value can be considered as a measure for the repulsive interaction between oxygen and CO on Pd. Second, the difference between CO binding energies on Pd(111) and Pd nanoparticles of about 10–15 kJ mol^{-1} remains preserved even on the O-covered surfaces. Previously, the decreasing CO adsorption energy with decreasing particle size on clean Pd particles was attributed primarily to the contraction of the Pd lattice in small clusters. The presented data demonstrate that the same trend is valid also for Pd surfaces covered with oxygen with approximately the same magnitude.

Conclusions

In conclusion, we investigated the interaction of CO and O_2 molecules with well-defined Pd nanoclusters and with the extended Pd(111) surface by a direct calorimetric method. The main focus of this study lies in a detailed understanding of how a nanometer scale confinement of matter affects the binding strength of gaseous adsorbates relevant to CO oxidation.

CO and oxygen adsorption were investigated both on clean and O-precovered Pd nanoparticles and Pd(111) and corresponding sticking coefficients and adsorption energies were determined as a function of adsorbate coverages.

The CO adsorption kinetics on Pd nanoparticles investigated by the King–Well method show a strong enhancement of the adsorbate flux onto Pd clusters due to a "capture zone" effect, which involves weak trapping of the adsorbates on the support and diffusion to the metal particles. The CO adsorption rate on nanoparticles was estimated to be enhanced by factor of 1.4–1.9 depending on the assumptions on the initial sticking coefficient on metallic Pd. For oxygen adsorption, the "capture zone effect" was found to also play a role; however, its exact quantification was difficult because of possibly higher activity of Pd nanoparticles towards dissociative O_2 adsorption as compared to Pd(111).

The initial binding energy of carbon monoxide was found to be lower on Pd nanoparticles than on Pd(111) (134 *vs.* 149 kJ mol^{-1}, correspondingly) despite the high fraction of low-coordinated surface sites present on nanoclusters. This observation can be rationalized as a consequence of Pd lattice contraction in small particles leading to weakening of chemisorptive interaction. Additionally, a reduced polarizability of small Pd nanoparticles can result in a reduction of van der Waals attraction. On the O-precovered Pd surfaces, the CO binding energy was found to be reduced by about 40–45 kJ mol^{-1} as compared to the corresponding pristine Pd surfaces in the full range of CO coverages. Also in the presence of a monolayer of oxygen, CO binds weaker on Pd nanoclusters than on Pd(111) by about 10–15 kJ mol^{-1}. This result suggests that the strain effect observed for pristine Pd nanoclusters remains preserved also on the particles pre-covered with adsorbates, particularly oxygen.

Table 1 Summary of the adsorption energies of the CO and oxygen adsorbates on the clean O-precovered Pd surfaces obtained in this study

	Adsorption enthalpy, kJ mol^{-1} (low coverage $\rightarrow$ quasi saturation[a])	
	Pd(111)	Pd nanoparticles (~4 nm)/Fe_3O_4/Pt(111)
CO on pristine Pd, 300 K	$149 \pm 4 \rightarrow 75$	$134 \pm 4 \rightarrow 90$
O_2 on pristine Pd (per O_2 molecule), 300 K	$205 \pm 7 \rightarrow 100$	$280 \pm 13 \rightarrow 100$
CO on O (1 ML)/Pd, 100 K	$104 \pm 4 \rightarrow 30$	$94 \pm 2 \rightarrow 30$

[a] Note that the absolute accuracy of the enthalpy values in the quasi saturation regime is lower than for the lowest adsorbate coverage because of the experimental difficulties in determining very low sticking coefficients.

The measured initial oxygen adsorption heat on Pd(111) amounts to 205 ± 7 kJ mol^{-1}. On Pd nanoparticles, this values increases by ~75 kJ mol^{-1}; this energy can be attributed to strong oxygen binding at the low coordinated adsorption sites of Pd nanoparticles such as edges as corners, which was also confirmed by spectroscopic data. The unexpectedly strong binding energy of oxygen at the particles edges exceeds all literature values of oxygen adsorption energies at stepped single crystal surfaces. Apparently, the local adsorption environment of Pd–O bond plays a very important role in determining the binding strength in this adsorption system. This result is in a sharp contrast to the CO–Pd case, where the presence of the low-coordinated surface sites on a Pd nanoparticle was observed to not result in stronger CO–Pd binding energy. Instead, weakening of CO interaction energy with Pd nanoparticles was detected, which can be most likely attributed to the contraction of the Pd lattice. Both Pd surfaces exhibit very strong coverage dependence of O binding energy covering the range up to 180 kJ mol^{-1}.

In summary, the reduced dimensions of the Pd substrate can affect the binding strength of the adsorbates involved in CO oxidation in two principle ways: (i) *via* the change of the local adsorption environment that can result *e.g.* in stronger adsorbate bonding *e.g.* at the particle edges and (ii) *via* contraction of the Pd lattice in small clusters and a concomitant weakening of chemisorptive interaction. The latter effect is expected to play a role not only for CO as shown in this study but also for other adsorbates (provided that the local adsorption environment remains preserved). The presented experimental data (summarized in Table 1) provide important benchmarks for upcoming theoretical calculation.

Acknowledgements

Support by the Fonds der Chemischen Industrie is gratefully acknowledged. The authors thank Hans-Joachim Freund, Charles T. Campbell (UW) and Beatriz Roldan (UCF) for valuable discussions.

References

1 G. Ertl, H. Knoezinger and J. Weitkamp, in *Handbook of Heterogeneous Catalysis*, ed. G. Ertl, H. Knoezinger and J. Weitkamp, VCH, Weinheim, 1997.
2 J. M. Thomas and W. J. Thomas, *Principle and Practice of Heterogeneous Catalysis*, VCH, Weinheim, 1997.
3 T. Engel and G. Ertl, *Adv. Catal.*, 1979, **28**, 41.
4 S. M. McClure and D. W. Goodman, *Chem. Phys. Lett.*, 2009, **469**, 1.
5 J. Libuda and H.-J. Freund, *Surf. Sci. Rep.*, 2005, **57**, 157.
6 J. Wintterlin, S. Völkening, T. V. W. Janssens, T. Zambelli and G. Ertl, *Science*, 1997, **278**, 1931.

7 C. R. Henry, *Surf. Sci. Rep.*, 1998, **31**, 231.
8 I. Stará and V. Matolín, *Surf. Sci.*, 1994, **313**, 99.
9 G. Zheng and E. I. Altman, *Surf. Sci.*, 2002, **504**, 253.
10 H. Conrad, G. Ertl, J. Küppers and E. E. Latta, *Surf. Sci.*, 1977, **65**, 245.
11 R. Westerström, J. Gustafson, A. Resta, A. Mikkelsen, J. N. Andersen, E. Lundgren, N. Seriani, F. Mittendorfer, M. Schmidt, J. Klikovits, P. Varga, M. D. Ackermann, J. W. M. Frenken, N. Kasper and A. Stierle, *Phys. Rev. B*, 2007, **76**, 155410.
12 F. P. Leisenberger, G. Koller, M. Sock, S. Surnev, M. G. Ramsey, F. P. Netzer, B. Klötzer and K. Hayek, *Surf. Sci.*, 2000, **445**, 380.
13 G. Zheng and E. Altman, *Surf. Sci.*, 2000, **462**, 151.
14 A. Vlad, A. Stierle, R. Westerström, S. Blomberg, A. Mikkelsen and E. Lundgren, *Phys. Rev. B*, 2012, **86**, 35407.
15 B. Klötzer, K. Hayek, C. Konvicka, E. Lundgren and P. Varga, *Surf. Sci.*, 2001, **482–485**, 237.
16 P. Nolte, A. Stierle, N. Kasper, N. Y. Jin-Phillipp, N. Jeutter and H. Dosch, *Nano Lett.*, 2011, **11**, 4697.
17 R. Westerström, M. E. Messing, S. Blomberg, A. Hellman, H. Grönbeck, J. Gustafson, N. M. Martin, O. Balmes, R. van Rijn, J. N. Andersen, K. Deppert, H. Bluhm, Z. Liu, M. E. Grass, M. Hävecker and E. Lundgren, *Phys. Rev. B*, 2011, **83**, 115440.
18 J. Han, G. Zhu, D. Y. Zemlyanov and F. H. Ribeiro, *J. Catal.*, 2004, **225**, 7.
19 E. Lundgren, G. Kresse, C. Klein, M. Borg, J. N. Andersen, M. de Santis, Y. Gauthier, C. Konvicka, M. Schmidt and P. Varga, *Phys. Rev. Lett.*, 2002, **88**, 246103.
20 T. Schalow, M. Laurin, B. Brandt, S. Schauermann, S. Guimond, H. Kuhlenbeck, D. E. Starr, S. K. Shaikhutdinov, J. Libuda and H.-J. Freund, *Angew. Chem., Int. Ed.*, 2005, **44**, 7601.
21 W. A. Brown, R. Kose and D. A. King, *Chem. Rev.*, 1998, **98**, 797.
22 J. Stuckless, N. Frei and C. T. Campbell, *Rev. Sci. Instrum.*, 1998, **69**, 2427.
23 C. T. Campbell and O. Lytken, *Surf. Sci.*, 2009, **603**, 1365.
24 J.-H. Fischer, J. Hartmann, J. A. Farmer, J. M. Flores-Camacho, C. T. Campbell, S. Schauermann and H.-J. Freund, *Rev. Sci. Instrum.*, 2011, **82**, 24102.
25 D. King and M. Wells, *Surf. Sci.*, 1972, **29**, 454.
26 T. Schalow, B. Brandt, D. E. Starr, M. Laurin, S. K. Shaikhutdinov, S. Schauermann, J. Libuda and H.-J. Freund, *Phys. Chem. Chem. Phys.*, 2007, **9**, 1347.
27 T. Schalow, B. Brandt, D. E. Starr, M. Laurin, S. Schauermann, S. K. Shaikhutdinov, J. Libuda and H.-J. Freund, *Catal. Lett.*, 2006, **107**, 189.
28 C. Lemire, R. Meyer, V. Henrich, S. K. Shaikhutdinov and H.-J. Freund, *Surf. Sci.*, 2004, **572**, 103.
29 T. Mitsui, M. K. Rose, E. Fomin, D. F. Ogletree and M. Salmeron, *Phys. Rev. Lett.*, 2005, **94**, 036101.
30 A. M. Bradshaw and F. M. Hoffmann, *Surf. Sci.*, 1978, **72**, 513.
31 J. Szanyj, W. K. Kuhn and D. W. Goodman, *J. Vac. Sci. Technol., A*, 1993, **11**, 1969.
32 S. Surnev, M. Sock, M. G. Ramsey, F. P. Netzer, M. Wiklund, M. Borg and J. N. Andersen, *Surf. Sci.*, 2000, **470**, 171.
33 T. Gießel, O. Schaff, C. J. Hirschmugl, V. Fernandez, K.-M. Schindler, A. Theobald, S. Bao, R. Lindsay, W. Berndt, A. M. Bradshaw, C. Baddeley, A. F. Lee, R. M. Lambert and D. P. Woodruff, *Surf. Sci.*, 1998, **406**, 90.
34 H. Conrad, G. Ertl, J. Kocha and E. E. Latta, *Surf. Sci.*, 1974, **43**, 462.
35 T. Engel, *J. Chem. Phys.*, 1978, **69**, 373.
36 A. Ortega, F. M. Hoffmann and A. M. Bradshaw, *Surf. Sci.*, 1982, **119**, 79.
37 K. Tsu and M. Boudart, *2nd Actes du Congr. Int. Cat. (Technip, Paris)*, 1961, **1**, 593.
38 Y. Kim and M. Boudart, *Langmuir*, 1991, **7**, 2999.
39 E. Gillet, S. Channakhone, V. Matolin and M. Gillet, *Surf. Sci.*, 1985, **152–153**, 603.
40 V. Matolin and E. Gillet, *Surf. Sci.*, 1986, **166**, L115.
41 M. Bowker, P. Stone, R. Bennett and N. Perkins, *Surf. Sci.*, 2002, **497**, 155.
42 H. Conrad, G. Ertl, J. Koch and E. E. Latta, *Surf. Sci.*, 1974, **43**, 462.
43 G. Ertl and J. Koch, *Z. Naturforsch.*, 1970, **25a**, 1906.
44 R. D. Ramsier, K.-W. Lee and J. T. Yates, Jr., *Surf. Sci.*, 1995, **322**, 243.
45 T. Engel, *J. Chem. Phys.*, 1978, **69**, 373.
46 J.-H. Fischer-Wolfarth, J. A. Farmer, J. M. Flores-Camacho, A. Genest, I. V. Yudanov, N. Rösch, C. T. Campbell, S. Schauermann and H.-J. Freund, *Phys. Rev. B*, 2010, **81**, 241416.
47 I. V. Yudanov, R. Sahnoun, K. M. Neyman, N. Rösch, J. Hoffmann, S. Schauermann, V. Johánek, H. Unterhalt, G. Rupprechter, J. Libuda and H.-J. Freund, *J. Phys. Chem. B*, 2003, **107**, 255.
48 C. T. Campbell, G. Ertl, H. Kuipers and J. Segner, *Surf. Sci.*, 1981, **107**, 207.

49 S. A. Nepijko, M. Klimenkov, M. Adelt, H. Kuhlenbeck, R. Schlögl and H.-J. Freund, *Langmuir*, 1999, **15**, 5309.
50 R. Lamber, S. Wetjen and N. I. Jaeger, *Phys. Rev. B*, 1995, **51**, 10968.
51 S. Krüger, S. Vent, F. Nötermann, M. Staufer and N. Rösch, *J. Chem. Phys.*, 2001, **115**, 2082.
52 P. Nava, M. Sierka and R. Ahlrichs, *Phys. Chem. Chem. Phys.*, 2003, **5**, 3372.
53 O. Häberlen, S.-C. Chung, S. Krüger and N. Rösch, *J. Chem. Phys.*, 1997, **106**, 5189.
54 I. V. Yudanov, R. Sahnoun, K. M. Neyman and N. Rösch, *J. Chem. Phys.*, 2002, **117**, 9887.
55 I. V. Yudanov, M. Metzner, A. Genest and N. Rösch, *J. Phys. Chem. C*, 2008, **112**, 20269.
56 E. Shustorovich and H. Sellers, *Surf. Sci. Rep.*, 1998, **31**, 1.
57 R. A. van Santen and G. J. Kramer, *Chem. Rev.*, 1995, **95**, 637.
58 B. Hammer, Y. Morikawa and J. K. Nørskov, *Phys. Rev. Lett.*, 1996, **76**, 2141.
59 S. Sandel, J. Libuda, P. Brühwiler, S. Andersson, A. J. Maxwell, M. Bäumer, N. Mårtensson and H.-J. Freund, *J. Vac. Sci. Technol., A*, 1996, **14**, 1546.
60 K. Watanabe, Y. Matsumoto, M. Kampling, K. Al-Shamery and H.-J. Freund, *Angew. Chem., Int. Ed.*, 1999, **38**, 2192.
61 D. Mulugeta, K. H. Kim, K. Watanabe, D. Menzel and H.-J. Freund, *Phys. Rev. Lett.*, 2008, **101**, 146103.
62 To be published.
63 P. W. Davies and R. M. Lambert, *Surf. Sci.*, 1981, **110**, 227.
64 K. Hųjrup Hansen, Ž. Šljivančanin, E. Lęgsgaard, F. Besenbacher and I. Steinsgaard, *Surf. Sci.*, 2002, **505**, 25.
65 K. Wolter, O. Seiferth, H. Kuhlenbeck, M. Bäumer and H.-J. Freund, *Surf. Sci.*, 1998, **399**, 190.
66 M. Peter, J. M. Flores Camacho, S. Adamovsky, L. K. Ono, K.-H. Dostert, C. P. O'Brien, B. Roldan Cuenya, S. Schauermann and H.-J. Freund, *Angew. Chem., Int. Ed.*, 2013, DOI: 10.1002/anie.201209476.
67 P. Nolte, A. Stierle, N. Kasper, N. Y. Jin-Phillipp, N. Jeutter and H. Dosch, *Nano Lett.*, 2011, **11**, 4697.
68 T. Gießel, O. Schaff, C. J. Hirschmugl, V. Fernandez, K.-M. Schindler, A. Theobald, S. Bao, R. Lindsay, W. Berndt, A. M. Bradshaw, C. Baddeley, A. F. Lee, R. M. Lambert and D. P. Woodruff, *Surf. Sci.*, 1998, **406**, 90.
69 D. Loffreda, D. Simon and P. Sautet, *Surf. Sci.*, 1999, **425**, 68.
70 L. Surnev, M. Sock, M. G. Ramsey, F. P. Netzer, M. Wiklund, A. Borg and J. N. Andersen, *Surf. Sci.*, 2000, **470**, 171.
71 M. K. Rose, T. Mitsui, J. C. Dunphy, A. Borg, D. F. Ogletree, M. Salmeron and P. Sautet, *Surf. Sci.*, 2002, **512**, 48.
72 J. Méndez, S. H. Kim, J. Cerdá, J. Wintterlin and G. Ertl, *Phys. Rev. B*, 2005, **71**, 085409.

Influence of hot carriers on catalytic reaction; Pt nanoparticles on GaN substrates under light irradiation

Sun Mi Kim,[a] Dahee Park,[a] Youngji Yuk,[a] Sang Hoon Kim[b] and Jeong Young Park*[a]

Received 29th November 2012, Accepted 21st December 2012
DOI: 10.1039/c2fd20133j

We report the hot carrier-driven catalytic activity of two-dimensional arrays of Pt nanoparticles on GaN substrate under light irradiation. In order to elucidate the effect of a hot carrier in a catalytic chemical reaction, the CO oxidation reaction was carried out on Pt nanoparticles on p- and n-type GaN under light irradiation. Metal catalysts composed of Pt nanoparticles were prepared using two different preparation methods: the one-pot polyol reduction and arc plasma deposition methods. Under light irradiation, the catalytic activity of the Pt nanoparticles supported on GaN exhibited a distinct change depending on the doping type. The catalytic activity of the Pt nanoparticles on the n-doped GaN wafer decreased by 8–28% under light irradiation, compared to no irradiation (*i.e.*, in the dark), while the Pt nanoparticles on the p-doped GaN wafer increased by 11–33% under light irradiation, compared to no irradiation. The catalytic activity increased on the smaller Pt nanoparticles, compared to the larger nanoparticles, presumably due to the mean free path of hot carriers. Based on these results, we conclude that the flow of hot carriers generated at the Pt–GaN interface during light irradiation is responsible for the change in catalytic activity on the Pt nanoparticles.

I Introduction

Understanding the mechanisms of electronic excitation in exothermic chemical processes on surfaces is a long-standing question of both fundamental and practical importance. Recent experimental and theoretical studies have demonstrated nonadiabatic electronic excitations that result from chemisorption and physisorption of gases at surfaces and from chemical reactions at surfaces.[1–3] Electronic excitation in exothermic catalytic reactions involves the flow of hot electrons (1–3 eV), assuming that most of the chemical energy is converted to electron flow.[4,5] Once the electrons on the metal surface become energetic (or hot) through nonadiabatic electronic excitation, the hot electrons move into the bulk of the metal catalyst and eventually dissipate energy, turning into low-energy electrons through lattice atom relaxation within the length scale of the electron mean free path. Hellberg *et al.* proposed that an exothermic surface reaction ($K + Cl_2$ in their case) induces electron transfer from the metal surface to the molecular orbital of the adsorbed diatomic molecule.[6] A catalytic nanodiode was reported that detected a steady-state flux of hot (>1 eV) conduction electrons generated during a gas-phase exothermic reaction on the surface of Pt metal.[7–10] Furthermore, it has been demonstrated that hot electron flows can influence atomic or molecular processes, including

[a]*Graduate School of EEWS (WCU) and NanoCentury KI, KAIST (Korea Advanced Institute of Science and Technology), Daejeon, 305-701 South Korea. E-mail: jeongypark@kaist.ac.kr*
[b]*Centre for Materials Architecturing, KIST, Seoul, 136-791, Republic of Korea*

desorption,[11] diffusion,[12] and isomerization of molecules.[13] Shen *et al.* reported that incident electrons released from scanning tunneling microscopy could be utilized for atomic-scale hydrogen desorption through electronic and vibrational excitation mechanisms.[11] Wadayama *et al.* found that hot electrons induce the reductive coupling reaction of *p*-nitrobenzoic acid adsorbed on the Ag top electrode surface.[14]

Many research groups also discovered the so-called strong metal–support interaction (SMSI) effect that generates newly created, unique properties when the metal is dispersed in support materials; although in most cases, there is no significant interaction between the metal and the support.[15–19] G.-M. Schwab *et al.* first reported this phenomenon in methane oxidation on metal-semiconductor mixed catalysts, ZnO/Ag, which display higher catalytic activity than other catalysts, ZnO and Ag, giving an extremely high yield of products at even lower temperatures. For this catalytic promoter effect, they assumed that it originated from an electron exchange between the support and catalyst. Tauster *et al.* reported that the adsorption of hydrogen on iridium supported on transition metal oxides such as TiO_2, V_2O_3, Nb_2O_5 and Ta_2O_5 show the SMSI effect, while iridium supported on transition metal oxides such as Al_2O_3, SiO_2, MgO, Sc_2O_3, Y_2O_3, ZrO_2 and HfO_2 showed no change.[20,21]

Recently, Schäfer *et al.* observed that the type of gallium nitride (GaN) doping affected the reactivity of Pt nanoparticles. They suggested that this is attributed to the strong electronic interaction between the nanoparticles and the n- or p-type GaN substrates, showing a large influence on the chemical composition and oxygen affinity of supported nanoparticles under X-ray irradiation.[22] Kim *et al.* carried out CO oxidation on Pt–CdSe–Pt nanodumbbells and showed that the catalytic activity of the nanodumbbells increases by a factor of two during irradiation by light with energy higher than the bandgap of CdSe.[23] These experiments indicate that the hot electrons generated by photon absorption on the semiconductor influence the catalytic activity. In this study, we carried out the CO oxidation reaction on Pt nanoparticles on n- and p-doped GaN substrates with and without light to identify the role of hot carriers in the chemical reaction, as shown in Fig. 1(a). We also studied the effect of the size of the Pt nanoparticles and the presence of capping layers on the change of catalytic activity due to hot carrier flows.

II Experimental details

1 Preparation of n- or p-doped GaN substrates as the support and Pt nanoparticles as the metal catalyst

As a support, n- or p-doped GaN thin films were grown on a sapphire substrate *via* the metal–organic chemical vapor deposition (MOCVD) method. Monosilane (SiH_4) and bis-cyclopentadienylmagnesium (Cp_2Mg), as acceptor dopants, were

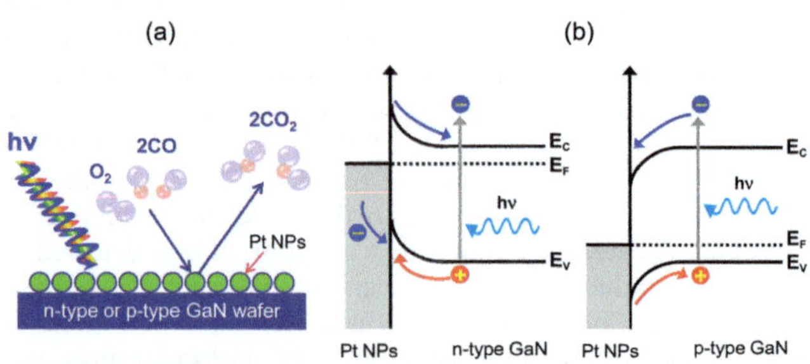

Fig. 1 (a) Schematic drawing representing the CO oxidation reaction on Pt nanoparticles on the n- or p-type gallium nitride (GaN) substrate under light irradiation and (b) energy band alignment of the Pt–GaN model systems showing two plausible routes of hot carrier flows according to the doping type of the GaN under light irradiation.

used for n-type and p-type doping, respectively. MOCVD-grown GaN exhibits Ga-face crystal polarity along the [0001] axis, resulting in high mechanical and chemical stability due to low surface defect concentrations.[24–26] However, as-grown Mg-doped GaN films have a high resistivity and do not have any p-type conduction. Post-treatment is generally required to activate the dopants, and the GaN was annealed using rapid thermal annealing at 823 K for 5 min in pure oxygen at ambient pressure prior to nanoparticle deposition. Through 4-probe measurements, we confirmed that the resistance of the p-doped GaN wafer decreased by a factor of $\sim$10.

The Pt nanoparticles were synthesized by two different preparation routes, the one-pot polyol reduction and arc plasma deposition (APD) methods, in order to study the effect of the organic capping agent on catalytic activity. Colloidal nanoparticles are usually prepared in the presence of an organic capping agent, such as polymers or surfactants, that prevent the aggregation of nanoparticles in solution.[27] The organic capping agent is known to partially affect the catalytic activity by blocking reactive sites of the metallic nanoparticles.[28] In other words, the functional groups on the organic capping agents can potentially shield the surface of the catalyst atom, which in turn alters the bonding modes of the adsorbents during catalytic reactions, thereby affecting the catalytic activity. For this reason, plasma processing, including the APD method, or electrical explosion have attracted a lot of attention because it is possible to not only prepare the nanoparticles or thin films without using an organic capping agent to but also synthesize nanocatalysts at a large scale.[29–33] This is mainly due to the high reactivity of the medium *via* which the energetic transfer occurs through various forms: excited atoms or molecules, radicals, ions, electrons, *etc.*

The Pt nanoparticles were prepared *via* the one-pot polyol reduction method, as reported by Song *et al.* and Tsung *et al.*[34,35] First, 2.1 nm Pt nanoparticles were synthesized using the same method described in Song *et al.*[34] and details of the synthesis procedure are as follows: 5 mL of 0.5 M NaOH solution in ethylene glycol was put into an ethylene glycol solution containing 250 mg $H_2PtCl_6 \cdot 6H_2O$ in a 50 mL round-bottom flask. After stirring at room temperature for 20 min, the mixture was heated at 433 K for 3 h with N_2 bubbling in a silicon oil bath. After the reaction, the particles were precipitated by adding 1 mL of 2 M HCl and dispersed in ethanol containing 12.2 mg of poly(vinylpyrrolidone) (PVP, $M_w = 29{,}000$). Secondly, 4.3 nm Pt nanoparticles were synthesized using the same method described in Tsung *et al.*[35] Here, the size of the Pt nanocrystals is determined by controlling the concentration ratios of Pt(II) to Pt(IV) in the initial solution because the oxidation state of the metal ion precursors affects the reducing rate of the Pt precursor and, consequently, the number of nuclei. The details of the synthesis procedure are as follows: 20 mg of ammonium tetrachloroplatinate (II) ($(NH_4)_2PtCl_6$), 100 mg of PVP, and 120 mg of tetramethylammonium bromide ($N^+(CH_3)_4Br^-$) were dissolved into 10 mL of ethylene glycol in a 25 mL round-bottom flask at room temperature. The mixed solution was heated to 453 K in a silicon oil bath and was held for 20 min under argon protection and magnetic stirring, resulting in a dark brown solution. After the reaction, the solution was cooled to room temperature. The size, shape and size distribution of the as-synthesized 2.1 and 4.3 nm Pt nanoparticles were characterized by transmission electron microscopy (TEM), as shown in Fig. 2 (a-b) and (c-d), respectively.

2 Two technical approaches to fabricate two-dimensional (2D) arrays of Pt nanoparticles on the n- or p-doped GaN substrates: the Langmuir–Blodgett (LB) technique and arc plasma deposition (APD) method

To systematically investigate the characteristics of Pt/GaN model catalysts based on the strong metal–support interaction (SMSI) effect and to identify the effect of the doping type of semiconductor on Pt nanoparticle reactivity, the Pt nanoparticles were deposited on the n- or p-doped GaN wafer using two different preparation

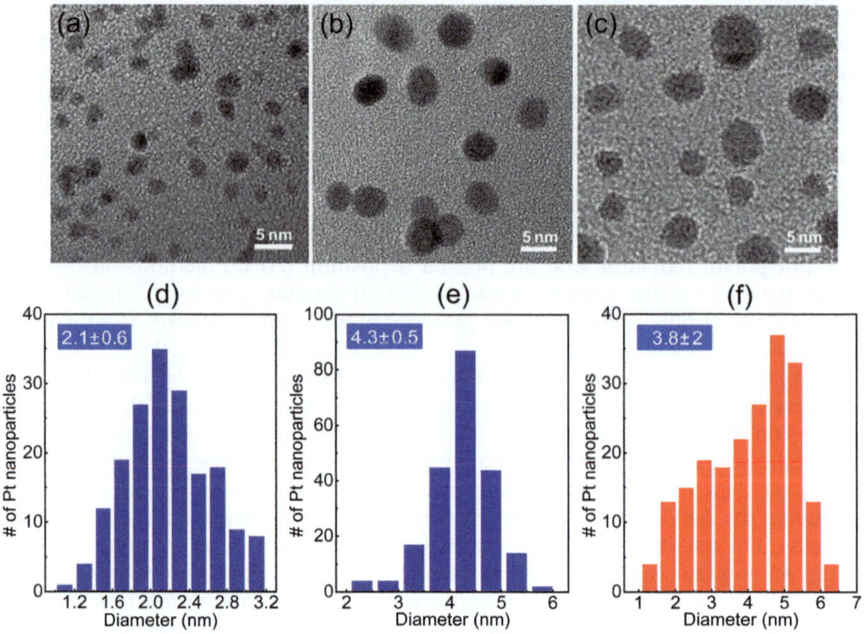

Fig. 2 TEM images (a) 2.1 nm and (b) 4.3 nm Pt nanoparticles prepared using the one-pot polyol reduction method on a copper TEM grid. (c) TEM images Pt nanoparticles prepared using the APD method (100 V) on a silicon oxide TEM. Size distribution histograms of (d) 2.1 nm Pt and (e) 4.3 nm Pt nanoparticles prepared using the one-pot polyol reduction method, and (f) Pt nanoparticles prepared using the APD method.

methods: the Langmuir–Blodgett (LB) technique and APD method. First of all, the Langmuir–Blodgett technique is a very versatile tool to fabricate a massive, parallel, controlled organization of nanostructures, such as nanocrystals or nanowires, over a large area.[27] The assembly process was carried out in a water-filled trough (611 Nima Technology) equipped with a mobile barrier and a pressure sensor (Wilhelmy plate) at room temperature. The Pt nanoparticles synthesized by the one-pot polyol reduction method were dispersed in chloroform, a volatile solvent that is immiscible. The surface pressure was monitored with a Wilhelmy plate, and was adjusted to zero before spreading the nanoparticles. The solution is then spread dropwise onto the water surface where it spreads to equilibrium surface pressure and evaporates, leaving behind a water-supported thin film layer of nanoparticles. The resulting layer of nanoparticles floating on the water was compressed at a rate of 15 cm^2 min^{-1} using the mobile barrier, and two-dimensional (2D) arrays of nanoparticles on the silicon wafer substrate were created by lifting the submerged silicon wafer from the water.

Pt nanoparticles were also deposited onto the n- or p-doped GaN wafer by a coaxial-pulsed APD system (ULVAC, ARL-300) at room temperature under a vacuum of 10^{-6} Torr. The size of the Pt nanoparticles deposited onto the substrates was tuned depending on the arc voltage, with an arc voltage of 100 V used here. The APD system is largely comprised of three parts: a cathode, a trigger, and an anode electrode. Structurally, a cathode rod made of Pt and a trigger electrode were placed at the center and a cylindrical anode was coaxially mounted, wrapping the cathode. A trigger pulse induced an arc discharge on the surface of the target rod, generating a highly ionized Pt plasma.[36] The Pt plasma was deposited onto the n- or p-doped GaN surface. The Pt nanoparticles made by arc plasma were deposited on a silicon oxide TEM windows grid (SiMPore Inc.) and were characterized by TEM, as shown in Fig. 2 (c) and (f) to analyze the size, shape, and size distribution. The morphology of the prepared nanocatalyst 2D arrays was characterized using scanning electron microscopy (SEM, Magellan 400).

3 Measurement of the catalytic activity of the CO oxidation reaction on 2D-Pt nanoparticles/n- or p-doped GaN substrates

CO oxidation on the active Pt metal nanoparticles/n- or p-doped GaN substrates prepared *via* the various methods was carried out in a batch reactor under 40 Torr CO and 100 Torr O_2 as reactant gases and 620 Torr He as the balancing gas, equipped with a recirculation pump, similar to that used for the studies of single-crystal metal samples. The gases were circulated through the reaction line by a Metal Bellows recirculation pump. A gas chromatograph equipped with a thermal conductivity detector and a $6' \times 1/8''$ SS molecular sieve 5A was used to separate the products for analysis. The measured reaction rates are reported as turnover frequencies (TOF) and are measured in units of product molecules of CO_2 produced per metal surface site per second of reaction time. The number of metal sites is calculated by geometrical considerations based on SEM measurements of the surface area of the nanoparticle arrays. A halogen lamp (Dolhan-Janner D150) was used as the light source. With the distance between the sample and halogen lamp set at ~8 cm, the light intensity was measured to be 49.4 mW cm^{-2} by an optical power meter (ADCMT 8230).

III Results and discussion

In order to identify the role of hot carriers in the surface chemical reaction according to the doping type of semiconductor, we carried out the CO oxidation on Pt nanoparticles/n- or p-doped GaN wafers with light irradiation and in the dark (without light irradiation). Fig. 3 shows the catalytic activity of the 2.1 and 4.3 nm Pt nanoparticles on the n- or p-doped GaN substrates with and without light. As shown in Fig. 3 (a) and (c), the TOF of the 2.1 nm Pt nanoparticles on the n- or p-doped GaN was two times higher than that of the 4.3 nm Pt nanoparticles on the n- or p-doped GaN substrates with and without light. It is also clear that the CO oxidation reactivity of the Pt nanoparticles shows two different trends according to the doping type of semiconductor: the catalytic activity of the nanoparticles on the n-doped GaN wafer decreased by 8–15% under light irradiation while that on the p-doped GaN wafer increased by 16–33% under light irradiation, compared to no light irradiation, although the size of the nanoparticles affected the extent to which the catalytic activity increased or decreased. The activation energies for each Pt/GaN catalyst were obtained from Arrhenius plots of the catalytic activity. As shown in Fig. 3 (b), the activation energy increased from 24.7 kcal mol^{-1} to 25.5 kcal mol^{-1} for the 2.1 nm Pt nanoparticles on the n-doped GaN substrate, and decreased from 24.1 kcal mol^{-1} to 23.6 kcal mol^{-1} for the nanoparticles on the p-doped GaN substrate, depending on the light irradiation. For the 4.3 nm Pt nanoparticles/n- or p-doped GaN, a similar tendency was observed, as shown in Fig. 3 (d): the activation energy increased from 25.4 kcal mol^{-1} to 25.8 kcal mol^{-1} for the 4.3 nm Pt nanoparticles/n-doped GaN, and decreased from 25.7 kcal mol^{-1} to 25.2 kcal mol^{-1} for the Pt nanoparticles/p-doped GaN.

Fig. 4 (a) summarizes the catalytic activity of the 2.1 and 4.3 nm Pt nanoparticles on the n- or p-doped GaN supports with and without light at 523 K, respectively. The catalytic activity shows a remarkable difference according to the type of GaN doping. First, the catalytic activity of the Pt nanoparticles on the n-type GaN is higher than that on the p-type GaN. We suppose that this result is related to the higher oxidation states of the Pt nanoparticles on the n-type GaN, compared to that on the p-type GaN. Another interesting point we noticed in Fig. 4(a) is that the catalytic activity of the smaller (2.1 nm) Pt nanoparticles is higher than that of the larger (4.3 nm) nanoparticles. Because the smaller nanoparticles have a larger

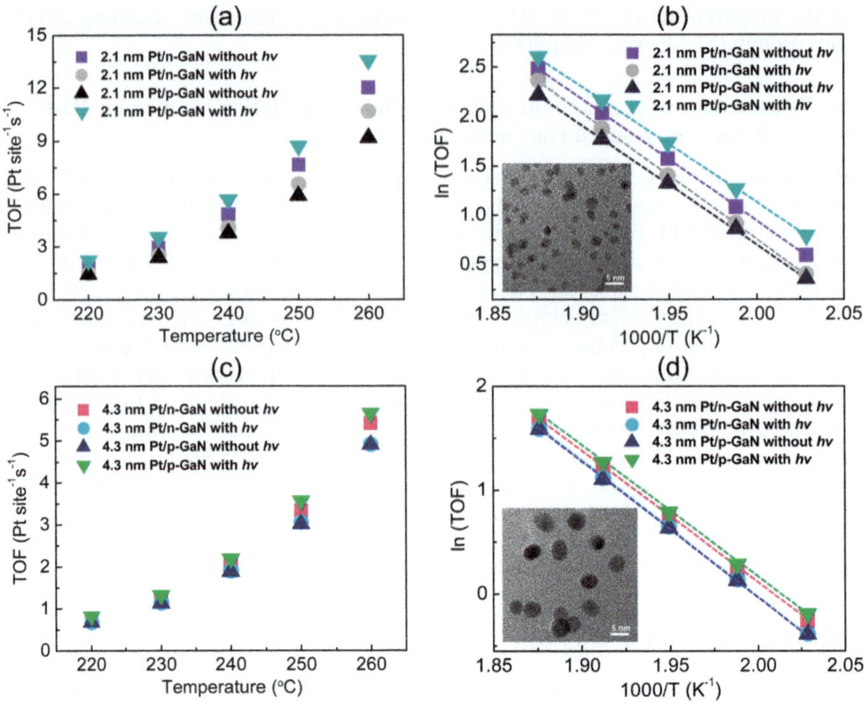

Fig. 3 Catalytic activity of the CO oxidation reaction on two-dimensional arrays of two different-sized Pt nanoparticles on the n- or p-type GaN substrate with and without light: (a) catalytic turnover rate with temperature; (b) Arrhenius plots of 2.1 nm Pt nanoparticles on the n- or p-type GaN substrate; (c) catalytic turnover rate with temperature and (d) Arrhenius plots of 4.3 nm Pt nanoparticles on the n- or p-type GaN substrate with and without light.

interface area making contact with the GaN, the result implies a strong metal–support interaction the in Pt/GaN system.

Under light irradiation, the activity decreased when the nanoparticles were deposited on the n-doped GaN wafer while the activity increased when the nanoparticles were deposited on the p-doped GaN wafer. Another appealing aspect of the catalytic activity of the Pt/n- or p-doped GaN catalysts is that the percentage change of the activity based on light irradiation depends on the size of the nanoparticles. Fig. 4(b) shows the percentage change of TOF of the 2.1 nm and 4.3 nm Pt nanoparticles/GaN during light irradiation. It is clear that the smaller nanoparticles show a higher change of TOF compared to the larger nanoparticles, presumably due to the mean free path of the hot carriers. In other words, hot carriers can be injected to the surface of smaller Pt nanoparticles more easily, inducing the more prominent change of catalytic activity. Fig. 5 shows SEM images (a) before and (b) after CO oxidation of two-dimensional (2D) 4.3 Pt nanoparticle arrays on the n-doped GaN substrates prepared using the LB technique. These SEM images indicate that the agglomeration of Pt nanoparticle during CO oxidation can be ignored.

In order to investigate the effect of organic capping agents in addition to identifying the role of hot carriers, Pt nanoparticles were deposited on the n- or p-doped GaN wafers *via* the arc plasma deposition method and CO oxidation was carried out with and without light irradiation. Fig. 6 shows the results of catalytic activity measured on APD Pt nanoparticles on the n- or p-doped GaN substrates with and without light. The inset of Fig. 6 (b) shows TEM image of the Pt nanoparticles prepared using APD. As shown in Fig. 6 (a), the TOF of the APD Pt nanoparticles

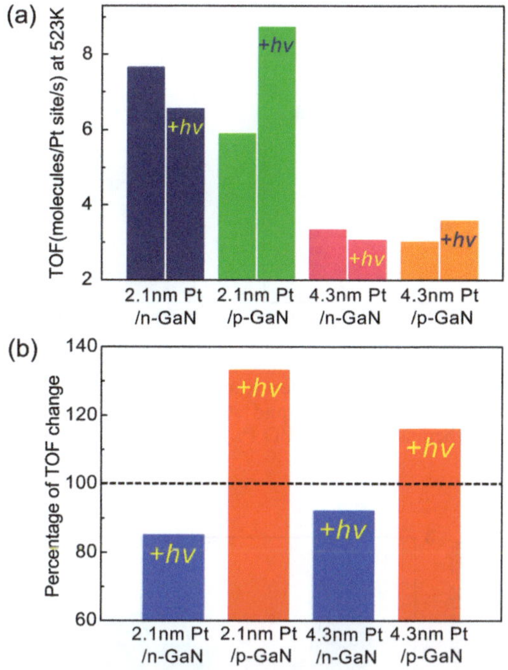

Fig. 4 Comparison of the TOF of the CO oxidation reaction measured on Pt nanoparticles/n- or p-type GaN at 523 K as a function of light: TOF of (a) Pt nanoparticles arrayed on the n- or p-type GaN using the LB technique and (b) the percentage change of TOF of the 2.1 nm and 4.3 nm Pt nanoparticles/ GaN during light irradiation.

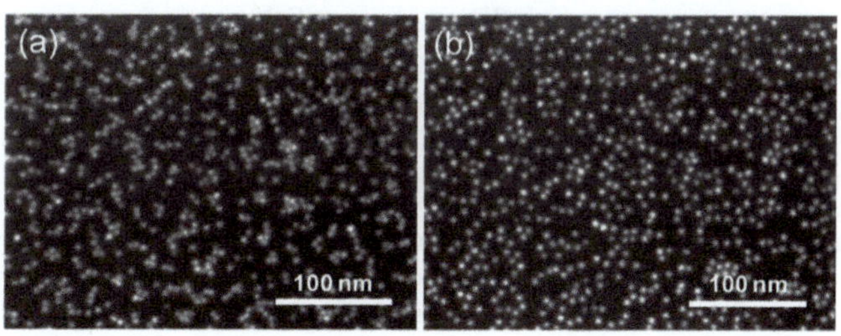

Fig. 5 SEM images (a) before and (b) after CO oxidation of two-dimensional (2D) 4.3 Pt nanoparticle arrays on the n-doped GaN substrates prepared *via* the LB technique.

on the n-doped GaN substrate was somewhat higher than that of the p-doped GaN without light. It is also clear that the activity of the APD Pt nanoparticles on the n-doped GaN wafer decreased by ~28% under light irradiation while the activity of the p-doped GaN increased by ~12% under light irradiation, compared to no irradiation. The activation energies for the APD Pt nanoparticles on the n- or p-doped GaN were obtained from Arrhenius plots of the catalytic activity. The activation energy increased from 26.3 kcal mol^{-1} to 26.5 kcal mol^{-1} for the APD Pt nanoparticles on the n-doped GaN and decreased from 26.3 kcal mol^{-1} to 26.0 kcal mol^{-1} for the Pt nanoparticles on the p-doped GaN, depending on the light irradiation, as shown in Fig. 6 (b).

Fig. 7 summarizes the catalytic activity of the APD Pt nanoparticles on the n- or p-doped GaN supports with and without light at 523 K, respectively. Regardless of the preparation method of the Pt nanoparticles, the activity shows a remarkable

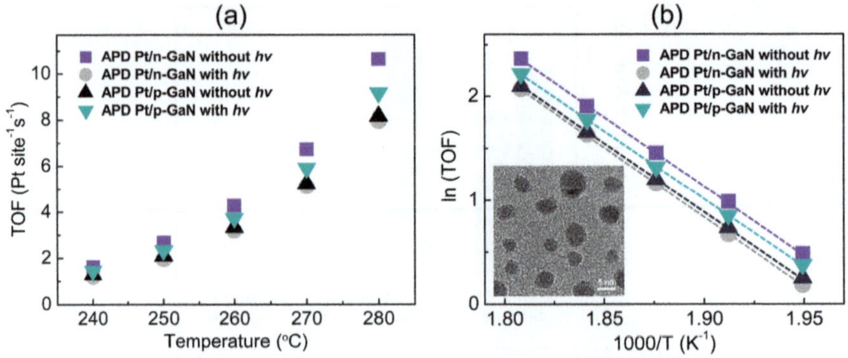

Fig. 6 Catalytic activity of the CO oxidation reaction on Pt nanoparticles deposited on the n- or p-type GaN substrate using APD with and without light: (a) catalytic turnover rate with temperature and (b) Arrhenius plots of APD Pt nanoparticles deposited on the n- or p-type GaN substrate.

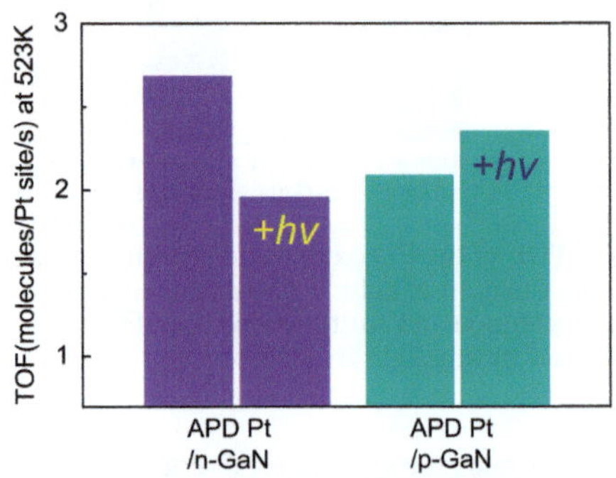

Fig. 7 Comparison of the TOF of the CO oxidation reaction measured on APD Pt nanoparticles/n- or p-type GaN at 523 K with and without light.

difference according to the type of GaN doping. The activity decreased when the nanoparticles were deposited on the n-doped GaN wafer, while the activity increased when the nanoparticles were deposited on the p-doped GaN wafer, where both samples were under light irradiation. Fig. 1 (b) shows a possible mechanism for how the type of semiconductor doping influences catalytic activity. With the given energetic alignment, the recombination of electrons from the Pt nanoparticles with holes from the n-type GaN semiconductor under illumination is expected. Likewise, electron transfer to the Pt nanoparticles is expected for the p-type GaN supports, as shown on the right of Fig. 1 (b). Such interfacial charge transfer processes would lead to a reduced net charge of the Pt nanoparticles on the n-type GaN and an increased net charge of the Pt nanoparticles on the p-type GaN. This change in net charge according to the type of GaN doping will influence the catalytic activity. Schäfer *et al.* also observed enhanced oxidation of the Pt nanoparticles on the p-type GaN substrate.[22] The lower catalytic activity of CO oxidation on the Pt/n-type semiconductor is also consistent with the earlier result by Baker *et al.* who found that catalytic activity decreased on Pt/n-type Si upon exposure to light.[37]

As shown in Fig. 4(b), the percentage of change in catalytic activity under light irradiation is higher on the smaller nanoparticles than on the larger nanoparticles. On smaller nanoparticles, the ballistic transport of hot carriers is facilitated more because of the constant value of the mean free path. Generally, after the hot carriers are generated at the interfaces between the Pt metal nanoparticles and the GaN semiconductor, it is predicted that the smaller the size of particles, the faster the hot carriers can approach the surface of particles, and then participate in the surface chemical reaction, such as CO oxidation, to reduce the charge on the surface. We can conclude, therefore, that the hot carriers are generated at the metal–semiconductor (Pt–GaN) interface under light irradiation, which affects the catalytic activity. This result implies an intrinsic relationship between heterogeneous catalysis and electronic excitation on the surface, with the potential for applications as catalytic actuators using metal-semiconductor hybrid catalysts.

IV Conclusions

CO oxidation on Pt nanoparticles on GaN substrates was carried out to identify the generation of hot carriers and their role in the surface catalytic chemical reaction with and without light. Metal catalysts made up of Pt nanoparticles were prepared using two different preparation methods, the one-pot polyol reduction and arc plasma methods, to investigate the effect of organic capping agents on catalytic activity. As a support, n- or p-type GaN wafers were grown on a sapphire substrate *via* the metal–organic chemical vapor deposition (MOCVD) method. The catalytic activity of the Pt nanoparticles supported on GaN displays a distinct difference according to the doping type of the semiconductor, no matter how the nanoparticles were prepared. The activity decreased when the Pt nanoparticles were deposited on the n-doped GaN, while the activity increased when the Pt nanoparticles were deposited on the p-doped GaN during light irradiation. Based on these results, we conclude that the hot carriers are generated at the Pt–GaN interface during light irradiation and that the two different behaviors of the hot carriers, according to the type of GaN doping, will influence the catalytic activity.

Acknowledgements

This work was supported by the WCU (World Class University) program through the National Research Foundation of Korea (31-2008-000-10055-0), KRF-2012-009249 and by a grant from the Fundamental R&D Program for Core Technology of Materials funded by the Ministry of Knowledge Economy, Republic of Korea.

References

1 J. W. Gadzuk, *J. Phys. Chem. B*, 2002, **106**, 8265–8270.
2 H. Nienhaus, *Surf. Sci. Rep.*, 2002, **45**, 3–78.
3 G. A. Somorjai, K. M. Bratlie, M. O. Montano and J. Y. Park, *J. Phys. Chem. B*, 2006, **110**, 20014–20022.
4 H. Nienhaus, H. S. Bergh, B. Gergen, A. Majumdar, W. H. Weinberg and E. W. McFarland, *Phys. Rev. Lett.*, 1999, **82**, 446–449.
5 J. Y. Park, H. Lee, J. R. Renzas, Y. W. Zhang and G. A. Somorjai, *Nano Lett.*, 2008, **8**, 2388–2392.
6 L. Hellberg, J. Stromquist, B. Kasemo and B. I. Lundqvist, *Phys. Rev. Lett.*, 1995, **74**, 4742–4745.
7 A. Hervier, J. R. Renzas, J. Y. Park and G. A. Somorjai, *Nano Lett.*, 2009, **9**, 3930–3933.
8 X. Z. Ji and G. A. Somorjai, *J. Phys. Chem. B*, 2005, **109**, 22530–22535.
9 J. Y. Park, J. R. Renzas, A. M. Contreras and G. A. Somorjai, *Top. Catal.*, 2007, **46**, 217–222.
10 J. Y. Park and G. A. Somorjai, *ChemPhysChem*, 2006, **7**, 1409–1413.
11 T. C. Shen, C. Wang, G. C. Abeln, J. R. Tucker, J. W. Lyding, P. Avouris and R. E. Walkup, *Science*, 1995, **268**, 1590–1592.
12 E. Fomin, M. Tatarkhanov, T. Mitsui, M. Rose, D. F. Ogletree and M. Salmeron, *Surf. Sci.*, 2006, **600**, 542–546.
13 J. Henzl, M. Mehlhorn, H. Gawronski, K. H. Rieder and K. Morgenstern, *Angew. Chem., Int. Ed.*, 2006, **45**, 603–606.
14 T. Wadayama and M. Yokawa, *Chem. Phys. Lett.*, 2006, **428**, 348–351.

15 J. R. Croy, S. Mostafa, J. Liu, Y. H. Sohn, H. Heinrich and B. R. Cuenya, *Catal. Lett.*, 2007, **119**, 209–216.
16 C. Hippe, R. Lamber, G. Schulz-Ekloff and U. Schubert, *Catal. Lett.*, 1997, **43**, 195–199.
17 T. Schalow, B. Brandt, D. E. Starr, M. Laurin, S. K. Shaikhutdinov, S. Schauermann, J. Libuda and H. J. Freund, *Angew. Chem., Int. Ed.*, 2006, **45**, 3693–3697.
18 J. Stubenrauch and J. M. Vohs, *Catal. Lett.*, 1997, **47**, 21–25.
19 S. J. Tauster, S. C. Fung, R. T. K. Baker and J. A. Horsley, *Science*, 1981, **211**, 1121–1125.
20 S. J. Tauster, *Acc. Chem. Res.*, 1987, **20**, 389–394.
21 S. J. Tauster, S. C. Fung and R. L. Garten, *J. Am. Chem. Soc.*, 1978, **100**, 170–175.
22 S. Schafer, S. A. Wyrzgol, R. Caterino, A. Jentys, S. J. Schoell, M. Havecker, A. Knop-Gericke, J. A. Lercher, I. D. Sharp and M. Stutzmann, *J. Am. Chem. Soc.*, 2012, **134**, 12528–12535.
23 S. M. Kim, S. J. Lee, S. H. Kim, S. Kwon, K. J. Yee, H. Song, G. A. Somorjai and J. Y. Park, *Nano Lett.*, 2013, **13**, 1352–1358.
24 S. S. Dhesi, C. B. Stagarescu, K. E. Smith, D. Doppalapudi, R. Singh and T. D. Moustakas, *Phys. Rev. B: Condens. Matter*, 1997, **56**, 10271–10275.
25 B. J. Kowalski, R. J. Iwanowski, J. Sadowski, J. Kanski, I. Grzegory and S. Porowski, *Surf. Sci.*, 2002, **507–510**, 186–191.
26 M. Stutzmann, O. Ambacher, M. Eickhoff, U. Karrer, A. L. Pimenta, R. Neuberger, J. Schalwig, R. Dimitrov, P. J. Schuck and R. D. Grober, *Phys. Status Solidi B*, 2001, **228**, 505–512.
27 G. A. Somorjai and J. Y. Park, *Top. Catal.*, 2008, **49**, 126–135.
28 J. Y. Park, C. Aliaga, J. R. Renzas, H. Lee and G. A. Somorjai, *Catal. Lett.*, 2009, **129**, 1–6.
29 Z. R. Ismagilov, O. Y. Podyacheva, O. P. Solonenko, V. V. Pushkarev, V. I. Kuz'min, V. A. Ushakov and N. A. Rudina, *Catal. Today*, 1999, **51**, 411–417.
30 C. J. Liu, G. P. Vissokov and B. W. L. Jang, *Catal. Today*, 2002, **72**, 173–184.
31 H. Shim, J. Phillips, I. M. Fonseca and S. Carabineiro, *Appl. Catal., A*, 2007, **316**, 259–259.
32 A. L. Thomann, J. P. Rozenbaum, P. Brault, C. Andreazza, R. Andreazza, B. Rousseau, H. Estrade-Szwarckopf, A. Berthet, J. C. Bertolini, F. J. C. S. Aires, F. Monnet, C. Mirodatos, C. Charles and R. Boswell, *Pure Appl. Chem.*, 2002, **74**, 471–474.
33 J. Y. Yun, A. S. Reddy, S. Yang, H. J. Kim, H. Y. Koo, H. M. Lee, C. H. Jung, K. Qadir, S. Kim and J. Y. Park, *Catal. Lett.*, 2012, **142**, 326–331.
34 H. Song, R. M. Rioux, J. D. Hoefelmeyer, R. Komor, K. Niesz, M. Grass, P. D. Yang and G. A. Somorjai, *J. Am. Chem. Soc.*, 2006, **128**, 3027–3037.
35 C. K. Tsung, J. N. Kuhn, W. Y. Huang, C. Aliaga, L. I. Hung, G. A. Somorjai and P. D. Yang, *J. Am. Chem. Soc.*, 2009, **131**, 5816–5822.
36 K. Qadir, S. H. Kim, S. M. Kim, H. Ha and J. Y. Park, *J. Phys. Chem. C*, 2012, **116**, 24054–24059.
37 L. R. Baker, A. Hervier, G. Kennedy and G. A. Somorjai, *Nano Lett.*, 2012, **12**, 2554–2558.

Switching-off toluene formation in the solvent-free oxidation of benzyl alcohol using supported trimetallic Au–Pd–Pt nanoparticles

Qian He,[a] Peter J. Miedziak,[b] Lokesh Kesavan,[b] Nikolaos Dimitratos,[b] Meenakshisundaram Sankar,[b] Jose Antonio Lopez-Sanchez,[b] Michael M. Forde,[b] Jennifer K. Edwards,[b] David W. Knight,[b] Stuart H. Taylor,[b] Christopher J. Kiely*[a] and Graham J. Hutchings*[b]

Received 17th December 2012, Accepted 2nd January 2013
DOI: 10.1039/c2fd20153d

Trimetallic Au–Pd–Pt nanoparticles have been supported on activated carbon by the sol-immobilisation method. They are found to be highly active and selective catalysts for the solvent-free aerobic oxidation of benzyl alcohol. The addition of Pt promotes the selectivity to the desired product benzaldehyde at the expense of toluene formation. Detailed aberration corrected STEM-XEDS analysis confirmed that the supported particles are indeed Au–Pd–Pt ternary alloys, but also identified composition fluctuations from particle-to-particle which vary systematically with nanoparticle size.

Introduction

The oxidation of primary alcohols to aldehydes is an industrially important process, as aldehydes are valuable as intermediates and as high value components for the perfume industry. Conventionally this reaction is carried out using expensive and toxic oxidants such as chromate or permanganate, causing serious environmental and economical concerns.[1–3] Hence, catalytic oxidation of primary alcohols to aldehydes using molecular oxygen/air, or other greener oxidants such as hydrogen peroxide, becomes an attractive alternative. We have previously demonstrated that Au–Pd bimetallic nanoparticles can be very active for the alcohol oxidation reaction, and provide high selectivity toward aldehydes even under solvent-free reaction conditions.[4–8] Au–Pd catalysts are generally prepared *via* co-impregnation[4] or sol-immobilization[5–8] methods, the latter of which provides much better control over the particle size and allows the possibility of tuning the particle morphology at will, to generate core–shell structures or homogeneous random Au–Pd alloys.[9] However previously, in order to achieve high selectivity towards the aldehyde, either a low reaction temperature approach was chosen resulting in a low turnover frequency (TOF), or the conversion was kept at a low level. Herein, using benzyl alcohol oxidation as an example, we demonstrate that when a small amount of Pt metal is alloyed into the Au–Pd sols, a high selectivity toward benzaldehyde can be achieved in the solvent-free oxidation of benzyl alcohol reaction while still preserving high conversion levels.

[a]*Department of Materials Science and Engineering, Lehigh University, 5 East Packer Avenue, Bethlehem, Pennsylvania, PA 18015, USA. E-mail: chk5@lehigh.edu*
[b]*Cardiff Catalysis Institute, School of Chemistry, Cardiff University, Main Building, Park Place, Cardiff, CF10 3AT, UK. E-mail: hutch@cardiff.ac.uk*

Experimental details

Catalyst preparation – Sol-immobilisation method

Au-only, Pd-only, Pt-only monometallic catalysts, as well as Au–Pd, Au–Pt, Pt–Pd bimetallic catalysts and Au–Pd–Pt trimetallic catalysts supported on activated carbon were prepared using a colloidal method. For the synthesis of the trimetallic Au–Pd–Pt supported catalysts, an aqueous solution of $PdCl_2$, $HAuCl_4 \cdot 3H_2O$ and H_2PtCl_6 (Johnson Matthey, 99.9% purity) of the desired concentration was prepared. To this solution, polyvinyl alcohol (PVA) (1 wt % solution, Aldrich, weight average molecular weight M_w = 9000–10 000 g mol^{-1}, 80% hydrolysed) was added (PVA/Au (wt/wt) = 1.2). Subsequently, 0.1 M of freshly prepared $NaBH_4$ solution (>96% purity, Aldrich, $NaBH_4$/Au (mol/mol) = 5) was then introduced to form a dark-brown sol. After 30 min of sol generation, the colloid was immobilised by adding the activated carbon (G60 Aldrich) support material (acidified to pH 1 using sulfuric acid) under vigorous stirring conditions. The amount of support material required was calculated so as to give a total final metal loading of 1% wt. After 2 h the slurry was filtered, the catalyst washed thoroughly with distilled water and then dried at 120 °C overnight.

Catalyst preparation – Impregnation method

For comparative purposes Au-only, Pd-only and Au–Pd, Pt–Pd and Au–Pd–Pt catalysts were also prepared supported on titania (P25 Degussa) or carbon (G60 Aldrich), using a conventional wet impregnation technique. The trimetallic Au–Pd–Pt material was prepared by impregnation of the support using aqueous solutions of $PdCl_2$, $HAuCl_4 \cdot 3H_2O$ and $PtCl_2$ (Johnson Matthey). For instance, the detailed procedure for the preparation of 2 g of the 5 wt% (2.27Au : 2.27Pd : 0.45Pt)/TiO_2 catalyst is as follows: $PdCl_2$ (75.7 mg) was dissolved in a stirred and heated aqueous solution (4.55 ml) of $HAuCl_4 \cdot 3H_2O$ (5 g in 250 ml water) and an aqueous solution (0.09 ml) of $PtCl_2$ (1 g in 10 ml water). The resultant solution was added to TiO_2 (1.9 g) and the slurry formed was dried at 110 °C for 16 h. The recovered powder was ground and calcined (1 g, 6 inch quartz boat) in static air at 400 °C for 3 h at a ramp rate of 20 °C min^{-1}.

Catalytic testing – Benzyl alcohol oxidation

Benzyl alcohol oxidation was carried out in a stirred reactor (100 ml, Parr reactor). The vessel was charged with alcohol (40 ml) and catalyst (0.05 g). The autoclave was then purged 5 times with oxygen leaving the vessel at 10 bar gauge. The stirrer was set at 1500 rpm and the reaction mixture was raised to the required temperature. Samples from the reactor were taken periodically *via* a sampling system. For the analysis of the products, a GC-MS and GC (a Varian star 3400 cx with a 30 m CP-Wax 52 CB column) were employed. The products were identified by comparison with known standards. For the quantification of the amounts of reactants consumed and products generated, the external calibration method was used.

Catalyst characterisation

The aqueous sols, contained in a quartz cuvette, were optically characterised using a UV-vis spectrometer (V-570, JASCO) operating over the 200 to 900 nm wavelength range. Samples for examination by scanning transmission electron microscopy (STEM) were prepared by dispersing the dry catalyst powder onto a holey carbon film supported by a 300 mesh copper TEM grid. STEM high angle annular dark field (HAADF) images of the metallic particles were obtained using an aberration corrected JEOL 2200FS (S)TEM operating at 200 kV. X-ray energy dispersive (XEDS) spectra were acquired from individual nanoparticles larger than 2 nm in size while rastering the beam over the entire particle, using a Thermo Noran

Si(Li) XEDS detector. STEM-XEDS spectrum imaging experiments were carried out at 200kV using an aberration corrected JEM-ARM 200F microscope equipped with a JEOL Centurio SDD XEDS spectrometer.

Results and discussion

Supported Au–Pd–Pt trimetallic catalysts were synthesized by the sol-immobilization method as described above. This process involved the co-reduction of three metal precursors ($HAuCl_4$, $PdCl_2$ and H_2PtCl_6) at the desired concentration in aqueous solutions, using sodium borohydride as chemical reductant and polyvinyl alcohol (PVA) as the protective ligand. The synthesised Au–Pd–Pt colloids were analysed by UV-vis spectrometry (Fig. 1) and the spectrum obtained did not show a characteristic plasmon resonance band for gold, suggesting the formation of trimetallic alloy particles. The colloids were then immobilized onto an activated carbon support, followed by a washing and 120 °C drying process to yield the final catalysts with 1 wt% total metal loading. For comparative purposes Au-only, Pd-only, Pt-only catalysts, as well as Au–Pd, Au–Pt and Pt–Pd bimetallic catalysts were also prepared using a similar sol-immobilization procedure.

Oxidations were carried out in a stirred autoclave reactor using solvent-free conditions (40 ml of benzyl alcohol), at 120–140 °C, and $pO_2 = 10$ bar. The catalytic results are given in Table 1, and Table 2 presents the catalytic data at iso-conversion levels. The oxidation of benzyl alcohol with O_2 under solvent-free conditions proceeds efficiently with very high catalytic activity using C supported Au–Pd nanoparticles, with TOF values of 50 000 to 60 000 h^{-1} being reached. The main product is benzaldehyde and the major by-product is toluene, which primarily comes from the disproportionation of benzyl alchohol[10–12] as well as *via* the hydrogenolysis pathway.[13] It is considered that Pd sites are responsible for the disproportionation reaction.[10–12] However, we have recently reported that this disproportionation reaction can be controlled to some extent by the choice of catalyst support when using Au–Pd catalysts, as the disproportionation of benzyl alcohol can be switched-off simply by using basic support materials (*e.g.* ZnO, MgO).[8] While carbon supported

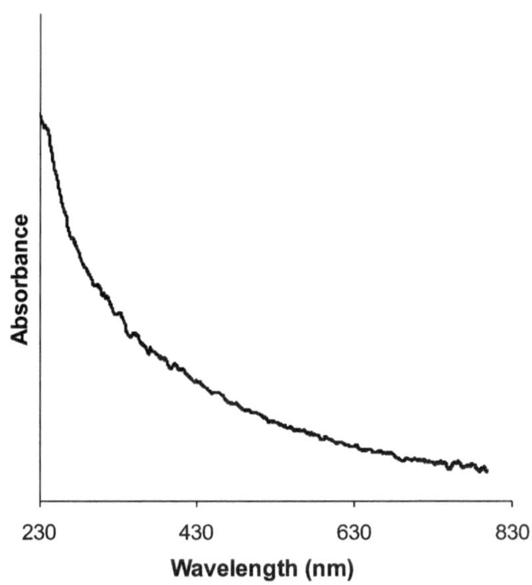

Fig. 1 Typical UV-vis spectrum from a Au–Pd–Pt aqueous sol preparation (0.45 : 0.45 : 0.1 by weight in this case).

Table 1 Benzyl alcohol oxidation results after 2 h reaction with Pd-only, Pt-only, Au–Pd, Au–Pt, Pt–Pd and various Au–Pd–Pt catalysts supported on carbon prepared by the sol-immobilisation method[a]

Catalyst	Conversion (%)	Selectivity (%) Toluene	Benzaldehyde	Benzoic acid	Benzyl benzoate	TOF/ h^{-1}[b]
1 wt% (0.65Au + 0.35Pd)/C	71.1	4.0	69.8	19.9	6.3	52 700
1 wt% (0.5Au + 0.5Pd)/C	80.7	3.4	67.0	23.1	6.5	63 800
1 wt% (0.3Au + 0.4Pd + 0.3Pt)/C	35.4	0.9	83.3	7.9	7.9	16 000
1 wt% (0.4Au + 0.4Pd + 0.2Pt)/C	36.7	0.6	81.7	8.0	9.7	22 100
1 wt% (0.45Au + 0.45Pd + 0.1Pt)/C	53.9	0.1	80.2	13.1	6.6	31 900
2 wt% (1Au + 1Pt)/C	11.8	2.0	79.9	8.4	9.7	4770
1 wt% (0.5Au + 0.5Pt)/C	9.7	1.1	84.9	3.1	10.9	8450
1 wt% (0.5Pd + 0.5Pt)/C	14.8	0.7	84.7	5.7	8.9	9580
1 wt% Pt/C	2.8	0.0	90.7	0.7	8.6	2910
1 wt% Pd/C	59.3	6.8	74.7	10.4	8.1	12 200

[a] *Reaction conditions*: benzyl alcohol, 0.05g of catalyst, $T = 120$ °C, $pO_2 = 150$ psi, stirring rate 1500 rpm. [b] Calculation of TOF (h^{-1}) value was done after 0.5 h of reaction, and was based on the total metal loading.

Table 2 Comparison of benzyl alcohol oxidation at iso-conversion of 30% with Pd-only, Au–Pd and various Au–Pd–Pt catalysts supported on carbon prepared by the sol-immobilisation method[a]

Catalyst	Selectivity (%) Toluene	Benzaldehyde	Benzoic acid	Benzyl benzoate
1% (0.65Au + 0.35Pd)/C	8.1	80.5	2.6	8.8
1% (0.5Au + 0.5Pd)/C	6.9	81.5	3.6	8.0
1% (0.3Au + 0.4Pd + 0.3Pt)/C	0.8	85.7	6.5	7.0
1% (0.4Au + 0.4Pd + 0.2Pt)/C	0.6	83.3	6.8	9.3
1% (0.45Au + 0.45Pd + 0.1Pt)/C	0.4	86.7	5.9	7.0
1% Pd/C	5.6	78.7	5.8	9.9

[a] *Reaction conditions*: benzyl alcohol, 0.05g of catalyst, $T = 120$ °C, $pO_2 = 150$ psi, stirring rate 1500 rpm.

Au–Pd nanoparticles were always found to catalyse the disproportionation reaction, we observed a very low selectivity towards toluene for carbon supported monometallic Pt and bimetallic Au–Pt (1.1%) and Pd–Pt (0.7%) catalysts. However these desirable selectivities were accompanied by low benzyl alcohol conversion levels of 2.8%, 9.7% and 14.8% respectively. Taking into account the fact that the highest conversion level was observed with the Au–Pd/C catalysts, it was natural to consider the possibility of maintaining high activity while simultaneously keeping the selectivity to toluene at a very low level, simply by incorporating Pt into the Au–Pd supported catalyst. Consequently, we synthesised a series of trimetallic Au–Pd–Pt nanoparticles supported on carbon in which the amount of Pt metal added was systematically varied. By comparing three trimetallic catalyst formulations, namely (0.3Au + 0.4Pd + 0.3Pt)/C, (0.4Au + 0.4Pd + 0.2Pt)/C, (0.45Au + 0.45Pd + 0.1Pt)/C, it is clear from Table 1 that, by decreasing the amount of Pt incorporated, an increase in benzyl alcohol conversion was observed, whereas the selectivity to toluene remained below 1% in all cases. The highest selectivity toward benzaldehyde, measured at iso-conversion levels, was found for the 1 wt% (0.45Au + 0.45Pd + 0.1Pt)/C catalyst where the formation of toluene was almost completely switched-off. These results clearly demonstrate that only a small amount of Pt is needed for the effective termination of toluene formation. The best of our knowledge, this is the first report of the effective blocking of toluene formation under solvent-free conditions using Au–Pd nanoparticles modified with the optimum amount of a third alloying component.

Selected catalysts were characterized by aberration corrected scanning transmission electron microscopy (AC-STEM) in order to evaluate their microstructure. Fig. 2 shows representative low magnification STEM-HAADF images and the corresponding measured particle size distribution of various trimetallic (Au–Pd–Pt/C), bimetallic (Au–Pd/C, Au–Pt/C and Pt–Pd/C) and monometallic (Pt/C) catalyst samples analysed in this study. To produce each histogram 500–1000 particles were analysed in each survey. It was found that the trimetallic (Fig. 2a and b) and bimetallic (Fig. 2c–e) samples analyzed all have quite similar particle size distributions with mean diameters in the 2–3 nm range. Although some minor coarsening effects were noted as a result of the 120 °C drying process, especially for the 1% (0.5Au + 0.5Pt)/C (Fig. 2e) and 1% Pt/C samples (Fig. 2f), the vast majority of the particles in each sample resided in the 1–3 nm size range. For these reasons, particle size variation issues from sample-to-sample should not be too significant when comparing the catalytic activities of these samples.

It proved possible to collect reasonable quality X-ray energy dispersive point spectra (XEDS) from individual nanoparticles larger than about 2 nm in size in our JEM 2200FS microscope which was equipped with a standard Si(Li) XEDS detector.[14] Representative STEM-HAADF and XEDS data from small (~2 nm), intermediate size (~5 nm), and larger (~10 nm) metal particles are shown in Fig. 3a–c respectively for the most active 1 wt% (0.45Au + 0.45Pd + 0.1Pt)/C catalyst. From the relative intensity ratios between the overlapping Au and Pt M series and Pd L series, a clear size-dependent composition variation can be seen. The larger (~10 nm) particles tend to be Pd-rich, while the smaller (~2 nm) particles were consistently found to be Pd-deficient. This same general size/composition trend was also noted for all the Pd-containing alloys analyzed in the current study; (*i.e.* the 1 wt% (0.3Au + 0.4Pd + 0.3Pt)/C trimetallic sample (Fig. 4); the 1 wt% (0.5Pd + 0.5Pt)/C bimetallic sample (Fig. 5); and the 1 wt% (0.5Au + 0.5Pd)/C bimetallic sample (Fig. 6). It is interesting to note that no comparable composition/size variation was found for the particles in the Au–Pt binary system (Fig. 7). It has been reported previously by Braidy *et al.*[15] that similar size-dependent composition variations in bimetallic Au–Ag nanoparticles can be generated as an artifact in the electron microscope if one of the constituent elements has a tendency to be depleted preferentially from the particle by electron beam bombardment. Since only Pd, but not Au or Pt, has a knock-on damage threshold below 200 keV,[16] we examined the possibility of beam damage occurring by taking two consecutive XEDS spectra from a single Au–Pd–Pt particle about 2 nm in size. As shown in Fig. 8, even though significant structural modification can be clearly seen from the consecutive HAADF images, the two sequentially obtained XEDS spectra gave similar composition results within the limits of experimental error. This suggests that the composition variation noted in our study is inherent to the material itself and not an artifact of electron beam damage. This particle size/composition trend with respect to Pd concentration is also consistent with XEDS composition measurements presented in our previous study of the sol-immobilized bimetallic Au–Pd system.[14] Interestingly, as shown in Fig. 3 and 4

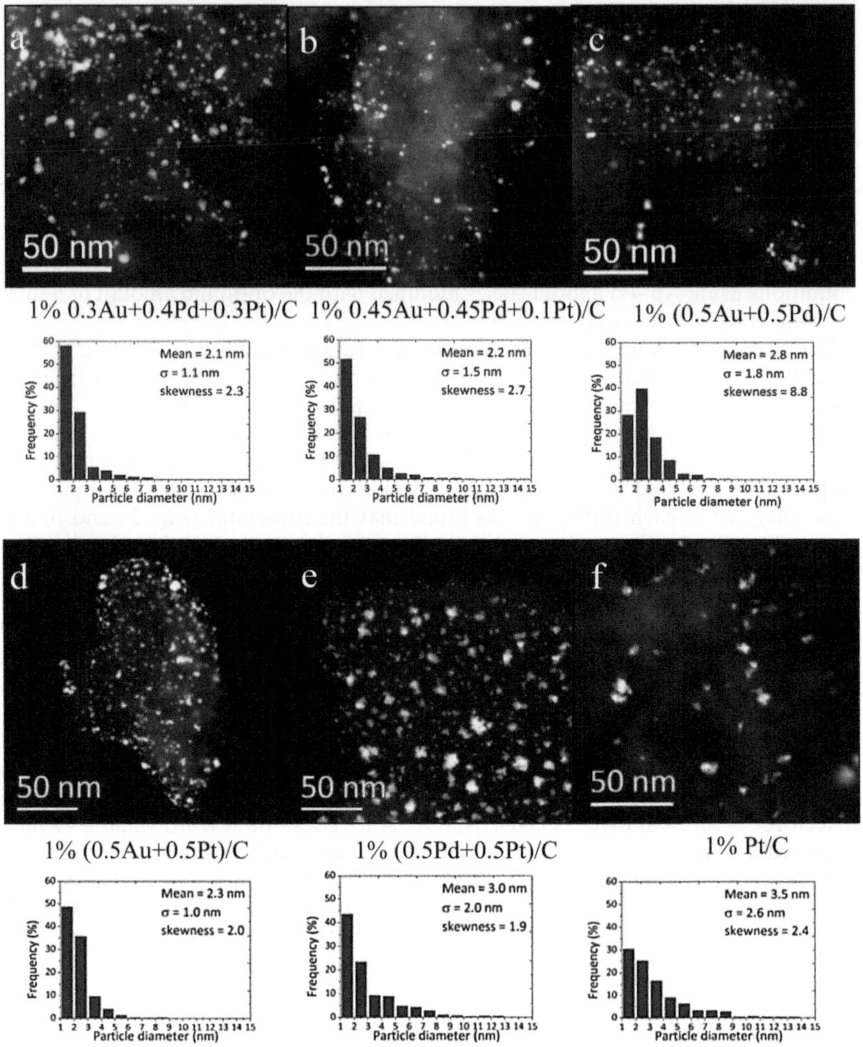

Fig. 2 Representative low magnification STEM-HAADF images and corresponding particle size distributions of selected trimetallic, bimetallic and monometallic catalysts characterized in this study; namely; (a) 1 wt% (0.3Au + 0.4Pd + 0.3Pt)/C; (b) 1 wt% (0.45Au + 0.45Pd + 0.1Pt)/C; (c) 1 wt% (0.5Au + 0.5Pd)/C; (d) 1 wt% (0.5Au + 0.5Pt)/C; (e) 1 wt% (0.5Pd + 0.5Pt)/C and (f) 1 wt% Pt/C.

for the Au–Pt–Pd trimetallic samples, and Fig. 7 for the bimetallic 1 wt% 0.5Au + 0.5Pt/C sample, the Pt L series peak can be detected in all spectra and the relative intensities of the Au L_α (9.71 keV) to Pt L_α (9.44 keV) peaks was found to be relatively constant with particle size, suggesting that no systematic particle size/composition variation develops for this particular pair of elements.

The elemental distributions within individual particles, which have a profound effect on the resultant catalytic activities, can also be probed using STEM. It has been demonstrated by our group[9] and others[17–19] that Pd (Z = 46) segregation with respect to Au (Z = 79) or Pt (Z = 78) in multi-metallic nanoparticles can be easily distinguished by STEM-HAADF imaging, which provides mass-thickness contrast approximately proportional to Z^2. From the HAADF images presented

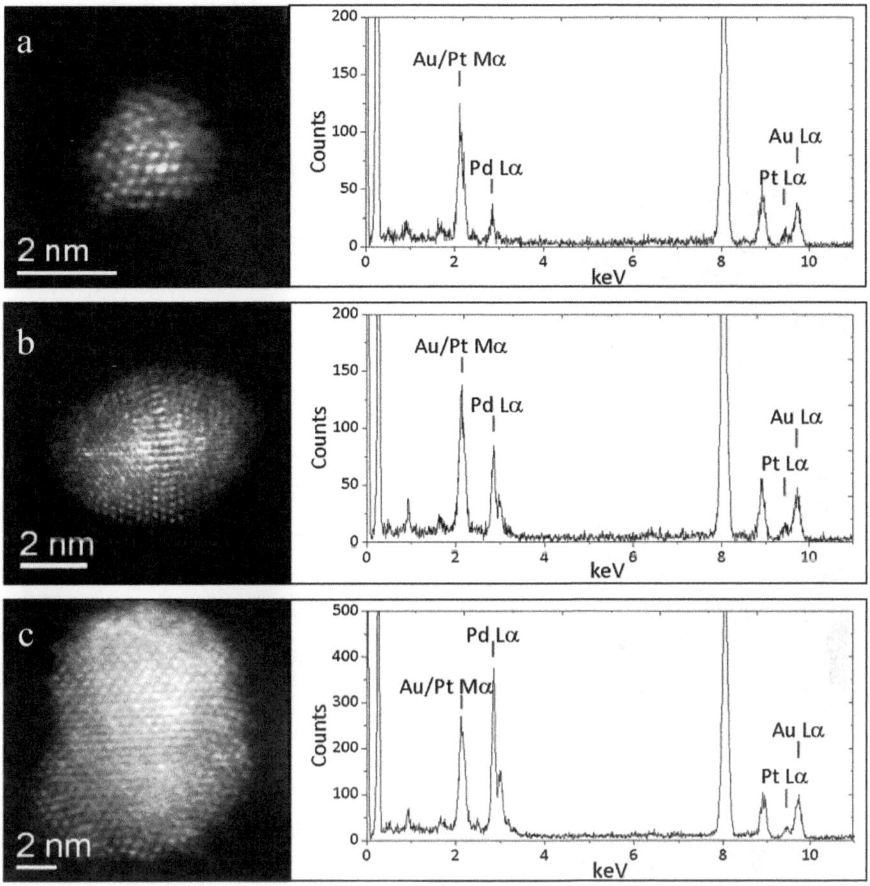

Fig. 3 High magnification STEM-HAADF images and the corresponding X-ray energy dispersive spectra acquired from representative (a) 2 nm, (b) 5 nm and (c) 10 nm particles in the sol-immobilized 1 wt% (0.45Au + 0.45Pd + 0.1Pt)/C catalyst dried at 120 °C.

in Fig. 3, 4, 5, 6 and 7 no Pd segregation or chemical ordering was detected within trimetallic or bimetallic alloy nanoparticles containing Pd, suggesting that the Pd atoms were intimately mixed with the Au and Pt components. Indeed, this homogeneous alloy morphology is to be expected for materials generated by the co-reduction colloidal route and is consistent with our previous studies.[9,14] Although the HAADF *Z*-contrast between Au and Pt is minimal, the relative distribution of Au and Pt within individual alloy nanoparticles can in principle be probed by STEM-XEDS mapping. However, in practice, small nanoparticles are rapidly damaged when subjected to an intense electron beam, whereas the use of low probe currents provides insufficient X-ray generation for XEDS analysis. Furthermore, conventional Si(Li) XEDS detectors only have a very small solid angle of collection (*i.e.* ~0.13 sr) which captures less than 1% of the total characteristic X-rays emitted. For these two reasons, STEM-XEDS mapping on nanoparticles smaller than about 6 nm is extremely challenging in our aberration corrected JEOL 2200FS (S)TEM. This situation has dramatically improved recently with the installation of our aberration corrected JEM-ARM 200F (S)TEM which is equipped with a Centurio SDD XEDS spectrometer, having a collection solid angle of 0.9 sr, which represents a 5–10 fold improvement in X-ray collection efficiency.[20] Therefore it has now become feasible to routinely acquire STEM-XEDS mapping data on nanoparticles

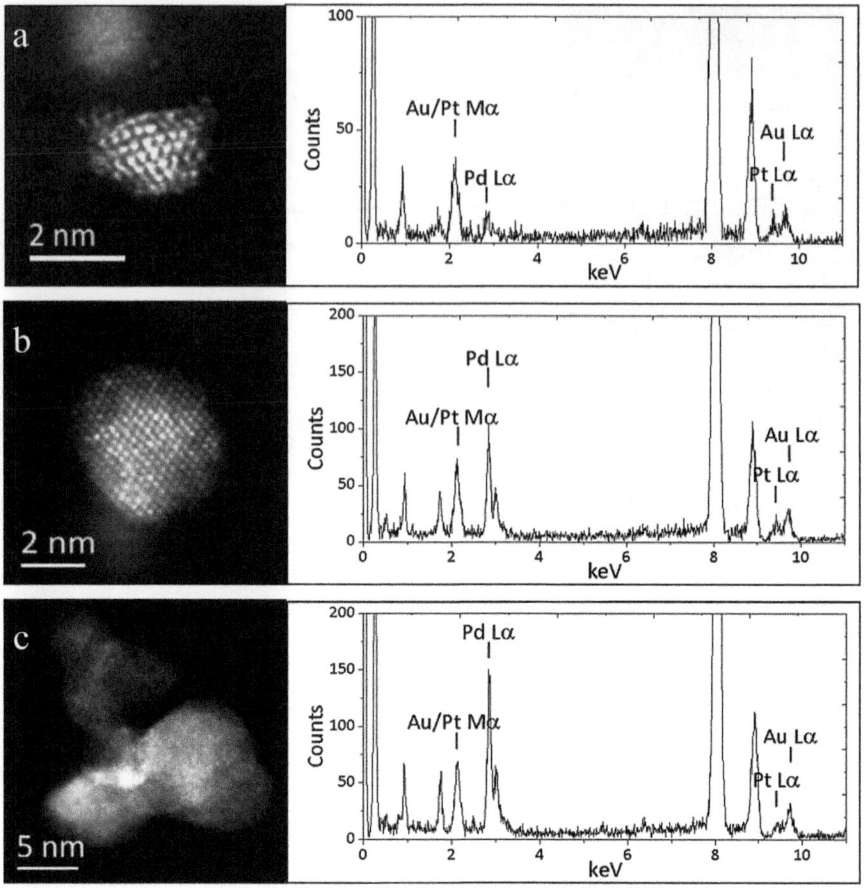

Fig. 4 Representative high magnification STEM-HAADF images and the corresponding X-ray energy dispersive spectra acquired from representative (a) 2 nm, (b) 5 nm and (c) 10 nm particles in the sol-immobilized 1 wt% (0.3Au + 0.4Pd + 0.3Pt)/C trimetallic catalyst dried at 120 °C.

as small as 2–3 nm. Fig. 9 shows representative STEM-XEDS spectrum images of 3 nm, 5 nm and 10 nm diameter alloy particles in the 1 wt% (0.3Au + 0.4Pd + 0.3Pt)/C sol-immobilized catalyst taken using this new instrument. This sample was deliberately chosen, instead of the most active 1 wt% (0.45Au + 0.45Pd + 0.1Pt)/C catalyst, because it had the highest Pt loading allowing the distribution of all three elements to be more clearly visualized. These results conclusively demonstrate that regardless of size, the particles are all homogenous trimetallic alloys which show no apparent signs of elemental segregation. It is also reasonable to extrapolate that even for the catalyst with the most dilute Pt loading (*i.e.* 1 wt% (0.45Au + 0.45Pd + 0.1Pt)/C), the Pt atoms get randomly incorporated into the Au–Pd nanoparticles. Even though a wide miscibility gap exists in the ‘bulk’ binary phase diagram for the Au–Pt binary system,[21] both experimental[22] and theoretical[23,24] investigations now suggest that a wide range of homogeneous alloy compositions between Pt, Au and Pd can be prepared and stabilized at low temperature in the nano-regime.

The beneficial effects of adding some Pt into Au–Pd nanoparticles, in terms of activity and selectivity for the benzyl alcohol oxidation reaction, are quite remarkable. The incorporated Pt atoms seem to act as either as an electronic or strain modifier to the Au–Pd nanoparticles, which nullifies the Pd sites that are normally active

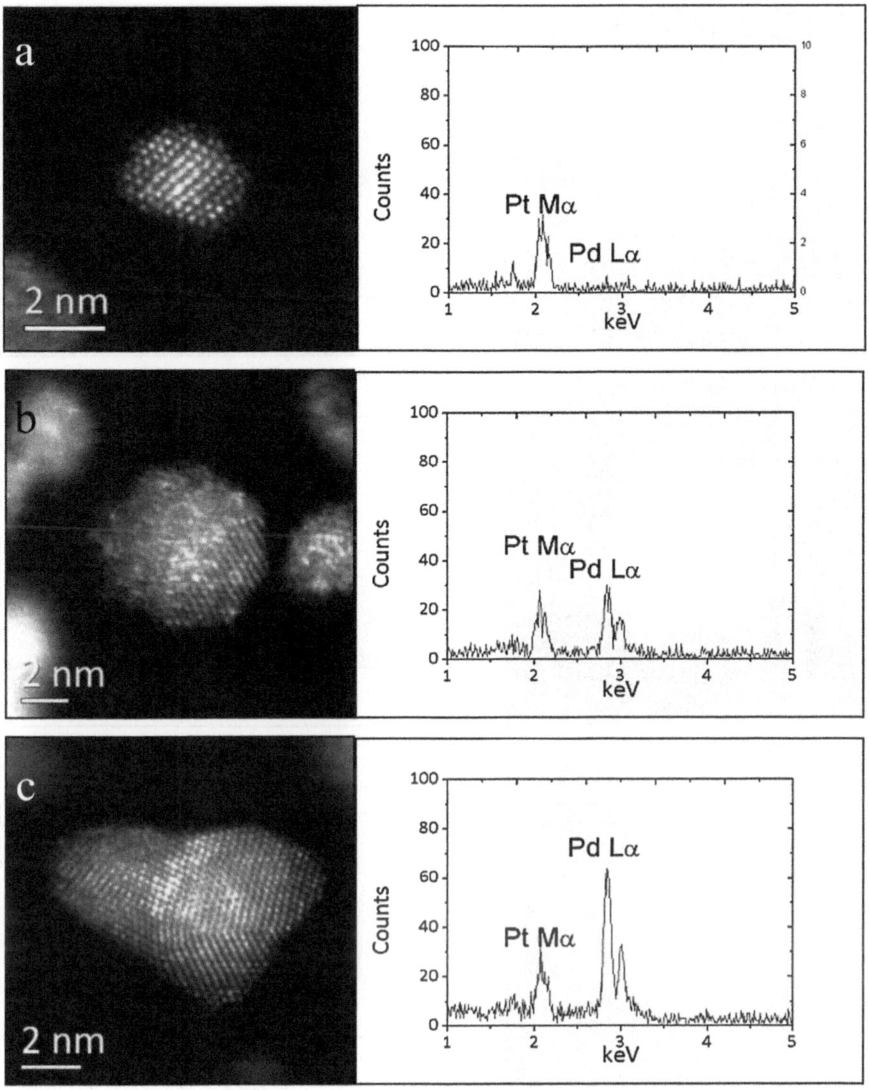

Fig. 5 Representative high magnification STEM-HAADF images and the corresponding X-ray energy dispersive spectra acquired from representative (a) 2 nm, (b) 5 nm and (c) 10 nm particles in the sol-immobilized 1 wt% (0.5Pd + 0.5Pt)/C bimetallic catalyst dried at 120 °C.

for disproportionation reactions. One issue that hinders a more detailed mechanistic understanding of this process is that the co-reduction colloidal preparation route employed provides only limited composition control as the larger particles are consistently Pd-rich, while the smaller ones are always Pd-deficient. Therefore although the nominal composition of the trimetallic catalysts can be modified by adjusting the molar ratios of the three metal precursors, the resultant materials still contain a complex distribution of particle sizes and compositions. One possible cause of such size dependent composition variations in nanoalloys was proposed by Alloyeau *et al.*,[25] who found that in Pt–Co nanoparticles, the Co has a higher rate of Ostwald ripening than the Pt component on the surface of the carbon

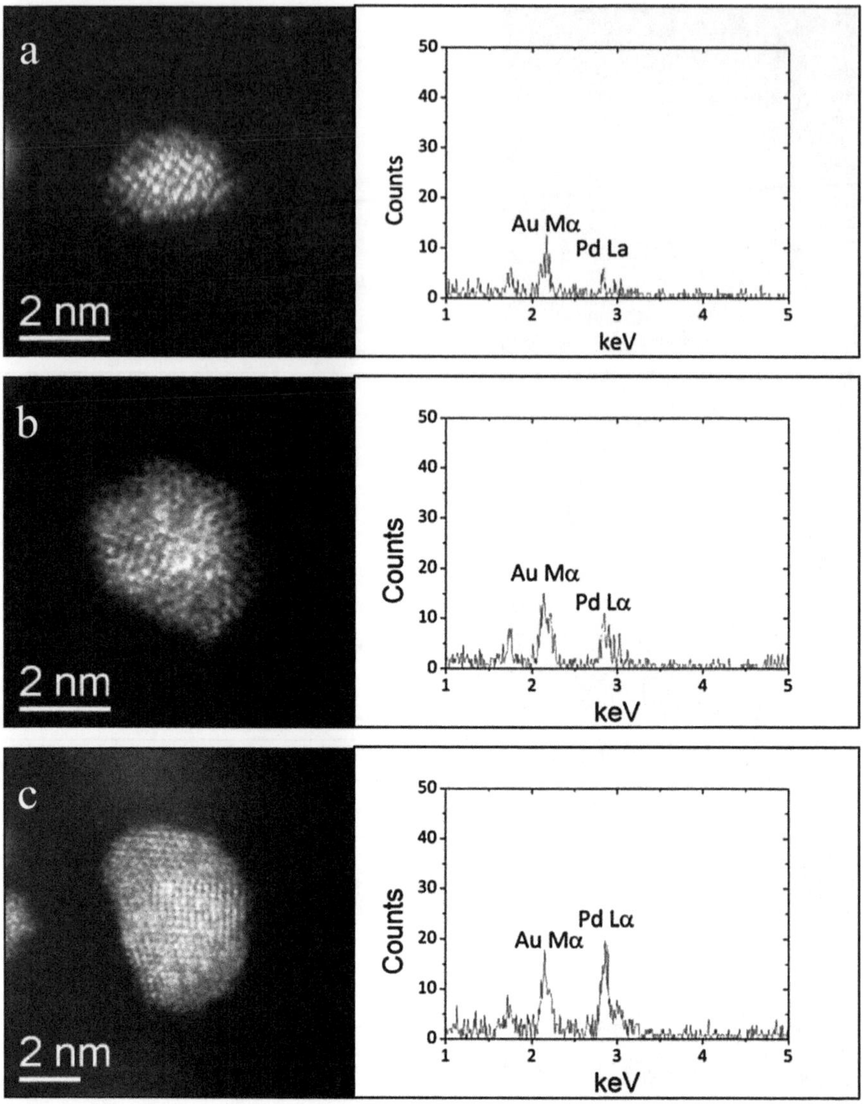

Fig. 6 Representative high magnification STEM-HAADF images and the corresponding X-ray energy dispersive spectra acquired from representative (a) 2 nm, (b) 5 nm and (c) 10 nm particles in the sol-immobilized 1 wt% (0.5Au + 0.5Pd)/C bimetallic catalyst dried at 120 °C.

support, causing the Co to preferentially segregate onto larger particles after high temperature (*c.a.* 700 °C) heat treatments. If this type of effect were happening in our situation, we suspect that the selective Ostwald ripening process for Pd is more likely to occur in the solution phase during the sol synthesis process instead of while on the carbon support after immobilization, because the catalyst only underwent a very mild drying treatment at 120 °C. Hence, the ability to synthesize trimetallic Au–Pd–Pt nanoparticles where there is a tighter control over the composition from particle-to-particle, while preserving the mean particle size below 5 nm, still remains as a significant technical challenge.

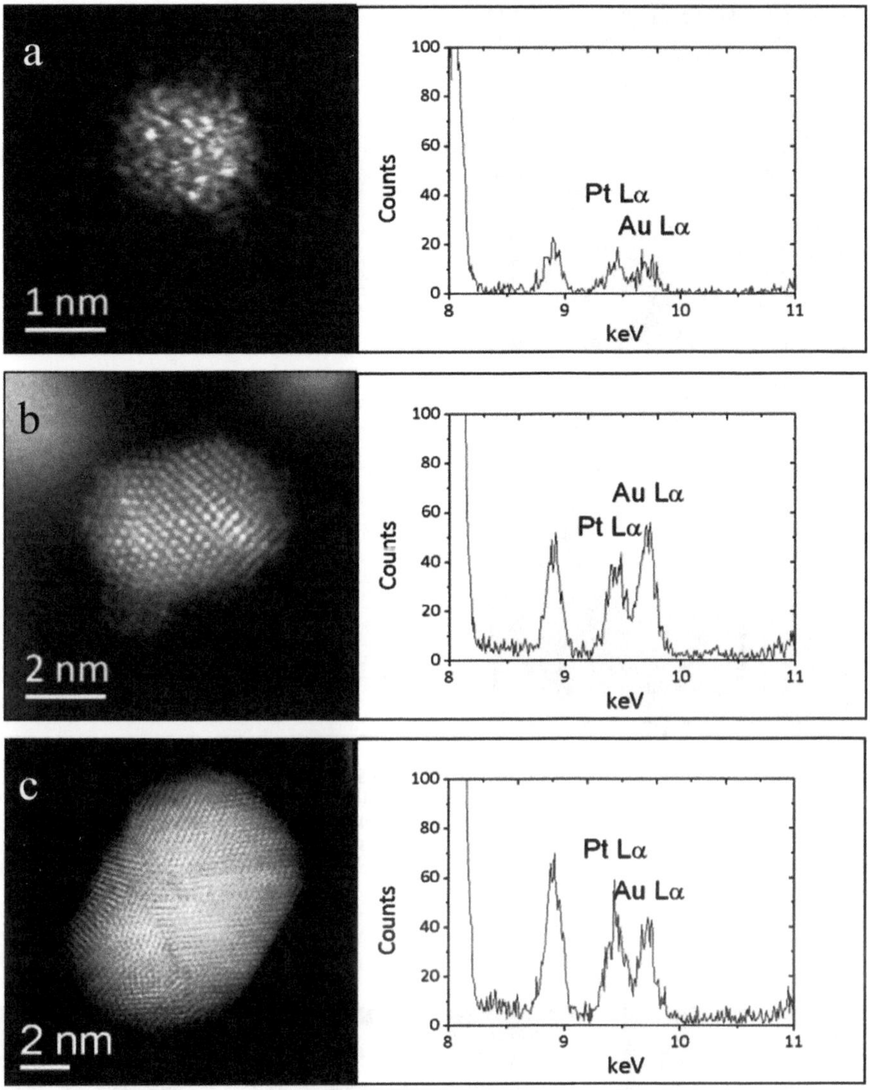

Fig. 7 Representative high magnification STEM-HAADF images and the corresponding X-ray energy dispersive spectra acquired from representative (a) 2 nm and (b) 5 nm and (c) 10 nm particles in the sol-immobilized 1 wt% (0.5Au + 0.5Pt)/C bimetallic catalyst dried at 120 °C.

Finally, we have also extended the scope of using Pt as a third alloying element by synthesizing trimetallic Au–Pd–Pt nanoparticles supported on C and TiO_2 using a conventional impregnation method. This synthesis method would be preferred by industry due to its greater simplicity, so we investigated if the same catalytic trends could be obtained with impregnation. The catalytic results for the impregnated catalysts are presented in Table 3, and it is evident that the impregnated trimetallic catalysts also significantly decreased the formation of toluene. However, it should be noted that the catalytic activity of the impregnated supported Au–Pd–Pt catalysts are much lower than the corresponding sol-immobilized catalysts, probably due to (i) a significant increase in mean particle size and (ii) wider composition variations as compared to the colloid derived materials.

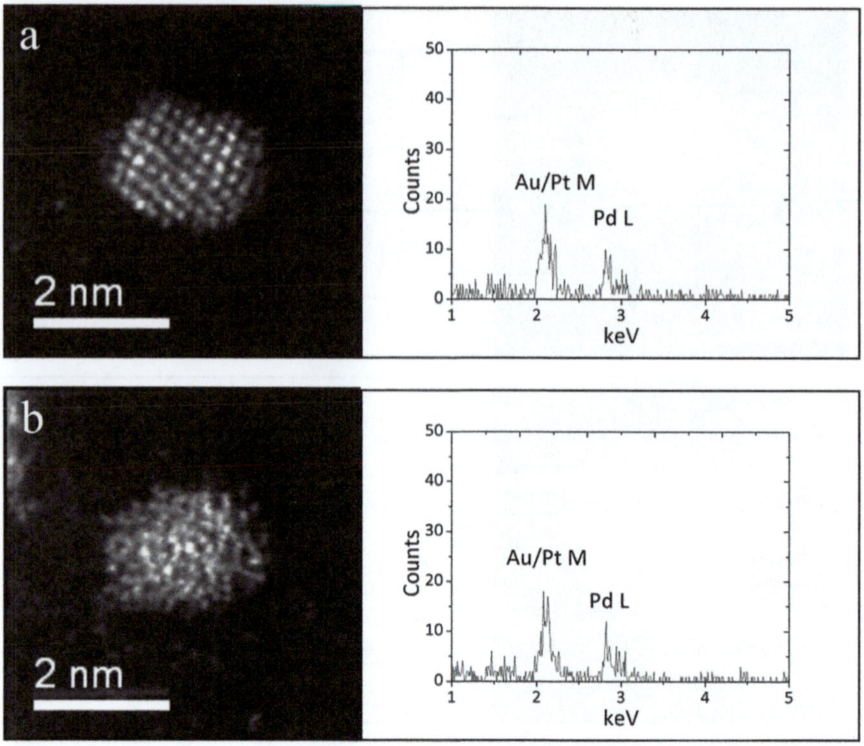

Fig. 8 (a) An HAADF image and the corresponding XEDS spectra from a 2 nm Au–Pd–Pt particle in a 1 wt% (0.3Au + 0.4Pd + 0.3Pt)/C trimetallic catalyst prepared by the sol-immobilization method; (b) A second consecutive HAADF image and corresponding XEDS spectra of the same particle. A comparison of these two datasets indicates that the composition quantification from XEDS spectra was not significantly affected by electron beam damage under the experimental conditions employed.

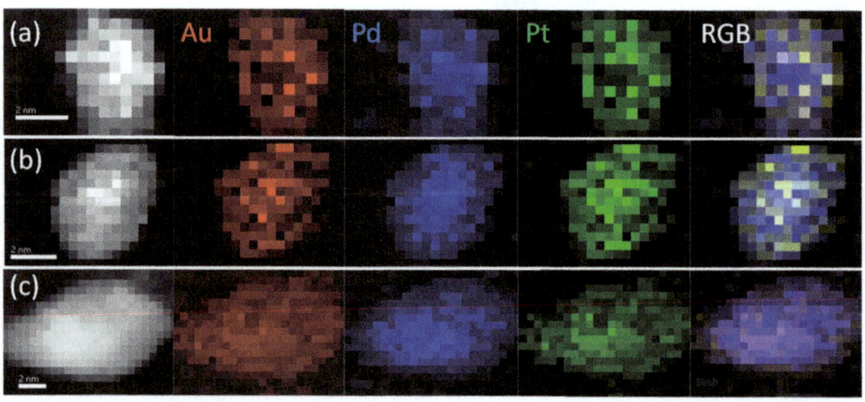

Fig. 9 STEM-XEDS spectrum image analysis of the 1 wt% (0.3Au + 0.4Pd + 0.3Pt)/C trimetallic sample. The first column shows HAADF images from representative particles that are (a) ~3 nm, (b) ~5 nm, and (c) ~10 nm in size. The scale bar represents 2 nm. The next three columns show the corresponding XEDS maps derived from the Au L (9.64–9.84 keV), Pd L (2.75–3.05 keV) and Pt L (9.35–9.55 keV) peaks. The final column shows the corresponding RGB overlay maps where Au is red, Pd is green and Pt is blue.

Table 3 Benzyl alcohol oxidation results after 3 h reaction with, Au–Pd, Pt–Pd and various Au–Pd–Pt catalysts supported on carbon and titania as prepared by the impregnation method (calcined at 400 °C in air)[a]

			Selectivity (%)				
Entry	Catalyst	Conversion (%)	Toluene	Benzaldehyde	Benzoic acid	Benzyl benzoate	TOF/ h^{-1}[b]
1	5% (2.5Au + 2.5Pd)/TiO_2	45.1	13.6	74.2	9.4	2.8	6420
2	5% (1.67Au + 1.67Pd + 1.67Pt)/ TiO_2	8.95	0.4	94.5	2.8	2.2	1410
3	5% (2.5Au + 2.5Pd + 0.5Pt)/TiO_2	18.6	2.7	86.9	3.3	7.1	2470
4	5% (2.27Au + 2.27Pd + 0.45Pt)/ TiO_2	10.6	1.8	86.1	5.1	7.0	1150
5	5% (2.5Pd + 2.5Pt)/TiO_2	8.6	1.1	91.3	4.3	3.3	1220
6	5% (2.5Au + 2.5Pd)/C	55.0	16.6	79.7	1.8	1.9	7830
7	5% (2.27Au + 2.27Pd + 0.45Pt)/C	6.2	2.7	87.9	4.8	4.6	910

[a] *Reaction conditions*: benzyl alcohol, 0.025g of catalyst, $T = 140$ °C, $pO_2 = 150$ psi, stirring rate 1500 rpm. [b] Calculation of TOF (h^{-1}) value was done after 0.5 h of reaction, and was based on the total metal loading.

Conclusions

In summary, Au–Pd–Pt/C trimetallic catalysts were successfully synthesized using the sol-immobilization method. Aberration corrected STEM characterization reveals that the nanoparticles were tightly controlled in size, with mean diameters around 2–3 nm, and they were all random ternary alloys. However, a systematic size-dependant composition variation with respect to Pd was observed, as larger particles were always found to be Pd-rich and smaller particles were consistently Pd-deficient. It was shown that the addition of a small amount of Pt can significantly enhance the selectivity of benzaldehyde in the solvent-free oxidation of benzyl alcohol reaction, while still maintaining a high alcohol conversion.

References

1 W. J. Mijs and C. R. H. I. de Jonge, *Organic syntheses by oxidation with metal compounds*, New York, Plenum Press, 1986.
2 R. A. Sheldon and J. K. Kochi, *Metal-catalyzed oxidations of organic compounds: mechanistic principles and synthetic methodology including biochemical processes*, Academic Press, New York, 1981.
3 G. Cainelli and G. Cardillo, *Chromium oxidations in organic chemistry*, Springer-Verlag, Berlin; New York, 1984.
4 D. I. Enache, J. K. Edwards, P. Landon, B. Solsona-Espriu, A. F. Carley, A. A. Herzing, M. Watanabe, C. J. Kiely, D. W. Knight and G. J. Hutchings, *Science*, 2006, **311**, 362–365.
5 J. A. Lopez-Sanchez, N. Dimitratos, P. Miedziak, E. Ntainjua, J. K. Edwards, D. Morgan, A. F. Carley, R. Tiruvalam, C. J. Kiely and G. J. Hutchings, *Phys. Chem. Chem. Phys.*, 2008, **10**, 1921–1930.
6 N. Dimitratos, J. A. Lopez-Sanchez, D. Morgan, A. F. Carley, R. Tiruvalam, C. J. Kiely, D. Bethell and G. J. Hutchings, *Phys. Chem. Chem. Phys.*, 2009, **11**, 5142–5153.

7 J. C. Pritchard, Q. He, E. N. Ntainjua, M. Piccinini, J. K. Edwards, A. A. Herzing, A. F. Carley, J. A. Moulijn, C. J. Kiely and G. J. Hutchings, *Green Chem.*, 2010, **12**, 915–921.
8 M. Sankar, E. Nowicka, R. Tiruvalam, Q. He, S. H. Taylor, C. J. Kiely, D. Bethell, D. W. Knight and G. J. Hutchings, *Chem.–Eur. J.*, 2011, **17**, 6524–6532.
9 R. C. Tiruvalam, J. C. Pritchard, N. Dimitratos, J. A. Lopez-Sanchez, J. K. Edwards, A. F. Carley, G. J. Hutchings and C. J. Kiely, *Faraday Discuss.*, 2011, **152**, 63–86.
10 S. Meenakshisundaram, E. Nowicka, P. J. Miedziak, G. L. Brett, R. L. Jenkins, N. Dimitratos, S. H. Taylor, D. W. Knight, D. Bethell and G. J. Hutchings, *Faraday Discuss.*, 2010, **145**, 341–356.
11 S. S. Hladyi, M. K. Starchevsky, Y. A. Pazdersky, M. N. Vargaftik and I. I. Moiseev, *Mendeleev Commun.*, 2002, **12**, 45–46.
12 A. Villa, D. Wang, N. Dimitratos, D. S. Su, V. Trevisan and L. Prati, *Catal. Today*, 2010, **150**, 8–15.
13 J. Chen, Q. H. Zhang, Y. Wang and H. L. Wan, *Adv. Synth. Catal.*, 2008, **350**, 453–464.
14 J. Pritchard, L. Kesavan, M. Piccinini, Q. He, R. Tiruvalam, N. Dimitratos, J. A. Lopez-Sanchez, A. F. Carley, J. K. Edwards, C. J. Kiely and G. J. Hutchings, *Langmuir*, 2010, **26**, 16568–16577.
15 N. Braidy, Z. J. Jakubek, B. Simard and G. A. Botton, *Microsc. Microanal.*, 2008, **14**, 166–175.
16 R. F. Egerton, P. Li and M. Malac, *Micron*, 2004, **35**, 399–409.
17 D. Ferrer, D. A. Blom, L. F. Allard, S. Mejia, E. Perez-Tijerina and M. Jose-Yacaman, *J. Mater. Chem.*, 2008, **18**, 2442–2446.
18 D. Ferrer, A. Torres-Castro, X. Gao, S. Sepulveda-Guzman, U. Ortiz-Mendez and M. Jose-Yacaman, *Nano Lett.*, 2007, **7**, 1701–1705.
19 M. R. Ward, T. Hyde, E. D. Boyes and P. L. Gai, *ChemCatChem*, 2012, **4**, 1622–1631.
20 M. Watanabe, A. Yasuhara and E. Okunishi, *Microsc. Microanal.*, 2012, **18**, 974–975.
21 Z. M. Peng and H. Yang, *Nano Today*, 2009, **4**, 143–164.
22 N. Braidy, G. R. Purdy and G. A. Botton, *Acta Mater.*, 2008, **56**, 5972–5983.
23 H. B. Liu, U. Pal and J. A. Ascencio, *J. Phys. Chem. C*, 2008, **112**, 19173–19177.
24 K. Yun, Y. H. Cho, P. R. Cha, J. Lee, H. S. Nam, J. S. Oh, J. H. Choi and S. C. Lee, *Acta Mater.*, 2012, **60**, 4908–4916.
25 D. Alloyeau, G. Prevot, Y. Le Bouar, T. Oikawa, C. Langlois, A. Loiseau and C. Ricolleau, *Phys. Rev. Lett.*, 2010, **105**, 255901.

General discussion

Professor Asakura opened the discussion of the paper by Professor Anderson: You showed interesting results on size dependence that is the activity increases with particle size until 7 nm and then drops. Do you have information about the structure and morphology? For example 2D growth until 7 nm and then 3D?

Professor Anderson replied: We are planning to use ion scattering to examine the as-deposited morphology of these clusters. At this point, I have no evidence about cluster morphology, however, you suggestion seems quite reasonable to me.

Professor Dr Suzer remarked: The number of atoms designated presumably correspond to the mass selected particle in the gas phase. How does that number translate to the numbers after landing on the surfaces? Any way to prove that those numbers are conserved?

Professor Anderson said: This is an important point. We do not have any direct way of imaging the clusters, to see if they remain intact or not. Perhaps the strongest evidence that the clusters are not sintering or fragmenting extensively, is that both the chemical reactivity and Pd 3d binding energies are non-monotonically dependent on deposited cluster size. This does not rule out some sintering or fragmenting, but it does show that the catalyst retains significant "memory" of the deposited cluster size. The exception in this case may be for deposition of Pd atoms. In that case there is some sign that atoms may be sintering.

Professor Behm commented: You referred in your presentation to the different reactivities of the different Pd clusters, which was determined *via* the integrated areas of the TPR runs. To my understanding, this area results from the competition between reaction and (non-reactive) desorption of oxygen. In that case it should be very hard to extract real activities from these data.

Dr Nilius asked: Is it not surprising that the overall Pd activity on a non-reducible oxide support such as alumina is so much higher than on the reducible titania studied before? What conclusions can be drawn on the crystal quality and defect structure of your alumina films from this finding?

Professor Anderson answered: I am not sure that I know what to expect when comparing activity of such small Pd clusters on different supports. There is no reason to expect that they necessarily will behave like the much larger Pd nanoparticles that most authors have studied. What is found is that O_2 binding to Pd clusters on $TiO_2(110)$ is clearly activated (see ref. 1), with cluster size-dependent activation energies, such that none of the Pd/TiO_2 samples are active if the oxidation step is carried out below room temperature. For the same size Pd clusters on alumina films, O_2 binds to Pd even at cryogenic temperatures, and oxidizes the Pd, at least as far as can be determined by XPS. This is true both for the thick, and probably relatively defective alumina films we have grown on Ta(110), and the much thinner, crystalline films grown on NiAl(110) in a earlier study (see ref. 2).

1 Kaden *et al.*, *J. Am. Chem. Soc.*, 2010, **132**, 13097–13099.
2 Wu *et al.*, *Surf. Sci.*, 2009, **603**, 2764–2770.

Dr Yang said: The paper has presented very good experimental results on TPR and TPD. My questions are: (a) What is the reproducibility of these experiments? (b) Can you control the pattern of clusters deposited on surface?

Professor Anderson responded: The reproducibility of integrated desorption/product yield in repeated runs on samples with the same size clusters, is generally pretty good – something like 10% of the reported value (see error bars). As discussed in the paper, it is essential to make sure that the alumina film thickness and growth rate are controlled, so that the support is reproducible. The cluster deposition, itself, is really very reproducible.

Unfortunately, we are not able to pattern the clusters on the surface, although this could, in principle, be done by depositing through a resist or mask. Another approach has been adopted by Friedrich Esch and Ueli Heiz is to deposit on a graphene layer grown on a metal substrate, where lattice mismatch results in a Moiré pattern that gives shallow potential minima. Because the cluster–graphene interaction is weak, the clusters diffuse into the minima, and become localized at low temperatures. This particular approach to patterning is probably not useful at catalytically interesting temperatures, because the wells are too shallow.

Professor Hutchings queried: Does the oxidation state of Pd change as the temperature is increased in your experiments?

Perhaps during heating the nanoclusters change in size, is it possible to do an initial heat treatment experiment and then cool the nanoparticles and repeat the treatment. It would be interesting to see if the same profile was observed.

Professor Anderson answered: We have looked at the oxidation state, *i.e.*, the Pd 3d core level energy, before and after heating and CO and O_2 exposures, but have never actually looked during heating. Our typical protocol is to deposit clusters at room temperature, cool to our TPD starting temperature, and then flash the sample to 560 K to desorb any adventitious CO prior to doing TPR experiments. We have looked with both XPS and ISS at the effects of this quick flash to 560 K, and don't see any significant changes. The question of whether clusters are growing during heating is not one that we can address directly, however, it is observed that the (size-dependent) CO_2 production activity is very stable in repeated TPR experiments, suggesting that the changes in size induced by a cycle of TPR are modest.

Professor Bowker said: I imagine it must be a disappointment to find so little apparent structure sensitivity in CO oxidation on these varying cluster sizes. Why is there so little variation?

Professor Anderson replied: I think that the weak dependence of activity on cluster size in this system is simply a consequence of the high activity in this system. All the Pd_n/alumina model catalysts were highly active, even at low temperatures. In other cases, for example Pd_n/TiO_2 or Au_n/TiO_2, where activity is low due to low O_2 activation efficiency, there is a much stronger cluster size dependence. In the case of Pd_n/TiO_2, the size dependence was clearly shown to result from size-dependent activation energetics and kinetics for O_2 adsorption.[1–3]

1 W. Kaden, T. Wu, W. Kunkel and S. Anderson, *Science*, 2009, **326**, 826.
2 S. Lee, C. Fan, T. Wu and S. Anderson, *J. Chem. Phys.*, 2005, **123**, 124710.
3 W. Kaden, W. Kunkel, M. Kane, F. Roberts and S. Anderson, *J. Am. Chem. Soc.*, 2010, **132** 13097–13099.

Professor Bowker then asked: Is it possible that for all these cluster sizes they flatten out on the surface to make rafts rather than nanoparticles?

Professor Anderson answered: It certainly is possible. For the Pd/alumina system we do not (yet) have structural characterization of the as-deposited samples, however, for a number of metals on $TiO_2(110)$, we have used ion scattering to probe morphology, and as you might expect, the transition between 2D and 3D structures is quite variable. For example, Ir clusters deposited on TiO_2 adopt 3D structures already at 5 atoms, and by 15 atoms, only ~20% of the Ir atoms are in the surface layer, suggesting a nearly spherical structure. In contrast, both Ni and Pd clusters on TiO_2 appear to deposit as 2D islands for sizes up to 10 atoms, then go multilayer for larger sizes.[1–3]

1 M. Aizawa, S. Lee and S. Anderson, *Surf. Sci.*, 2003, **542**, 253–275.
2 M. Aizawa, S. Lee and S. Anderson, *J. Chem. Phys.*, 2002, **117**, 5001–5011.
3 W. Kaden, T. Wu, W. Kunkel and S. Anderson, *Science*, 2009, **326**, 826.

Professor Ertl questioned: How do you quantify your reactivity? Apart from the complication by the competition with desorption, can you relate this quantity to the number of surface atoms?

Professor Anderson replied: In all the different model catalysts I compared, the total number of Pd atoms deposited was identical, and the catalysts differ only in the size cluster the deposited atoms were packaged in. We do not yet have any morphology information for these catalysts, but it would not surprise me if the drop in activity between Pd_6 and Pd_7 reflected a transition from 2D to 3D island structure, reducing the number active surface sites.

Professor Campbell opened the discussion of the paper by Dr Schauermann: In your explanation of the strong particle size dependence for CO adsorption energy for Pd particles between 3 and 8 nm in diameter; it is due to the change in Pd–Pd lattice constant and the manifestation of that on CO bonding to the same type sites? My memory of the size range over which lattice contraction occurs is that it is only significantly lower than the bulk value below 3 nm particle diameter. Can you please add a few citations that verify that lattice parameters significantly smaller than the bulk are seen above 3 nm.

Dr Schauermann replied: The available literature data on Pd nanoparticles show a pronounced decrease of the lattice parameter below 3 nm as measured by TEM.[1] However, this particle size needs to be considered as the lowest limit for the onset of contraction. In fact, the relaxation of the uppermost metal layers is well known even on the single crystals and is it very likely that the uppermost layers on the large Pd nanoparticles can be also significantly contracted. However, such kind of effects would be very difficult to clearly show experimentally (at least with TEM) since a large fraction of the deeper lying atoms in large clusters will still exhibit the lattice separation similar to the bulk value, which will be averaged with the smaller lattice parameter of the uppermost layers. To judge about the possibility and the magnitude of this effect, one would need to determine the lattice parameter of a nanoparticle as the function of its depth. To the best of my knowledge, such depth-resolved experimental data are not available for Pd nanoparticles supported on oxides so far. However, it is generally well known that the long range relaxation usually occurs on a large variety of metals, including nanoparticles and core–shell nanoparticles. Thus, contraction of the nanoparticles was reported for Pt (size range 3.8–12.2 nm), Au (size range 1.8–12.5 nm), Ag (size range 3–17.8 nm), Cu (size range 2.4–9.2 nm), and Bi (size range 3.5–10 nm) (summary in ref. 2). Therefore we believe that partial contraction of the uppermost Pd layers in the large clusters might occur and with this cause the observed lowering of the CO binding energy.

1 S. A. Nepijko, M. Klimenkov, M. Adelt, H. Kuhlenbeck, R. Schlögl and H.-J. Freund, *Langmuir*, 1999, **15**, 5309–5313.
2 P. M. Diehm, P. Agoston and K. Albe, *ChemPhysChem*, 2012, **13**, 2443.

Professor Freund added in response to Professor Campbell: There is no question in my mind that small particles contract with respect to the interatomic distances and the amount of contraction depends strongly on the system. At what size this starts is less clear, however Nepijko[1] have measured Moiré patterns of particles supported on alumina films which reveal that at 4 nm average size the effect is already clearly visible. Even if there is not a large contraction on average, there will be a stronger contraction of the surface layers of the particles which, in turn, will lead *via* a bound-order conservation argument, to weaker adsorbate bonding.

As far as calculations are concerned, those are performed for very small particles not containing the sufficient number of atoms, and, even more importantly, the calculations do not take the strain between particles and support into account, which will also strongly influence the particle response to adsorption. In general, I believe, that the flexibility of nanoparticles is so much higher than that of (even stepped) single crystals, so that a proper description of adsorption on them based on slab calculations is not appropriate.

1 Nepijko *et al.*, *Langmuir*, 1999, **15**, 5309.

Professor Campbell thanked Dr Schauermann and Professor Freund for these very good points.

Professor Bowker remarked: These results have considerable similarity to our previous work on Pd nanoparticles on titania[1] which also shows that the CO is more weakly bound on the Pd nanoparticles than on Pd(111) and (110). However, we also found that the CO adsorption was strongly affected by the temperature to which the supported sample was annealed in vacuum. Annealing above 300 °C resulted in reduced sticking probability and heating above 450 °C resulted in complete poisoning of adsorption. This was because TiO_2 is a reducible support and annealing in vacuum is effectively like a reducing atmosphere, and encapsulation of the Pd particles by TiO occurs (the SMSI effect[2]) which poisons adsorption. Have you done annealing experiments to examine whether such encapsulation occurs for the iron oxide supported Pd?

1 M. Bowker, P. Stone, R. A. Bennett and N. Perkins, *Surf. Sci.*, 2002, **497**, 155–165.
2 M. Bowker, P. Stone, P. Morrall, R. Smith, R. A. Bennett, N. Perkins, R. Kvon, C. Pang, E. Fourre and M. Hall, *J. Catal.*, 2005, **234**, 172–181.

Dr Schauermann responded: These systems have been extensively studied for possible encapsulation effects both with STM and CO sticking coefficient measurements in 2006–2008 (*e.g.* refs. 1 and 2 and other unpublished results from Sh. Shaikhutdinov's group). In absence of oxygen treatment at elevated temperatures (up to 600 K), neither particles restructuring nor encapsulation were observed for Pd nanoparticles. Oxygen treatment was found to induce and/or facilitate the Ostwald ripening of the particles resulting in bigger Pd aggregates but no encapsulation was observed either. Additionally, the CO sticking coefficient measurements in our own studies showed the same CO adsorption capacity as for the clean Pd(111) and Pd(100) single crystals. Based on these observations, we can very safely exclude any encapsulation of Pd by iron oxide.

1 T. Schalow *et al.*, *Phys. Chem. Chem. Phys.*, 2007, **9**, 1347.
2 T. Schalow *et al.*, *Catal. Lett.*, 2006, **107**, 189.

Professor Freund added in response to Professor Bowker: While for Pt on magnetite this phenomenon of encapsulation and reduced binding energy of adsorbed molecules has been observed after heat treatment at 850 K, for the case of Pd/F_3O_4(111) this has not been observed. Pt–support interaction is considerably, stronger (surface energy from Wulff–Kaicev construction: 4.15 ± 0.1 J m^{-2})[1] than for Pd–magnetite (surface energy from Wulff–Kaicev construction: 3.25 ± 0.3 J m^{-2}).[2,3]

1 Z.-H. Qin *et al.*, *J. Phys. Chem. C*, 2008, **112**, 10209.
2 R. Meyer *et al.*, *Z. Phys.*, 2004, **218**, 905.
3 T. Schalow *et al.*, *Surf. Sci.*, 2006, **600**, 2528.

Dr Nilius asked: Why does the saturation of step edges on Pd(111) single crystals with CO not produce a more pronounced kink in the heat of adsorption *versus* coverage curve?

Dr Schauermann answered: Our experimental results suggest very similar binding energies of CO at the low-coordinated sites, such as edges and corners, and the regular (111) terraces as evidenced by a rather smooth decrease of the CO binding energy as a function of CO coverage. This observation is in a good experimental agreement with the numerous results of temperature-programmed experiments showing very similar desorption temperatures from the stepped and low-Miller-index Pd surfaces. This issue is extensively discussed on pages 7 and 8 of our manuscript.

Professor Campbell noted: The temperature at which FeO_x begins to spread over the Pd particle was measured for rather large size Pd particles, if my memory is correct. Is it possible that the temperature at which some small amount of Fe moves onto or into the Pd particle is lower than that temperature? Can you rule out that some small amount of Fe has moved onto the 2–8 nm particles, with more Fe on/in the smaller particles, and this explains why they bind CO more weakly than bulk Pd? By this I am thinking of such small amounts that it does not decrease the number of CO binding sites, just the strength with which they bind CO.

Dr Schauermann responded: In our experimental studies we never observed a spread of Fe over Pd supported on the Fe_3O_4 films – neither on the small nor on the large Pd nanoclusters. Such kind of encapsulation was detected for *e.g.* Pt, were one can observe formation of a Moiré pattern of FeO over Pt nanoclusters after the high temperature (~800 K) treatment in oxygen.

The reported initial binding energies are related to the very low coverage limit, where only a few, typically 3 to 5 CO molecules or oxygen atoms were adsorbed per Pd cluster, containing on the surface a few hundreds of Pd atoms. Since the experimental time scale is much longer than the time needed for diffusion to the strongest adsorption site on the Pd cluster, these initial binding energies are related to the adsorption on the strongest binding site on Pd nanoparticle. If your assumption were true, then essentially all binding sites – including the strongest binding sites – should have been modified by co-adsorbed iron. This scenario implies a nearly complete saturation of the Pd surface with iron, which most likely should result both in the lowering of the total adsorption capacity of the cluster and also formation of an FeO layer after the high temperature treatment with oxygen. Since neither of these effects were observed in our experiments, we can rule out any significant Fe incorporation into Pd particles, that might affect the initial binding energies of the gaseous molecules at the strongest adsorption sites.

Professor Attard opened a general discussion of Professor Anderson's and Dr Schauermann's papers: I would like to go back (a long time) to when I was a PhD student and the concern was about bridging the so-called "pressure" and "materials" gap between fundamental surface science and heterogeneous catalysis. In the light of this Faraday Discussions meeting so far, would you say that these problems have now been addressed and solved?

Professor Bowker replied: I believe we have gone a long way to bridging that gap. Surface science began by using very perfect samples (*e.g.* single crystal metals), but such basic information is essential (*e.g.* Dr Schauermann's paper, which requires such data to understand the behaviour of nanoparticle forms of Pd). So we have greater understanding from the basis of well-defined materials, which has now advanced to the imaging and reactivity of nanoparticles on flat crystal surfaces. There is a kind of pincer movement being applied to the materials gap, one side being this surface science approach, but on the other, catalysis is now becoming a surface science (as we'll see in Professor Hutching's talk), and so the relationship between the particle and its support now much clearer. The pincer in nearly closed.

Professor Hutchings agreed with Professor Bowker: With the advent of aberration-corrected STEM together with *in situ* spectroscopies the amount of information we can gain on complex real heterogeneous catalysts has advanced tremendously in recent years. In this way the approaches of surface science and catalysis are certainly drawing very close together.

Professor Somorjai addressed Dr Schauermann: Can you follow the surface structure as you change CO coverage, or as a function of pressure? Step surfaces with 90% coverage of CO alone with no O_2, restructures the step surface, with high pressure, at lower pressure steps reappear. High pressure STM studies clearly indicate reversible restructuring of the step structures of metals as the CO pressure is altered.

Dr Schauermann responded: The highest pressures that we can reach in our experimental setup are on the order of magnitude 10^{-6} to 10^{-5} mbar. To the best of my knowledge, such pressures are still too low to induce some irreversible restructuring of the low coordinated sites with CO.

Professor Attard asked a general question addressed to the catalytic chemists at the Discussion: Is the fundamental surface characterisation knowledge being discussed here useful/capable of helping you synthesise better heterogeneous catalysts?

Professor Hutchings responded: Yes indeed. This meeting has shown that the surface science experiments are moving towards studying materials that are relevant to the more complex real heterogeneous catalysts and *vice versa*. There has been immense progress in the last decade and the two communities are now converging due to the dialogue created by meetings such as this. A point raised by Professor Somorjai is relevant to this discussion. He suggested that we should select reactions that can be carried out at the gas–solid interface as well as the liquid–solid interface as this would provide a means to explore the difference in reactant concentration on the reactions. The oxidation of alcohols could be a suitable model system to work on in this respect as this is amenable to study by both subject communities.

Professor Ertl commented: The kinetics of catalytic CO oxidation had been investigated already by Langmuir many years ago, and the dependence on the partial pressures of O_2 and CO has been well established before.

Dr Schauermann agreed with the comment by Professor Ertl: I think that the major contribution of this study to the CO oxidation issue consists in an exact quantification of the CO binding energies on the particles of different sizes. This parameter has a pivotal importance for the establishing the "CO-poisoning" conditions. Lower CO binding energies result in higher CO desorption rates and, consequently, later onset of the CO-poisoning regime with the growing CO pressure. The reported strong particle size dependence of the CO binding energy might be used to tune the overall rate of CO oxidation by tuning the particle size.

Professor Hutchings returned to the discussion point raised by Professor Attard concerning designing catalysts on the basis of surface science studies, addressing Dr Schauermann: It is clear that the morphology of the nanoparticles is very important. In your case you consider the edge sites to be important but what about the interfacial sites? If the edge sites are important for reactivity we will need experimental approaches that minimize the interfacial sites, *i.e.* we would need to minimize the interaction of the nanoparticles with the surface. Can you comment on this?

Dr Schauermann answered: We are not generally claiming that the edge sites are important for catalysis and the interfacial sites are not. In contrast, the interfacial sites, that often bind the gaseous adsorbates rather weakly, might be more important in certain surface reactions that require a weak bonding of adsorbates. I think it is not possible to make any general statement on the importance of the particular sites for the reactivity. The design of the nanoparticles should rather depend on the particular surface process and its demands with respect to the binding energies of the adsorbates.

Professor Anderson closed the general discussion remarking: We observe that for small Pd clusters on TiO_2, there is strong dependence on cluster size of the activation energy for O_2 dissociative binding, and that this appears to be correlated with the particle's electronic structure.[1] On the other hand, the dependence of CO desorption energy on cluster size is relatively weak. For CO binding on top of the clusters, there are non-monotonic size-dependent variations of up to 20 °C in the peak desorption temperature, and for CO binding around the cluster periphery, there is a monotonic decrease of similar magnitude in the peak desorption temperature as size increases from 1 to 20 atoms.[2]

1 Kaden *et al.*, *J. Am. Chem. Soc.*, 2010, **132**, 13097–13099.
2 Kaden *et al.*, *J. Chem. Phys.*, 2012, **136**, 204705.

Dr Su opened the discussion of the paper by Professor Park: How stable are your samples? Have you tried to measure the recyclability of the photocatalyst? What is the quantum efficiency of the material? The loading of the metal nanoparticles is very important parameter, did the authors characterized the loading of the *meta* NPs?

Professor Park replied: For Pt–CdSe–Pt nanodumbbells, we checked the recyclability of the catalysts as the temperature or the presence of the light source changed.[1] Regarding the metal loading, because the sample is a two-dimensional catalyst, we can obtain the loading of the metal nanoparticles based on the estimation of size of a nanoparticle (from TEM images) and surface coverage (from SEM images).

1 S. M. Kim, S. J. Lee, S. H. Kim, S. Kwon, K. J. Yee, H. Song, G. A. Somorjai and J. Y. Park, *Nano Lett.*, 2013, **13**, 1352.

Professor Anderson enquired: In these samples the turnover frequency seems to increase with temperature for both light and dark experiments, but the relative and absolute magnitudes of the photo effect also seems to increase with temperature, which surprised me since there is obviously more thermal (nonphotochemical) reaction at higher temperatures. Can you comment?

Professor Park responded: The activation energies for the Pt nanoparticles on the n- or p-doped GaN were obtained from Arrhenius plots of the catalytic activity. The activation energies remained unchanged, depending on the light irradiation. This means that the relative magnitudes of the photo effect are constant, while the absolute magnitude increases with temperature. On the other hand, we found both relative and absolute magnitudes for the Pt–CdSe–Pt nanodumbbells change with temperature.[1] This temperature effect would be associated with the lower energy barrier of Pt–CdSe compared to Pt–GaN. Because of the higher energy barrier between Pt and GaN, the effect of temperature on light-induced catalytic activity would be less prominent than for the Pt–CdSe–Pt nanodumbbells. Another possible mechanism is associated with trapping the photocarrier in the surface states of the CdSe nanorods. Thermal activation of the photocarrier would facilitate the transport of electrons from the CdSe to the Pt nanoparticles or the transport of holes from the CdSe to silicon at high temperature. The effect of the surface state for the GaN substrate would not be as prominent as for the CdSe nanorods. In order to determine the exact mechanism, *in situ* microscopic and spectroscopic studies on these nanocatalysts under light irradiation would be required.

1 S. M. Kim, S. J. Lee, S. H. Kim, S. Kwon, K. J. Yee, H. Song, G. A. Somorjai and J. Y. Park, *Nano Lett.*, 2013, **13**, 1352.

Professor Ertl said: These are very exciting results. I am wondering about the underlying mechanism. Have you any information about the dependence of the rate on the partial pressures of the reactants which would provide some indication about the rate-determining step?

Professor Park answered: For the CO oxidation reaction on Pt nanoparticles on GaN substrate, we used a partial pressure of 40 torr CO and 100 torr O_2 as the reactant gases and the temperature between 220–260 °C. The catalytic behavior of the Pt-based nanoparticles under this pressure and temperature range follows the Langmuir–Hinshelwood mechanism, resulting in a positive first order in O_2 and negative first order in CO pressure.[1] However, we haven't studied the dependence of the partial pressure on the hot electron effect. We plan to carry out this study in order to ascertain the rate-determining step of the hot electron effect.

1 J. Y. Park, Y. Zhang, S. H. Joo, Y. Jung and G. A. Somorjai, *Catal. Today*, 2012, **181**, 133.

Professor Ertl then asked: How does the effect depend on the intensity of light?

Professor Park replied: We carried out CO oxidation on Pt–CdSe–Pt nanodumbbells as a function of light intensity to understand the correlation between the catalytic activity and power density. This result is published elsewhere.[1] The measured catalytic activity of CO oxidation as a function of light intensity reveals a linear relationship between catalytic enhancement and photon flux. This result could be associated with an increase in the number of hot carriers created at the interface of the Pt metal and CdSe nanorod as the power density increases.

1 S. M. Kim, S. J. Lee, S. H. Kim, S. Kwon, K. J. Yee, H. Song, G. A. Somorjai and J. Y. Park, *Nano Lett.*, 2013, **13**, 1352.

Professor Dr Suzer queried: In the paper, it is stated that photoillumination measurements were carried out at only one power density or lamp intensity. Have you checked, or considering to check, the power dependence as a way to shed some light into the mechanism(s) of these processes?

Professor Park answered: We haven't investigated the dependence of power density yet. We plan to carry out this aspect to understand the detailed mechanism of the hot electron effect on the catalytic activity.

Professor Campbell remarked: You mentioned that the rate of CO oxidation is negative first order in CO. Under these conditions, the surface is poisoned by CO and the rate goes up whenever you do anything to increase the rate of CO desorption. Since adsorbed CO is well known to be desorbed by hot electrons (created by either photons or electrons hitting the surface), I suggest that your increase in rate with photons is simply due to the well known effect of excited electrons on CO desorption.

Professor Park responded: I appreciate your comments on the possible mechanism of hot electron effects. Desorption of adsorbed CO by hot electrons is quite plausible. We plan to carry out *in situ* spectroscopy on Pt nanoparticles/GaN substrate to confirm this point.

Professor Asakura said: I would like to ask the particle size dependence of activity and reaction site. You said it could be explained by mean free path of hot electrons. Hot electrons are created in the bulk of GaN. Pt has a hemispherical shape and aspect ratio is constant. If so, the hot electrons travel through the bulk to the surface of GaN and go through the interface Pt and GaN and then reach the surface of the Pt particle. So your active site was the top of the Pt particle. Otherwise there would be no particle size dependence. How about the edge and peripheral activities? If the active site is located at the periphery of the Pt particle, we cannot explain the particle size effect by mean free path. What do you think? Also the hot electrons might be created near the surface region. Might the band bending be affected by Pt nanoparticle size?

Professor Park answered: The particle size dependence in the change of catalytic activity can be explained by several factors including the mean free path of hot electrons. We think that the edge and periphery of Pt nanoparticles can be the active sites that affect the catalytic activity. If this is the case, because the area of the peripheral region scales with $\sim 1/d$, where d is the size of the nanoparticle, we can expect higher catalytic activity with smaller nanoparticles. Because band bending can be affected by the size of the nanoparticles, we can't rule out the influence of local band bending. Nevertheless, we think the argument about the role of hot electrons affecting the catalytic activity is still valid. In other words, the change of catalytic activity by light irradiation can be explained by the injection of hot electrons. Smaller nanoparticles may have a larger area of active sites or a shorter travel length of hot electrons, leading to the enhancement of catalytic activity.

Professor Bowker enquired: The reaction probability is around 10^{-6}, so I wonder how that number equates to the lifetime of the "hot" electrons which are produced. Is there a mismatch between excited state lifetime and turnover frequency?

Professor Park replied: The light intensity used in our experiment is 49.4 mW cm^{-2}. The number of photons incident to the Pt nanoparticle on GaN substrate is $\sim 8.7 \times 10^{16}$ cm^{-2} s^{-1}. Assuming that an absorbed photon creates an electron–hole pair and the surface atomic density is 10^{15} cm^{-2}, the maximum number of hot electrons generated on the surface is 87 surface atom s^{-1}. The turnover frequency (TOF). which is the number of product molecule per metal surface site per second, for the 4.3 nm Pt nanoparticles/p-GaN substrate with and without light at 523 K is ~3.57 and 3.01 molecules per Pt site per second, respectively. Therefore, the difference in TOF is 0.56 molecules per Pt site per second, which is associated with the effect of hot electrons on catalytic activity. Therefore, the rough estimation of the quantum efficiency, which is the ratio of the number of chemical reaction events to the number of generated hot electrons, is less than 1%.

Professor Menzel remarked: The suggested mechanism is related to the well-known TNI (transient negative ion) mechanism or its inverse, the TPI (transient positive ion) mechanism which have been shown to operate also on MNPs.[1] The difference from the present suggestion is – if I understand correctly – that here light absorption is assumed to happen in the GaN substrate and the hot carriers then transferred (or not) to the MNP (Fig. 1 in the paper), while usually it is found that the absorption and hot carrier formation take place in the MNP (possibly enhanced by plasmon action), and the hot carrier efficiency for the TNI can be enhanced by the fact that the hot electron mean free path is longer than the NP size. In order to test this a wavelength-dependent study would be very helpful, since the energy of the relevant LUMO or HOMO is decisive and can be sampled by the photon energy. Even if the primary absorption should take place in the MNP, a doping (and therefore band-bending) dependent effect could still result if the band bending at the interface would confine the hot carriers after transfer in the one and not the other case.

1 D. Mulugeta *et al.*, *J. Chem. Phys.*, 2011, **134**, 164702.

Professor Park responded: In case of Pt–CdSe–Pt nanodumbbells, we have carried out the wavelength-dependent experiment.[1] To identify the influence of photon energy on catalytic enhancement, we used two types of energy filters, a short-wave pass filter (SWF, $h\nu = 2.0–3.0$ eV) and a long-wave pass filter (LWF, $h\nu = 1.0–2.0$ eV). We found that the Pt–CdSe–Pt nanodumbbells irradiated by light with a higher photon energy (2.0 eV $< h\nu <$ 3.0 eV) were two times more active for CO oxidation, compared to when irradiated by light with a lower photon energy (1.0 eV $< h\nu <$ 2.0 eV). In the latter case, the catalytic activity was very similar to CO oxidation in the dark. Thus, this experiment indicates that hot electrons are generated when irradiated with light having a photon energy above the bandgap energy (E_g) of the CdSe rods, and that the hot electrons promote the CO oxidation reaction on the Pt metal. We are planning to carry out the wave-dependent experiment for the Pt nanoparticles/GaN support.

1 S. M. Kim, S. J. Lee, S. H. Kim, S. Kwon, K. J. Yee, H. Song, G. A. Somorjai and J. Y. Park, *Nano Lett.*, 2013, **13**, 1352.

Professor Hutchings enquired: Have you considered using PVA as a capping ligand in place of PVP? The reason is that PVA can be readily removed by treatment with water at 90 °C but PVP is much more difficult to remove from the surface of nanoparticles.[1] Using PVA and treating with water could enable a facile comparison between liganded and non-liganded surfaces.

1 J. A. Lopez-Sanchez, N. Dimitratos, S. White, G. Brett, L. Kesavan, P. Miedziak, R. Tiruvalam, R. L. Jenkins, A. F. Carley, D. Knight, C. J. Kiely and G. J. Hutchings, *Nat. Chem.*, 2011, **3**, 551.

Professor Park answered: Understanding the role of capping layers in catalytic activity is quite important in our study. We will consider using PVA in place of PVP, which may allow us to compare liganded with non-liganded nanoparticle surfaces.

Professor Ertl commented: For adsorbed CO the lowest lying unoccupied level (derived from the CO 2p orbital) is located about 5 eV above the Fermi level which is presumably much too high in energy to be populated by hot electrons arising from your photon energies.

Professor Park responded: Currently, we don't understand the exact mechanism for how hot electrons influence the CO oxidation process. Earlier theoretical DFT studies showed that hot electron generation is mediated *via* a negatively-charged CO_2^- or O_2^- transition state, implying the important role of negatively-charged reaction intermediates. Another plausible mechanism involves the activation of Pt–O bonds by hot electrons, resulting in the reaction with coadsorbed CO. Bonn *et al.* showed that oxidation of CO was initiated by hot electrons generated by femtosecond infrared laser pulses on a Ru (0001) surface. Based on these results, we assume that the hot electrons created at the interface of the Pt nanoparticles and GaN mediate the facile formation of a negatively charged CO_2^- or O_2^- transition state, leading to a change in catalytic activity.

Professor Attard asked: Have you investigated the impact on catalytic activity of the level of doping in the semiconductor support?

Professor Park answered: No, we haven't. So far, we have focused on the hot electron effect as the doping type of GaN substrate changes. We plan to investigate the influence of the doping level on the catalytic activity in the future.

Mr Zhao opened the discussion of the paper by Professor Hutchings: You used PVA as capping agent to prepare Au–Pd–Pt nanoparticles, which were then immobilized on activated carbon. Where was the location of PVA in the resulting catalyst? Was it on the carbon support or on the metallic nanoparticles? If on the nanoparticles, did it attach with all the three metal atoms? How did you determine this? You recommend using hot water refluxing method to remove capping agent, but in your paper here, you didn't use it. Is there any specific reason for this? In one previous publication, you declared different morphology of Au–Pd colloid on different support, where wetting angle of the colloid on activated carbon was extremely bigger than on oxide support. Does this mean that the colloid has weak interaction with activated carbon and can be easily washed away together with PVA by hot water?

Professor Hutchings communicated in reply: We consider that as the PVA is present is excess in the sol-immobilisation preparation method it will be deposited on both the support and the nanoparticles. However, PVA is soluble in benzyl alcohol and so during the catalysed reaction some of the PVA will be removed from the surface on the nanoparticles thereby facilitating reaction. We have previously shown that the removal/presence of PVA on the catalyst surface can be followed using laser Raman spectroscopy.[1] Of course complete removal may not be necessary nor desirable since residual ligands could play a role in deactivating sites responsible for non-desired side reactions. As yet we do not consider that this aspect of catalyst design has been sufficiently explored.

1 J. A. Lopez-Sanchez, N. Dimitratos, S. White, G. Brett, L. Kesavan, P. Miedziak, R. Tiruvalam, R. L. Jenkins, A. F. Carley, D. Knight, C. J. Kiely and G. J. Hutchings, *Nat. Chem.*, 2011, **3**, 551.

Mr Zhao said: I agree with your comment on the important role of capping agent on the nanoparticle surface if complete removal is not possible. In our study, we also observed the different influence of capping agent in the same catalyst working in different phase. For example, PVA could cover and block surface of nanoparticles in gas phase; however, in aqueous phase, the blocking effect was less significant. This may due to the different morphology of PVA in aqueous phase than in gas phase. Even more, the capping agent possibly also influences reaction selectivity.

Professor Hutchings responded: I certainly agree with these ideas. Appropriate choice of the stabilizing ligand and the solvent for the reaction could play a vital role in controlling or engendering catalytic activity. With PVA-stabilized AuPd nanoparticles we have previously shown that for gas phase reactions, in our case CO oxidation, the PVA could be removed by solvent extraction with water.[1] It is possible that this approach can be extended to other systems.

1 J. A. Lopez-Sanchez, N. Dimitratos, S. White, G. Brett, L. Kesavan, P. Miedziak, R. Tiruvalam, R. L. Jenkins, A. F. Carley, D. Knight, C. J. Kiely and G. J. Hutchings, *Nat. Chem.*, 2011, **3**, 551.

Dr Whiston addressed Professor Hutchings: The results reported indicate that supported Pt nanoparticles do not catalyse the benzaldehyde disproportionation however they are not useful oxidation catalysts. The addition of platinum to the Au–Pd alloy nanoparticles principally inhibits the disproportionation and formation of toluene. One hypothesis for why this is the case is that the presence of platinum in the alloy nanoparticles could stabilise a hydride intermediate from the disproportionation reaction on the metal surface thus inhibiting the hydrocarbon formation.

One method of testing this hypothesis might be to look at the sensitivity of toluene selectivity to "pH" in the reaction mixture. Addition of a small amount of sulfonic acid (such as $MeSO_4H$ or $PhSO_4H$ for example) might further stabilise the hydride and further inhibit this route. It might have other reactivity consequences as well of course. Is the reaction acid sensitive?

Professor Hutchings replied: You raise an interesting point concerning the stabilization of a surface hydride as the origin of the effect when Pt is added to the nanoparticles. A hydrogen transfer process is key to this chemistry and we agree that anything than can hinder or aid this process could exert an effect. This is therefore an aspect that requires further consideration. Concerning the effect of acidity; yes, the disproportionation reaction is highly sensitive to the acidity/basicity of the reaction medium and the catalyst. We have reported on this aspect in two previous publications.[1,2] Whenever the Au–Pd particles were supported on an acidic metal oxide some toluene formation was observed. However, if the Au–Pd particles were supported a basic metal oxide, the toluene formation was completely switched-off.

1 M. Sankar, E. Nowicka, R. Tiruvalam, Q. He, S. H. Taylor, C. J. Kiely, D. Bethell, D. W. Knight and G. J. Hutchings, *Chem.–Eur. J.*, 2011, **17**, 6524–6532.
2 S. Meenakshisundaram, E. Nowicka, P. J. Miedziak, G. L. Brett, R. L. Jenkins, N. Dimitratos, S. H. Taylor, D. W. Knight, D. Bethell and G. J. Hutchings, *Faraday Discuss.*, 2010, **145**, 341–356.

Dr Philip Davies asked: In your paper you indicate that you believe the role played by the Pt is as an electronic modifier. Could you expand on your reasons for reaching that conclusion?

Professor Hutchings answered: We observe that the different nanoparticles compositions do not exhibit any marked changes in their morphology yet give very different activities and selectivities. We have observed this previously in the bimetallic compositions of supported AuPd and AuPt nanoparticles.[1,2] We therefore consider that the addition of the second or third metal influences the electronic nature of the nanoparticles.

1 J. Pritchard, L. Kesavan, M. Piccinini, Q. A. He, R. Tiruvalam, N. Dimitratos, J. A. Lopez-Sanchez, A. F. Carley, J. K. Edwards, C. J. Kiely and G. J. Hutchings, *Langmuir*, **26**, 2010, 16568.
2 G. Brett, Q. He, C. Hammond, P. J. Miedziak, N. Dimitratos, M. Sankar, A. A. Herzing, M. Conte, J. A. Lopez-Sanchez, C. J. Kiely, D. W. Knight, S. H. Taylor and G. J. Hutchings, *Angew. Chem., Int. Ed.*, 2011, **50**, 10136.

Dr Hughes noted: Fig. 3b and c and Fig. 5b and c appear to shows changes in lattice direction within individual particles; are these related to faceting or defect structures such as twinning or grain (or subgrain) boundaries?

Given that structure sensitivity of many reactions depends on the detailed nature of the site (*e.g.* B5 site on Ru for N_2 dissociation) could the sites between these different regions be important sites for some reaction?

Professor Hutchings replied: Nanoparticles of FCC metals often tend to form multiply twinned icosahedral structures in order to expose low energy {111}-type surfaces at the cost of introducing some extra lattice strain. We have previously shown that support identity plays an important role in determining the particle morphology.[1] Activated C supports give rise to more multiply twinned particles with only {111}-type surfaces, where particles on TiO_2 supports tend to show more untwinned particles with cuboctahedral structures which are less strained, but expose both{111}- and {200}-type facet terminations. To the best of our knowledge, it still remains a big challenge to control the particle morphology on specific supports such that *only* multiply twinned icosahedral or only untwinned cuboctahedral particles exist. Therefore it is difficult to experimentally test your hypothesis that the line traces where twin defects intersect the particle surface may be particularly active catalytic sites. This is however, a very interesting point that deserves further investigation. Maybe first principle calculations could provide some insight into this intriguing possibility.

1 L. Kesavan, R. Tiruvalam, M .I. Bin Saiman, D. I. Enache, R. Jenkins, N. Dimitratos, J. A. Lopez-Sanchez, S. H. Taylor, D. W. Knight, C. J. Kiely and G. J. Hutchings, *Science*, 2011, **331**, 195–199.

Dr Robert Davies said: Your results indicate that the presence of Pd in the trimetallic catalyst decreases with decreasing nanoparticle size, and also show no segregation of the three species. Do you think that for smaller particles the Pd encounters a more significant ligand effect such as a change in electron density distribution?

Professor Hutchings answered: It is apparent that when Pd is present in the nanoparticles formed by the sol-immobilisation method there is a marked composition range with the nanoparticles size. It is observed for AuPd bimetallic nanolloys[1] and so this trend is continued for these trimetallic compositions. It is possible that Pd^{2+} interacts more strongly with the PVA stabilizing ligand and in this case your suggestion is that it would be more difficult to reduce on addition of the $NaBH_4$. Certainly, Au and Pt will be more readily reduced and so it is feasible that this difference could give rise to the composition variation. However, we have found that enhancing the amount of halide ligand in a modified impregnation method can remove this compositional variation. In this case we consider the effect is due to enhanced dispersion of the Pd due to the presence of the halide.[2]

1 J. Pritchard, L. Kesavan, M. Piccinini, Q. A. He, R. Tiruvalam, N. Dimitratos, J. A. Lopez-Sanchez, A. F. Carley, J. K. Edwards, C. J. Kiely and G. J. Hutchings, *Langmuir*, **26**, 2010, 16568.
2 M. Sankar, Q. He, M. Morad, J. Pritchard, S. J. Freakley, J. K. Edwards, S. H. Taylor, D. J. Morgan, A. F. Carley, D. W. Knight, C. J. Kiely and G. J. Hutchings, *ACS Nano*, **6**, 2012, 6600.

Dr Torrente enquired: As indicated in the experimental section, metallic nanoparticles are formed by their reduction with $NaBH_4$ therefore, it is expected that initially, reduced metallic nanoparticles are present in the catalyst. However, is it possible that the oxidation state of the NP changes during the reaction under O_2 atmosphere? The Au–Pd–Pt catalysts reported present an extraordinarily narrow particle size distribution with a high proportion of the NP below 2 nm (Fig. 2). The smaller the particle size, the more likely the re-oxidation to happen according to previous studies (*e.g.* ref. 1). Is it possible that the presence of Pt in the ternary alloy modifies (probably by decreasing it) the re-oxidation of metal under an O_2 atmosphere explaining the high selectivity observed towards aldehyde?

1 Miller *et al.*, *J. Catal.*, 2006, **240**, 222.

Professor Hutchings replied: It is possible that during the catalytic reaction that the surface of the catalyst becomes oxidized. This is particularly true for Pd and the effect becomes dominant at high temperatures as the formation of PdO on the surface of catalysts can become a driving force in the formation of core–shell structures.[1] Hence it is possible that this will have an effect. However, in the present case Pd is present in both the unmodified and Pt-modified compositions, hence we do not consider this to be the dominant origin of the effect in this case.

1 J. K. Edwards, A. F. Carley, A. A. Herzing, C. J. Kiely and G. J. Hutchings, *Faraday Discuss.*, 2008, **138**, 225.

Dr Robert Davies asked: How does the addition of a third metal species affect the choice of support?

Professor Hutchings answered: There are two aspects to designing a catalyst for the oxidation of alcohols, namely, the composition of the metals and the nature of the support. These will each have a contributing factor to the nature of the interfacial sites which are critical for some aspects of this reaction. We have previously shown that by choosing basic supports the formation of the unwanted toluene by-product can be eliminated.[1,2] We now show that this can also be achieved by fine-tuning the metal composition. In view of this it is apparent that by adding a third metal to the catalyst we have opened up the possibility of using a wider range of supports.

1 M. Sankar, E. Nowicka, R. Tiruvalam, Q. He, S. H. Taylor, C. J. Kiely, D. Bethell, D. W. Knight and G. J. Hutchings, *Chem.–Eur. J.*, 2011, **17**, 6524–6532.
2 S. Meenakshisundaram, E. Nowicka, P. J. Miedziak, G. L. Brett, R. L. Jenkins, N. Dimitratos, S. H. Taylor, D. W. Knight, D. Bethell and G. J. Hutchings, *Faraday Discuss.*, 2010, **145**, 341–356.

Dr Yang remarked: In the paper, Au–Pd–Pt trimetallic alloy particles have shown high selectivity and activity for oxidation of benzyl alcohol. I wonder if you can speculate what will happen in the case of Au/Pd/Pt core–shell nanoparticles?

Professor Hutchings responded: When we set out on this study our aim was to make core–shell structures with these trimetallic particles. However, we have concentrated to date on the study of the homogeneous alloys. We consider that core–shell structures when they form either using impregnation or sol-immobilisation do not comprise exclusively one metal at the surface although one metal may be dominant. In this way we expect that the catalyst activity can be fine-tuned by using such structures. However, shell compositions that are rich in Pd would need to be avoided since these can be expected to favour toluene formation in the oxidation of benzyl alcohol.

Professor Anderson said: You state that the particles are uniform alloys, however the STEM/EDS results shown in one of the figures in your presentation give the distinct impression that the larger particles are not uniform, with significant enrichment of, I believe it was Pt, in the surface layers. Can you comment on this point and what effect it might have on toluene suppression.

Professor Hutchings replied: In the work presented we have used two methods, namely STEM-HAADF and STEM-XEDS spectrum imaging, to identify possible elemental segregation in our nanoparticles. We have previously shown that any Pd segregation within individual Au–Pd particles can easily be seen by *z*-contrast.[1] No significant Pd segregation in Au–Pd–Pt particles was detected by this technique in the current study. STEM-XEDS spectrum imaging is used to probe if there is any segregation between Pt and Au, which are too close to each other in the periodic table for HAADF imaging to distinguish. However compared to HAADF, XEDS spectrum imaging requires an electron dose that is several orders of magnitude higher, mainly because of the low X-ray collection efficiency of the instrument. A consequence of this technicality is it that it will introduce more damage to the particle, as shown in Fig. 8 in the paper. Therefore the STEM-XEDS map shown in Fig. 9 should only be considered as a qualitative demonstration that all three elements co-exist in the particle, since we do not yet know how the elements might re-distribute themselves during XEDS spectrum image acquisition. If we accept the proposition that smaller particles (~2 nm) are homogeneous alloys and larger particles (>5 nm) have a Pd-deficient surface, we still believe that the electronic ligand effect is more significant than Pt surface segregation. Statistically the larger particles only account for a very small fraction of the total particle size distribution and therefore contribute even less to the total surface area.

1 R. C. Tiruvalam, J. Pritchard, N. Dimitratos, J. A. Lopez-Sanchez, J. K. Edwards, A. F. Carley, G. J. Hutchings and C. J. Kiely, *Faraday Discuss.*, 2011, **152**, 63–86.

Professor Attard opened a general discussion of Professor Park's and Professor Hutchings' papers: Although somewhat disparate in terms of the nature of the subject matter, I believe both contributions have a couple of common factors: to improve catalytic properties by either tuning systematically the electronic properties of the support or *via* the bulk composition of the metal catalyst itself. The former I think is quite novel and I would be interested if correlations between the level of doping of the semiconductor and catalytic activity would exist. Another thread common to both presentations (and indeed to almost all of the talks) was the use of *ex situ* techniques such as TEM and XPS to characterise a surface post reaction. In electrochemistry, there were concerns about removing an electrode into UHV and trying to deduce anything about the electrochemical interface since important components of the electrochemical interface are "missing" (water for example). Should we be worried about deductions made about catalytic surfaces working at ambient or elevated pressures after they too have been transferred to UHV? Does the TEM method suffer too much from perturbation of the catalyst or are we sufficiently vigilant concerning these matters?

Professor Hutchings said: You raise a very interesting point. The use of *in situ* spectroscopic techniques is of crucial importance. We have used these to try to understand reaction mechanisms alongside the *ex situ* methods we have been describing and discussing in this session. However, this becomes quite difficult even for such simple reactions where there can be multiple pathways. We have recently found this to be the case in studying the reaction of benzyl alcohol oxidation using three *in situ* spectroscopies.[1]

1 G. J. Hutchings *et al.*, *Phys. Chem. Chem. Phys.*, 2013, DOI: 10.1039/c3cp50710f.

Professor Hutchings then said: Following on from a discussion point raised earlier in this meeting by Professor Somorjai, it could be important to study the same reaction in the gas and liquid phases and contrast the rates observed. In this case the oxidation of alcohols using supported nanoparticles could provide useful insights.

Professor Attard responded: I think that this is a very good point and may lead to further insights concerning the role of the solvent.

Professor Attard asked a general question: Referring to the title of Faraday Discussions 162, one needs to understand the nature of how the nanoparticles are actually anchored to the support. Are there better methods of probing this interface more explicitly? I can envisage great future progress in relation to the support, the metal and the perimeter of the particle sitting on the support but the region directly below the particle seems a difficult area to address experimentally except in a slightly indirect way.

Professor Park responded: I think probing of the interface between anchored nanoparticles and the support is quite important but challenging because it is a buried interface. I think the surface chemistry community should look into this issue to find suitable spectroscopic or microscopic techniques for investigation of the nature of the metal–oxide interface. I am planning to implement conductive probe atomic force microscopy to probe charge transport between nanoparticles and support under a gaseous environment.

Professor Attard replied to Professor Park: I agree with this comment and refer you back to my previous comment to Dr Yang concerning deposition of small nanoparticles onto oxide-coated SERS active gold core nanoparticles.

Dr Yang made a general comment is regarding the selectivity of catalysts: In my opinion, selectivity is never a concern if the catalysts are optimised for a specific reactions, one can hardly expect that they are optimised for other reactions in the mean time. In another word, if we can optimise the catalytic activities for each reaction systems, the catalysts will automatically have the catalytic selectivity.

Professor Campbell closed the general discussion in response to Dr Yang: In general, I disagree with this for the following reason. A common approach to improve selectivity in industrial catalysts is the addition of some ingredient (either in the catalysts itself or as a minor additive to the feed) which poisons the desired reaction but poisons the undesired product more rapidly. For example, we have proven that this is the case for chlorine promoters in ethylene epoxidation over Ag (see ref. 1) Here, chlorine adatoms suppress the rate of epoxide formation but suppress the rate of complete combustion faster. If one "optimized the catalyst surface for the desired product rate", as you suggest, it would be much less selective than the slower catalyst whose surface is partially poisoned by chlorine adatoms. The difference in this case would be worth over a billion dollars per year in wasted ethylene feed costs due to its complete combustion.

1 C. T. Campbell and B. E. Koel, *J. Catal.*, 1985, **92**, 272–283.

Concluding remarks

Gabor A. Somorjai*

Received 9th May 2013, Accepted 10th May 2013
DOI: 10.1039/c3fd00081h

The 162 Faraday Discussion Meeting in Berlin under the title "Fabrication, Structure and Reactivity of Anchored Nanoparticles" was held on April 10–12, 2013. Twenty papers were presented and lively discussion followed all of them. Seven of the papers were focused on bimetallic nanoparticles (NPs), six on oxide metal structure and reactivity, and one on oxide–oxide interface chemistry, indicating that the chemistry of bimetallic NPs is the preoccupation of research studies in many laboratories in the field. Seventeen of the NP systems were studied at the solid–gas interfaces at relatively low pressures (less than one bar), while three papers focused on solid–liquid interface reactivity. Twelve of the papers reported restructuring of the reactive systems that were studied, indicating the acceptance of surface restructuring under reaction conditions as a dominant phenomena by most of the participants. Five theoretical papers that investigated important surface reactions and catalysts indicated that theory plays important roles to help atomic and molecular level interpretation of chemical data and mechanisms.

An important trend in research is the synthesis and characterization of nanoparticle systems combined with reactivity or catalytic performance studies in the same group or the same laboratory. NP synthesis by colloid chemistry techniques (Fig. 1) provides the precise size control as compared to the classical impregnation

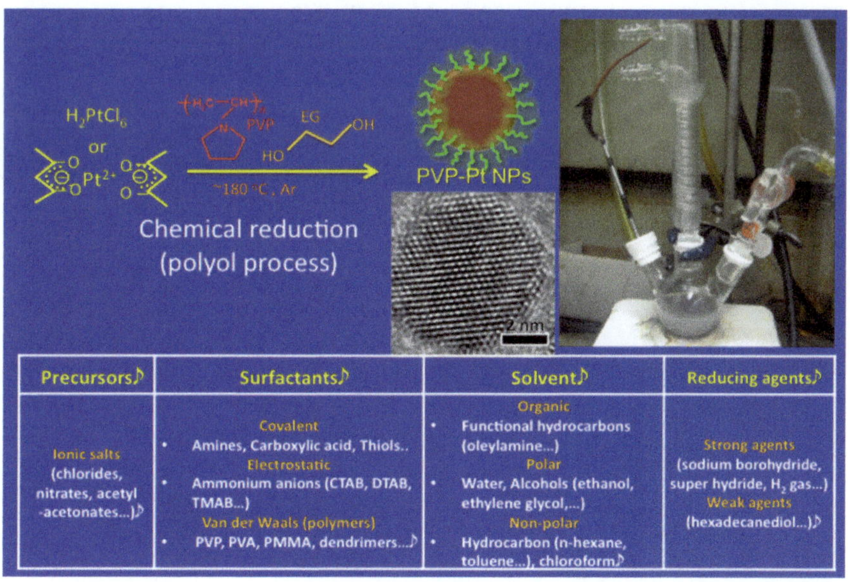

Fig. 1 Colloidal synthesis of nanoparticles.

Department of Chemistry, University of California, Berkeley, CA 94720, USA. E-mail: somorjai@berkeley.edu

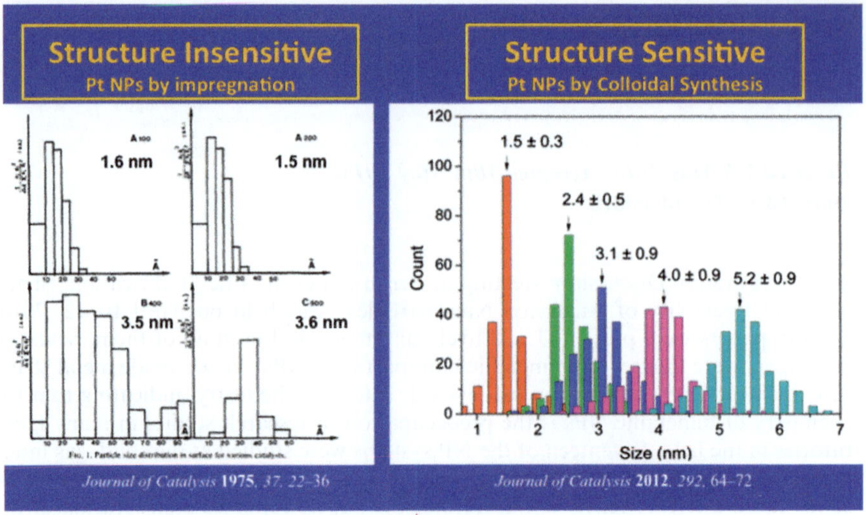

Fig. 2 Benzene hydrogenation over supported Pt NP catalysts.

method of metal salt precursors followed by oxidation–reduction treatment (Fig. 2). As a result, the catalyst size sensitivity of the turnover rates of reactions has been determined as the dominant effect as shown by hydrogenation of benzene, toluene and many other molecules. The metal and bimetallic size range between 1–10 nm that overlaps the catalyst size range utilized in most studies and in the catalyst technologies are readily achievable by colloid synthesis techniques (Fig. 3). The NP shape control can be obtained by selective adsorption of atoms or molecules on one crystal face that leads to reduced growth rates at certain crystal faces (Fig. 4). The paper by Beaumont[1] listed fifteen bimetallic NPs with variable compositions in the 1–10 nm size range.Most NPs are capped by organic films that stop their

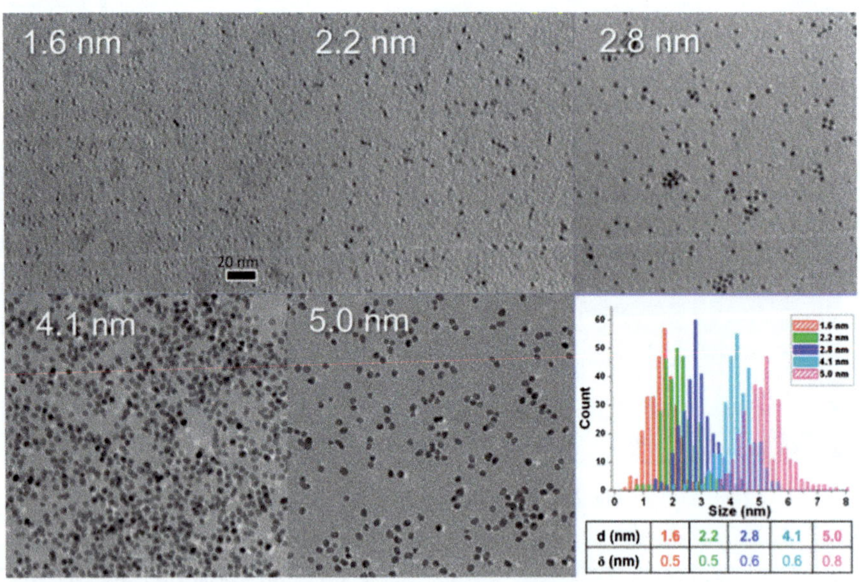

Fig. 3 Monodispersed PVP-capped Pt NPs.

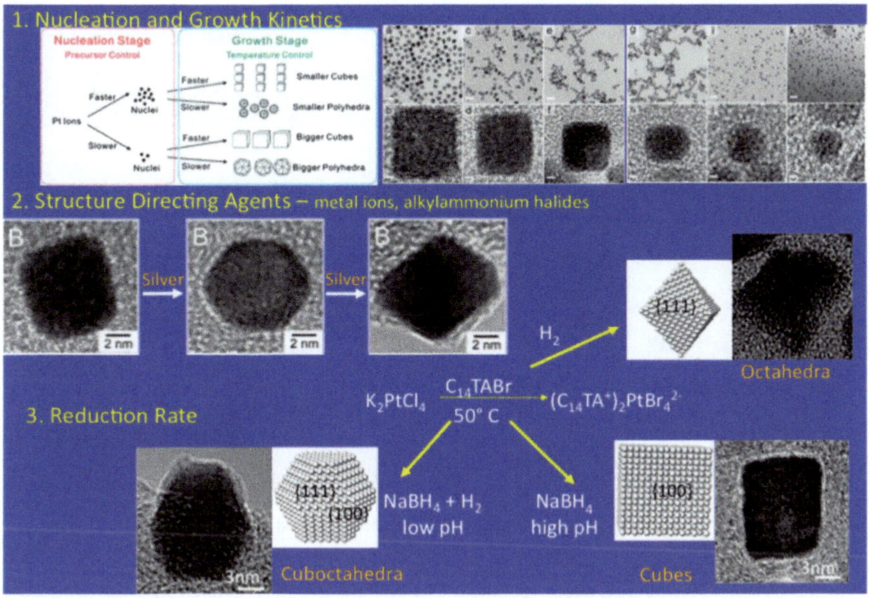

Fig. 4 Shape control of Pt NPs.

growth, and aggregation that can be removed by UV irradiation or made porous to incoming reactants or desorbing products by hydrogen treatments. Core–shell NPs with inorganic shells such as mesoporous silica impart high thermal stability, which is essential for catalytic reactions at high temperatures, like combustion, to prevent the melting of metal NPs that have lower melting points with decreasing size (Fig. 5). The NP catalysts can be utilized as 2D films by deposition with the Langmuir–Blodgett techniques, or in 3D mode by encapsulation in mesoporous oxide supports (Fig. 6). Techniques practiced in H. J. Freund's laboratory in Berlin can produce small NPs like palladium on single crystal oxide films such as $Fe_3O_4(111)$ for use as catalysts. The so-called hard templating technique deposits metal oxide films

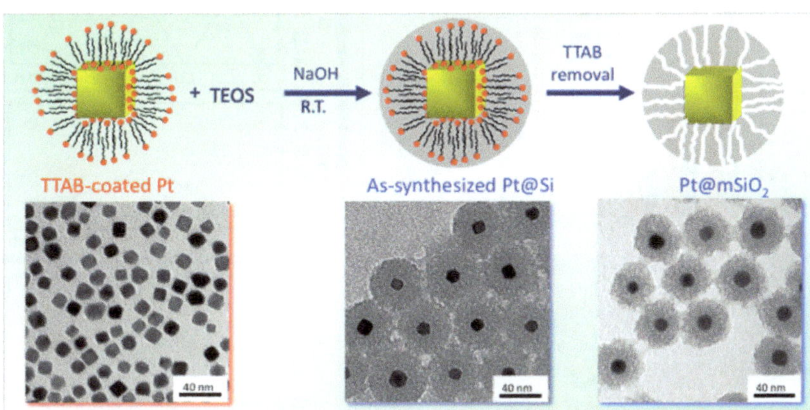

Fig. 5 $Pt@mSiO_2$ from TTAP capped Pt. Silica shell/platinum (or other metal) core nanoparticles have exceptionally high thermal stability. The inorganic oxide shell enables high temperature catalytic oxidation studies, which would not be possible with bare platinum nanoparticles.

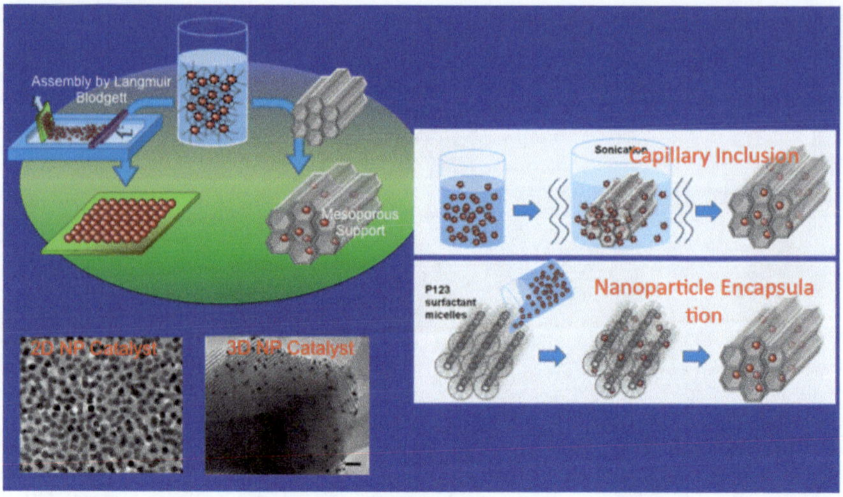

Fig. 6 Nanoparticle-based catalysts.

on mesoporous silica and after leaching out the silica to leave behind a mesoporous high surface area oxide that is used as supports such as CO_3O_4, MnO_2 or CeO_2 for deposition of metal NPs (Fig. 7).

The main conclusion of catalytic studies using transition metal NPs in 2D or 3D configurations is that the "**size and shape of metal NPs control both catalytic reaction rates and selectivities**".

Fig. 8 and 9 show examples of how the size and shape of platinum NPs control the selectivity of catalytic reactions like furfural and methyl cyclopentane ring opening.

Why are catalytic selectivity and turnover rates size and shape dependent? Instead of relying on techniques, which characterize catalysts only before and after reaction

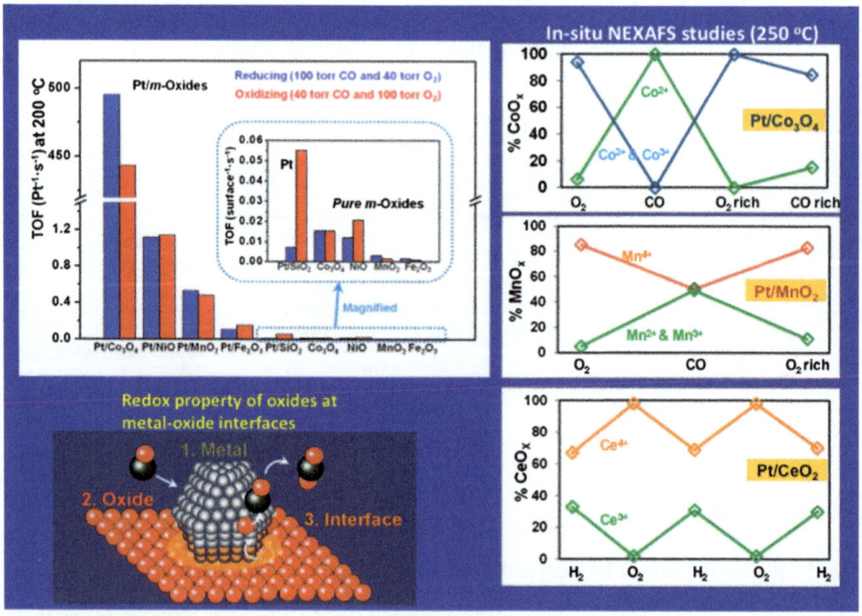

Fig. 7 Strong metal–support interaction for CO oxidation.

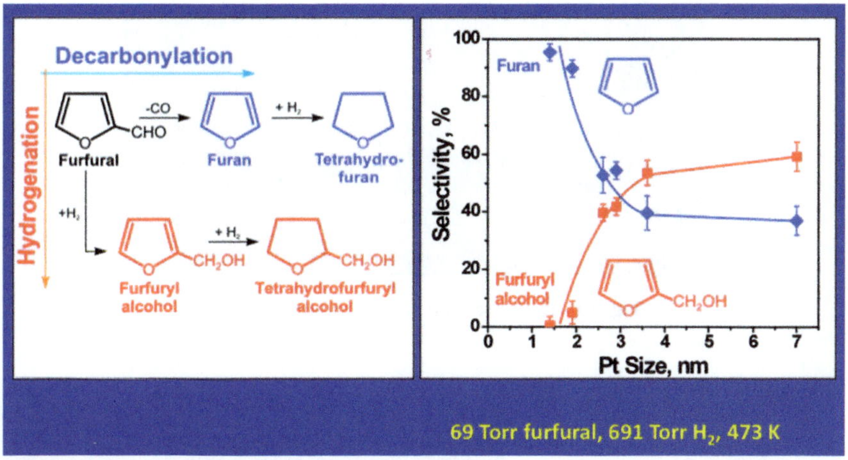

Fig. 8 Platinum nanoparticle size dependence of furfural decarbonylation and hydrogenation.

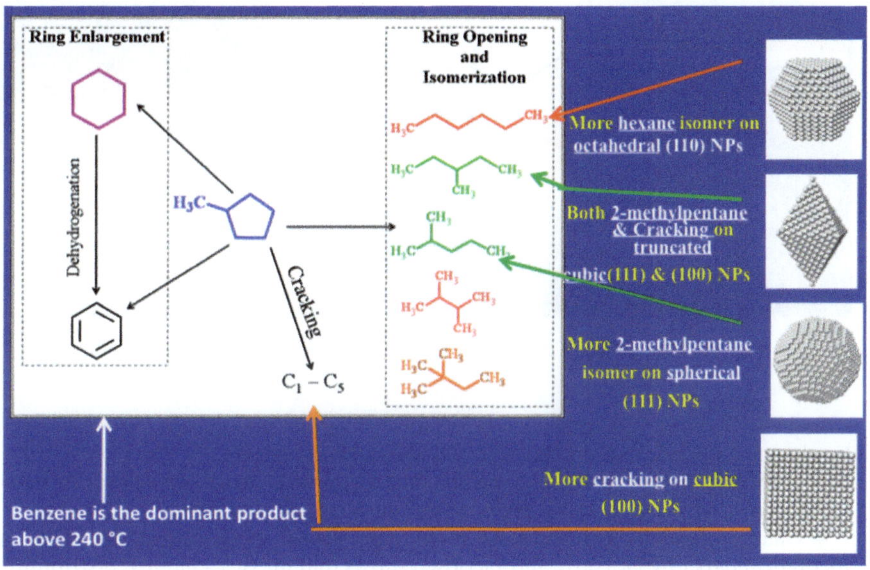

Fig. 9 Platinum nanoparticle shape dependence of methylcyclopentane ring opening.

studies, high pressure *in situ* techniques are required that can monitor the dynamics of molecular catalysis under reaction conditions to understand the size- and shape-dependent effects.

The high pressure *in situ* surface techniques that I use in my laboratory are listed in Fig. 10.[2,3] Sum frequency generation (SFG) vibrational spectroscopy, high pressure scanning tunneling microscopy (HPSTM) and synchrotron-based techniques of ambient pressure X-ray photoelectron spectroscopy (APXPS), X-ray Absorption Near Edge Structure (EXAFS) and Extended X-ray Absorption Fine Structure (NEXAFS) spectroscopies. Examples of these studies that can be carried out at high pressures under reaction conditions are described in several publications.

To summarize:

1. The reaction intermediates detected by SFG are changing with NP size and shape.

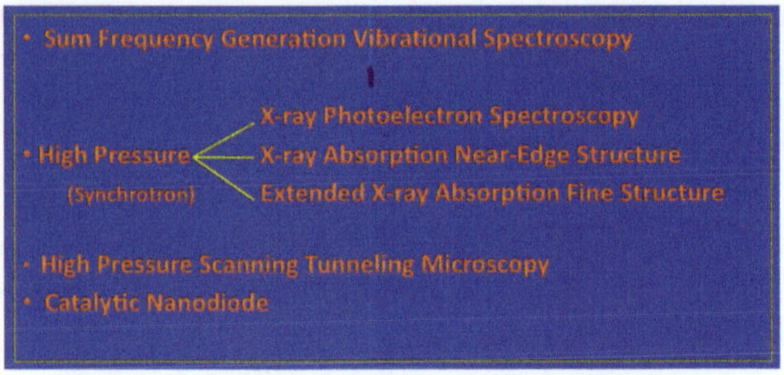

Fig. 10 High pressure *in situ* surface techniques.

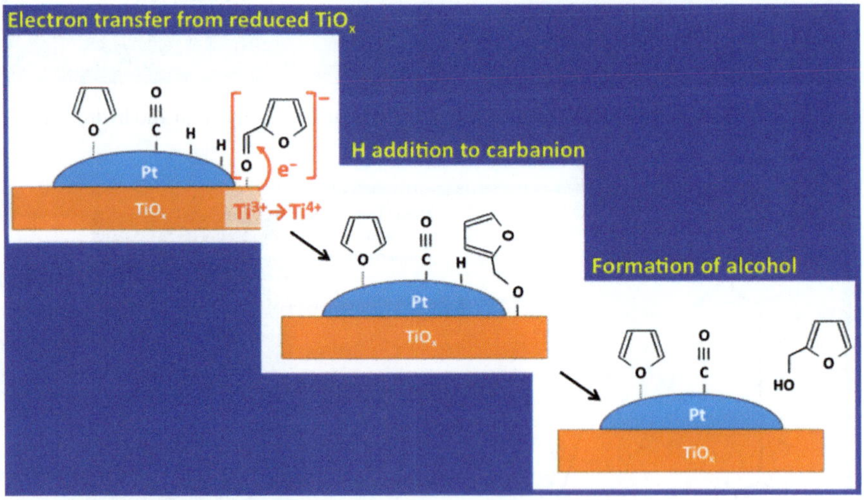

Fig. 11 The platinum–titanium oxide interface produces furfuryl alcohol.

2. APXPS shows that bimetallic NPs undergo surface composition change driven by changing chemical (oxidizing and reducing) environments.

3. HPSTM shows adsorbate mobility under reaction conditions and adsorbate-induced restructuring of metal surfaces at high reactant pressures.

4. Change of metal oxidation state by decreasing the size of metal nanoparticles. A theoretical paper by Nørskov and Jacobsen shows that massive charge redistribution takes place on metal clusters upon oxygen adsorption.[4]

5. Oxide–metal interfaces exhibit hot electron flow under exothermic reaction conditions. The mean free path of these hot electrons is of the order of 4–5 nm after being generated on the femtosecond scale before they equilibrate with the phonons. For metal NPs of this size (4–5 nm) the electrons can overcome the Schottky barrier at the oxide–metal NP interface and ionize the reacting molecules such as CO, O_2 or H_2 in the transition states to carry out what is commonly called acid–base catalysis. Fig. 11 shows how the furfural becomes furfuryl alcohol at the platinum–TiO_2 interface by the charge transfer as Ti^{3+} oxidizes to Ti^{4+} during the hydrogenation reaction.

Future directions

The future of catalysis science is very bright as we assert precise control of nanoparticle size and shape in reactions. We have the instruments to characterize the catalyst NPs under reaction conditions and in the atomic and molecule scales.

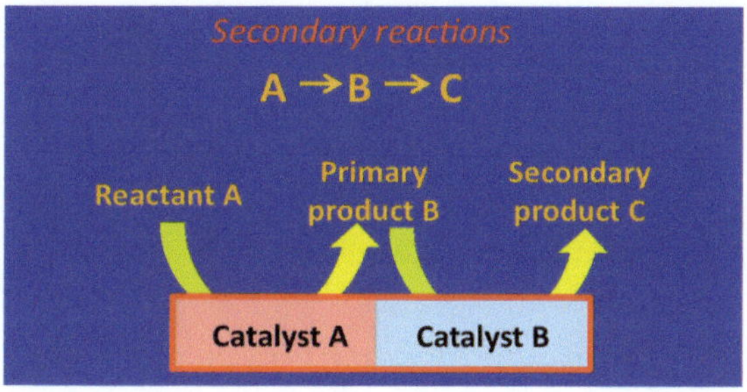

Fig. 12 Architectural design of multifunctional catalysts.

Heterogeneous, homogeneous and enzyme catalysts are all nanoparticles. All three systems show the catalytic importance of low coordination sites. It appears that the molecular mechanisms of chemistry might be the same although one is more frequently utilized in the gas phase, the other in organic solvents and the third in water. We should be able to produce hybrid catalysts as we find already that heterogenizing homogeneous catalysts is possible. We should correlate gas phase and liquid phase catalysis to understand the effects of the 1000-fold difference of molecular density at the interface.

We could combine charge and covalent bond catalysis, commonly called tandem catalysis.

Finally, we should attempt to construct catalytic architecture instead of reactions by single catalysts. Enzyme catalysts are slow in turnover and produce complex molecules that require catalytic architecture with many different catalytic sites (Fig. 12).

References

1 S. K. Beaumont, S. Alayoglu, V. V. Pushkarev, Z. Liu, N. Kruse and G. A. Somorjai, *Faraday Discuss.*, 2013, DOI: 10.1039/c2fd20145c.
2 G. A. Somorjai and C. Aliaga, *Langmuir*, 2010, **26**, 16190.
3 G. A. Somorjai, S. K. Beaumont and S. Alayoglu, *Angew. Chem., Int. Ed.*, 2011, **50**, 10116.
4 J. Kleis, J. Greeley, N. A. Romero, V. A. Morozov, H. Falsig, A. H. Larsen, J. Lu, J. J. Mortensen, M. Dułak, K. S. Thygesen, J. K. Nørskov and K. W. Jacobsen, *Catal. Lett.*, 2011, **141**, 1067.

Poster titles

Ag- and Au-nanoparticle submonolayers stabilized by high-K sol–gel–ZrO_2, **Y. Joseph, T. Nestler, F. Gueth, D. Gao, J. Heitmann and R. Otto**, *TU Bergakademie Freiberg, Germany*

An STM/XPS study of the oxychlorination of Cu(111), Cu(100) and Cu(110) surfaces, **H. Altass, P. Davies, A. Carley and R. J. Davies**, *Cardiff University and Catalysis Institute, UK*

Visualising electrocatalysis at the single nanoparticle level, **S. Lai, P. Dudin, S. Kleijn, M. Koper and P. Unwin**, *University of Warwick, UK*

Biotemplated cobalt doped nanomagnetic arrays, **S. Bird, J. Galloway, J. Bramble, A. Rawlings, S. Evans and S. Staniland**, *University of Leeds, UK*

Chemoselectivity in hydrogenation of the α,β-unsaturated carbonyl compound acrolein on Pd model catalysts, **K.-H. Dostert, C. O'Brien, S. Schauermann and H.-J. Freund**, *Fritz-Haber-Institut der Max-Planck-Gesellschaft, Germany*

Chlorine suppression stabilizer poisoning on supported Pd nanoparticles, **Y. Zhao, J. Medrano and L. Lefferts**, *University of Twente, The Netherlands*

CO oxidation and defect creation by Au nanoparticles on α-Fe_2O_3 from DFT+U calculations, **S. W. Hoh and D. J. Willock**, *Cardiff University, UK*

Elucidating migration pathways of small vanadium oxide clusters at the CeO_2(111) surface by DFT/nudged elastic band calculations, **J. Paier, T. Kropp, C. Penschke and J. Sauer**, *Humboldt University Berlin, Germany*

Redox chemistry of supported Pd with hydroxyls on MgO(001), **Y. Fujimori, F. Ringleb, M. A. Brown, B. Roldan-Cuenya, M. Sterrer and H.-J. Freund**, *Fritz-Haber-Institut der Max-Planck-Gesellschaft, Germany*

Ni@Ag/Au core–shell nanoparticles fabricated in superfluid helium droplets, **A. Boatwright, D. Spence, E. Latimer and S. Yang**, *University of Leicester, UK*

Observation of adsorbed formate, CO and solution CO_2 during formic acid oxidation at supported Pt nanoparticles: an *in situ* ATR-IR spectroelectrochemical approach, **K. Vincent, I. McPherson, P. A. Ash, J. Liu and R. M. J. Jacobs**, *University of Oxford, UK*

Preparation and optical properties of sol–gel matrixes doped with lanthanides, **S. Gutzov and N. Danchova**, *Sofia University "St Kliment Ohridski", Bulgaria*

Preparation of a model system for extra-framework Fe in zeolites, **J. A. Boscoboinik, X. Yu, S. Shaikhutdinov and H.-J. Freund**, *Fritz-Haber-Institut der Max-Planck-Gesellschaft, Germany*

Preparation of triethylamine stabilized silver nanoparticles, **S. L.-C. Hsu and J.-T. Wang**, *National Cheng-Kung University, Taiwan*

Sintering of cobalt nanoparticles at Fischer–Tropsch synthesis conditions, **D. Kistamurthy, C. J. Weststrate, A. M. Saib and J. W. Niemantsverdriet**, *Eindhoven University of Technology, The Netherlands*

Strain and substrate effects on nanoparticle adsorption, **C. M. Yim, C. L. Pang, C. Muryn, R. Perez and G. Thornton**, *University College London, UK*

Structure and activity of Pt on $CeO_2(111)$ surface using first principles thermodynamics, **A. Khetan, H. Pitsch and B. Han**, *RWTH Aachen, Germany*

Supported titanium nitride Pt nanofilm. Electrolysis and CO electrooxidation in acidic and alkaline media, **G. García, M. Roca-Ayats, O. Guillén-Villafuerte, E. Pastor and Mª V. Martínez-Huerta**, *Instituto de Catálisis y Petroleoquímica-CSIC, Spain*

Tuning the reactivity of a Cu/ZnO nanocatalyst *via* gas phase pressure and methanol synthesis on ZnO from MD, **J. Frenzel, L. Martínez-Suárez, D. Marx and B. Meyer**, *Ruhr-University Bochum, Germany*

Synthesis and structural investigation of "well-defined" bimetallic Fe–MnO Fischer–Tropsch model catalysts, **M. Dad and H. Freddikson**, *Eindhoven University of Technology, Netherlands*

The effect of acid treatment on the surface chemistry and topography of graphite, **R. Davies, E. Bouleghlimat, P. Davies, J. Kulhavy and D. Morgan**, *Cardiff University, UK*

TiO_2 based photocatalysis – the influence of metal decoration, **R. Su, R. Tiruvalam, Q. He, N. Dimitratos, R. Bechstein, C. Kiely, G. Hutchings and F. Besenbacher**, *University of Aarhus, Denmark*

The formation of –SRAuSR– staple in thiolate protected gold clusters from first-principles, **D. Pichugina, A. Majouga, M. Askerka, S. Nikolaev, A. Shestakov and N. Kuz'menko**, *Lomonosov Moscow State University, Russia*

The modification of tetrahexahedral Pt nanocrystal catalysts and their enhanced activity for small organic molecules electrooxidation, **H. Liu, Z. Huangfu, N. Tian, Z. Zhou and S. Sun**, *Xiamen University, China*

Titania-supported vanadia and tungsten oxide model catalysts, **C. Davies, J. Granestrand, M. Bowker and S. Golunski**, *Cardiff Catalysis Institute, UK*

Towards multiscale modelling of Ag particle sintering, **M. Garcia Mota, M. Rieger and K. Reuter**, *BASF, Germany*

The Skinner Prize for the best poster was jointly awarded to Ian McPherson of the University of Oxford, UK, for his poster on *in situ* ATR-IR on supported Pt nanoparticles and Soon Wen Hoh of Cardiff University, UK, for his poster on CO oxidation and defect creation by Au nanoparticles using DFT calculations.

List of participants

Mr Abdulmohsen Alshehri, *Cardiff University, United Kingdom*
Professor Scott Anderson, *University of Utah, USA*
Professor Kiyotaka Asakura, *Hokkaido University, Japan*
Dr Philip A. Ash, *University of Oxford, United Kingdom*
Professor Gary Attard, *Cardiff University, United Kingdom*
Dr Florian Ausfelder, *Deutsche Bunsen-Gesellschaft für Physikalische Chemie e.V., Germany*
Dr Aude Bailly, *Institut Néel, CNRS, France*
Professor Dr Matthias Ballauff, *Helmholtz-Zentrum Berlin für Materialien und Energie, Germany*
Dr Simon Beaumont, *Durham University, United Kingdom*
Dr Alisa Becker, *Royal Society of Chemistry, United Kingdom*
Professor Dr Rolf Jürgen Behm, *Ulm University, Germany*
Dr Malte Behrens, *Fritz-Haber-Institut der Max-Planck-Gesellschaft, Germany*
Professor Sophia Belfer, *Ben-Gurion University of the Negev, Israel*
Mr Scott Bird, *University of Leeds, United Kingdom*
Dr Adrian Boatwright, *University of Leicester, United Kingdom*
Dr J. Anibal Boscoboinik, *Fritz-Haber-Institut der Max-Planck-Gesellschaft, Germany*
Professor Michael Bowker, *Cardiff University, United Kingdom*
Professor Dr Charles Campbell, *University of Washington, USA*
Mr Chunhua Cui, *Technische Universität Berlin, Germany*
Mr Mohammadhassan Dad, *Eindhoven University of Technology, Netherlands*
Dr Robert Davies, *Cardiff University, United Kingdom*
Dr Philip Davies, *Cardiff University, United Kingdom*
Miss Catherine Davies, *Cardiff University, United Kingdom*
Dr Huiqiu Deng, *Hunan Agricultural University, China*
Mr Karl-Heinz Dostert, *Fritz-Haber-Institut der Max-Planck-Gesellschaft, Germany*
Dr Jennifer Edwards, *Cardiff University, United Kingdom*
Professor Gerhard Ertl, *Fritz-Haber-Institut der Max-Planck-Gesellschaft, Germany*
Dr Johannes Frenzel, *Ruhr-University Bochum, Germany*
Professor Hans-Joachim Freund, *Fritz-Haber-Institut der Max-Planck-Gesellschaft, Germany*
Mr Yuichi Fujimori, *Fritz-Haber-Institut der Max-Planck-Gesellschaft, Germany*
Dr Cornelia Futter, *hte Aktiengesellschaft, Germany*
Dr Monica Garcia Mota, *BASF, Germany*
Dr Gonzalo García, *Instituto de Catálisis y Petroleoquímica-CSIC, Spain*
Dr Juan Maria Gonzalez Carballo, *Sasol Technology (UK) Ltd, United Kingdom*
Miss Helen Gray, *Royal Society of Chemistry, United Kingdom*
Dr Stoyan Gutzov, *Sofia University "St Kliment Ohridski", Bulgaria*
Professor Dr Klaus Hermann, *Fritz-Haber-Institut der Max-Planck-Gesellschaft, Germany*
Mr Soon Wen Hoh, *Cardiff University, United Kingdom*
Professor Dr Steve Lien-chung Hsu, *National Cheng-Kung University, Taiwan*
Dr Anthony Hughes, *Commonwealth Scientific and Industrial Research Organisation (CSIRO), Australia*
Professor Graham Hutchings, *Cardiff University, United Kingdom*
Professor Dr Yvonne Joseph, *Technische Universität Bergakademie Freiberg, Germany*
Dr Margarethe Kampling, *SPECS Surface Nano Analysis GmbH, Germany*

Dr Imran Khan, *Atomic Weapons Establishment (AWE), United Kingdom*
Mr Abhishek Khetan, *RWTH Aachen University, Germany*
Mr Deshen Kistamurthy, *Eindhoven University of Technology, Netherlands*
Mr Thomas Kropp, *Humboldt University Berlin, Germany*
Dr Stanley Lai, *University of Warwick, United Kingdom*
Miss Haixia Liu, *Xiamen University, China*
Dr Yan Lu, *Helmholtz-Zentrum Berlin für Materialien und Energie, Germany*
Dr Alexander Majouga, *Lomonosov Moscow State University, Russia*
Dr Maria Victoria Martinez-Huerta, *Consejo Superior de Investigaciones Cientificas, Spain*
Mr Ian McPherson, *University of Oxford, United Kingdom*
Mr Marko Melander, *Aalto University, Finland*
Professor Dr Dietrich Menzel, *Fritz-Haber-Institut der Max-Planck-Gesellschaft, Germany*
Dr Peter Miedziak, *Cardiff University, United Kingdom*
Dr Henrik Junge Mortensen, *Haldor Topsøe A/S, Denmark*
Ms Tina Nestler, *Technische Universität Bergakademie Freiberg, Germany*
Dr Matthew Neurock, *University of Virginia, USA*
Dr Niklas Nilius, *Fritz-Haber-Institut der Max-Planck-Gesellschaft, Germany*
Mr Hyung-suk Oh, *Technische Universität Berlin, Germany*
Dr Joachim Paier, *Humboldt University Berlin, Germany*
Professor Jeong Y. Park, *Korea Advanced Institute of Science and Technology (KAIST), South Korea*
Mr Michael Parkin, *Royal Society of Chemistry, United Kingdom*
Mr Christopher Penschke, *Humboldt University Berlin, Germany*
Dr Daria Pichugina, *Lomonosov Moscow State University, Russia*
Dr Julia Rinck, *Karlsruhe Institute of Technology, Germany*
Dr Swetlana Schauermann, *Fritz-Haber-Institut der Max-Planck-Gesellschaft, Germany*
Professor Gabor A. Somorjai, *University of California, Berkeley, USA*
Mr Weiyu Song, *Eindhovenn University of Technology, Netherlands*
Professor Dr Peter Strasser, *Technische Universität Berlin, Germany*
Dr Ren Su, *University of Aarhus, Denmark*
Professor Dr Sefik Suzer, *Bilkent University, Turkey*
Mrs Rachel Thompson, *Royal Society of Chemistry, United Kingdom*
Professor Geoffrey Thornton, *University College London, United Kingdom*
Dr Na Tian, *Xiamen University, China*
Dr Laura Torrente, *University of Bath, United Kingdom*
Professor Dr Rutger van Santen, *Eindhoven University of Technology, Netherlands*
Dr Kylie Vincent, *University of Oxford, United Kingdom*
Professor Kazuo Watanabe, *Tokyo University of Science, Japan*
Dr Keith Whiston, *Invista Textiles UK Ltd, United Kingdom*
Dr David Willock, *Cardiff University, United Kingdom*
Dr Shengfu Yang, *University of Leicester, United Kingdom*
Dr Chi Ming Yim, *University College London, United Kingdom*
Mr Yingnan Zhao, *University of Twente, Netherlands*

Index of contributors*

*The page numbers in **bold** type indicate papers submitted for discussions.

